Study & Master
Study Guide 10
Mathematics

Noleen Jakins | Deirdre Yeo

CAMBRIDGE
UNIVERSITY PRESS

CAMBRIDGE
UNIVERSITY PRESS

The Water Club, Beach Road, Granger Bay, Cape Town 8005, South Africa

Cambridge University Press is part of the University of Cambridge.

It furthers the University's mission by disseminating knowledge in the pursuit of education, learning and research at the highest international levels of excellence.

www.cambridge.org
Information on this title: www.cambridge.org/9781107470811

First published 2015
Reprinted 2016

Printed in South Africa by Creda Communications

ISBN 978-1-107-47081-1 Paperback

Additional resources for this publication at www.educators.co.za

Editors: Alison Paulin and Christine de Nobrega
Proofreaders: Magda de Jager, Amy Moore, Loraine Paulin
Typesetter: Maryke Garifallou
Illustrators: Maryke Garifallou, Laura Brecher, Claudia Eckard
Proj ID: 20327

....

Contents

Solutions

TOPIC 1: Real numbers and surds

In Grades 8 and 9 you worked mostly with rational numbers and were introduced briefly to irrational numbers. In Grade 10 we extend our knowledge and understanding of irrational numbers and work more specifically with surds.

Knowledge and skills for this topic

If you struggle with any of the work listed below, revise it before continuing with this Topic:

* working knowledge of numbers up to and including rational numbers
* working knowledge of composite and prime factors, perfect squares and cubes.

Content of final exam

* Classify real numbers as rational or irrational.
* Work with rational and irrational numbers.
* Establish between which two integers a given surd lies.
* Round numbers off to an appropriate degree of accuracy.

Rational and irrational numbers

The real number system consists of both rational and irrational numbers.

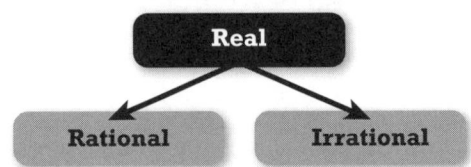

Rational numbers

Rational numbers include all numbers that can be expressed as fractions, so rational numbers include all numbers that can be written in the form $\frac{a}{b}$, where a and b are integers and $b \neq 0$.

- All positive and negative fractions are rational numbers, e.g. $\frac{7}{3}$, $-\frac{2}{11}$ and $7\frac{3}{8}$.
- All terminating decimal numbers are rational numbers. This is because any terminating decimal number can be written as a fraction, e.g. $0{,}56 = \frac{56}{100}$ and $-4{,}6 = -4\frac{6}{10}$.
- All recurring decimal numbers are rational numbers, because they can be expressed as fractions, e.g.; $0{,}\dot{6} = \frac{6}{9} = \frac{2}{3}$, $0{,}\dot{2}\dot{4} = \frac{24}{99} = \frac{8}{33}$ and $0{,}\dot{3}7\dot{8} = \frac{378}{999} = \frac{14}{37}$.
- Rational numbers include all integers because all integers can be expressed as fractions, e.g. the rational number 21 can be expressed as $\frac{42}{2}$.

Irrational numbers

Irrational numbers include all numbers that cannot be written as fractions.

- Irrational numbers are infinite/non-terminating decimal numbers, e.g.:
  - $\sqrt{3}$ because the value of $\sqrt{3} = 1{,}73205080\ldots$
  - $\sqrt[3]{71}$ because the value of $\sqrt[3]{71} = 4{,}14081774\ldots$
  - π because the value of $\pi = 3{,}14159265\ldots$

Surds

- Surds include all numbers that can only be expressed using the root symbol, e.g. $\sqrt{}$ or $\sqrt[n]{}$ where $n \in \mathbb{N}$. So surds can be square roots, cube roots, fourth roots, fifth roots, etc.
- Note that $\sqrt{9}$ is **not** a surd, because it can be expressed as an exact whole number, 3.
- However, $\sqrt[3]{9}$ is a surd, because it cannot be written as an exact decimal number: $\sqrt[3]{9} \approx 2{,}08008382\ldots$
- Division by 0 is **undefined**, therefore $\frac{\sqrt{7}}{0}$ is undefined and is neither rational or irrational.
- The square root, or any root, of a negative number has no answer. We say $\sqrt{-25}$ is **non-real** and therefore neither rational or irrational.
- The cube root, or any uneven root, of a negative number has either a rational or irrational answer, e.g. $\sqrt[3]{-27} = -3$, $\sqrt[3]{-11} \approx -2{,}223\ldots$

Approximate values for surds

Worked examples

1. Between which two whole numbers does $\sqrt{31}$ lie?
2. Without a calculator find the approximate value of $\sqrt{31}$ correct to one decimal place.

Solutions

1. Write as surds the perfect squares that lie on either side of the required surd:

   $$\sqrt{25} \qquad \sqrt{31} \qquad \sqrt{36}$$

   Now convert these perfect square surds into whole numbers:

   $$5 \qquad \sqrt{31} \qquad 6$$

   Therefore $\sqrt{31}$ lies between 5 and 6.

2. $\sqrt{31}$ lies closer to $\sqrt{36}$ than to $\sqrt{25}$, so a safe approximation of its value would be 5,6.

Note

Using a calculator:
$\sqrt{31} \approx 5,5677 \ldots$
$\approx 5,6$

Using a calculator to find approximate values for surds

A calculator may be used to find the approximate value of any given surd, rounded off to an appropriate degree of accuracy.

Worked examples

Use a calculator to find the value for each of these surds. Round off your answers to two decimal places.

1. $\sqrt{360}$
2. $\sqrt[3]{360}$
3. $\sqrt[4]{360}$
4. $3\sqrt{72} + 2\sqrt{34}$
5. $-3\sqrt{65} + 2\sqrt{30}$
6. $-3\sqrt[3]{3}$
7. $\left(-3\sqrt[3]{3}\right)^3$

Solutions

1. $\sqrt{360} = 6\sqrt{10} \approx 18,973665\ldots \approx 18,97$

2. $\sqrt[3]{360} \approx 7,1137866\ldots \approx 7,11$

3. $\sqrt[4]{360} \approx 4,355877\ldots \approx 4,36$

4. $3\sqrt{72} + 2\sqrt{34} \approx 37,117747\ldots \approx 37,12$

5. $-3\sqrt{65} + 2\sqrt{30} \approx -13,2323220\ldots \approx -13,23$

6. $-3\sqrt[3]{3} \approx -4,3267487\ldots \approx -4,33$

7. $\left(-3\sqrt[3]{3}\right)^3 = -81$

Note

Press the general root key ⌄▣ and then press 4 to find the fourth root.

Exercise 1

Give the value of each number, rounding off to three decimal places where necessary. Also classify each number as rational, irrational, or neither. (Use your calculator to check.)

1. 12
2. $\sqrt{12}$
3. $\sqrt[3]{12}$
4. $-0,36$
5. $-\sqrt{0,36}$
6. $-\sqrt[4]{0,36}$
7. $\frac{9}{25}$
8. $\sqrt{\frac{9}{25}}$
9. $\sqrt[6]{\frac{9}{25}}$
10. $0,45$
11. $0,4\dot{5}$
12. $3\sqrt{81}$
13. $-3\sqrt{81}$
14. $-3\sqrt{-81}$
15. $-3\sqrt[3]{-81}$
16. $25 + 144$
17. $\sqrt{25 + 144}$
18. $\sqrt{25} + \sqrt{144}$
19. $36 - 169$
20. $\sqrt{36 - 169}$
21. $\sqrt{36} - \sqrt{169}$
22. 9π
23. $\sqrt[3]{9\pi}$
24. $4\pi + 5\pi$

25. Choose the correct answer: $21\sqrt{3} =$ _____

   A $\sqrt[21]{3}$ **B** $21 \div \sqrt{3}$ **C** $21 \times \sqrt{3}$

   D $21 - \sqrt{3}$ **E** $21 + \sqrt{3}$

Simplifying surds by expressing in simplest form

Using a calculator to find the approximate value of an irrational number is easy. But remember that this will give an approximation of the value. Throughout your study of mathematics you are often required in answers to give an irrational number in its most accurate simplified form.

Rules for simplifying surds

In general, where a and b are both positive numbers:

$\sqrt{a} \times \sqrt{b} = \sqrt{ab}$ e.g. $\sqrt{3} \times \sqrt{2} = \sqrt{3 \times 2} = \sqrt{6}$

$\sqrt{a} \times \sqrt{a} = \sqrt{a^2} = a$ e.g. $\sqrt{3} \times \sqrt{3} = \sqrt{3 \times 3} = 3$ or $\sqrt{3^2} = 3$ or $\left(\sqrt{3}\right)^2 = 3$

$\sqrt{a} \div \sqrt{b} = \sqrt{ab}$ e.g. $\sqrt{16} \div \sqrt{9} = \sqrt{\frac{16}{9}} = \frac{4}{3}$

Expressions can be simplified by adding or subtracting terms that contain the same surd.

$\sqrt{a} + \sqrt{a} = 2\sqrt{a}$ e.g. $3\sqrt{5} + 4\sqrt{5} = 7\sqrt{5}$

$3\sqrt{a} - \sqrt{a} = 2\sqrt{a}$ e.g. $10\sqrt{2} - 6\sqrt{2} = 4\sqrt{2}$

Note the following:
- Surds can be cancelled with identical surds: $\frac{\sqrt{5}}{\sqrt{5}} = 1$, but $\frac{\sqrt{5}}{5}$ cannot be simplified.

- It is not acceptable to leave a surd as a denominator in your answer. To change the surd to a rational number, multiply it by an identical surd. You therefore need to multiply the numerator by the same surd, for example:

  - $\frac{5}{\sqrt{3}} = \frac{5}{\sqrt{3}} \times \frac{\sqrt{3}}{\sqrt{3}} = \frac{5\sqrt{3}}{3}$

  - $\frac{2}{\sqrt{8}} = \frac{2}{\sqrt{4 \times 2}} = \frac{2}{2\sqrt{2}} = \frac{1}{\sqrt{2}} \times \frac{\sqrt{2}}{\sqrt{2}} = \frac{\sqrt{2}}{2}$

Worked examples

Simplify, without using a calculator:

1. $\sqrt{75}$ **2.** $\sqrt{75} \times 2\sqrt{3}$ **3.** $\sqrt{75} + 2\sqrt{27}$ **4.** $\frac{12\sqrt{30}}{3\sqrt{5}}$

Solutions

1. Surds can be simplified by expressing the number under the root sign as a product of its factors, one of which must be a perfect square. So for question 1. follow these steps:

   Step 1: Find the factors of 75: $\sqrt{75} = \sqrt{5 \times 5 \times 3} = \sqrt{25 \times 3}$

   Step 2: Express this as a product of two surds: $\sqrt{25} \times \sqrt{3}$

   Step 3: Simplify the perfect square: $5 \times \sqrt{3}$

   Step 4: Write the surd in its simplest form: $\sqrt{75} = 5\sqrt{3}$

2. $\sqrt{75} \times 2\sqrt{3} = 5\sqrt{3} \times 2\sqrt{3} = 5 \times 2 \times \sqrt{3} \times \sqrt{3} = 10 \times 3 = 30$

3. $\sqrt{75} + 2\sqrt{27} = \sqrt{25 \times 3} + 2\sqrt{9 \times 3} = 5\sqrt{3} + (2 \times 3)\sqrt{3} = 5\sqrt{3} + 6\sqrt{3} = 11\sqrt{3}$

4. If the factors of the number under the root sign do not result in a perfect square, it is useful to use the multiplication law to write each factor as a separate surd:

   $\frac{12\sqrt{30}}{3\sqrt{5}} = \frac{12\sqrt{2 \cdot 3 \cdot 5}}{3\sqrt{5}} = \frac{12\sqrt{2}\sqrt{3}\sqrt{5}}{3\sqrt{5}} = 4\sqrt{2}\sqrt{3} = 4\sqrt{6}$

Roots other than square or cube roots

When dealing with roots other than square or cube roots, write the number in the root sign as the product of a prime number to the same power as the root, multiplied by whatever prime number is left.

Worked example
Simplify $\sqrt[4]{243}$.

Solution
$\sqrt[4]{243} = \sqrt[4]{3 \cdot 3 \cdot 3 \cdot 3 \cdot 3} = \sqrt[4]{3^4 \cdot 3} = 3\sqrt[4]{3}$

Exercise 2
Simplify each of the following without using a calculator.

1. $\sqrt{50}$
2. $\sqrt{125}$
3. $\sqrt{56}$
4. $\sqrt{128}$
5. $3\sqrt{48}$
6. $-2\sqrt{32}$
7. $2\sqrt{18} \times 3\sqrt{12}$
8. $2\sqrt{18} + 3\sqrt{12}$
9. $2\sqrt{18} \div 3\sqrt{12}$
10. $\sqrt{\frac{24}{6}}$
11. $3\sqrt{\frac{24}{6}}$
12. $\sqrt{\frac{24}{6}} + \frac{3}{4}\sqrt{\frac{16}{9}}$
13. $\sqrt{8} + \sqrt{162} - \sqrt{18} + \sqrt{50}$
14. $8\sqrt{50} - 3\sqrt{24} - \sqrt{96}$
15. $\sqrt[3]{8}$
16. $\sqrt[3]{125}$
17. $\sqrt[4]{16}$
18. $\sqrt[5]{128}$
19. $\sqrt[3]{8} - 3\sqrt[3]{27} + 5\sqrt[5]{32}$

Test A: Knowledge and routine procedures

1. Classify each number as rational or irrational.

    1.1 $\sqrt{44}$ 1.2 $-\sqrt{144}$ 1.3 $\frac{\sqrt{9}}{2}$

    1.4 $\sqrt{\frac{21}{3}}$ 1.5 $0{,}123444444\ldots$ (5)

2. Choose the correct answer.

    2.1 $\sqrt{48}$ written in its simplest form is:

        A $\quad 48\sqrt{1}$ B $\quad 16\sqrt{3}$ C $\quad 6\sqrt{8}$

        D $\quad 4\sqrt{3}$ E $\quad 4\sqrt{12}$ (2)

    2.2 $\sqrt{432}$ written in its simplest form is:

        A $\quad 16\sqrt{27}$ B $\quad 32\sqrt{3}$ C $\quad 12\sqrt{3}$

        D $\quad 3\sqrt{12}$ E $\quad 72\sqrt{6}$ (2)

3. Simplify each of the following without using a calculator.

   3.1 $5\sqrt{16}$ (1)

   3.2 $-8\sqrt{8}$ (1)

   3.3 $3\sqrt{6} \times 5\sqrt{3}$ (3)

   3.4 $3\sqrt{6} \div 5\sqrt{3}$ (3)

4. Simplify each of the following without using a calculator.

   4.1 $3\sqrt{12} + 5\sqrt{3}$ (2)

   4.2 $3\sqrt{12} - 5\sqrt{27}$ (3)

   4.3 $\dfrac{-3\sqrt{40}}{\sqrt{5}}$ (2)

   4.4 $\sqrt{72} - \sqrt{162} - 3\sqrt{75}$ (4)

5. Use your calculator to give the value of each expression to three decimal places where necessary.

   5.1 $\sqrt{2} - \sqrt[3]{3}$ (2)

   5.2 $\dfrac{\sqrt{41}}{3}$ (2)

   5.3 $\dfrac{4}{3} \times \sqrt{21^2}$ (3)

6. **6.1** Between which two whole numbers does $\sqrt{78}$ lie?   (3)

   6.2 Without a calculator find the approximate value of $\sqrt{78}$, correct to one decimal place.   (2)

7. From the given set of numbers $0,\dot{4}$; $\frac{4}{0}$; $\frac{0}{4}$; $\sqrt{4}$; $\sqrt{-4}$; $\sqrt[3]{4}$; -4π, write down:

   7.1 two irrational numbers   (2)

   7.2 a number that is undefined   (1)

   7.3 two whole numbers.   (2)

                                                                  Total 45

Test B: Complex procedures and problem solving

1. $\sqrt[5]{64}$ written in its simplest form is equal to:

   A $8\sqrt{8}$ **B** $4\sqrt[5]{2}$ **C** $2\sqrt{2}$

   D $4\sqrt[5]{8}$ **E** $2\sqrt[5]{2}$ (2)

2. Simplify each of the following without using a calculator:

   2.1 $5\sqrt[3]{16}$ (2)

   2.2 $-8\sqrt[3]{8}$ (2)

3. **3.1** Between which two whole numbers does $\sqrt{470}$ lie?   (3)

   3.2 Without a calculator find the approximate value of $\sqrt{470}$ correct to two decimal places.   (1)

4. **4.1** Between which two whole numbers does $\sqrt[3]{225}$ lie? (3)

    4.2 Without a calculator find the approximate value of $\sqrt[3]{225}$ correct to two decimal places. (1)

5. Simplify the following.

    5.1 $3\sqrt[3]{27}$ (2)

    5.2 $-3\sqrt[3]{-27}$ (2)

    5.3 $-3\sqrt[4]{81}$ (2)

6. Give the value of each of the following without using a calculator.
    Use your solution to classify each as rational, irrational or neither.

    6.1 $\sqrt{225 - 289}$ (2)

    6.2 $\sqrt{225} - \sqrt{289}$ (2)

7. Simplify each of the following, giving your answer in simplified surd form:

    7.1 $4\sqrt[3]{16} - 7\sqrt[4]{32}$ (4)

    7.2 $3\sqrt[3]{54} + 5\sqrt{8} - \sqrt{32} + \sqrt[3]{8}$ (6)

8. From the set of numbers $\sqrt[3]{-27}$; $\sqrt[8]{-8}$; $\sqrt[5]{10}$; $\frac{\pi^2}{2}$; $12 \times 0,\dot{3}$, write down:

    8.1 two irrational numbers (2)

    8.2 one non-real number (1)

    8.3 two integers. (2)

9. Give the value of each number, rounding off to three decimal places where necessary. Also classify each number as rational, irrational or neither.

    9.1 $-\sqrt{\pi^2}$ (3)

    9.2 $\sqrt{(1 + 2\pi)^2}$ (3)

Total 45

Test C: Content and breakdown as for exam

1. Choose the correct answer.

    1.1 $19\sqrt{5} = \underline{\qquad}$

         A $19 + \sqrt{5}$ **B** $19 \div \sqrt{5}$ **C** $19 \times \sqrt{5}$

         D $19 - \sqrt{5}$ **E** $\sqrt[19]{5}$ (2)

    1.2 $\sqrt[4]{48}$ written in its simplest form is:

         A $48\sqrt[4]{1}$ **B** $2\sqrt[4]{3}$ **C** $6\sqrt[4]{2}$

         D $4\sqrt{3}$ **E** $3\sqrt[4]{2}$ (2)

2. Simplify as far as possible and then classify as rational or irrational.

    2.1 $3\sqrt{27}$ (1)

    2.2 $-\frac{\sqrt{27}}{3}$ (2)

    2.3 $3,78888$ (2)

3. Simplify without using a calculator.

    3.1 $\sqrt{56}$ (1)

    3.2 $-3\sqrt{128}$ (2)

    3.3 $2\sqrt{18} \times 3\sqrt{12}$ (3)

    3.4 $2\sqrt{18} - 3\sqrt{12}$ (4)

4. Simplify each of the following without using a calculator:

    4.1 $\sqrt[3]{27}$ (1)

    4.2 $\sqrt[4]{2\,592}$ (4)

5. **5.1** Between which two whole numbers does $\sqrt[3]{33}$ lie? (3)

    5.2 Without a calculator find the approximate value of $\sqrt[3]{33}$ correct to one decimal place. (1)

6. Give the value of each number, and classify as rational, irrational or neither.

    6.1 $36 + 169$ (2)

    6.2 $\sqrt{36 + 169}$ (2)

    6.3 $\sqrt{36} + \sqrt{169}$ (2)

7. Simplify, and then classify as rational, irrational or neither.

    7.1 π^2 (2)

    7.2 $\sqrt{\pi^2}$ (2)

    7.3 $\sqrt[3]{\pi^2}$ (2)

8. Indicate whether the statements given below are true or false. If false, correct the statement.

    8.1 All real numbers are rational numbers.

    8.2 All recurring decimal numbers can be written as fractions.

    8.3 An undefined number can be written as zero. (5)

                                            Total 45

TOPIC 2: Algebraic expressions

You were introduced to formal algebra in Grade 8. In Grade 9 your knowledge of algebra was extended to include multiplication, factorisation and the simplification of basic algebraic fractions. Grade 10 algebra builds on and extends this foundation, taking multiplication, factorisation and algebraic fractions to a higher level.

Knowledge and skills for this topic

If you struggle with any of the work listed below, revise it before continuing with this Topic:

- simplification, including addition, subtraction, multiplication and division of basic polynomials
- multiplication of a binomial by a binomial
- factorisation, including common factors, common brackets, difference of two squares and basic trinomials
- addition, subtraction, multiplication and division of simple algebraic fractions
- cubes and cube roots, squares and square roots.

Content of final exam

- All algebra covered in Grades 8 and 9.
- Multiply a binomial by a trinomial.
- Factorise, including work from Grade 9 and:
  - trinomials
  - the difference and the sum of two cubes
  - factorising by grouping in pairs.
- Simplify, add and subtract algebraic fractions up to and including cubic denominators (limited to the sum and difference of two cubes).

Vocabulary and terminology

Terms in algebraic expressions are separated by + and – signs.

- *mono* = one: a monomial has one term
- *bi* = two: a binomial has two terms
- *tri* = three: a trinomial has three terms
- *poly* = many: a polynomial has four or more terms

Rules of signs

$$(+) \times (+) = + \qquad (+) \times (-) = - \qquad (-) \times (-) = + \qquad (-) \times (+) = -$$

Polynomials: revision

Multiplying and simplifying polynomials that contain monomials

To multiply and simplify polynomials that contain monomials, you will need the following skills:

- how to work and with simplify polynomials
- how to use the rules of signs
- how to multiply a monomial by a binomial
- how to multiply a binomial by a binomial.

Worked example

Simplify $x(x^2 - 2x + 4) - 2x^2 + (x - 3)3x$.

Remember

$a \times b = b \times a$

$\therefore (x - 3)3x$

$= 3x(x - 3)$

Solution

$x(x^2 - 2x + 4) - 2x^2 + (x - 3)3x$

$= x(x^2 - 2x + 4) - 2x^2 + 3x(x - 3)$ Simplify by multiplying to remove the brackets.

$= x^3 - 2x^2 + 4x - 2x^2 + 3x^2 - 9x$ Simplify by adding/subtracting all like terms.

$= x^3 - x^2 - 5x$ We usually arrange an algebraic answer in descending order of powers.

Multiplying binomials by binomials

Worked examples

Simplify the following:

1. $(2x - 1)(x + 4)$ **2.** $\left(\frac{1}{x} + 2x\right)\left(\frac{2}{x} - 3x\right)$

Solutions

1. Start by multiplying out.

Note

If you can, do steps ② and ③ in your head, and write down only the answer, e.g. 7x.

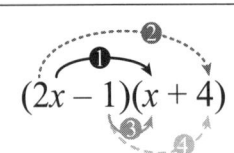

❶ $2x \times x = 2x^2$

❷ $2x \times (+4) = +8x$

❸ $-1 \times (+x) = -x$

❹ $-1 \times (+4) = -4$

$= 2x^2 + 8x - x - 4$

$= 2x^2 + 7x - 4$ Simplify by adding together the like terms.

2. When working with fractions, it helps to express all the terms as fractions.

$\left(\frac{1}{x} + 2x\right)\left(\frac{2}{x} - 3x\right)$

$= \left(\frac{1}{x} + \frac{2x}{1}\right)\left(\frac{2}{x} - \frac{3x}{1}\right)$

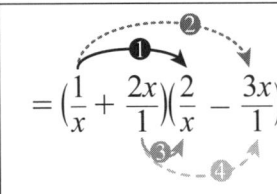

❶ $\frac{1}{x} \times \left(+\frac{2}{x}\right) = \frac{2}{x^2}$

❷ $\frac{1}{x} \times \left(-\frac{3x}{1}\right) = -\frac{3}{1}$

❸ $\frac{2x}{1} \times \left(+\frac{2}{x}\right) = +4$

❹ $\frac{2x}{1} \times \left(-\frac{3x}{1}\right) = -\frac{6x^2}{1}$

$= \frac{2}{x^2} - \frac{3}{1} + \frac{4}{1} - \frac{6x^2}{1}$

$= \frac{2}{x^2} + 1 - 6x^2$ Simplify by adding together the like terms.

Multiplying a perfect square

$(2x - 3)^2$ is an example of a perfect square binomial. Practise doing this multiplication by inspection, in other words do it in one step. This will speed up your working and help you to recognise and factorise perfect square trinomials. However, note that 'squaring' a binomial results in a product that has **three** terms, as shown in the worked example below.

Worked example

Simplify $(2x - 3)^2$.

Remember

$(2x - 3)^2 \neq 4x^2 + 9$

Solution

Long method:

$(2x - 3)^2$

$= (2x - 3)(2x - 3)$

$= 4x^2 - 6x - 6x + 9$

$= 4x^2 - 12x + 9$

Steps for multiplying a perfect square by inspection

1. Square the first term: $(2x)^2 = 4x^2$

2. To get the middle term you multiply the two terms in the brackets and double the result: $2x \times -3 = -6x$ and $-6x \times 2 = -12x$.

3. Square the last term: $(-3)^2 = +9$

By inspection:

$(2x - 3)^2$

$= 4x^2 - 12x + 9$

The product is the difference of two squares

The only time the result of multiplying two binomials is a product that has **two** terms is when the binomials are identical, **but** one has a plus sign and the other one has a minus sign.

Worked example

Multiply $(3x - 4)(3x + 4)$.

Solution

Long method:

$(3x - 4)(3x + 4)$

$= 9x^2 + 12x - 12x - 16$

$= 9x^2 - 16$

By inspection:

By recognising that adding the middle terms will result in zero, you will be able to write the answer only.

$(3x - 4)(3x + 4)$

$= 9x^2 - 16$

Multiplying two binomials or a binomial and a trinomial by a monomial

If there is a factor that is single term (called a monomial) in front of the brackets, first multiply out the brackets, and then multiply the product by the monomial. Remember BODMAS: do brackets before multiplication.

Worked example

Simplify $2x(x - 3)(3x - 5)$.

Solution

$2x(x - 3)(3x - 5)$
$= 2x(3x^2 - 14x + 15)$
$= 6x^2 - 28x + 30x$

Mixed problems involving multiplication

Worked example

Multiply and simplify $(3p - y)^2 + (2p - 3y)(2p + 3y) - (p - 2y)(p + y)$.

Note

In this sort of problem, work carefully, and write down all your steps.

Solution

It will help you if you recognise the difference of two squares pattern, which has no middle term. This is in this part of the expression: $(2p - 3y)(2p + 3y) = 4p^2 - 9y^2$.

$(3p - y)^2 + (2p - 3y)(2p + 3y) - (p - 2y)(p + y)$
$= (9p^2 - 6py + y^2) + (4p^2 - 9y^2) - (p^2 - py - 2y^2)$
$= 9p^2 - 6py + y^2 + 4p^2 - 9y^2 - p^2 + py + 2y^2$
$= 12p^2 - 5py - 6y^2$ Simplify fully by adding/subtracting all like terms.

Remember

When multiplying out, a **minus sign** in front of a bracket **changes the signs** of the terms inside the bracket, so
$-(p^2 - py - 2y^2)$
$= -p^2 + py + 2y^2$.

Exercise 1

Simplify the following:

1. $f^2(f - 2g)(f + g)$
2. $-(x - 3y)(x + 5y)$
3. $-5a^3(a^2 - 1)(a^2 + 1)$
4. $\frac{1}{2}(\frac{1}{x} - 2x)^2$
5. $(9 - a)(a + 3) - (2a - 5)^2$
6. $y(xy - z^2) + xy(x - y)^2 - (x^2 - z^2)y$
7. $2(3a - \frac{1}{a})(2a + \frac{3}{a}) - (-\frac{2}{a})^2$
8. $(p - 2)^2 - p^2(2p + 1) + (3p - 2)(-2p^2) - (4p + 3) - p^2$
9. $(a^2 - 4)^2 + (a^2 + 1)(a^2 - 1) - (2a^2)^2$

Multiplying binomials by trinomials

Multiplying binomials by trinomials can get quite complicated. There are six specific steps you need to follow – approach the problem one step at a time. Stay focused and keep track of your working to avoid errors.

Worked examples

1. Multiply $(x - 2)$ by $(3x^2 - 4x + 2)$.
2. Simplify $-3x(3 - x)(2x^2 - x - 5)$.

Solutions

1. $(x - 2)(3x^2 - 4x + 2)$

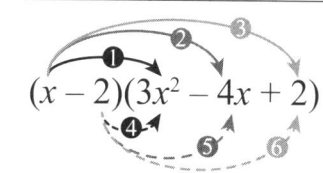

❶ $x \times (+3x^2) = 3x^3$	❹ $-2 \times (+3x^2) = -6x^2$
❷ $x \times (-4x) = -4x^2$	❺ $-2 \times (-4x) = +8x$
❸ $x \times (+2) = +2x$	❻ $-2 \times (+2) = -4$

$= 3x^3 - 4x^2 + 2x - 6x^2 + 8x - 4$

$= 3x^3 - 10x^2 + 10x - 4$ Simplify by adding together the like terms.

2. $-3x(3 - x)(2x^2 - x - 5)$

$= -3x(6x^2 - 3x - 15 - 2x^3 + x^2 - 5x)$ First multiply the brackets, i.e. the binomial by the trinomial.

$= -18x^3 + 9x^2 + 45x + 6x^4 - 3x^3 - 15x^2$ Then multiply by the monomial.

$= 6x^4 - 21x^3 - 6x^2 + 45x$ Add or subtract like terms and give answer in descending order of the exponents.

Exercise 2

1. Given that A = $2x^2 - 3x + 4$, B = $3x - 7$ and C = $3 - 2x$, find:

   1.1 B – A   **1.2** B^2 **1.3** BC

   1.4 AB   **1.5** AB – C   **1.6** A + 2B – BC

   1.7 AC – C^2 – 3C

2. Multiply out and simplify fully:

   2.1 $(3x^2 - 3x)(x^2 - 3x - 4)$ **2.2** $-2x(5x - 2x^3)(4x^2 + 2x - 7)$

   2.3 $(2x - 3y)^3$

3. Subtract $2x^3 - x + 4$ from the product of $(3x - 1)$ and $(5x^2 - 2x - 1)$.

4. Multiply out and simplify fully:

   4.1 $\left(\frac{1}{2}x - 3\right)\left(\frac{1}{4}x^2 + \frac{3}{2}x + 9\right)$ **4.2** $-\left(\frac{1}{2}x - 3\right)^3$

Methods for factorising

Factorisation is a tool you can use to simplify algebraic expressions and fractions in order to solve mathematical problems, so make sure you are able to use it well.

In Grade 9 you learnt to factorise using these methods:

- common factors
- common brackets, with and without sign changes
- the difference of two squares
- trinomials in the form $ax^2 + bx + c$ where $a = 1$ and $b, c \in \mathbb{Z}$.

Worked examples

Factorise the following expressions:

1. $5x^4p^2 - 15p^3x^5 - 20p^3x^2$ 2. $x(a - 2b) - y^2(a - 2b)$

3. $2x(a - 2p) - 3(2p - a)$ 4. $4a^2 - 25b^4$

5. $x^2 + 2x - 15$ 6. $x^2 - 18x + 81$

Solutions

The method used for factorising is indicated next to each solution.

Common factors

1. $5x^4p^2 - 15p^3x^5 - 20p^3x^2$
   $= 5x^2p^2(x^2 - 3px^3 - 4p)$

Common brackets

2. $x(a - 2b) - y^2(a - 2b)$
   $= (a - 2b)(x - y^2)$

Common brackets with sign change

3. $2x(a - 2p) - 3(2p - a)$
   $= 2x(a - 2p) + 3(a - 2p)$
   $= (a - 2p)(2x + 3)$

Difference of two squares

4. $4a^2 - 25b^4$
   $= (2a - 5b^2)(2a + 5b^2)$

Trinomial

5. $x^2 + 2x - 15$
   $= (x + 5)(x - 3)$

Trinomial: perfect square

6. $x^2 - 18x + 81$
   $= (x - 9)(x - 9)$

Combining factorisation methods

It is sometimes necessary to use a combination of methods to factorise an expression fully. In the examples below you have to use more than one method or the expression will not be factorised fully.

Worked examples

Factorise fully:

1. $4(a - 2b) - x^2(a - 2b)$
2. $2x(a - 2p) - 6x^2(2p - a)$
3. $8a^3 - 50ab^8$
4. $3x^3 - 30x^2 + 72x$

Solutions

Common brackets, followed by difference of two squares

1. $4(a - 2b) - x^2(a - 2b)$
   $= (a - 2b)(4 - x^2)$
   $= (a - 2b)(2 - x)(2 + x)$

Common brackets with sign change, followed by separating out common factor

2. $2x(a - 2p) - 6x^2(2p - a)$
   $= 2x(a - 2p) + 6x^2(a - 2p)$

   A sign change is necessary in this case.

   $= (a - 2p)(2x + 6x^2)$
   $= (2x)(a - 2p)(1 + 3x)$

   It is common practise to put the common factor in front of the factorised expression.

   $= 2x(a - 2p)(1 + 3x)$

Common factors, followed by difference of two squares

3. $8a^3 - 50ab^8$
   $= 2a(4a^2 - 25b^8)$
   $= 2a(2a - 5b^4)(2a + 5b^4)$

STUDY & MASTER MATHEMATICS STUDY GUIDE GRADE 10

4. $3x^3 - 30x^2 + 72x$

$= 3x(x^2 - 10x + 24)$

$= 3x(x - 6)(x - 4)$

Exercise 3

Factorise the expressions fully:

1. $4p^2y - 16py^2$

2. $p^2(x + y) - 16(y + x)$

3. $25x^2 - 9y^2$

4. $3a(a - b) + 9b(b - a)$

5. $x^2 - 2xy - 3y^2$

6. $\frac{9}{16}x^2 - \frac{1}{4}y^2$

7. $4x^2(2x - 1) + y^2(1 - 2x)$

8. $2x^2 + 10x + 12$

9. $12y^2 - 7xy + x^2$

10. $\frac{x^2}{2} - \frac{1}{2}$

Factorising by grouping two by two

When factorising polynomial expressions that have two or more terms, you need to group the terms together so you can use a common bracket to factorise the expression.

Steps for grouping

1. Group the expression into two groups of two.

2. Take out the common factor from each group.

3. Write down the common bracket as the first factor, and group the remaining terms as the second factor.

4. If necessary, factorise each bracket further to ensure that the expression is fully factorised.

Worked examples

Factorise fully:

1. $ab + 5b + 3a + 15$

2. $9x^2 - x^2y^2 - 36 + 4y^2$

Solutions

1. $ab + 5b + 3a + 15$

$= (ab + 5b) + (+3a + 15)$

$= b(a + 5) + 3(a + 5)$

$= (a + 5)(b + 3)$

2. $9x^2 - x^2y^2 - 36 + 4y^2$

$= (9x^2 - x^2y^2) + (-36 + 4y^2)$

$= x^2(9 - y^2) + 4(-9 + y^2)$

$= x^2(9 - y^2) - 4(9 - y^2)$ Use a sign change to create a common bracket.

$= (9 - y^2)(x^2 - 4)$

$= (3 - y)(3 + y)(x - 2)(x + 2)$ You need to factorised further using the difference of two squares.

Exercise 4

Factorise fully:

1. $3y + 9z + by + 3bz$

2. $40p + 60r - 20pq - 30qr$

3. $ac - bc - ad + bd$

4. $b^3 + 6 + 3b^2 + 2b$

5. $6ac - 15ax + 10xb - 4bc$

6. $5(a + b) - (a + b)^2$

7. $3ax^2 + 2ab^2 - 2x^3 - 4xb^2$

8. $1 - p - p^2 + p^3$

9. $x^2(p - q) - 2x(p - q) - p + q$

Factorising trinomials

In Grade 9 you learnt to factorise quadratic expressions of the form $ax^2 + bx + c$ and $ax^2 + bxy + cy^2$, where $a = 1$ and $b, c \in \mathbb{Z}$. In these expressions the coefficient of the squared term is 1.

Note

The methods described here only work when the expression is written in descending powers of x.

Worked example

Factorise $x^2 + 3x - 10$.

Solution

> **Steps for factorising trinomials**
>
> 1. Think of two numbers which when multiplied give -10, and when added give $+3$.
>
> 2. Rewrite the middle term as the sum of the two numbers you found in step 1.
>
> 3. Now use your knowledge of grouping to factorise the expression.
>
> 4. Check your solution by finding the product of the two factors. This is just a check, so don't write down this step.

$-2 \times 5 = -10$	or	$2 \times -5 = -10$
$(-2) + (+5) = +3$	or	$(+2) + (-5) = -3$	Step 1
Correct values		Incorrect values

$\therefore x^2 + 3x - 10$

$= x^2 - 2x + 5x - 10$ Step 2

$= (x^2 - 2x) + (+5x - 10)$ Step 3

$= x(x - 2) + 5(x - 2)$

$= (x - 2)(x + 5)$

> **Steps for factorising trinomials where $a \neq 1$**
>
> Factorising a trinomial where $a \neq 1$ becomes more complicated.
>
> 1. Multiply the coefficient of the squared term by the last term, and call this value K.
>
> 2. Now use trial and error to find two numbers that are **factors** of K that **add up** to the value of the **middle term** of the ordered trinomial expression.
>
> 3. Rewrite the **middle term** as the **sum of the two numbers** you found in step 2.
>
> 4. Use your knowledge of grouping to factorise the expression.
>
> 5. Check your solution by finding the product of the two factors. Don't show this step.

Worked example

Factorise $6x^2 + 11x + 3$.

Solution

Step 1: This method is essentially the same as when the coefficient of the squared term is 1, except that you start by multiplying the coefficient of the squared term by the last term.	$6x^2 + 11x + 3$ so $6 \times 3 = 18$
Step 2: Look for factors of 18 that add up to $+11$. Use trial and error until you find the correct numbers.	$1 \times 18 = 18$ but $18 + 1 = 19$, so these numbers are incorrect $6 \times 3 = 18$ but $6 + 3 = 9$, so these numbers are incorrect $2 \times 9 = 18$ and $2 + 9 = 11$, so these numbers are correct
Step 3: Rewrite the trinomial expression replacing the middle term with the two numbers you found.	$\therefore 6x^2 + 11x + 3$ $= 6x^2 + 2x + 9x + 3$
Step 4: Use grouping to factorise the expression.	$= (6x^2 + 2x) + (+9x + 3)$ $= 2x(3x + 1) + 3(3x + 1)$
Step 5: Check that your factors are correct by mentally calculating the product of your two factors.	$= (3x + 1)(2x + 3)$

Worked example

Factorise $3y^2 - 10x^2 - xy$.

Solution

$3y^2 - 10x^2 - xy$

$= -10x^2 - xy + 3y^2$ Rearrange the trinomial in descending powers.

$= -1(10x^2 + xy - 3y^2)$ Take out a common factor of -1 to simplify your working.

Use the steps to find the factors. Steps 1 and 2 should be done in rough. Show only steps 3 and 4.

$K = 10 \times (-3) = -30$ Step 1

$6 \times (-5) = -30$ and $6 + (-5) = +1$ Step 2

$\therefore$ the numbers are 6 and -5.

$= -1(10x^2 + 6xy - 5xy - 3y^2)$ Step 3

$= -1[(10x^2 + 6xy) + (-5xy - 3y^2)]$

$= -1[2x(5x + 3y) - y(5x + 3y)]$ Step 4

$= -1(5x + 3y)(2x - y)$

Exercise 5

Factorise fully:

1. $x^3 + 2x^2y - 15xy^2$ **2.** $3x^2 + 24x + 45$ **3.** $2x + 15 - x^2$

4. $30 + 2x^2 - 16x$ **5.** $3a^2 + 2ab - 8b^2$ **6.** $12x^2 - 14x + 4$

7. $11x + 6 + 3x^2$ **8.** $27x + 6 - 15x^2$ **9.** $22a^2 - 37a + 6$

10. $78x^2 - 111xy - 9y^2$

Sum or difference of two cubes

This type of expression consists of:

• only two terms

• both terms are perfect cubes.

Some examples of perfect cubes are $8x^3$, 27, $\frac{8}{x^3}$, $64p^3$.

The general form for this pattern is summarized below.

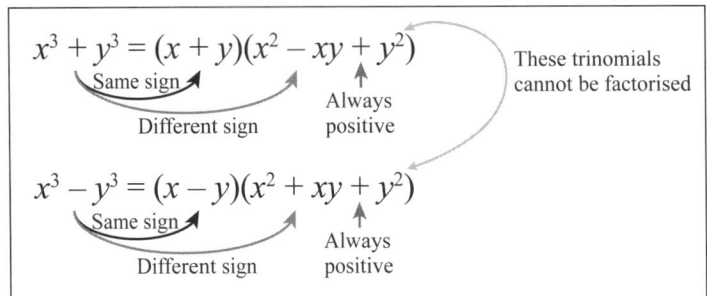

Work out the cube roots and the squares of the terms carefully, for example:

$$\sqrt[3]{125x^3} = 5x \quad \sqrt[3]{8p^6} = 2p^2 \quad (5x)^2 = 25x^2 \quad 5x \times 2p^2 = 10xp^2 \quad (2p)^2 = 4p^2$$

Steps for factorising the sum or difference of two cubes

Example: $27 - x^3$

1. In the binomial first bracket, write the cubed root of each term: $(3 - x)(\ )$

2. For the trinomial second bracket, first square the term: $3^2 = 9$

3. Then for the middle term of the trinomial, multiply the two terms in the binomial, and write the product with the **opposite sign**: $3 \times (-x) = -3x$ so write $+3x$

4. Finally, square the last term: $(-x)^2 = +x^2 \therefore 27 - x^3 = (3 - x)(9 + 3x + x^2)$

Worked example

Factorise fully: $1\,000x^3 + 64p^6$

Solution

$1\,000x^3 + 64p^6$

$= 8(125x^3 + 8p^6)$ Always take out the common factor if there is one.

$= 8(5x + 2p^2)(25x^2 - 10xp^2 + 4p^4)$

Exercise 6

Factorise fully:

1. $8b^3 - 343$ **2.** $64b^3 + 27c^3$ **3.** $81ab^4 + 24a^7b$

4. $-1\,000x^6 + 125y^3$ **5.** $64b^3 - 125c^3$ **6.** $(x - 1)^3 - 27$

7. $\frac{a^3}{8} - \frac{b^3}{27}$ **8.** $\frac{1\,000}{x^6} + \frac{y^3}{27}$

How to recognise the factorising method

The secret to factorising correctly is about recognising which method to use, especially when more than one method is needed. You need to identify the type of expression and then apply the factorising method that applies to the type of expression. Use this flow chart to help you recognise methods.

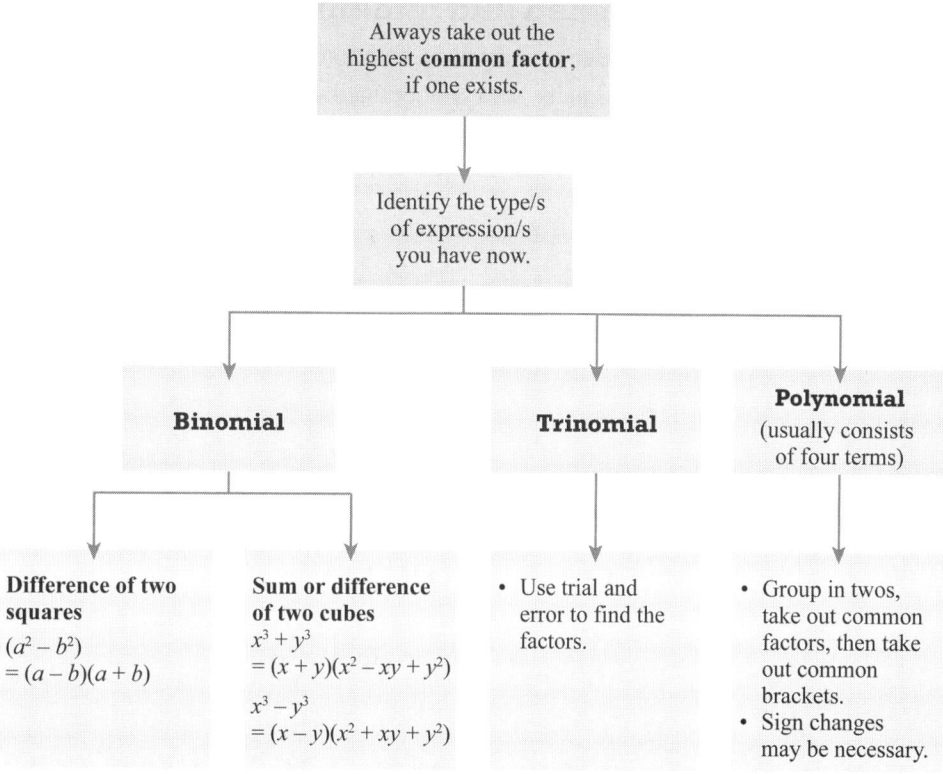

Always take out the highest **common factor**, if one exists.

Identify the type/s of expression/s you have now.

Binomial **Trinomial** **Polynomial** (usually consists of four terms)

Difference of two squares
$(a^2 - b^2)$
$= (a - b)(a + b)$

Sum or difference of two cubes
$x^3 + y^3$
$= (x + y)(x^2 - xy + y^2)$
$x^3 - y^3$
$= (x - y)(x^2 + xy + y^2)$

- Use trial and error to find the factors.

- Group in twos, take out common factors, then take out common brackets.
- Sign changes may be necessary.

Steps for recognising methods needed to factorise

1. Take out common factors.

2. Identify the expression, i.e. is it a binomial, a trinomial or a polynomial?

3. Decide which factorising method to use and apply it correctly.

4. Consider your answer carefully:
   - Is it fully simplified?
   - Can it be factorised further? If so start the process again.

Exercise 7

Factorise each expression:

1. $3x^2 - 7x - 6$

2. $6x^2 - 9ax + 8cx - 12ac$

3. $4x^4 - 36$

4. $x^4 - 13x^2 + 36$

5. $\frac{2}{5}x^2y^4 - \frac{4}{5}py^3$

6. $8 - f^3$

7. $6x^2 + 4x - 16$

8. $4x^2 - 12xy + 9y^2$

9. $18a^2 - 32b^2$

10. $(2z - 1)^2 - 7(2z - 1)$

11. $(5c + 1)(a - b) - 2c(a - b)$

12. $3ay - 5y + y^2 - 15a$

13. $\frac{1}{2}f^3 - 4g^3$

14. $2x^2 - 3 - \frac{2}{x^2}$

15. $3x^6 - 3y^6$

Algebraic fractions

Algebraic fractions can be simplified, multiplied, divided, added and subtracted in much the same way as arithmetic fractions. Factorising the numerator and the denominator plays an important part in the simplification of algebraic fractions.

Defined and undefined fractions

The denominator of a fraction may not be zero. We cannot divide by zero because division by zero is undefined or meaningless. So the examples of fractions given below are all undefined.

$$\frac{5}{0} \qquad \frac{5x-1}{0} \qquad \frac{5}{x} \text{ if } x = 0$$

$$\frac{5}{x-3} \text{ if } x = 3 \text{ because } \frac{5}{3-3} = \frac{5}{0}$$

Worked example

For which values of x will the following fractions be undefined?

1. $\frac{5}{x+3}$ 2. $\frac{x+1}{x-1}$ 3. $\frac{x+1}{x(x-1)}$ 4. $\frac{x+1}{3x^2-x-2}$

Solutions

1. $\frac{5}{x+3}$ is undefined if $x = -3$ because $\frac{5}{-3+3} = \frac{5}{0}$

   Or $x + 3 = 0 \therefore x = -3$ Make the denominator equal to 0 to work out the values of x for which the fraction is undefined.

   $\therefore \frac{5}{x+3}$ is undefined for $x = -3$

2. $x - 1 = 0 \therefore x = 1$

   $\therefore \frac{x+1}{x-1}$ is undefined for $x = 1$

3. $x(x - 1) = 0$

   $\therefore x = 0 \text{ or } x = 1$

   $\therefore \frac{x+1}{x(x-1)}$ is undefined for $x = 0$ or $x = 1$

   Check:

   If $x = 0$ then $\frac{0+1}{0(0-1)} = \frac{1}{0}$ which is undefined.

   If $x = 1$ then $\frac{1+1}{1(1-1)} = \frac{2}{0}$ which is undefined.

4. $3x^2 - x - 2 = 0$

   $\therefore (3x + 2)(x - 1) = 0$

   $\therefore x = -\frac{2}{3} \text{ or } x = 1$

   $\therefore \frac{x+1}{3x^2-x-2}$ is undefined for $x = -\frac{2}{3}$ or $x = 1$

Manipulating and simplifying fractions

Your ability to factorise and to recognise when an expression is fully factorised is vitally important when manipulating fractions.

> **Steps for simplifying fractions**
> 1. Factorise both the denominator and the numerator.
> 2. Cancel out the factors that are common to both the numerator and the denominator.

Worked example

Simplify:

1. $\dfrac{9g^2h}{18gh^3}$ 2. $\dfrac{6x^2 - 3x}{2x^2 - 7x + 3}$ 3. $\dfrac{8x^3 + 4x + 2x}{16x^3 - 2}$

Solutions

Note

You should be able to cancel out mentally.

1. $\dfrac{9g^2h}{18gh^3}$

$= \dfrac{\cancel{9} \times g \times g \times \cancel{h}}{2 \times \cancel{9} \times g \times \cancel{h} \times h \times h}$

$= \dfrac{g}{2h^2}$

Remember

You CANNOT cancel across plus or minus signs! So:

$\dfrac{6x^2 - 3x}{2x^2 - 7x + 3} \neq \dfrac{6 - 3}{2 - 7 + 3}$

2. $\dfrac{6x^2 - 3x}{2x^2 - 7x + 3}$

It is very important to factorise both the numerator and the denominator fully before cancelling.

$= \dfrac{3x\cancel{(2x - 1)}}{\cancel{(2x - 1)}(x - 3)}$

Factorise the quadratic trinomial in the denominator, then cancel common factors.

$= \dfrac{3x}{(x - 3)}$

3. $\dfrac{8x^3 + 4x + 2x}{16x^3 - 2}$

$= \dfrac{2x(4x^2 + 2x + 1)}{2(8x^3 - 1)}$

First factorise.

$= \dfrac{\cancel{2}x\cancel{(4x^2 + 2x + 1)}}{\cancel{2}(2x - 1)\cancel{(4x^2 - 2x + 1)}}$

Cancel common factors

$= \dfrac{x}{(2x - 1)}$

Multiplying and dividing fractions

When working with fractions, the instruction 'simplify fully' means that you must manipulate the expression and express it as a single fraction in its simplest form.

Worked examples

Simplify fully:

1. $\dfrac{18pq}{15p^2} \times \dfrac{5p^2}{9q}$ 2. $\dfrac{2x^2 - 4x + 2}{2x^2 - 5x + 3} \div \dfrac{x^3 - 1}{4x^2 - 9}$

Solutions

1. $\dfrac{^2\cancel{18pq}}{^3\cancel{15p^2}} \times \dfrac{\cancel{5p^2}}{\cancel{9q}} = \dfrac{2p}{3}$

2. $\dfrac{2x^2 - 4x + 2}{2x^2 - 5x + 3} \div \dfrac{x^3 - 1}{4x^2 - 9}$

$= \dfrac{2(x^2 - 2x + 1)}{2x^2 - 5x + 3} \times \dfrac{4x^2 - 9}{x^3 - 1}$

Use the rule 'turn the divisor upside down and multiply' to convert division into multiplication.

$= \dfrac{2\cancel{(x - 1)}\cancel{(x - 1)}}{\cancel{(2x - 3)}\cancel{(x - 1)}} \times \dfrac{\cancel{(2x - 3)}(2x + 3)}{\cancel{(x - 1)}(x^2 + x + 1)}$

difference of two squares

difference of two cubes

quadratic trinomials

$= \dfrac{2}{1} \times \dfrac{(2x + 3)}{x^2 + x + 1}$

$= \dfrac{2(2x + 3)}{x^2 + x + 1}$

Express as a single fraction.

Using a sign change to simplify fractions

It will sometimes be necessary to use a legal sign change to simplify algebraic fractions.

Worked examples

Simplify the following expressions.

1. $\dfrac{x - y}{y - x}$

2. $\dfrac{x + y - z}{z - y - x}$

Solutions

1. $\dfrac{x - y}{y - x}$

   $= \dfrac{(x - y)}{-(x - y)}$ $(y - x) = -(-y + x) = -(x - y)$

   $= \dfrac{1}{-1} = -1$

2. $\dfrac{x + y - z}{z - y - x}$

   $= \dfrac{x + y - z}{-x - y + z}$

   $= \dfrac{x + y - z}{-(x + y - z)}$ $(-x - y + z) = -(x + y - z)$

   $= \dfrac{1}{-1} = -1$

Steps for simplifying, multiplying and dividing fractions

1. Factorise both the numerator and the denominator.

2. Cancel out common factors that occur in both the numerator and the denominator (using a sign change if necessary).

3. Simplify fully.

Exercise 8

1. For which values of x will the following expressions be undefined?

   1.1 $\dfrac{1}{x}$ **1.2** $\dfrac{x}{x + 8}$ **1.3** $\dfrac{x + 1}{2x + 3}$ **1.4** $\dfrac{2x - 1}{(x + 3)(x - 5)}$

Simplify each of the fractions given below.

2. $\dfrac{x^2 - 4}{x^2 - 5x + 6}$

3. $\dfrac{x - y}{x + y} \times \dfrac{xy + y^2}{x^2 - xy}$

4. $\dfrac{x}{x - y} \div \dfrac{y}{y - x}$

5. $\dfrac{3z - 2}{z^2 - 4z} \times \dfrac{3z}{3z^2 + 10z - 8}$

6. $\dfrac{x^3 - 4x}{21x} \times \dfrac{7x^2}{x^4 - 8x}$

Adding and subtracting fractions

Fractions can be added and subtracted only when they have a common denominator.

Because the solution is required in its simplest form, always use the lowest common denominator.

Worked example

Express as a single fraction in its simplest form: $\dfrac{x}{3} - \dfrac{x - 4}{2x}$.

Solution

$\dfrac{x}{3} - \dfrac{x-4}{2x}$

$= \dfrac{2x^2 - 3(x-4)}{6x}$

$= \dfrac{2x^2 - 3x + 12}{6x}$

LCD $= 6x$

Using brackets is essential when adjusting the numerator.

Be careful with the signs.

> **Steps for adding and subtracting fractions**
> 1. Identify the lowest common denominator.
> 2. Adjust the numerator(s). Be careful with your signs!
> 3. Simplify fully.

Worked example

Express as a single fraction in its simplest form: $\dfrac{2}{x-2} - \dfrac{3}{2x-1}$.

Solution

$\dfrac{2}{x-2} - \dfrac{3}{2x-1}$

$= \dfrac{2(2x-1) - 3(x-2)}{(x-2)(2x-1)}$

$= \dfrac{4x - 2 - 3x + 6}{(x-2)(2x-1)}$

$= \dfrac{x+4}{(x-2)(2x-1)}$

Step 1: Identify the lowest common denominator: $(x-2)(2x-1)$

Step 2: Adjust the numerators.

Step 3: Simplify fully.

Using factorisation to find the common denominator

Worked example

Simplify: $\dfrac{2ab}{a^2 - b^2} + \dfrac{a}{a+b} + \dfrac{b}{b-a} + 1$.

Solution

$\dfrac{2ab}{a^2 - b^2} + \dfrac{a}{a+b} + \dfrac{b}{b-a} + 1$

$= \dfrac{2ab}{(a-b)(a+b)} + \dfrac{a}{(a+b)} + \dfrac{b}{(b-a)} + \dfrac{1}{1}$

$= \dfrac{2ab}{(a-b)(a+b)} + \dfrac{a}{(a+b)} - \dfrac{b}{(a-b)} + \dfrac{1}{1}$

$= \dfrac{2ab + a(a-b) - b(a+b) + 1(a-b)(a+b)}{(a-b)(a+b)}$

$= \dfrac{2ab + a^2 - ab - ba - b^2 + a^2 - b^2}{(a-b)(a+b)}$

$= \dfrac{2a^2 - 2b^2}{(a-b)(a+b)}$

$= \dfrac{2(a^2 - b^2)}{(a-b)(a+b)}$

$= \dfrac{2(a-b)(a+b)}{(a-b)(a+b)}$

$= 2$

Use a sign change when finding the LCD: $+(b-a) = -(a-b)$

LCD $= (a-b)(a+b)$

Carefully adjust the numerator to the new denominator.

Simplify the numerator by adding together like terms.

Factorise to simplify.

Cancel common factors.

Exercise 9

Simplify:

1. $\frac{1}{a} + \frac{1}{3a} + \frac{1}{2a}$

2. $\frac{2y-1}{3y} - \frac{1}{6}$

3. $\frac{1}{b+1} + \frac{1}{b-1}$

4. $\frac{6}{2w-3} - \frac{3}{3-2w}$

5. $\frac{1}{x-2} + \frac{2x}{x} - \frac{x-1}{x+2}$

6. $\frac{3}{1-c} + \frac{4}{1-2c+c^2}$

7. $\frac{4}{r^2+2r} - \frac{3r}{r+2} + \frac{3r-2}{r}$

8. $\frac{1}{2z-4} + \frac{1}{z^2-5z+6} + \frac{1}{(z-2)(z^2-5z+6)}$

Fractions involving addition/subtraction and multiplication/division

It can become confusing when operations are combined or mixed up all in one exercise. Use the steps given above for each type of operation. First use the steps for multiplication and division, and only then use the steps for adding and subtracting. Sometimes you may have to factorise more than once in order to simplify fully.

Worked example

Simplify fully: $\left(1 - \frac{1}{x}\right) \div \left(1 + \frac{1}{x}\right)$.

Solution

$\left(1 - \frac{1}{x}\right) \div \left(1 + \frac{1}{x}\right)$

$= \left(\frac{x-1}{x}\right) \div \left(\frac{x+1}{x}\right)$ Work with one bracket at a time, adding and subtracting by forming common denominators.

$= \left(\frac{x-1}{x}\right) \times \left(\frac{x}{x+1}\right)$ Invert and multiply.

$= \frac{x-1}{x+1}$ Simplify.

Exercise 10

Given that $x = \frac{t-2}{t+2}$ and $y = \frac{t+2}{t-2}$, simplify:

1. $x + y$

2. $x - y$

3. xy

4. $x \div y$

5. Use your results to express $(x + y) \div (x - y)$ in terms of t.

6. Use your answer for 1.5 to show that $(x + y) \div (x - y) \neq x + y \div x - y$. Support your answer with a fully worked solution.

Factorising: complex procedures and problem solving

In this section we will:

- use factorisation to simplify multiplication
- work with more involved factorisation including:
    - using a suitable substitution
    - more complicated problems involving grouping
- work with more complicated fractions.

Using factorisation to simplify multiplication

Your knowledge of factorisation can make multiplication easier.

Worked example

Use your knowledge of factorisation to simplify the following expressions:

1. $(a - b - c)(a - b + c)$ **2.** $(x + y)(2x - y)(4x^2 + 2xy + y^2)$

Solutions

1. $(a - b - c)(a - b + c)$

$= [(a - b) - c][(a - b) + c]$

$= (a - b)^2 - c^2$

$= a^2 - 2ab + b^2 - c^2$

> **Note**
>
> Multiplying a trinomial by a trinomial is a lengthy procedure, which can be simplified by rewriting this expression as the difference of two squares.

2. $(x + y)(2x - y)(4x^2 + 2xy + y^2)$

$= (x + y)(8x^3 - y^3)$

$= 8x^4 - xy^3 + 8x^3y - y^4$

> **Note**
>
> Recognising that the second and the third factors represent a difference of two cubes makes this multiplication problem a lot easier.

K substitution

Using a suitable substitution when factorising a complicated expression can help to simplify your working. In the examples below the variable K is used to replace sections of an expression, which makes the pattern easier to recognise and the working easier.

Worked example

Factorise:

1. $16x^2 - 4(5x - 1)^2$ **2.** $(2m + 1)^2 - 8(2m + 1) + 15$

Solutions

1. $16x^2 - 4(5x - 1)^2$

Let K $= (5x - 1)$, then $16x^2 - 4(5x - 1)^2 = 16x^2 - 4K^2$

$= 4(4x^2 - K^2)$ Common factor

$= 4(2x - K)(2x + K)$ Difference of two squares

K $= (5x - 1)$

$\therefore 4(2x - K)(2x + K) = 4[2x - (5x - 1)][2x + (5x - 1)]$

$= 4(2x - 5x + 1)(2x + 5x - 1)$

$= 4(-3x + 1)(7x - 1)$

> **Note**
>
> Substitute and factorise in terms of K.

> **Note**
>
> Change K back into the original expression, and simplify each factor.

2. $(2m + 1)^2 - 8(2m + 1) + 15$

Let K $= (2m + 1)$, then $(2m + 1)^2 - 8(2m + 1) + 15 = K^2 - 8K + 15$

$K^2 - 8K + 15 = (K - 5)(K - 3)$ This is a trinomial.

K $= (2m + 1)$

$\therefore (K - 5)(K - 3) = [(2m + 1) - 5][(2m + 1) - 3]$

$= (2m + 1 - 5)(2m + 1 - 3)$

$= (2m - 4)(2m - 2)$

$= (2)(m - 2)(2)(m - 1)$ Simplify fully by taking out the common factors.

$= 4(m - 2)(m - 1)$

Grouping: complex procedures

Sometimes a problem does not seem to fit any of the recognised patterns. This type of problem requires more thought and some creative thinking. One method is to order the terms differently and group differently.

Worked examples

Factorise fully:

1. $4x^2 - px^2 - 8x + 2px + 4 - p$

2. $4x^2 - 25p^2 - 4x + 1$

Solutions

Note

Grouping in two groups of two may not always work when factoring polynomials. In this case there are three groups, and each group has two terms.

1. $4x^2 - px^2 - 8x + 2px + 4 - p$

$= (4x^2 - px^2) + (-8x + 2px) + (4 - p)$

$= x^2(4 - p) + 2x(-4 + p) + (4 - p)$

$= x^2(4 - p) - 2x(4 - p) + (4 - p)$ Use a legal sign change to change the middle bracket.

$= (4 - p)(x^2 - 2x + 1)$ Expression has a binomial and a trinomial, so factorise the trinomial if possible.

$= (4 - p)(x - 1)(x - 1)$

Note

You may need to order and group a polynomial differently from the traditional two by two pattern.

2. $4x^2 - 25p^2 - 4x + 1$

$= 4x(x - 1) + (1 - 25p^2)$

$= 4x^2 - 4x + 1 - 25p^2$

$= (4x^2 - 4x + 1) - (5p)^2$ If we rearrange this expression, it is made up of a perfect square trinomial and a monomial.

$= (2x - 1)^2 - (5p)^2$ The resulting pattern is now a difference of two squares.

$= [(2x - 1) - 5p][(2x - 1) + 5p]$

$= (2x - 1 - 5p)(2x - 1 + 5p)$ You are not allowed to leave brackets within brackets, so you must remove the inner brackets.

Fractions: complex procedures

More complicated fractions problems can involve addition, subtraction and division all in the same problem.

Worked example

Note

This is known as a complex fraction.

Simplify $\dfrac{\frac{x}{y} - \frac{y}{x}}{\frac{2}{y} + 2 + \frac{y}{x}}$.

Solution

$\dfrac{\frac{x}{y} - \frac{y}{x}}{\frac{2}{y} + 2 + \frac{y}{x}}$

$= \dfrac{\frac{x^2 - y^2}{xy}}{\frac{2x + 2xy + y^2}{xy}}$ Work with the numerator and the denominator separately. Find and adjust for the LCD for each fraction.

$= \dfrac{x^2 - y^2}{xy} \times \dfrac{xy}{2x + 2xy + y^2}$ Invert and multiply.

$= \dfrac{x^2 - y^2}{2x + 2xy + y^2}$ It does not help to factorise the numerator because this fraction cannot be simplified any further.

Exercise 11: Complex procedures

1. Use your knowledge of factorisation to simplify the expressions given below.

    1.1 $(a - b - 3 - x)(a - b + 3 + x)$

    1.2 $(8x^3 + 1)(2x - 1)(4x^2 + 2x + 1)$

    1.3 $(a - b)(a + b)(a^2 + b^2)(a^4 + b^4)$

2. Factorise fully.

    2.1. $x^2 - 2x - y^2 - 2y$ **2.2** $a^2(1 + b) - b^2(1 - a)$

    2.3 $p^2 + 4m - 4m^2 - 1$ **2.4** $a^2 - ab + 2a + b - 3$

    2.5 $16 - 2(p - 3)^3$ **2.6** $a^2 - 16 - 2ab + b^2$

    2.7 $6(2x - 1)^2 + (2x - 1) - 2$ **2.8** $(2x - 1)^2 - 4(a - 1)^2$

    2.9 $2(a - b)^2 - 2(a - b) - 12$ **2.10** $2(2x - y)^2 - 8(a + b)^2$

3. Simplify fully: $\left(\frac{2(3 + x)}{2 - x} + 1\right) \div \left(\frac{3 + x}{2 - x} - 1\right)$.

4. Simplify fully: $\dfrac{\frac{1}{y - 1} + \frac{1}{y}}{\frac{2}{y} - \frac{3}{y} - 1}$.

Test A: Knowledge and routine procedures

1. Simplify:

    1.1 $(3 - x)(3 + 2x)$ (2)

    1.2 $(2a + 3b)(a^2 - 4ab + 3b^2)$ (4)

    1.3 $\left(\frac{2}{x} + \frac{x}{2}\right)^2$ (3)

2. What must be added to $(z + 2t)^2$ to give $z^2 + 4t^2$?   (2)

3. Given that A $= \frac{1}{m + 2}$ and B $= \frac{1}{m + 1}$, calculate:

    3.1 A $-$ B   (4)

    3.2 A $\div$ B   (2)

    3.3 For what value(s) of m is A $= \frac{1}{m + 2}$ undefined?   (1)

4. Simplify fully:

    4.1 $\frac{(m - 3)^2}{m^2 - 9}$ (3)

    4.2 $\frac{2x - 3y}{xy - y^2} \div \frac{9xy - 6x^2}{2x - 2y}$ (5)

    4.3 $\frac{6x}{(x + 2)(x - 2)} - \frac{2}{x + 2} + \frac{3}{2 - x}$ (6)

    4.4 $\left(1 + \frac{1}{r}\right) \div (1 + r)$ (3)

5. Factorise fully:

    5.1 $5a^3 - 15a^2 - 20a$ (3)

    5.2 $F^3 - 125$ (2)

    5.3 $5 - \frac{45}{z^2}$ (3)

    5.4 $x^2(a - 3) + 7(3 - a) - 10(3 - a)$ (3)

    5.5 $6ab - 15ac + 4b^2 - 25c^2$ (4)

Total 50

Test B: Complex procedures and problem solving

1. Use your knowledge of factorisation to simplify:

   1.1 $(3a + b)(3a - b)(81a^4 + 9a^2b^2 + b^4)$ (3)

   1.2 $(x - 2)(x + 2)(x^2 + 4)(x^4 + 16)$ (3)

2. Factorise fully:

   2.1 $5a + 5b - a^2 - 2ab - b^2$ (3)

   2.2 $4x^2(3y + 1) - 9y^2(2x + 1)$ (5)

   2.3 $6(7x - 3)2 - 7(7x - 3) - 3$ (5)

   2.4 $p^7 - 27p(x + y)^3$ (6)

   2.5 $a^2 - 2ab + b^2 - m^2 + 2mn - n^2$ (5)

3. Simplify fully:

   3.1 $\dfrac{3x - 4}{3x^2 - x - 4} - \dfrac{4x}{x^2 - 2x - 3}$ (8)

   3.2 $\left(2x - 3 + \dfrac{7}{x + 3}\right) \div \left(x + 1 - \dfrac{3}{2x + 1}\right)$ (6)

4. If $9x^2 - 30x + k$ is a perfect square, what is the value of k?   (2)

5. What is the value of f if $(4 - 3x)$ is a factor of $9x^2 + fx + 8$?   (2)

6. Show that $(2x - 5)^2 - (3 - x)^2$ can be written as $(3x - 8)(x - 2)$.   (2)

Total 50

Test C: Content and breakdown as for exam

1. Are the following statements true or false? If false, change the right hand side of the equation to make the statement true.

   1.1 $p - q = -(q - p)$ for all values of p and q (1)

   1.2 $(x - 2)^2 = x^2 + 4$ for all values of x (1)

   1.3 $(x - y)(x + y) = (x^2 - y^2)$ for all values of x and y (2)

   1.4 $(x - 3)(x + 2) = x^2 - 5x - 6$ (2)

2. Determine the following products and simplify fully:

   2.1 $\left(3 - \dfrac{x}{2}\right)^2$ (3)

   2.2 $(x - y)(2x^2 - x + 2)$ (3)

   2.3 Show that $2(2p - 1)^2 + 5(2p - 1) - 3 = 2(4p - 3)(p + 1)$.   (6)

3. Given that $A = \dfrac{n}{1 - n^2}$ and $B = \dfrac{1}{1 - 2n + n^2}$, find:

   3.1 $A + B$ (5)

   3.2 $B \div A$ (4)

   3.3 For which value(s) of n is $\dfrac{1}{1 - 2n + n^2}$ undefined?   (2)

4. If $x + \dfrac{1}{x} = 7$, determine the value of $x^2 + \dfrac{1}{x^2}$.   (3)

5. Factorise fully:

    5.1 $(x - y)^3 - 125$ (4)

    5.2 $c^2 - b^2 - 2ab - 2ac$ (4)

    5.3 $x^2(p + q) - 2x(p + q) + p + q$ (4)

6. Simplify fully: $\left(1 - \frac{1}{u}\right) \div (1 - u)$.   (4)

7. For what value(s) of u is the expression in question 6 undefined?   (1)

Total 50

Grade 10 we revise the laws of exponents learnt in Grade 9 and introduce exponential equations.

Knowledge and skills for this topic

If you struggle with any of the work listed below, revise it before continuing with this Topic:

- prime factors
- exponential notation
- using the exponential laws and definitions to simplify expressions.

Content of final exam

- Simplify expressions using the laws of exponents for exponents that are integers.
- Use the laws of exponents to simplify expressions and solve equations, accepting that the rules also hold for $m, n \in \mathbb{Q}$.

Definitions and laws of exponents

Definition

$x^n = x \cdot x \cdot x \cdot x \ldots$ until there are n factors of x, where $x \in \mathbb{R}$, and $n \in \mathbb{N}$.

Terminology

x^n is the nth power of x, or you can say 'x to the nth'.

n is called the exponent.

x is the base of the power x^n.

Law where $x, y > 0$ and $m, n \in \mathbb{Z}$	Application	Description
$x^m \times x^n = x^{m+n}$	$10^5 \times 10^3 = 10^{5+3} = 10^8$ $2^3 \times 2^7 = 2^{3+7} = 2^{10}$ NB: The base stays the same! $2^3 \times 2^7 \neq 4^{3+7} \neq 4^{10}$ $2x^2y \times 3xy^3 = 2 \times 3x^{2+1}y^{1+3} = 6x^3y^4$ $x^n \times x^{n+3} = x^{n+n+3} = x^{2n+3}$	When we **multiply** two powers that have the **same base**, we **add the exponents**.
$x^m \div x^n = x^{m-n}$	$10^5 \div 10^3 = 10^{5-3} = 10^2$ $2^3 \div 2^7 = 2^{3-7} = 2^{-4}$ $x^n \div x^{n+3} = x^{n-(n+3)} = x^{n-n-3} = x^{-3}$ $\dfrac{5py^3}{15p^3y^2} = \dfrac{y^{3-2}}{3p^{3-1}} = \dfrac{y}{3p^2}$	When we **divide** two powers that have the **same base**, we **subtract the exponents**.
$(x^m)^n = x^{mn}$	$(10^5)^3 = 10^{5 \times 3} = 10^{15}$ $(2^3)^7 = 2^{3 \times 7} = 2^{21}$ $(x^n)^{n+3} = x^{n(n+3)} = x^{n^2+3n}$	When we have a **power raised to another power**, we **multiply the exponents**.

Law where $x, y > 0$ and $m, n \in \mathbb{Z}$	Application	Description
$x^m \times y^m = (xy)^m$ $\left(\frac{x}{y}\right)^m = \left(\frac{x^m}{y^m}\right)$ NB: If the bases are the same, i.e. $x = y$, then $x^m \times x^m = (x^2)^m$	$10^5 \times 5^5 = (10 \times 5)^5 = 50^5$ $2^3 \times 3^3 = (2 \times 3)^3 = 6^3$ NB: $2^3 \times 3^3 \neq (2 \times 3)^6 \neq 6^6$ $2^3 \times 3^4 \neq (2 \times 3)^7 \neq 6^7$ $(2x^2)^3 = 2^3 x^{2 \times 3} = 2^3 x^6$ $\left(\frac{2x}{3y^4}\right)^3 = \frac{2^3 x^3}{3^3 y^{4 \times 3}} = \frac{8x^3}{27y^{12}}$ $3^3 \times 3^3 = (3 \times 3)^3 = 9^3$ or $3^3 \times 3^3 = (3 \times 3)^3 = 3^{2 \times 3} = 3^6$	When we **multiply or divide** two powers that have **different bases** but have the same power, we **only multiply the bases**.
Also by definition:		
$x^{-n} = \frac{1}{x^n}, x \neq 0$	$10^{-5} = \frac{1}{10^{+5}} = \frac{1}{10^5}$ $2^{-3} = \frac{1}{2^{+3}} = \frac{1}{8}$ $\frac{1}{x^{-2}} = \frac{x^{+2}}{1} = x^2$ $\frac{2^{-1}}{3a^{-2}} = \frac{a^2}{2 \cdot 3} = \frac{a^2}{6}$	This definition allows us to move terms from the numerator to the denominator and vice versa.
$x^0 = 1, x \neq 0$	$10^0 = 1$ $2^0 = 1$ $(2ab^2)^0 = 1$ $(w - 4x)^0 = 1$ NB: $-10^0 = -1$, but $(-10)^0 = +1$ $-(2^3)^0 = -1$ but $(-2^3)^0 = +1$ $-5(2^3)^0 = -5 \times 1 = -5$	Any base (except zero) to the power of zero is equal to 1.

Exercise 1

Use the laws of exponents to simplify each of the expressions given below. Always give your answers using positive exponents.

1. $2x^2 \times 3x^3 \times 4x$
2. $2^2 \times 3^3 \times 2 \times 3^{-2}$
3. $\frac{3x^5 y^2}{6xy^4}$

4. $-5f^7 \div 10ef^9$
5. $(-2x^3 y)^3$
6. $\left(\frac{21a^3}{7a^2}\right)^2$

7. $\frac{1}{(3)^4}$
8. $2^{-3} \times 5 \times 2^4 \times 5^{-2}$
9. $(-xy^3)^0$

10. $\left(\frac{3a}{v}\right)^0 + 2b^0$
11. $2^3 \times 5^3$
12. $\frac{4^2 \cdot 3^2}{2^3 \cdot 5^3}$

13. $\frac{2ab^{-3}}{c^{-3}}$
14. $\left(\frac{(3x)^{-2}}{3x^{-2}}\right)^{-2}$
15. $\left(\frac{a^2 b^{-3}}{a^4 b^2}\right)^{-2}$

16. $\left(\frac{p}{2}\right)^2 \times \left(\frac{p^{-2}}{t^{-3}}\right)^2$
17. $\left(\frac{1}{x^{-1}} + \frac{2}{x^{-1}} - \frac{10}{x^{-1}}\right)^2$
18. $\frac{q}{p^2} \times \left(\frac{q^3}{p}\right)^2 \times p^{-4}$

Prime numbers and composite numbers

Expressing a number as a product of its prime factors

Expressing individual terms as a product of their prime factors is a useful technique often used to simplify expressions.

Worked examples

1. Express the numbers 8 and 32 as a product of their prime factors.

2. Express the numbers 12 and 36 as a product of their prime factors.

Solutions

1. $8 = 2 \times 2 \times 2 = 2^3$

   $32 = 2 \times 2 \times 2 \times 2 \times 2 = 2^5$

2. $12 = 4 \times 3 = 2 \times 2 \times 3 = 2^2 \cdot 3$

   $36 = 4 \times 9 = 2 \times 2 \times 3 \times 3 = 2^2 \cdot 3^2$

Using prime factors to find the value of exponential expressions

Steps for using prime factors to find the value of exponential expressions

1. Begin by expressing each base as a product of its prime factors.
2. Then use the exponential laws to simplify fully. Be careful when working with exponents that are fractions.
3. You may check your answer using a calculator.

Worked examples

Find the value of each of the following without using a calculator. You need to show all your working.

1. $8^{-\frac{1}{3}}$

2. $\left(\frac{64}{125}\right)^{\frac{2}{3}}$

3. $\left(32^{-\frac{3}{5}}\right)^2$

Solutions

1. $8^{-\frac{1}{3}} = (2^3)^{-\frac{1}{3}} = 2^{-1} = \frac{1}{2}$

   Remember: $3 \times -\frac{1}{3} = -1$

2. $\left(\frac{64}{125}\right)^{\frac{2}{3}} = \left(\frac{2^6}{5^3}\right)^{\frac{2}{3}}$

   Remember: $6 \times \frac{2}{3} = 4$ and $3 \times \frac{2}{3} = 2$

   $= \frac{2^4}{5^2} = \frac{16}{25}$

3. $\left(32^{-\frac{3}{5}}\right)^2 = [(2^5)^{-\frac{3}{5}}]^2$

   Remember: $5 \times -\frac{3}{5} = -3$

   $= (2^{-3})^2 = 2^{-6} = \frac{1}{2^6} = \frac{1}{64}$

Using prime factors to simplify expressions involving exponents that are variables

Worked examples

Simplify fully, always giving your answers in positive exponents:

1. $8^n \cdot 2^{n+2} \cdot 16^{-n}$

2. $2^{2x+1} \cdot 3^{2x} \cdot 12^{1-x}$

Solutions

Steps for using prime factors to simplify exponential expressions

1. $8^n \cdot 2^{n+2} \cdot 16^{-n}$
2. $2^{2x+1} \cdot 3^{2x} \cdot 12^{1-x}$

   Step 1: Express each base as a product of its primes.

   $= (2^3)^n \cdot 2^{n+2} \cdot (2^4)^{-n}$

   $= 2^{2x+1} \cdot 3^{2x} \cdot (2^2 \cdot 3)^{1-x}$

Step 2: Use $(x^m)^n = x^{mn}$ to work out the new exponents.

$= 2^{3n} \cdot 2^{n+2} \cdot 2^{-4n}$ 　　　　　　　　$= 2^{2x+1} \cdot 3^{2x} \cdot 2^{-2-2x} \cdot 3^{-1-x}$

Step 3: Use $x^m \times x^n = x^{m+n}$ to add exponents that are like terms.

$= 2^{3n+n+2-4n}$ 　　　　　　　　　$= 2^{2x+1-2-2x} \cdot 3^{2x-1-x}$

　　　　　　　　　　　　　　　　Note that $(2^2)^{-1-x} \neq 2^{-2-x}$

Step 4: Use the relevant exponential laws to express in simplest form.

$= 2^2 = 4$ 　　　　　　　　　　$= 2^{-1} \cdot 3^{x-1}$

Step 5: To give your answer in positive exponents only, use the definition $x^{-n} = \frac{1}{x^n}$.

$$= \frac{3^{x-1}}{2}$$

Exercise 2

1. Give the value of the following without using a calculator. Show all working steps.

   1.1 $32^{\frac{4}{5}}$ 　　　　　　**1.2** $-(16)^{\frac{3}{4}}$ 　　　　　　**1.3** $64^{-\frac{2}{3}}$

   1.4 $\left(\frac{16}{9}\right)^{\frac{1}{2}}$ 　　　　　**1.5** $(16^{\frac{1}{2}})(16^{-\frac{1}{2}})$ 　　　**1.6** $(-32)^{\frac{3}{5}} \div (-32)^{\frac{2}{5}}$

   1.7 $81^{\frac{3}{4}} \times 27^{-\frac{2}{3}}$

2. Simplify fully using the exponential laws, giving your answer in positive exponents.

   2.1 $2^n \cdot 3^n \cdot 4^{n+1} \cdot 3^{n-2}$ 　　　　**2.2** $6^{2n-1} \cdot 9^{1-n} \cdot 3^{2n+1}$

   2.3 $25^n \cdot 5^{1-n} \cdot 15^n \cdot 9^{n-2}$ 　　　**2.4** $27^{n+1} \cdot 18^{2n} \cdot 4^{n-1}$

Exponential problems involving fractions

Exponential fractions involving multiplication and division

To simplify an exponential expression that is a fraction containing multiplication and division, rewrite all the bases as products of their prime factors, and then use the exponential laws to simplify the result.

Worked examples

Simplify:

1. $\frac{3^n \, 9^{n-3}}{27^{n-1}}$

2. $\frac{2^{4x+1} \cdot 9^x \cdot 6^{2x-1}}{12^{3x} \cdot 3^x}$

Solutions

Steps for using prime factors to simplify exponential fractions

1. $\frac{3^n \, 9^{n-3}}{27^{n-1}}$ 　　　　　　　　　2. $\frac{2^{4x+1} \cdot 9^x \cdot 6^{2x-1}}{12^{3x} \cdot 3^x}$

Step 1: Rewrite all the bases as products of their prime factors.

$= \frac{3^n \times (3^2)^{n-3}}{(3^3)^{n-1}}$ 　　　　　　　$= \frac{2^{4x+1} \cdot (3^2)^x \cdot (2 \cdot 3)^{2x-1}}{(2^2 \cdot 3)^{3x} \cdot 3^x}$

Step 2: Use the exponential laws to a power to another power.

Step 3: Multiply the terms carefully, and take care with the minus signs.

$$= \frac{3^n \times 3^{2n-6}}{3^{3n-3}}$$

$$= \frac{2^{4x+1} \cdot 3^{2x} \cdot 2^{2x-1} \cdot 3^{2x-1}}{2^{6x} \cdot 3^{3x} \cdot 3^x}$$

Step 4: To move terms from the bottom to the top of the fraction, apply the law for division.

$$\frac{1}{3^{3n-3}} = 3^{-3n+3}$$

Step 5: Add the exponents of different bases separately.

$$= 3^{n+2n-6-3n+3}$$

$$= 2^{4x+1+2x-1-6x} \cdot 3^{2x+2x-1-3x-x}$$

Step 6: Add like terms and simplify as necessary using the exponential laws.

$$= 3^{-3} = \frac{1}{3^3} = \frac{1}{27}$$

$$= 2^0 \cdot 3^{-1} = \frac{1}{3}$$

Exercise 3

Simplify fully:

1. $\dfrac{9^n \times 12^{n+1}}{4 \times 6^n}$ **2.** $\dfrac{2^{n-2} \cdot 4^{n+3}}{8^{n+2}}$ **3.** $\dfrac{2^n \cdot 8^{n+2} \cdot 4^{-3n}}{2^{-2n}}$ **4.** $\dfrac{2^{x+1} \cdot 2^{2x+1}}{(2^2)^{2x+1}}$

5. $\dfrac{5^{2x-1} \cdot (3^2)^{x-2}}{(5 \cdot 3)^{2x-3}}$ **6.** $\dfrac{4^{x+1} \cdot 8^{x-1}}{16^{x-2}}$ **7.** $\dfrac{45^{x-3} \cdot 3 \cdot 75^{4-x}}{25^{-x} \cdot 15^{x+2}}$ **8.** $\dfrac{(98^{x+1})^2 \cdot (2^x)^{-2}}{49^x}$

Simple exponential equations

There are two types of exponential equations:

- **Type 1** has the unknown variable as the base.
- **Type 2** has the unknown variable in the exponent.

Solving equations when the variable is the base

To solve for the variable when the variable is the base, the unknown base must be expressed as $1x^1$. To achieve this, raise the power of the variable to the inverse of its exponent. It will therefore be necessary to raise the number on the right hand side to the same power.

Worked examples

Solve for x:

1. $x^{\frac{1}{2}} = 5$ **2.** $2x^{\frac{3}{2}} = 54$ **3.** $2^2 x^{-2} = \frac{16}{100}$

Solutions

1. $x^{\frac{1}{2}} = 5$

Raise both sides of the equation by the power of two. Remember, $\frac{1}{2} \times \frac{2}{1} = 1$.

$(x^{\frac{1}{2}})^2 = 5^2$

$\therefore x = 25$

2. $2x^{\frac{3}{2}} = 54$

$x^{\frac{3}{2}} = 27$

First simplify both sides of the equation so that the unknown variable has a coefficient of 1. To do this, divide both sides of the equation by 2: $54 \div 2 = 27$.

$(x^{\frac{3}{2}})^{\frac{2}{3}} = (27)^{\frac{2}{3}}$

Then raise both sides to the power of $\frac{2}{3}$:

$\frac{3}{2} \times \frac{2}{3} = 1$.

Remember

A coefficient is the number used to multiply a variable, so in this expression the coefficient is 2.

$$x = (3^3)^{\frac{2}{3}}$$

Express the number on the RHS as the product of a prime number, and simplify.

$$x = 3^2$$

$$x = 9$$

3. $\quad 2^2 x^{-2} = \dfrac{16}{100}$

$\quad\quad 4x^{-2} = \dfrac{4}{25}$

First simplify both sides of the equation: $\dfrac{16}{100} = \dfrac{4}{25}$.

$\quad\quad\quad x^{-2} = \dfrac{1}{25}$

Make the coefficient of the unknown variable 1: $\dfrac{4}{25} \div 4 = \dfrac{4}{25} \times \dfrac{1}{4} = \dfrac{1}{25}$.

$\quad (x^{-2})^{-\frac{1}{2}} = (5^{-2})^{-\frac{1}{2}}$

Raise both sides to the power of $-\dfrac{1}{2}$: $-2 \times -\dfrac{1}{2} = 1$.

$$x = 5$$

Solving equations when variable is the exponent

To solve for the variable when it is an exponent, change both sides of the equation so that you have the same base on both sides, and then you can equate the exponents.

Worked examples

Solve for x:

1. $2^x = 2^4$

2. $2^x - 0,125 = 0$

3. $\left(\dfrac{9}{4}\right)2^{x-1} = 18$

4. $3 \cdot 2^{x^2 + 3} - 48 = 0$

Solutions

Note

The solution to this type of equation is based on the principle that if $a^x = a^b$, then $x = b$, as long as $a \neq 0$ and $a \neq \pm 1$.

1. $\quad 2^x = 16$

$\quad\quad 2^x = 2^4$

To get equal bases, rewrite 16 as a product of its prime factors.

$\quad \therefore x = 4$

Equate the exponents once the bases are equal.

2. $\quad 2^x - 0,125 = 0$

$\quad\quad\quad 2^x = 0,125$

$\quad\quad\quad 2^x = \dfrac{125}{1\,000}$

Express the decimal number as a fraction.

$\quad\quad\quad 2^x = \dfrac{1}{8}$

Simplify.

$\quad\quad\quad 2^x = 2^{-3}$

Express result as a product of its prime factors.

$\quad \therefore x = -3$

Drop the base and equate the exponents once the bases are equal.

3. $\quad \left(\dfrac{9}{4}\right)2^{x-1} = 18$

$\quad\quad 2^{x-1} = 18 \times \dfrac{4}{9}$

Divide 18 by $\dfrac{9}{4}$.

$\quad\quad 2^{x-1} = 8$

$\quad\quad 2^{x-1} = 2^3$

Express the result as a product of its prime factors.

$\quad \therefore x - 1 = 3$

Once the bases are equal, equate the exponents and solve for x.

$\quad \therefore x = 4$

4. $3 . 2^{x^2 + 3} = 48$

$2^{x^2 + 3} = \frac{48}{3}$

To get to the base 2 on its own on the left side, add 48 to the right side and divide the result by 3.

$2^{x^2 + 3} = 16$

$2^{x^2 + 3} = 2^4$

Express the result as a product of its prime factors.

$\therefore x^2 + 3 = 4$

Once the bases are equal, equate the exponents and solve for x.

$\therefore x^2 = 1$

$\therefore x = \pm 1$

Exercise 4

1. Solve for x:

   1.1 $x^{\frac{3}{2}} = 8$ **1.2** $-3x^{\frac{1}{3}} = 15$ **1.3** $10x^{\frac{3}{4}} = 270$

   1.4 $x^{\frac{1}{4}} - 2 = 0$ **1.5** $x^{-1} = \frac{4}{7}$

2. Solve for x:

   2.1 $16^x = 32$ **2.2** $16^{x-3} = 1$

   2.3 $3 . 5^x = 75$ **2.4** $2 . 3^{x+2} = 486$

   2.5 $4 \times 2^{x-1} = \frac{1}{4}$ **2.6** $4 \times 3^{2x+1} = 108$

   2.7 $5^{2x+1} = 125$ **2.8** $2^{x-1} = \frac{1}{16}$

   2.9 $2^{x+12} = 16^x$ **2.10** $3 . 9^{2x-3} - 81 = 0$

Exponential expressions, fractions and equations: complex procedures

Complex exponential expressions, fractions and equations involving addition and subtraction can usually be simplified using factorisation. You factorise in the same way as you would any other algebraic expression.

However, if a binomial exponent is involved, it is essential that, before factorising, you apply the law of multiplication of bases in 'reverse', e.g. $3^{x+1} = 3^x + 3^1$.

Worked examples

1. Use factors to simplify:

   1.1 $2^{x+2} + 2^x$ **1.2** $4^n + 2^{2n-1}$

2. Factorise fully:

   2.1 $3^{2n} - 3^n$ **2.2** $x^{n+1} - x^n + \frac{2}{x^{-n}}$

Solutions

1. **1.1** $2^{x+2} + 2^x$

      $= 2^x . 2^2 + 2^x$

Remember: $2^{x+2} = 2^x . 2^2$

Simplify the expression by taking out the common factor $2x$.

      $= 2^x(2^2 + 1)$

      $= 2^x(4 + 1)$

      $= 2^x(5)$

1.2 $4^n + 2^{2n-1}$

$\quad = (2^2)^n + 2^{2n-1}$ 　　　　　　　Express 4^n as a product of its prime factors.

$\quad = 2^{2n} + 2^{2n} \cdot 2^{-1}$

$\quad = 2^{2n}(1 + 2^{-1})$ 　　　　　　　Take out the common factor to simplify.

$\quad = 2^{2n}\left(1 + \frac{1}{2}\right)$

$\quad = 2^{2n}\left(\frac{3}{2}\right)$

2.1 $3^{2n} - 3^n$ 　　　　　　　　Remember: $3^{2n} = 3^n \cdot 3^n$

$\quad = 3^n \cdot 3^n - 3^n$

$\quad = 3^n(3^n - 1)$

2.2 $x^{n+1} - x^n + \dfrac{2}{x^{-n}}$

$\quad = x^{n+1} - x^n + 2x^n$

$\quad = x^{n+1} + x^n$

$\quad = x^n(x + 1)$

Exercise 5

Factorise fully:

1. $2^{2n+2} - 2^{2n}$ 　　　　　　　　2. $4^n \cdot 2^{2n+2} - 8^n$

3. $2 \cdot 2^x + 6 \cdot 2^{x-1}$ 　　　　　　4. $3^{x+1} + 3^{x-1}$

5. $3^{3n} - 3^{n+1}$ 　　　　　　　　6. $2^{2n} \cdot 3^n - 2^{n+1} \cdot 3^n + 2^n \cdot 3^{2n}$

Exponential fractions involving addition and subtraction

To simplify a fraction in which you need to add and/or subtract exponential terms:

1. Factorise both the numerator and the denominator if necessary.
2. Take out only the common factor that has the variable as its exponent.
3. Numerical common factors are written inside the bracket.
4. Cancel like factors.
5. Simplify the result.

Worked examples

Simplify fully:

1. $\dfrac{3^{n+2} - 3^n}{3^{n+1}}$ 　　　　　　　2. $\dfrac{2^{2n} \cdot 3^n - 2^{n+1} \cdot 3^n + 2^n \cdot 3^{2n}}{2^n - 2 + 3^n}$

Solutions

1. $\dfrac{3^{n+2} - 3^n}{3^{n+1}}$ 　　　　　　2. $\dfrac{2^{2n} \cdot 3^n - 2^{n+1} \cdot 3^n + 2^n \cdot 3^{2n}}{2^n - 2 + 3^n}$

$\quad = \dfrac{3^n \cdot 3^2 - 3^n}{3^n \cdot 3^1}$ 　　　　　　$= \dfrac{2^n \cdot 2^n \cdot 3^n - 2^n \cdot 2^1 \cdot 3^n + 2^n \cdot 3^n \cdot 3^n}{2^n - 2 + 3^n}$

First **factorise** both the numerator and the denominator if necessary.

$\quad = \dfrac{3^n(3^2 - 1)}{3^n \cdot 3^1}$ 　　　　　　$= \dfrac{2^n \cdot 3^n(2^n - 2 + 3^n)}{(2^n - 2 + 3^n)}$

Then 'cancel out' any factors common to both the numerator and the denominator.

$\quad = \dfrac{3^2 - 1}{3^1}$ 　　　　　　　$= \dfrac{2^n \cdot 3^n}{1}$

Simplify the result.

$\quad = \dfrac{8}{3}$ 　　　　　　　　$= 6^n$

Exercise 6: complex procedures

Simplify fully:

1. $\dfrac{2^{n+2} - 2^{2n}}{2^{2n}}$

2. $\dfrac{3^n \cdot 5 - 3^{n+1}}{3^n \cdot 4}$

3. $\dfrac{3^{3n} - 3^{n+1}}{3^{n+1}}$

4. $\dfrac{2^n \cdot 2^{2n+2} - 8^n}{8^{n+1}}$

5. $\dfrac{2 \cdot 2^x + 6 \cdot 2^{x-1}}{10^x}$

6. $\dfrac{3^n \cdot 2 + 2^2 \cdot 3^n}{3^n \cdot 5 - 3^n \cdot 2}$

7. $\dfrac{5^{n+1} + 5^{n-1}}{5^n \cdot 10 - 5^{n+1}}$

Exponential equations involving factorisation

To simplify an equation that involves adding and/or subtracting exponential terms, follow these steps:

1. First simplify both sides of the equation.

2. If there is more than one term, factorise.

3. If the unknown variable is an exponent, change the equation so that both sides have the same base.

Worked examples

Solve for x:

1. $2^{x+1} - 2^x = 16$

2. $3^{x+1} + 3^{x-1} = \dfrac{10}{9}$

Solutions

1. $2^{x+1} - 2^x = 16$

$\quad 2^x(2-1) = 16$ Factorise the left hand side.

$\quad\quad 2^x \cdot 1 = 16$ Simplify the result.

$\quad\quad\quad 2^x = 2^4$ Express numbers as the product of their prime factors.

$\quad \therefore x = 4$ Once the bases are equal, drop the bases, equate the exponents, and then solve for x.

2. $3^{x+1} + 3^{x-1} = \dfrac{10}{9}$

$\quad 3^x(3^1 + 3^{-1}) = \dfrac{10}{9}$ Factorise.

$\quad\quad 3^x\left(3\tfrac{1}{3}\right) = \dfrac{10}{9}$ Simplify.

$\quad\quad 3^x\left(\dfrac{10}{9}\right) = \dfrac{10}{9}$

$\quad\quad\quad 3^x = \dfrac{10}{9} \times \dfrac{9}{10}$

$\quad\quad\quad 3^x = 1$

$\quad\quad\quad 3^x = 3^0$ Remember: $1 = 3^0$.

$\quad \therefore x = 0$

Exercise 7: complex procedures

Solve for x in each of the equations given below:

1. $2x \cdot 2^{x+2} = 64$

2. $2^{x+2} + 2^x = 40$

3. $2^x \cdot 2^{x+2} = \dfrac{25}{100}$

4. $2^{x+2} - 2^x = \dfrac{3}{4}$

5. $8^{x+1} = 2^x \cdot 16^x$

6. $8^{x+1} - 2^x \cdot 4^x = 112$

7. $(2^x)^{x+1} = 4$

8. $(2^x)^x + 2^{x^2} = 32$

9. $3^x - 10 = -3^{x+2}$

10. $3 \cdot 5^{x-1} + 4 \cdot 5^x = \frac{23}{25}$

11. $3^{x+1} - \left(\frac{1}{3}\right)^{-x-3} = -11(3^{x+2}) + 25$

Test A: Knowledge and routine procedures

1. Match column A with column B: (3)

Column A	Column B
1.1 2^{-12}	**A** -24
1.2 $\frac{1}{8^3}$	**B** 8^{-3}
	C $\frac{1}{2^{12}}$
1.3 $\frac{1}{3^8}$	**D** 3^{-8}

2. Simplify each of the following, giving your answer in positive exponents:

   2.1 $r^4 \times r^{-3}$ (1)

   2.2 $2r^{-7} \times 3r^9$ (2)

   2.3 $\frac{12t^{-7}}{3t}$ (2)

   2.4 $(x^{-3})^3$ (1)

   2.5 $\frac{9^{a-2} \cdot 10^{a-2}}{6^{a-4} \cdot 15^a}$ (7)

3. Copy and complete the grid given below. Each space must contain an expression using an exponent, or the operation sign $\times$ or $\div$. (5)

3^6	$\times$		$=$	3^6
$\times$				
		3^2	$=$	
$=$		$=$		$=$
3^8	$\div$		$=$	3^6

4. Give the value of each of the expressions without using a calculator. Show all working steps.

   4.1 $\left(\frac{16}{81}\right)^{\frac{3}{4}}$ (4)

   4.2 $3^{\frac{5}{4}} \div 48^{\frac{1}{4}}$ (3)

   4.3 $(2^0 + 4^0 + 6^0)^{-2}$ (3)

5. Solve for x in each equation:

   5.1 $2^x = 32$ (2)

   5.2 $x^{-2} = \frac{1}{49}$ (2)

   5.3 $x^{\frac{3}{4}} = 27$ (3)

   5.4 $3^{3x-3} = 3^{x-4}$ (2)

   5.5 $81^{x-2} = 27^4$ (4)

6. In each statement below, a can be any integer, fraction or decimal except 0. Write down whether the statement is always true, sometimes true or never true. Give a reason for each answer.

   6.1 $a^0 \times a^0 \times a^0 > 2$ (2)

   6.2 $(2a)^{-3} = \dfrac{8}{a^3}$ (2)

   6.3 $2a^2 = a$ (2)

Total 50

Test B: Complex procedures and problem solving

1. Find the value of $\dfrac{3^2}{3^{-2} - 6^{-1}}$ without using a calculator. Show all working.   (4)

2. Simplify fully, using positive exponents in your answer:
$$\frac{(3x^3)^{-2}(-2x^{-1})^3}{\left(\frac{3x^2}{-2}\right)^{-2}}.$$
   (6)

3. Solve for x:

   3.1 $(0{,}5)^{x-1} = 4^{1,5}$ (4)

   3.2 $2^{2x} - 4^{x-1} = 6$ (5)

4. **4.1** Simplify fully: $\dfrac{3^{a+1} - 3^{a-1}}{3^a - 3^{a-2}}$.   (4)

   4.2 **4.2.1** Factorise: $9^x - 1$.   (2)

        4.2.2 Hence or otherwise, simplify the expression $\dfrac{9^x - 1}{3^x + 1}$.   (2)

5. State whether each of the following statements is true or false. If the statement is false, explain why in full.

   5.1. If $0 < x < 1$ then $x^{-4} < x^{-3}$.   (3)

   5.2 If $-1 < x < 1$ then $x^4 < x^3$.   (3)

6. Explain why there is no solution to the equation $2^x = 0$.   (2)

7. There are five children. Each child has five bags, each bag has five cats, and each cat has five kittens. If each kitten catches one mouse each day, and each cat catches five mice each day, how many mice will the cats and the kittens catch in seven weeks?   (4)

8. **8.1** Investigate the last digits of the powers of the form n^5, for example $1^5, 2^5, 3^5, \ldots$ Now find the last digit of 412^5.   (5)

   8.2 In a similar manner or otherwise, find the last digit of the number 9^{999}.   (6)

Total 50

Test C: Content and breakdown as for exam

1. From the list of terms given below, write down four matching pairs of terms.   (4)

   A 3^{-2} **B** 2^{-3} **C** 4^{-2} **D** 6^{-1} **E** -6

   F $\frac{1}{6}$ **G** $\frac{1}{16}$ **H** $\frac{1}{8}$ **I** $\frac{1}{9}$

2. Simplify fully, always giving answers in positive exponents:

    2.1 $x(-x)^{-1}$ (2)

    2.2 $\dfrac{(2+x)^0}{3(-3)^{-1}}$ (3)

3. Copy and complete the grid given below. Each space must contain an expression using an exponent or an operation sign $\times$ or $\div$.   (5)

	$\times$		$=$	x^3
$\times$	▓	$\times$	▓	
		x^{-3}	$=$	x^{-2}
$=$	▓	$=$	▓	$=$
1	$\div$		$=$	x^5

4. Solve for x in each of the following equations:

    4.1 $10^x = 1$ (2)

    4.2 $12^{-1} = \frac{1}{x}$ (2)

    4.3 $10^{x-1} = 100\,000$ (2)

    4.4 $4^x = 0{,}0625$ (3)

    4.5 $3 \cdot 5^{x-1} - 75 = 0$ (3)

    4.6 $3^{x-2} + 3^{x+1} = 28$ (4)

5. Simplify fully:

    5.1 $\dfrac{-3a^{-1}}{2a^{-2} - (a^3)^{-1}}$ (5)

    5.2 $\left(\frac{x}{y}\right)^{a+b} \times \left(\frac{y}{x}\right)^{a-b}$ (3)

    5.3 $\dfrac{3^{x+2} - 3^{x-2}}{4 \cdot 3^{x-3}}$ (4)

6. In each of the statements given below, n can be any integer. State whether each statement is always true, sometimes true, or never true. Give a reason for each answer.

    6.1 $(0{,}5)^n = 2^{-n}$ (2)

    6.2 n^3 is a prime number.   (2)

7. **7.1** Copy and complete the table:

n	1	2	3	4	5	6	7	8	9	10
2^n	2	4	8	16	32					
Last digit of 2^n	2	4	8	6	2					

    7.2 What is the last digit of 2^{100}?   (4)

Total 50

TOPIC 4: Number patterns

In mathematics a number pattern is a list of numbers or geometric shapes that can be described by a rule. Number patterns form the foundation of most mathematics, and they play an important role in both algebra and geometry.

Knowledge and skills for this topic

If you struggle with basic algebra, arithmetic or substitution, revise these before continuing with this Topic.

Content of final exam

- Investigate number patterns where the constant difference between consecutive terms is linear.
- Determine and apply the general term for a linear sequence of numbers.
- Investigate and work with simple quadratic and exponential number sequences.

Terminology

Look at this list of numbers:

$$5; \ 7; \ 9; \ 11; \ ...$$

- The list of numbers forms a pattern. A list of numbers that forms a pattern is called a **sequence**.
- Each number in the sequence is called a **term**.
- The general term, or the nth term is indicated by T_n, where $n = 1, 2, 3, ...$ indicates the position of the specific term in the sequence.
    - For example, the second term in the above sequence is 7. We write this as $T_2 = 7$.
    - The fourth term in the above sequence is 11. We write this as $T_4 = 11$.

Linear number patterns

Taking another look at the same sequence: 5; 7; 9; 11; ..., and notice that each term is obtained by adding two to the term before.

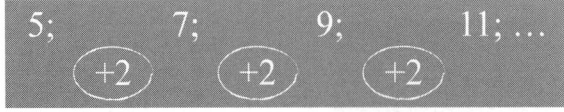

This means that there is a **common difference** between the terms. In this case the common difference is 2. A common difference can be positive or negative.

Number patterns that have a common difference are known as **linear number patterns**.

Study the sequence given below:

$$-15; \qquad -18; \qquad -21; \qquad -24; \ ...$$

You will subtract 3 from each term to get the next term, so this sequence is also linear because it has a common difference of -3.

Finding the rule/general term

Consider the number patterns given below:

2; 4; 6; 8; …

5; 7; 9; 11; …

 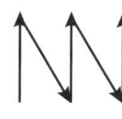 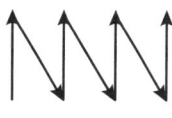 

Figure 1   Figure 2   Figure 3   Figure 4

Each of these sequences is linear. In each case the common difference is 2.

The rule used to generate each of these specific sequences is different because they start with different numbers, as shown in the table below.

2; 4; 6; 8; …	5; 7; 9; 11; …	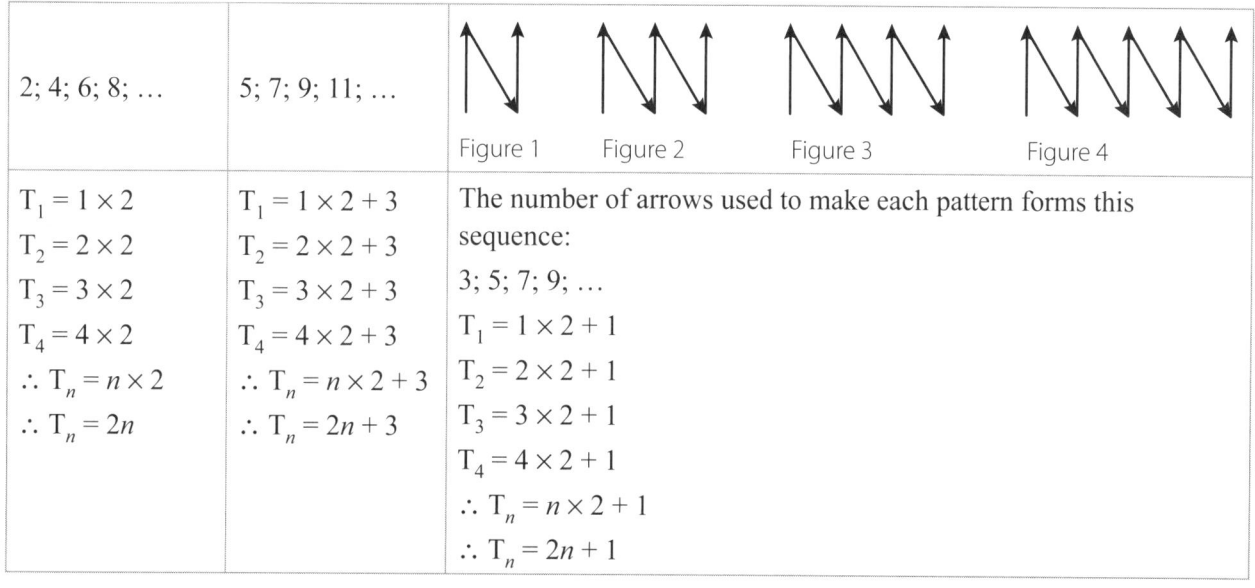 Figure 1　　Figure 2　　Figure 3　　Figure 4
$T_1 = 1 \times 2$ $T_2 = 2 \times 2$ $T_3 = 3 \times 2$ $T_4 = 4 \times 2$ $\therefore T_n = n \times 2$ $\therefore T_n = 2n$	$T_1 = 1 \times 2 + 3$ $T_2 = 2 \times 2 + 3$ $T_3 = 3 \times 2 + 3$ $T_4 = 4 \times 2 + 3$ $\therefore T_n = n \times 2 + 3$ $\therefore T_n = 2n + 3$	The number of arrows used to make each pattern forms this sequence: 3; 5; 7; 9; … $T_1 = 1 \times 2 + 1$ $T_2 = 2 \times 2 + 1$ $T_3 = 3 \times 2 + 1$ $T_4 = 4 \times 2 + 1$ $\therefore T_n = n \times 2 + 1$ $\therefore T_n = 2n + 1$

The rule used to generate all of these sequences contains $2n$, because they all have a common difference of 2.

The rule used to describe a number pattern is known as the **general term**, and it shows how the numbers in a pattern are obtained. We use T_n to indicate that we are giving the general term.

Exercise 1

Write down the general term used to obtain each of the linear sequences given below:

1. 3; 6; 9; 12; …

2. 1; 4; 7; 10; …

3. −5; −10; −15; −20; …

4. −2; −7; −13; −18; …

5. 1; 3; 5; 7; …

6. −7; −5; −3; −1; …

7. 1; 2; 3; 4; …

8. −5; −4; −3; −2; …

9. $\frac{1}{2}; \frac{1}{3}; \frac{1}{4}; \frac{1}{5}; …$

10. $1; \frac{1}{2}; \frac{1}{3}; \frac{1}{4}; …$

11.

Figure 1   Figure 2   Figure 3

12.

Figure 1   Figure 2   Figure 3

Using the general term

The general term, T_n, can be used to work out a number of things.

Worked examples

1. In the sequence, 5; 7; 9; 11; ... the general term is $T_n = 2n + 3$.

   1.1 What will T_{34} be?

   1.2 Which term will the number 103 be?

2. The number pattern below has the general term $T_n = 2n + 1$.

Figure 1   Figure 2   Figure 3   Figure 4

   2.1 Use the general term to calculate how many arrows will be needed to make Figure 25.

   2.2 If you have 203 arrows, and you use them to make only one of these patterns, which figure number would this be?

3. Write down the first three terms and the common difference for the linear sequence with the general term $T_n = -3n +4$.

Solutions

1. **1.1** $T_{34} = 2(34) + 3 = 71$

   1.2 $103 = 2n + 3$

   $100 = 2n$

   $\therefore n = 50$

2. **2.1** $T_{25} = 2(25) + 1 = 51$

   51 arrows will be needed to make Figure 25.

   2.2 $T_n = 2n + 1 = 203$

   $2n = 202$

   $\therefore n = 101$

   Figure 101 can be constructed using 203 arrows.

3. $T_1 = -3(1) + 4 = 1$

   $T_2 = -3(2) + 4 = -2$

   $T_3 = -3(3) + 4 = -5$

   The common difference is –3.

Exercise 2

1. Write down the first term and the common difference for the linear sequences with the general terms given below:

   1.1 $T_n = 8n + 5$ **1.2** $T_n = 7n - 1$

   1.3 $T_n = 6 - 2n$ **1.4** $T_n = -4n$

2. Use the first term and the common difference given to find the general term for each sequence:

   2.1 $T_1 = 4$, common difference is -7

   2.2 $T_1 = -1$, common difference is 12

   2.3 $T_1 = 0$, common difference is -23

   2.4 $T_1 = 1$, common difference is 0

   2.5 $T_1 = -13$, common difference is -5

   2.6 $T_1 = \frac{1}{10}$, common difference $= -\frac{1}{2}$

3. Write down the first term and the common difference for the linear sequences that have these general terms:

   3.1 $T_n = 3n + 2$ **3.2** $T_n = \frac{1}{3}n - 6$

   3.3 $T_n = 5 - n$ **3.4** $T_n = 8 - 3n$

   3.5 $T_n = -2n - 5$ **3.6** $T_n = 7 - \frac{n}{3}$

   3.7 $T_n = \frac{3}{4n} + \frac{5}{4}$ **3.8** $T_n = (n + 1)(n - 2) - n^2$

4. For each of the general terms below, find the term that is equal to the number that is given.

   4.1 $T_n = 7$ 294   **4.2** $T_n = -5n$ 195

   4.3 $T_n = 4 - 3n$ -149 **4.4** $T_n = \frac{n+5}{n-3}$ 2

Non-linear patterns

Not all sequences will have a common difference, for example 1; 4; 9; 16; 25 ...

This is a sequence of squares: 1^2; 2^2; 3^2; 4^2; 5^2; ...

This means that the general term for this sequence is $T_n = n^2$.

The sequence of squares is a simple example of a quadratic sequence. You will learn about quadratic sequences in Grade 11. At this stage, you should be able to identify sequences that are not linear and you should be able to write down the general terms for simple non-linear sequences. Some of the most common non-linear sequences are:

- **Quadratic sequences**

  1; 4; 9; 16; 25; ... 1^2; 2^2; 3^2; 4^2; 5^2; ... $T_n = n^2$

  3; 12; 27; 48; 75; ... 3×1^2; 3×2^2; 3×3^2; 3×4^2; 3×5^2; ... $T_n = n^3$

  You can check for a simple quadratic sequence by dividing by n^2, for example:

  -2; -8; -18; -32; ...

  $-2 \div 1^2 = -2$ $-8 \div 2^2 = -2$ $-18 \div 3^2 = -2$ $\therefore T_n = -2n^2$

- **Cubic sequences**

  1; 8; 27; 64; 125; ...   13; 23; 33; 43; 53; ... $T_n = n^3$

You can check for a simple cubic sequence by dividing by n^3, for example:

$\frac{1}{3}; \frac{8}{3}; 9; \frac{64}{3}; \frac{125}{3}; \dots$

$$\frac{1}{3} \div 1^3 = \frac{1}{3} \qquad \frac{8}{3} \div 2^3 = \frac{1}{3} \qquad 9 \div 3^3 = \frac{1}{3} \qquad \frac{64}{3} \div 4^3 = \frac{1}{3}$$

$\therefore T_n = \frac{1}{3}n^3$

- **The Fibonacci sequence:** 1; 1; 2; 3; 5; 8; 23; ...

  The general term for the Fibonacci sequence is unusual. It is $T_n = T_{n-2} + T_{n-1}$. This means that you add two previous terms together to get the next term.

Worked examples

For each of the sequences given, carry out the following:

(a) Identify the sequence as linear, quadratic, cubic or the Fibonacci sequence.

(b) Fill in the next two numbers.

(c) Write down the general term.

1. 1; –2; –5; –8; ...   **2.** –1; –8; –27; –64; ...   **3.** 5; 5; 10; 15; 25; ...

4. $\frac{1}{2}; \frac{1}{8}; \frac{1}{18}; \frac{1}{32}; \dots$ **5.** 1; 4; 7; 10; ...   **6.** $2; \frac{3}{4}; \frac{4}{9}; \frac{5}{16}; \dots$

Solutions

1. (a) 1;   –2;   –5;   –8; ...
       –3   –3   –3

Sequence is linear, with a common difference of –3.

(b) –11; –14

(c) $T_1 = 1 \times -3 + 4$
$T_2 = 2 \times -3 + 4$
$T_3 = 3 \times -3 + 4$
$T_4 = 4 \times -3 + 4$
$\therefore T_n = n \times -3 + 4$
$\therefore T_n = -3n + 4$

2. (a) –1;   –8;   –27;   –64; ...
       –7   –19   –37

The sequence is non-linear:

$\frac{T_1}{1^3} = \frac{-1}{1^3} = -1$

$\frac{T_2}{2^3} = \frac{-8}{2^3} = -1$

$\frac{T_3}{3^3} = \frac{-27}{3^3} = -1$

The sequence is cubic, because

$\frac{T_n}{n^3} = -1$

(b) –125; 216

(c) $T_1 = -1^3$
$T_2 = -2^3$
$T_3 = -3^3$
$T_4 = -4^3$
$T_4 = -4^3$
$\therefore T_n = -n^3$

3. (a) 5;   5;   10;   15; ...
       0   +5   +5

The sequence is the Fibonacci sequence, with each term multiplied by 5.

(b) 25; 40

(c) $T_n = T_{n-2} + T_{n-1}$. This means that you add the two previous terms together to get the next term.

4. (a) $\frac{1}{2}; \frac{1}{8}; \frac{1}{18}; \frac{1}{32}; \ldots$

The sequence is non-linear. Looking at the sequence of numbers generated by the denominators:

$$\frac{T_1}{1^2} = \frac{2}{1^2} = 2$$

$$\frac{T_2}{2^2} = \frac{8}{2^2} = 2$$

$$\frac{T_3}{3^2} = \frac{18}{3^2} = 2$$

The sequence is quadratic because $\frac{T_n}{n^2} = 2$.

(b) $2; \quad 8; \quad 18; \quad 32; \ldots$

$\quad +6 \quad +10 \quad +14$

$T_5 = 32 + 16 = 48$

$T_6 = 48 + 20 = 68$

The next two terms are 48 and 68.

(c) $T_1 = 1^2 \times 2$

$T_2 = 2^2 \times 2$

$T_3 = 3^2 \times 2$

$T_4 = 4^2 \times 2$

$T_4 = 4^2 \times 2$

$\therefore T_n = 2n^2$

5. (a) $1; \quad 4; \quad 7; \quad 10; \ldots$

$\quad +3 \quad +3 \quad +3$

The sequence is linear, with a common difference of 3.

(b) $13; 17$

(c) $T_2 = 2 \times 3 - 2$

$T_3 = 3 \times 3 - 2$

$T_4 = 4 \times 3 - 2$

$\therefore T_n = n \times 3 - 2$

$\therefore T_n = 3n \times 3 - 2$

6. 6.1 $\frac{2}{1}; \frac{3}{4}; \frac{4}{9}; \frac{5}{16}; \ldots$

The sequence represented by the numerators: $2; 3; 4; 5; \ldots$ is linear.

The sequence represented by the denominators: $1; 4; 9; 16; \ldots$ is quadratic.

6.2 $\frac{6}{25}; \frac{7}{36}$

6.3 The sequence of numbers generated by the numerators $2; 3; 4; 5; \ldots$ has the general term $T_n = n + 1$. The sequence of numbers generated by the denominators $1; 4; 9; 16; \ldots$ has the general term $T_n = n^2$. Putting the two sequences together: $T_n = \frac{n-1}{n^2}$.

Exercise 3

1. Use the given nth terms to find the missing terms in the sequences below.

1.1 $-7; \ldots; -21; \ldots; -35$ $-7n$

1.2 $0; 4; 18; \ldots; 100$ $n^3 - n^2$

1.3 $\ldots; 2; 8; \ldots; 26$ $n^2 + n - 4$

1.4 $2; \ldots; 12; \ldots; \ldots$ $n(n + 1)$

1.5 $\ldots; \frac{3}{2}; \ldots; \frac{5}{4}; \frac{6}{5}$ $\frac{n+1}{n}$

1.6 $0; -\frac{3}{2}; -\frac{8}{3}; \ldots; \ldots$ $-n + \frac{1}{n}$

2. Write down (a) the first three terms and (b) the 12th term for the number patterns given by the following general terms.

   2.1 $T_n = 12 - 3n$ **2.2** $T_n = -n + 7$

   2.3 $T_n = \frac{1}{2}n^2$ **2.4** $T_n = n(n + 1)(n - 1)$

   2.5 $T_n = \frac{2n}{1 + n}$ **2.6** $T_n = \frac{(n + 1)(n - 1)}{n(n + 2)}$

3. Write down the general term for each sequence.

   3.1 $-1; -2; -3; -4; -5; \ldots$ **3.2** $5; 10; 15; 20; 30; \ldots$

   3.3 $\frac{1}{2}; \frac{2}{3}; \frac{3}{4}; \frac{4}{5}; \frac{5}{6}; \ldots$ **3.4** $1; 8; 18; 32; 50; \ldots$

   3.5 $18; 19; 20; 21; 22; \ldots$ **3.6** $0; 2; 6; 12; 20; \ldots$

4. For each general term below, find the term that is equal to the number given after it.

   4.1 $T_n = n + 7$ 1 056   **4.2** $T_n = 6n$ 138

   4.3 $T_n = (n - 1)(n + 1)$ 255   **4.4** $T_n = \frac{3}{1 + n}$ $\frac{1}{2}$

5. Mpahlwa is told that a number sequence has an nth term of $15n - 2$. She has been asked to find which term will be equal to 96. She is stuck because she keeps getting an unexpected answer. Perform the calculations and then explain the answer.

Test A: Knowledge and routine procedures

1.
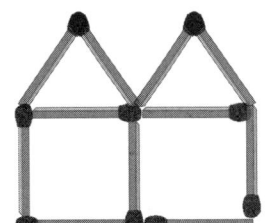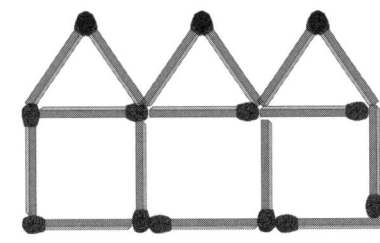

   Shape 1   Shape 2   Shape 3

   Study the shapes and use the repeating pattern to complete this table:

Shape number	1	2	3	4	15		n
Number of matchsticks	6	11	16			101	

   (8)

2. A number sequence has the general term $T_n = -5n + 7$.

   2.1 Write down the value of the 10th term.   (3)

   2.2 What is the first term of this sequence?   (2)

   2.3 Which term will be -103?   (3)

3. If the numbers $3, p$ and -6 are three consecutive terms in a linear sequence, find the value of p.   (4)

4. Write down the general terms for each sequence.

   4.1 $-6; -4; -2; 0; \ldots$ (3)

   4.2 $\frac{11}{3}; \frac{9}{3}; \frac{7}{3}; \frac{5}{3}; \ldots$ (4)

   4.3 $3; 12; 27; 48; \ldots$ (4)

$$1 + 3 + 5 = 0 + 3 + 6$$
$$2 + 4 + 6 = 1 + 4 + 7$$
$$3 + 5 + 7 = 2 + 5 + 8$$
$$4 + 6 + 8 = 3 + 6 + 9$$

5. Study the pattern on the left.

   5.1 Add one row at the beginning and one at the end of the pattern. (3)

   5.2 Write down a row of this pattern for which the first number is less than 0. (2)

   5.3 Rewrite the pattern given in the first line above in terms of n. (2)

   5.4 Simplify the expression you wrote down for 5.3. What can you conclude? (2)

6. Write down the first term and the common difference for the linear sequence with general term $T_n = 18n - 3$. (3)

7. Benny is packing chocolates into boxes. He packs eight chocolates in the first layer and then ten chocolates in each layer after that.

   7.1 Write down the values for (a) to (c). (4)

   7.2 If each box can hold seven layers, how many boxes will be completely filled if Benny has to pack 1 870 chocolates? (3)

Number of layers	Total number of chocolates
1	8
2	18
3	(a)
25	(b)
n	(c)

Total 50

Test B: Complex procedures and problem solving

1. Study the tile pattern given below. Each pattern consists of dark and light tiles.

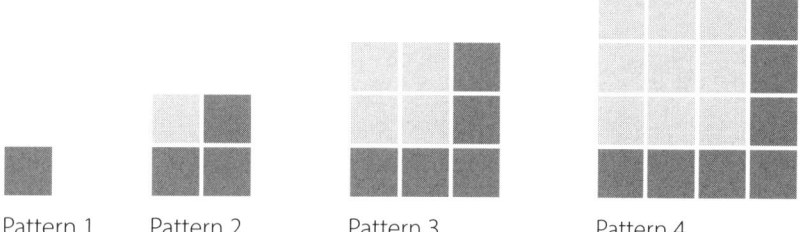

Pattern 1   Pattern 2   Pattern 3   Pattern 4

If the pattern of dark and light tiles continues:

   1.1 What number of dark tiles is there in the 8th pattern? (1)

   1.2 What number of light tiles is there in the 8th pattern? (1)

   1.3 Write down the general term for the number of dark tiles. (2)

   1.4 Write down the general term for the number of light tiles. (2)

   1.5 Which pattern number will have 841 light tiles? (5)

   1.6 How many dark tiles will there be in this pattern? (3)

2. Study the following pattern.

$$2^3 - 1^3 = 2^2 + 1^2 + 2 \times 1$$
$$3^3 - 2^3 = 3^2 + 2^2 + 3 \times 2$$
$$4^3 - 3^3 = 4^2 + 3^2 + 4 \times 3$$
$$5^3 - 4^3 = 5^2 + 4^2 + 5 \times 4$$

2.1 Add another two rows to the pattern. (3)

2.2 Rewrite the pattern algebraically, in terms of *n*. (6)

2.3 Simplify the left hand side and the right hand side individually.

2.4 What can you conclude? (2)

3. Write down the general term for each sequence.

    3.1 0,11; 0,22; 0,33; 0,44; … (3)

    3.2 $\frac{1}{8}$; $\frac{1}{27}$; $\frac{1}{64}$; $\frac{1}{125}$; … (3)

    3.3 1^2; 3^2; 5^2; 7^2; 9^2; … (3)

4. The general term of a number sequences is given by $T_n = n^2 + 4n - 6$

    4.1 Find the value of T_7. (3)

    4.2 Which term will be 15? (4)

5. The bacteria in a controlled experiment are increasing in number as shown in the sketch below:

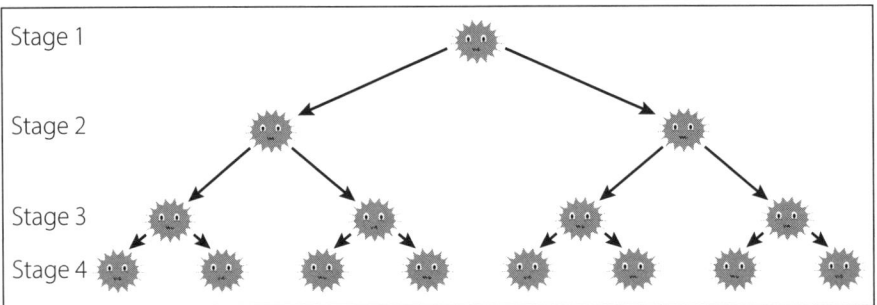

    5.1 If these bacteria continue to increase in this pattern, what type of growth is this? Write down the general term for this pattern. (2)

    5.2 How many individual cells of this bacteria will there be in the 18th stage? (2)

6. An odd number is of the form $2n - 1$. Letshaba is bored one day and chooses four consecutive odd numbers. He multiplies the middle two numbers and subtracts the product of the first and last number from his answer. He notices that he always gets the same answer. Use the general term for odd numbers to show what this answer will always be. (5)

Total 50

Test C: Content and breakdown as for exam

1. Determine the 4th term and the general term for this sequence of numbers:
    3; 10; 17; … (3)

2. Determine the next two terms if the sequences given below continue in the same pattern:

    2.1 2; 1; 4; 3; 6; 5; 8; … (2)

    2.2 2; 2; 4; 6; 10; 16; … (2)

3. The general term for a linear sequence is given by the rule $T_n = \frac{n}{2} + 3$.

    3.1 Determine the value of the first term in this sequence. (1)

3.2 Find the common difference. (2)

3.3 Find the sum of the 11th and the 13th terms, giving the answer in its simplest form. (2)

4. Choose four consecutive numbers, for example: 8; 9; 10; 11.

The product of the middle two numbers is: $9 \times 10 = 90$.

The product of the first and last numbers is: $8 \times 11 = 88$.

The difference between the two numbers is: $90 - 88 = 2$.

4.1 Choose any four consecutive numbers and repeat the same steps using your own numbers. (4)

4.2 Write a rule in words for what you find. (3)

4.3 Use algebra to prove that your rule will always work. (5)

5. $3x - 7$; $2x$; $3x + 1$; ... are the first three terms of a linear pattern.

5.2 If the pattern continues in this manner, determine the value of x. (4)

5.3 Which term in the sequence is the first to be greater than 31? (3)

6. Franco is working on a project for design. He develops a design in the following way:

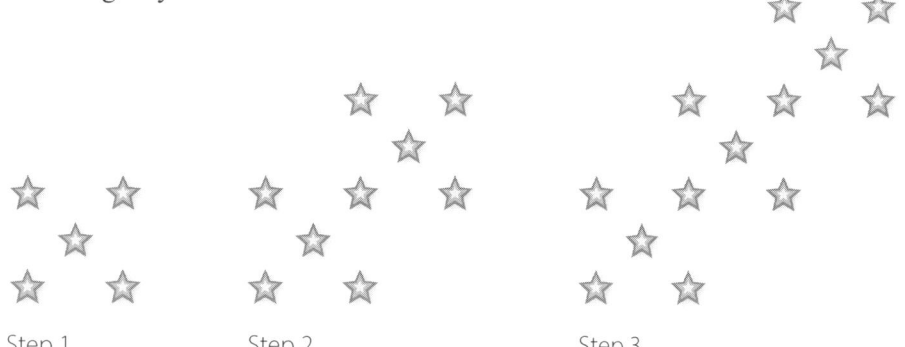

Step 1   Step 2   Step 3

Franco recognises that there are at least three number patterns in his design and he draws up the following table.

Step	Pattern 1 Number of rows with one star	Pattern 2 Total number of stars	Pattern 3 Number of horizontal rows
1	(a)	5	3
2	2	(b)	5
3	3	13	(c)
4	4	17	(d)

6.1 Write down the missing values (a to d). (4)

6.2 Show that each number pattern that Franco has found is a linear one. (3)

6.3 Work out the general term for each number pattern in the form $T_n = \ldots$ (3)

6.4 Determine T_{86} for each pattern. (2)

6.5 Franco would like to complete 59 rows of the design altogether. How many stars will he need? (2)

6.6 Franco counts a total of 286 stars in his design box. Does he have enough stars to develop his design to 105 rows? (5)

Total 50

TOPIC 5: Equations and inequalities

In Grades 8 and 9 you learnt how to solve basic linear equations. Grade 10 builds on and extends your knowledge to include more complicated algebraic equations involving factorisation, literal equations, simultaneous linear equations and inequalities.

Knowledge and skills for this topic

If you struggle with any of the work listed below, revise it before continuing with this Topic:

- a solid working knowledge of products, factors and fractions covered in Topic 2 of this book on algebraic expressions
- basic algebra:
  - distribution, i.e. eliminating brackets and multiplying binomials and trinomials by binomial expressions
  - simplifying, adding and subtracting algebraic fractions up to and including cubic denominators (limited to the sum and difference of two cubes)
  - basic equations and inequalities
- factorisation:
  - using common factors
  - trinomials
  - difference of two squares
  - difference and the sum of two cubes
  - factorising by grouping in pairs.

Content of final exam

Solve:

- linear equations
- quadratic equations by factorisation
- simultaneous linear equations with two unknowns
- word problems involving linear, quadratic or simultaneous linear equations
- literal equations (changing the subject of the formula)
- linear inequalities, showing the solution graphically (interval notation must be known).

Linear equations: revision

A linear equation is a first degree equation, which means that the highest power of the unknown variable (x) is 1. Linear equations usually have only one unique solution.

Linear equations with brackets

Steps for solving linear equations that contain brackets

1. Use distribution correctly to multiply out all brackets. (Watch your signs!)
2. Simplify by adding like terms.
3. Move all terms that contain the variable onto the left side of the equation, and move the constant terms onto the right side of the equation.
4. Simplify both sides fully until you have only one term on each side of the equation.
5. Divide both sides by the coefficient of the unknown.

Worked example

Solve for x: $4(x - 3) - 2(x + 2) = 3x - 2(5 - x)$.

Solution

$$4(x - 3) - 2(x + 2) = 3x - 2(5 - x)$$
$$4x - 12 - 2x - 4 = 3x - 10 + 2x$$
$$2x - 16 = 5x - 10$$
$$-3x = 6$$
$$\frac{-3}{-3}x = \frac{6}{-3}$$
$$x = -2$$

Identities: true for all values of x

Some (very few) equations have an unlimited number of solutions.

Worked example

Solve for x: $-2(x + 8) + 8 = 3 - 3(2 - x) - 5(x + 1)$.

Solution

$$-2(x + 8) + 8 = 3 - 3(2 - x) - 5(x + 1)$$
$$-2x - 16 + 8 = 3 - 6 + 3x - 5x - 5$$
$$-2x - 8 = -2x - 8$$

Remember

The symbol $\mathbb{R}$ stands for the set of all real numbers.

This equation is true for all values of x, $\therefore x \in \mathbb{R}$. An equation like this is called an **identity**, because the two sides are identical (the same).

If we tried to solve this equation further we would get:

$$-2x - 8 = -2x - 8$$
$$-2x + 2x = -8 + 8$$
$$0x = 0$$

We can see that replacing x with any number will make this equation true.

No solution: true for no values of x

Worked examples

Solve for x:

1. $(2x - 5)(3x + 2) = 2(3x^2 - 5x + 1) - x$

2. $-2[3(x - 2) + 2x] = 12$

Solutions

1.
$$(2x - 5)(3x + 2) = 2(3x^2 - 5x + 1) - x$$
$$6x^2 - 11x - 10 = 6x^2 - 10x + 2 - x$$
$$6x^2 - 6x^2 - 11x + 10x + x = 2 + 10$$
$$0x = 12$$

We will not find a number that when multiplied by zero will give us a 12, so there is **no** solution for x.

2. $-2[3(x - 2) + 2x] = 12$
$$-2(3x - 6 + 2x) = 12$$
$$-6x + 12 - 4x = 12$$
$$-10x = 12 - 12$$
$$-10x = 0$$
$$\therefore x = 0$$

Zero is a number, so this equation has **one unique solution**.

> **Summary: linear equations**
>
> - If $0x = 0$, then $x \in \mathbb{R}$, i.e. there are an infinite number of solutions.
> - If $0x =$ any number, then no solution exists.
> - If $x = 0$, then one unique solution exists, which is $x = 0$.

Linear equations with fractions

If the equation has binomial numerators that are not in brackets, it is wise to insert your own brackets as this will ensure that you multiply out correctly in the second step.

Also, remember to adjust any terms that are not fractions by multiplying them by the LCD as well.

Worked examples

Solve for x:

1. $\frac{x - 3}{2} - \frac{2x + 1}{3} = 4$

2. $\frac{2}{3}(x - 3) - \frac{1}{4}(3 - 2x) = \frac{1}{2}(3x + 2)$

Solutions

1. $\dfrac{x-3}{2} - \dfrac{2x+1}{3} = 4$

    $\dfrac{(x-3)}{2} - \dfrac{(2x+1)}{3} = 4$ Insert brackets.

    $\dfrac{3(x-3) - 2(2x+1)}{6} = \dfrac{24}{6}$ Find a common denominator and adjust numerators according to this new denominator.

    $3x - 9 - 4x - 2 = 24$ Drop the denominator and solve as you would any other equation involving brackets.

    $-x - 11 = 24$

    $-x = 35$

    $\therefore x = -35$

2. $\dfrac{2}{3}(x-3) - \dfrac{1}{4}(3-2x) = \dfrac{1}{2}(3x+2)$

    $\dfrac{2(x-3)}{3} - \dfrac{1(3-2x)}{4} = \dfrac{1(3x+2)}{2}$ First rewrite so that the denominators are clear.

    $\dfrac{2(4)(x-3) - 1(3)(3-2x)}{12} = \dfrac{1(6)(3x+2)}{12}$ Find a common denominator and adjust numerators according to this new denominator.

    $8(x-3) - 3(3-2x) = 6(3x+2)$ Drop the denominator and solve.

    $8x - 24 - 9 + 6x = 18x + 12$

    $14x - 33 = 18x + 12$

    $-4x = 45$

    $\therefore x = -\dfrac{45}{4}$

Exercise 1

Solve for x:

1. $5 - x - 2(x-5) = 3(x-3) - 3x$

2. $\dfrac{2x+1}{5} = -\dfrac{1}{7}(x-1)$

3. $6 - 3[2x - 4(x-2)] = 0$

4. $5\tfrac{1}{2} - 4(x+1) + \tfrac{1}{3}(x-2) = \tfrac{1}{3}\left(3x + 1\tfrac{1}{2}\right)$

5. $(x+3)^2 - 5x = (x+1)^2 + 6$

6. $2 - \dfrac{4x-3}{6} = \dfrac{3}{2}(x+3) - \dfrac{2(x-1)}{3}$

7. $\dfrac{1-5x}{6} = \dfrac{2x-1}{3} - \dfrac{3x-1}{2}$

8. $x - \dfrac{2}{3}(x-2) = \dfrac{3x+1}{4} - \dfrac{x+1}{6}$

Quadratic equations

Quadratic equations are second degree equations, which means that the highest power of the unknown variable (e.g. x) is **two**. Quadratic equations have two and sometimes three solutions.

- Solving quadratic equations depends largely on your ability to factorise.
- To solve a quadratic equation all terms **must** be moved onto the left side and the equation set equal to zero.

> **Steps for solving quadratic equations**
> 1. Use the correct algebraic rules (i.e. addition and subtraction) to move all terms onto the left side of the equation, so that the equation is set equal to zero.
> 2. Factorise fully.
> 3. Set each factor equal to zero.
> 4. Solve for the unknown in each of the resulting linear equations.

The solution of a quadratic equation is based on these principles:

- If $AB = 0$, then either $A = 0$ or $B = 0$.
- If $A = 0$, then $AB = 0 \times B = 0$.
- If $B = 0$ then $AB = A \times 0 = 0$.

Worked examples

Solve for x:

1. $6x^2 + 7x - 3 = 0$ **2.** $7x^2 = 12x + 4$ **3.** $8x^2 - 18 = 0$

Solutions

1.
$$6x^2 + 7x - 3 = 0$$
$$(3x - 1)(2x + 3) = 0$$

$3x - 1 = 0$	or	$2x + 3 = 0$
$3x = 1$	or	$2x = -3$
$\therefore x = \frac{1}{3}$	or	$x = -\frac{3}{2}$

2.
$$7x^2 = 12x + 4$$
$$7x^2 - 12x - 4 = 0$$
$$(7x + 2)(x - 2) = 0$$

$7x + 2 = 0$	or	$x - 2 = 0$
$7x = -2$	or	$x = 2$
$\therefore x = -\frac{2}{7}$	or	$x = 2$

3.
$$8x^2 - 18 = 0$$
$$2(4x^2 - 9) = 0$$
$$(2x - 3)(2x + 3) = 0$$

Divide both sides by 2, and remember that $\frac{0}{2} = 0$.

$2x - 3 = 0$	or	$2x + 3 = 0$
$2x = 3$	or	$2x = -3$
$\therefore x = \frac{3}{2}$	or	$x = -\frac{3}{2}$

Check whether equation is factorised or not

Don't be fooled by equations that appear to be factorised and those that are factorised. Study the two examples given below.

Worked examples

Solve for x in each of the equations given below:

1. $(x - 8)(x + 1) = 0$

2. $(x - 8)(x + 1) = -18$

Solution

1. $(x - 8)(x + 1) = 0$ This equation is fully factorised.

$x - 8 = 0$	or	$x + 1 = 0$
$\therefore x = 8$	or	$x = -1$

2. $(x-8)(x+1) = -18$

This equation is **not** fully factorised, so you need to multiply out and move all terms to the left side.

$$x^2 - 7x - 8 = -18$$
$$x^2 - 7x + 10 = 0$$
$$(x-5)(x-2) = 0$$

$x - 5 = 0$	or	$x - 2 = 0$
$\therefore x = 5$	or	$x = 2$

Exercise 2

Solve for x:

1. $(x-3)(6x+5) = 0$ **2.** $x(3x-4) = 0$

3. $(x^2 + 4) = 49$ **4.** $36 - 4x^2 = 0$

5. $5x^2 + 6 = 17x$ **6.** $2x^2 + 6 = -6 - 11x$

7. $(3x-2)(x-2) = 20$

Equations with fractions in which the variable is in the denominator

Restrictions on equations that contain fractions

All equations with unknowns (variables) in their denominators will have restrictions:

- Division by zero is undefined, because $\frac{5}{0}$ is meaningless, or undefined. In the same way, $\frac{5x-1}{0}$ is undefined.
- The fraction $x = 9 - \frac{8}{x}$ will be undefined for $x = 0$, because if $x = 0$, the equation $0 = 9 - \frac{8}{0}$ is undefined. This simply means that the solution to this equation may not be equal to zero.
- Similarly the equation $x - \frac{x-3}{x-1} = \frac{2x}{x-1}$ is undefined if the denominator $x - 1 = 0$. So if the solution of this equation results in $x = 1$, this solution must be discarded.

Worked examples

Solve for x:

1. $x = 9 - \frac{8}{x}$ **2.** $x - \frac{x-3}{x-1} = \frac{2x}{x-1}$

Solutions

1. $x = 9 - \frac{8}{x}$

Restriction: $x \neq 0$

$$\frac{x(x)}{x} = \frac{9(x) - 8}{x}$$
$$x^2 = 9x - 8$$
$$x^2 - 9x + 8 = 0$$
$$(x-8)(x-1) = 0$$
$$\therefore x = 8 \text{ or } x = 1$$

In this equation, $x = 0$ is not one of the solutions, so the restriction does not affect the answer.

Remember

This equation will have restrictions. The denominator of a fraction may not be zero, so $\frac{8}{0}$ will be undefined. Therefore this equation is undefined for $x = 0$.

2. $x - \frac{x-3}{x-1} = \frac{2x}{x-1}$

Restriction: $x \neq 1$

$\frac{x(x-1) - (x-3)}{x-1} = \frac{2x}{x-1}$

$\frac{x^2 - x - x + 3}{x-1} = \frac{2x}{x-1}$

$x^2 - 2x + 3 = 2x$

$x^2 - 4x + 3 = 0$

$(x-3)(x-1) = 0$

$x - 3 = 0$ or $x - 1 = 0$

$\therefore x = 3$ or $x = 1$ But the equation is undefined for $x = 1$, which means that this solution cannot exist, so you discard this solution, and only one solution exists.

$\therefore x = 3$

Steps for solving equations that contain fractions

1. Write down the restrictions.
2. Find a common denominator.
3. Adjust the numerators according to the new common denominator.
4. Drop the denominators.
5. Simplify.
6. Solve for the unknown variable and check against your restrictions.

Exercise 3

Solve for x:

1. $x - \frac{5}{2} = \frac{6}{x}$

2. $\frac{4}{3x-2} - \frac{3}{2x-3} = \frac{1}{2x-1}$

3. $\frac{2}{x-1} - \frac{1}{x} = \frac{1}{x+2}$

4. $\frac{x-2}{x-1} = \frac{2x-1}{x+7}$

5. $\frac{5}{x-2} - \frac{4}{x} = \frac{3}{x+6}$

Simultaneous linear equations

Simultaneous equations involve finding the values of two unknowns at the same time. To solve for two unknowns simultaneously, one needs two equations. Graphically the solution to two equations solved simultaneously is the point at which the two graphs intersect.

There are two methods you can use to solve simultaneous equations:

- substitution
- elimination.

You can choose which method you use. Sometimes choosing elimination results in simpler mathematics and vice versa. The mechanics involved in each of these methods is best described using examples.

Solving simultaneous equations using substitution

Using substitution requires that you rewrite one of the equations with one of the variables as the subject of the formula. Then substitute for this variable in the other equation, which allows you to solve for the other variable.

Generally it is easier to use substitution when one or both equations are given in the x or y form, i.e. when one of the variables is written as the subject of the formula.

Worked example

Find the values of x and y that satisfy the equations $2x + 3y = 7$ and $y - 2 = x$.

Solution

$$2x + 3y = 7 \qquad \text{①}$$
$$y - 2 = x$$
$$\therefore y = x + 2 \qquad \text{②} \qquad \text{Rewrite in the } y \text{ form to get equation ②.}$$

Substitute $x + 2$ for y in ①:
$$2x + 3(x + 2) = 7$$
$$2x + 3x + 6 = 7$$
$$2x + 3x = 7 - 6$$
$$5x = 1$$
$$x = \tfrac{1}{5}$$

Substitute $\tfrac{1}{5}$ for x in ①:
$$2\left(\tfrac{1}{5}\right) + 3y = 7$$
$$3y = 7 - \tfrac{2}{5}$$
$$y = \tfrac{33}{5} \div 3$$
$$y = \tfrac{11}{5}$$
$$\therefore x = \tfrac{1}{5} \text{ and } y = \tfrac{11}{5}$$

OR Substitute $\tfrac{1}{5}$ for x in ②:
$$y = x + 2$$
$$y = \tfrac{1}{5} + 2$$
$$y = \tfrac{11}{5}$$

> **Note**
>
> Number the equations ① and ② to help you to explain your method.

> **Note**
>
> You can substitute for this value of x in either equation to get y. Choose the simpler equation to substitute into, which in this case would be equation ②.

Solving simultaneous equations using elimination

This method requires the elimination of one of the variable using addition or subtraction.

Elimination is easier to use when the coefficients of the unknown variables are equal or additive inverses of each other, which makes elimination easier.

Worked example

Solve for x and y: $x + 2y = 10$ and $x - 5y = 3$.

Solution

This problem lends itself to elimination because the coefficients of x are both equal to 1.

$$x + 2y = 10 \qquad \text{①}$$
$$x - 5y = 3 \qquad \text{②}$$

Subtract ② from ① so that you eliminate x:

$$\begin{array}{r} x + 2y = 10 \\ - \ \underline{x - 5y = 3} \\ 0 + 7y = 7 \end{array} \qquad \text{Remember: } 2y - (-5y) = 2y + 5y = 7y$$

$$\therefore y = \tfrac{7}{7} = 1$$

> **Note**
>
> Subtracting one equation from the other creates one new equation that has only one unknown. In this case x has been eliminated so the unknown is y.

Substitute 1 for y in ①:

$x + 2(1) = 10$

$x = 8$

$\therefore x = 8$ and $y = 1$ Write out your solution in full.

Worked examples using both methods

Solve for x and y: $4x = 6 + 3y$ and $3x + 2y = 13$.

Solution

We will show that it is easier to use elimination in this problem.

$4x = 6 + 3y$ ①

$3x + 2y = 13$ ②

Solution using substitution

Rewrite one of the equation in terms of x or y. There are four choices for what we will substitute.

Choice 1:

Rewrite ① in the x or y form:

$4x = 6 + 3y$	or	$3y = 4x - 6$
$x = \frac{6 + 3y}{4}$	or	$y = \frac{4x - 6}{3}$

Choice 2:

Rewrite ② in the x or y form:

$3x + 2y = 13$	or	$3x + 2y = 13$
$3x = 13 - 2y$	or	$2y = 13 - 3x$
$x = \frac{13 - 2y}{3}$	or	$y = \frac{13 - 3x}{2}$

Each choice results in an algebraic fraction that we will have to substitute into the other equation.

In Choice 1 we have chosen to substitute $\frac{6 + 3y}{4}$ for x into ②:

$\therefore 3\left(\frac{6 + 3y}{4}\right) + 2y = 13$

$\frac{3(6 + 3y) + 4(2y)}{4} = \frac{13(4)}{4}$

$3(6 + 3y) + 4(2y) = 13(4)$

$18 + 9y + 8y = 52$

$17y = 34$

$\therefore y = 2$

Substitute 2 for y in $x = \frac{6 + 3y}{4}$:

$x = \frac{6 + 3(2)}{4} = \frac{12}{4} = 3$

$\therefore x = 3$ and $y = 2$

Solution using elimination

Now compare this with the simpler solution using elimination.

- Rewrite the equations one below the other, lining up the like terms:

  $4x - 3y = 6$ ①

  $3x + 2y = 13$ ②

- Either the x terms or the y terms need to have the same coefficient, in order to eliminate them.

- We have chosen to eliminate y, so we have used the LCD of 3 and 2, i.e. 6.

- We will need to multiply the entire equation by whatever is necessary to get the coefficient of both ys to be 6. So we multiply ① by 2 to get $6y$, and ② by 3 to get $6y$ (see the calculation below).

- The coefficients of y have opposite signs, so you add ① and ② in order to eliminate y.

Note

If you had chosen to eliminate x, you would multiply equation ① by 3, and equation ② by 4, to obtain $12x$ in each equation. Then you would subtract ② from ①.

$$8x - 6y = 12$$
$$+ \quad 9x + 6y = 39$$
$$17x + 0 = 51$$

Multiply equation ① by 2 to obtain $6y$.

Multiply equation ② by 3 to obtain $6y$.

$$\therefore x = \frac{51}{17} = 3$$

Note

You may substitute in ① if you prefer.

Substitute 3 for x in ②:

$$3(3) + 2y = 13$$
$$9 + 2y = 13$$
$$2y = 4$$
$$\therefore y = 2$$
$$\therefore x = 3 \text{ and } y = 2$$

Exercise 4

1. Use substitution to solve for x and y:

   1.1 $4x - 3y = 10$ and $4x + y = 2$ **1.2** $x + 2y = 4$ and $3x + y = 7$

   1.3 $2x + y + 3 = 0$ and $y = x + 1$

2. Use elimination to solve for x and y:

   2.1 $2x + 2y = 14$ and $5x - 2y = 21$ **2.2** $3x - 4y = -8$ and $3x - y = 10$

   2.3 $2x - 5y = -12$ and $4x + 5y = 6$

3. Use either substitution or elimination to solve for x and y:

   3.1 $3x + 5y = -1$ and $4x + 7y = 4$ **3.2** $y = 2x - 1$ and $y = -5x + 2$

   3.3 $\frac{2x + y}{3} = 11$ and $\frac{1}{6}x - 2y - 34 = 0$

   3.4 $x + 2y + 3 = 7x - y$ and $2x - 3y + 16 = 7x - y$

Literal equations (changing the subject of the equation)

Equations are called 'literal' equations when they have 'letters' as coefficients and not numbers. Literal equations are solved in the same way as any other linear equation.

The most important difference is that we cannot always add or subtract terms because the equation contains different variables as well as numbers. So we will sometimes have to factorise to isolate the subject of the equation. The process will seem more complicated than usual because you cannot simplify your answer as you go along.

The solution to a literal equation is in fact a formula that could be used to find a specific value for the subject of that formula.

Steps for solving a literal equation

The aim is to isolate the subject of the formula.

1. Begin by moving any terms containing the subject of the formula onto the left side of the equation.

2. Move all terms that do not contain the subject of the formula onto the right side.

3. If the subject of the formula appears in more than one term, factorise to isolate it.

4. Divide through by the coefficient of the subject of the formula.

5. Simplify further, using factorisation if necessary.

Worked examples

Solve for x in each of the literal equations given below:

1. $x + b = c - d$
2. $ax = b + cx$
3. $\frac{ax}{b} = a - b$

4. $\frac{1}{x} - \frac{1}{a} = \frac{1}{b}$
5. $a^2 + ax = b^2 + bx$

Solutions

1. $x + b = c - d$

$\therefore x = c - d - b$ Isolate x on the left side using addition and subtraction.

2. $ax = b + cx$

$ax - cx = b$ When x appears in more than one term, move all terms containing the variable x onto the left side.

$x(a - c) = b$ Factorise to isolate x.

$\therefore x = \frac{b}{a - c}$ Divide both sides by $(a - c)$, which is the coefficient of x.

3. $\frac{ax}{b} = a - b$

$ax = b(a - b)$ Multiply both sides of the equation by b.

$\therefore x = \frac{ab - b^2}{a}$ Divide both sides by a, the coefficient of x.

4. $\frac{1}{x} - \frac{1}{a} = \frac{1}{b}$

$\frac{1(ab) - 1(xb)}{xab} = \frac{1(ax)}{xab}$ Get rid of the fractions by making a common denominator.

$ab - xb = ax$ Adjust the numerator and drop the denominator.

$-xb - ax = -ab$ Move all terms containing x onto the left side.

$xb + ax = ab$

$x(b + a) = ab$ Factorise to isolate x.

$\therefore x = \frac{ab}{b + a}$ Divide both sides by $(a + b)$, the coefficient of x.

5. $a^2 + ax = b^2 + bx$

$ax - bx = b^2 - a^2$ Move all terms containing x onto the left side.

$x(a - b) = b^2 - a^2$ Factorise to isolate x.

$x = \frac{b^2 - a^2}{(a - b)}$ Divide both sides by $(a - b)$, the coefficient of x.

$x = \frac{(b - a)(b + a)}{(a - b)}$ Factorise and use a sign change to simplify further.

$x = \frac{-(a - b)(b + a)}{(a - b)}$

$\therefore x = -(b + a)$

Changing the subject of the formula

Literal equations can be used extensively when working with problems involving formulae. In the examples above we were instructed to solve for x. These same techniques are used when we are asked to change the subject of the formula to any letter.

Worked examples

Make the letter in the brackets the subject of the formula:

1. $\frac{1}{a} = \frac{1}{b} + \frac{1}{c}$ (b)   **2.** $r = \sqrt[3]{\frac{3v}{4\pi}}$ (v)   **3.** $v = m\sqrt{a^2 - b}$ (a)

Solutions

Rearrange the formula with the required subject on the left side, then continue as usual.

1.
$$\frac{1}{a} = \frac{1}{b} + \frac{1}{c} \quad (b)$$
$$\frac{bc}{abc} = \frac{ac + ab}{abc}$$
$$bc = ac + ab$$
$$bc - ab = ac$$
$$b(c - a) = ac$$
$$\therefore b = \frac{ac}{(c - a)}$$

2.
$$r = \sqrt[3]{\frac{3v}{4\pi}} \quad (v)$$
$$\left(\sqrt[3]{\frac{3v}{4\pi}}\right)^3 = r^3$$

To remove a cube root, raise the expression to the power 3. The right side must also be raised to the power 3.

$$\frac{3v}{4\pi} = r^3$$
$$3v = r^3 \times 4\pi$$
$$\therefore v = \frac{4\pi r^3}{3}$$

3.
$$v = m\sqrt{a^2 - b} \quad (a)$$
$$m\sqrt{a^2 - b} = v$$
$$\sqrt{a^2 - b} = \frac{v}{m}$$
$$a^2 - b = \left(\frac{v}{m}\right)^2$$

To get rid the square root, raise both sides to the power 2.

$$a^2 = \left(\frac{v}{m}\right)^2 + b$$
$$\therefore a = \pm\sqrt{\left(\frac{v}{m}\right)^2 + b}$$

Take the square root of both sides. Remember to write ± on the RHS for the square root.

Exercise 5

1. Solve for x in each of the equations:

    1.1 $ax + bx = c$ **1.2** $k - \frac{m}{x} = t$

    1.3 $\frac{x - a}{x + b} = \frac{x}{x + a}$ **1.4** $(x - a)^2 = (x - b)^2$

2. Given that $\frac{x - m}{2a} = 3$:

    2.1   rewrite the expression with x as the subject of the formula

    2.2   rewrite the expression with a as the subject of the formula.

3. Given that $xy = p + xt$:

    3.1   rewrite the expression with x as the subject of the formula

    3.2   rewrite the expression with t as the subject of the formula.

4. Given that $p = \frac{1}{x} + v + \frac{1}{y}$:

   4.1 rewrite the expression with x as the subject of the formula

   4.2 rewrite the expression with y as the subject of the formula.

Solving inequalities

In solving inequalities, the basic rules for solving linear equations are applied. Unless otherwise specified, your solution will come from the set of real numbers, i.e. $x \in \mathbb{R}$. There are therefore an infinite number of solutions, even if the values of x lie between two real numbers.

Showing solutions in interval notation, set builder notation or on number lines

Solutions to inequalities can be written in set builder notation or in interval notation.

For interval notation:

- the inequalities $<$ and $>$ are represented using round brackets ( ).
- the inequalities $\leq$ and $\geq$ are represented using square brackets [ ].

A linear inequality can also be represented on a number line graph. In number lines:

- the inequalities $<$ and $>$ are represented using open dots $\bigcirc$.
- the inequalities $\leq$ and $\geq$ are represented using closed or filled in dots $\bullet$.

Worked examples

For each of the inequalities given below:

(a) solve for x

(b) rewrite your answer in interval notation

(c) represent your answer graphically on a number line.

1. $5(x - 5) > 0$ 2. $5(5 - x) \geq 0$

Solutions

1. $5(x - 5) > 0$
   - **(a)** $5x - 25 > 0$
     $$5x > 25$$
     $$x > 5$$
   - **(b)** $(5; \infty)$
   - **(c)**

2. $5(5 - x) \geq 0$
   - **(a)** $25 - 5x \geq 0$
     $$-5x \geq -25$$
     $$x \leq 5$$
   - **(b)** $(-\infty; 5]$
   - **(c)**

Worked examples

1. Solve for x: $\frac{1}{3}(4 - x) > \frac{1}{4}x + 3$.

2. Rewrite your answer in interval notation and represent the solution graphically on a number line for $x \in \mathbb{R}$.

Solutions

1. $\frac{1}{3}(4-x) > \frac{1}{4}x + 3$

    $\frac{(4-x)}{3} > \frac{1}{4}x + \frac{3}{1}$

    $\frac{4(4-x)}{12} > \frac{3x + 3(12)}{12}$

    $16 - 4x > 3x + 36$

    $-7x > 20$

    $x < -\frac{20}{7}$

Find a common denominator, adjust the numerator and then drop the denominator.

When dividing by –7, the inequality sign **must** be reversed.

2. $\left(-\infty; -\frac{20}{7}\right)$

$-\frac{20}{7}$

Worked examples

1. Solve for x: $-2 \leq \frac{4-5x}{3} < 3$.

2. Rewrite your answer in interval notation and represent the solution graphically on a number line for $x \in \mathbb{R}$.

Solutions

1. $-2 \leq \frac{4-5x}{3} < 3$

Note

There are two methods of solving this type of inequality.

Method 1:
You can treat this inequality as two separate inequalities and solve each one:

$-2 \leq \frac{4-5x}{3} < 3$

$-2 \leq \frac{4-5x}{3}$ and $\frac{4-5x}{3} < 3$

$-6 \leq 4 - 5x$ $4 - 5x < 9$

$-10 \leq -5x$ $-5x < 5$

$2 \geq x$ $x > -1$

Method 2:
You can work with all three parts of the inequality simultaneously, until only x remains in the centre.

$-2 \leq \frac{4-5x}{3} < 3$

$-6 \leq [4 - 5x] < 9$ Multiply throughout by 3.

$-10 \leq -5x < 5$ Subtract 4 throughout.

$2 \geq x > -1$ Divide by –5. Reverse the signs.

i.e. $-1 < x \leq 2$

2. $(-1; 2]$

-1 2

Exercise 6

Solve for x for $x \in \mathbb{R}$, give the solution in interval notation, and illustrate your solution on a number line.

1. $5x - 3(x + 1) < 2 + 3x$

2. $x - \frac{x-3}{2} \geq 2,5 + \frac{5x}{6}$

3. $\frac{1}{3}(x - 2) \geq 1\frac{1}{2} - \frac{1-2x}{2}$

4. $\frac{x-2}{4} - \frac{x-4}{6} \geq 1\frac{2}{3}$

5. $0 < \frac{x}{3} + 1 \leq 3$

Equations with fractions: complex procedures and problem solving

Complex procedures on this work involves:

- linear and quadratic equations that contain complex denominators
- complex simultaneous equations
- complex literal equations
- word problems.

Linear equations with complex denominators

> **Steps for factorising denominators**
>
> 1. For more complex denominators, factorise each denominator separately to find the lowest common denominator.
> 2. The restrictions can only be stated once the denominators have been factorised.
> 3. Continue as you would for any other equation involving fractions.

Worked example

Solve for x: $\dfrac{x-2}{x^2-1} + \dfrac{1}{1-x} = \dfrac{1}{x^2+3x+2}$.

Solution

$$\frac{x-2}{x^2-1} + \frac{1}{1-x} = \frac{1}{x^2+3x+2}$$

$$\frac{x-2}{(x-1)(x+1)} - \frac{1}{(x-1)} = \frac{1}{(x+2)(x+1)}$$

Factorise the denominators to find the lowest common denominator.

Restrictions: $x \neq 1$, $x \neq -1$, $x \neq -2$.

Remember to write down the restrictions.

$$\frac{(x-2)(x+2) - 1(x+1)(x+2)}{(x-1)(x+1)(x+2)} = \frac{1(x-1)}{(x+2)(x+1)(x-1)}$$

$$(x^2-4) - (x^2+3x+2) = x-1$$

$$x^2 - 4 - x^2 - 3x - 2 = x-1$$

$$-4x = 5$$

This is now a linear equation.

$$\therefore x = -\frac{5}{4}$$

Quadratic equations with complex denominators

Worked example

Solve for x: $\dfrac{2}{x^2-1} + \dfrac{1}{1-x} = \dfrac{1}{x^2+3x+2}$.

Solution

$$\frac{2}{x^2-1} + \frac{1}{1-x} = \frac{1}{x^2+3x+2}$$

$$\frac{2}{(x-1)(x+1)} - \frac{1}{(x-1)} = \frac{1}{(x+2)(x+1)(x-1)}$$

Restrictions: $x \neq 1$, $x \neq -1$, $x \neq -2$.

$$\frac{2(x+2) - 1(x+1)(x+2)}{(x-1)(x+1)(x+2)} = \frac{1(x-1)}{(x+2)(x+1)}$$

$$2x + 4 - (x^2+3x+2) = x-1$$

$$2x + 4 - x^2 - 3x - 2 = x-1$$

$$-x^2 - x + 2 = x-1$$

$$-x^2 - 2x + 3 = 0$$
$$x^2 + 2x - 3 = 0$$
$$(x + 3)(x - 1) = 0$$

$x + 3 = 0$	or	$x - 1 = 0$
$\therefore x = -3$	or	$x = 1$ (not a solution)

$\therefore x = -3$

This equation is quadratic, but it has only one solution because of the restrictions.

Exercise 7: complex procedures

Solve for x:

1. $\dfrac{x}{x-3} - \dfrac{6}{3x - x^2} = 1$

2. $\dfrac{3}{x} + \dfrac{3}{x^2 - x} = \dfrac{1}{x^2 - 1}$

3. $\dfrac{1}{x+3} + \dfrac{x}{x-1} = \dfrac{x^2 + 7}{x^2 + 2x - 3}$

4. $\dfrac{x+2}{x^2 - 3x - 4} = \dfrac{3}{x-4} - \dfrac{1}{2x+2}$

5. $\dfrac{3+8}{x^2 + x - 2} - \dfrac{2x-3}{x^2 + 6x + 8} = \dfrac{x-4}{x^2 + 3x - 4}$

Complex simultaneous equations

More complicated simultaneous equations may require some simplification and rearranging before you decide which method to use.

Worked example

Solve for x and y: $\dfrac{1-3x}{7} = 2 - \dfrac{3y-1}{5}$ and $y = 9 - \dfrac{3x+y}{11}$.

Solution

Find a common denominator and simplify each equation:

$$\frac{1-3x}{7} = 2 - \frac{3y-1}{5}$$
$$\frac{5(1-3x)}{35} = \frac{2(35) - 7(3y-1)}{35}$$
$$5 - 15x = 70 - 21y + 7$$
$$21y - 15x = 72$$
$$7y - 5x = 24 \quad ①$$

$$y = 9 - \frac{3x+y}{11}$$
$$\frac{11y}{11} = \frac{9(11) - (3x+y)}{11}$$
$$11y = 99 - 3x - y$$
$$12y + 3x = 99$$
$$4y + x = 33 \quad ②$$

Examine the equations and decide which method to use. In this case substitution will be simpler.

$x = 33 - 4y \qquad ②$ Make x the subject of equation.

Substitute $33 - 4y$ for x in ①:

$$7y - 5(33 - 4y) = 24$$
$$7y - 165 + 20y = 24$$
$$27y = 189$$
$$y = 7$$

Substitute 7 for y in ②:

$$x = 33 - 4(7)$$
$$x = 5$$
$$\therefore x = 5 \text{ and } y = 7$$

Exercise 8

Solve for x and y in the following simultaneous equations.

1. $\frac{x+y}{3-y} = \frac{1}{2}$ and $\frac{x-y}{3+2y} = 4$

2. $\frac{2(x-y)}{3} + \frac{3(x+y)}{10} = \frac{10}{3}$ and $\frac{x-y}{4} + \frac{x+y}{5} = 3$

3. $3x - \frac{5-y}{2} = \frac{5x-2}{3}$ and $\frac{2x-3}{5} = y - \frac{6}{5}$

4. $\frac{6}{x} - \frac{1}{y} = 4$ and $\frac{9}{x} + 1 = -\frac{2}{y}$

Complex literal equations

Worked example

Rewrite $s = \frac{n}{2}[2a + (n-1)d]$ with d as the subject of the formula.

Solution

$$s = \frac{n}{2}[2a + (n-1)d]$$

$$\frac{n}{2}[2a + (n-1)d] = s$$

$$2a + (n-1)d = \frac{2s}{n}$$

$$(n-1)d = \frac{2s}{n} - 2a$$

$$d = \left(\frac{2s}{n} - 2a\right) \div (n-1)$$

$$d = \left(\frac{2s - 2an}{n}\right) \times \frac{1}{(n-1)}$$

$$d = \frac{2s - 2an}{n(n-1)} = \frac{2s - 2an}{n^2 - n}$$

Exercise 9

1. Solve for x:

   1.1 $\frac{x-3}{b} = 5 - \frac{2x-1}{2c}$ **1.2** $\frac{9ax}{b} - \frac{4bx}{a} = 3a - 2b$

   1.3 $(x-a)(x-b) = (x+a)(x-3b)$ **1.4** $(x-a)^2 - (x-b)^2 = a - b$

   1.5 $\frac{1-ax}{1+bx} = \frac{b}{a}$

2. Rewrite $t = \frac{m}{2}\sqrt{\frac{f}{g}}$ with f as the subject of the formula.

Word problems

In this section we use linear, quadratic and simultaneous equations to solve practical word problems.

Steps for approaching word problems

1. Read very carefully and make sure that you understand the given information.

2. Drawing your own diagram or creating a table can be really helpful.

3. Decide what you have to determine, and use a variable (usually x and/or y) to represent it.
   Hint: If more than one unknown is present, let the smallest value be equal to x.

4. You often have to write the other unknown in terms of x, in other words as '$x \pm$ a number'.

5. You can also choose to call the other unknown y, but then you must set up two equations to solve the two variables.

---

6. Use the remaining information to set up one or two equations.

7. Make sure that you do answer the question that was asked. Interpret your answer(s) and discard those that are not realistic.

8. Check your answers.

Useful information

1. Area of a rectangle = length × breadth.

    Perimeter of a rectangle = 2(length + breadth)

2. Profit = selling price – cost price

3. Distance = speed × time

    $\therefore$ Time $= \frac{\text{distance}}{\text{speed}}$

    $\therefore$ Speed $= \frac{\text{distance}}{\text{time}}$

4. x 10c coins have a value of $10x$ cents.

5. If I am x years old now, my age 3 years later will be $x + 3$ years.

Worked examples

1. There are two consecutive integers. One quarter of the first integer added to one seventh of the second integer is greater than 19. Determine the smallest possible values for the two integers.

2. The sum of the digits of a two digit number is 9. If the digits are switched around, the new number is 9 less than the original number. Determine the new number.

3. At present a man is nine times the age of his son. In three years' time the father's age will be five times the age of his son. What are the present ages of the man and his son?

4. The prices of tickets for a show are R6,75 for adults and R4,50 for children. There were twice as many adults as children. The concert raised R3 600. How many adults and how many children attended the concert?

5. An aeroplane completes a journey of 240 km in 4 hours less than a car that travels at one fifth of the speed of the aeroplane. Calculate the speed of the car.

6. Two cyclists are approaching each other at 20 km/h and 24 km/h respectively. At 09:00 they are 198 km apart. At what time will they meet, if they start riding towards each other simultaneously?

7. Nadia and Siraaj have two sons. The sum of their ages is 17 and the difference between their ages is 5. If Moegsien is the older brother and Nazim is the younger brother, determine their ages.

8. A rectangle has an area of 6 cm^2 and its breadth is 5 cm less than its length. Find the dimensions of the rectangle.

Solutions

1. Let the smaller number be x $\therefore$ the other number is $x + 1$:

$$\tfrac{1}{4}x + \tfrac{1}{7}(x + 1) > 19$$

$\times 28$: $7x + 4x + 4 > 532$

$$11x > 528$$

$$\therefore x > 48$$

The integers are 49 and 50.

2. Let the digits be x and y $\therefore$ the first number itself is $10x + y$ (because the x is the tens digit):

$$x + y = 9 \qquad \text{①}$$

The second number is $10y + x$ Because the digits are swapped.

$$10x + y = 10y + x + 9$$

$$9x - 9y = 9$$

$$x - y = 1 \qquad \text{②}$$

①+②:

$$\begin{aligned} x + y &= 9 \\ + \quad x - y &= 1 \\ \hline 2x &= 10 \end{aligned}$$

$$\therefore x = 5$$

Substitute 5 for x in ①:

$$y = 9 - 5 = 4$$

$\therefore$ The first number = 54 and the second number = 45.

3. Let the son be x years old, $\therefore$ the father is $9x$ years old. In 3 years' time the son will be $x + 3$.

Father: $9x + 3$

$$\therefore 9x + 3 = 5(x + 3)$$

Be careful to multiply the correct side of the equation in order to keep both sides equal. In this case you must multiply 5 times the age of the son.

$$9x + 3 = 5x + 15$$

$$4x = 12$$

$$\therefore x = 3$$

The present age of the man is 27 years old, and the son is 3 years old.

4. Let there be x children, $\therefore$ there are $2x$ adults:

$$450x + 675(2x) = 360\,000$$

$$450x + 1\,350x = 360\,000$$

$$1\,800x = 360\,000$$

$$\therefore x = 200$$

There were 200 children and 400 adults.

5. Use simultaneous equations to solve this problem. Let the speed of the car be x km/h:

    The aeroplane takes 4 hours less:

    $\therefore \frac{240}{5x} + 4 = \frac{240}{x}$

    $\times 5x$: $240 + 20x = 1\ 200$

    $\qquad\qquad 20x = 960$

    $\therefore x = 48$

    The speed of the car is 48 km/h.

	Aeroplane	Car
D	³ 240	³ 240
S	² $5x$	¹ x
T	⁴ $\frac{240}{5x}$	⁵ $\frac{240}{x}$

6. Let them meet after x hours.

    Combined, they cover 198 km:

    $\therefore 20x + 24x = 198$

    $44x = 198$

    $\therefore x = 4{,}5$

    The two cyclists meet after $4\frac{1}{2}$ hours, which will be at 13:30.

	Cyclist 1	Cyclist 2
D	⁴ $20x$	⁵ $24x$
S	² 20 km/h	³ 24 km/h
T	¹ x	¹ x

7. Let Moegsien's age be x and Nazim's age be y.

    Then: $x + y = 17$ ①

    And: $x - y = 5$ ②

    By elimination $2x = 22$

    $\therefore x = 11$

    By substituting into ①:

    $11 + y = 17$

    $\therefore y = 6$

    Moegsien is 11 years old and Nazim is 6 years old.

8. Let x = the length of the rectangle.

    Area = $l \times b$

    $\therefore x(x - 5) = 6$

    $\qquad x^2 - 5x = 6$

    $\quad x^2 - 5x - 6 = 0$

    $(x - 6)(x + 1) = 0$

    $\therefore x = 6$ or $x = -1$

    $\therefore x = 6$

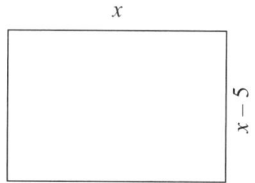

This is the only solution because a rectangle cannot have a side with a length of –1.

Dimensions of rectangle: length = 6 cm and breadth = 6 – 5 = 1 cm.

Exercise 10

1. One half of a number plus one third of the number plus one quarter of the number is 10 more than the number. Determine the number.

2. Twice a number reduced by half the original number is equal to 30. Determine the number.

3. The sum of two numbers is 50 and their difference is 10. Determine the numbers.

4. The sum of four consecutive even numbers is 68. Determine the numbers.

B. Ages

5. A mother is 30 years older than her daughter at present. Ten years ago she was twice the age of her daughter. What is the mother's age at present?

6. A mother is 3 times the age of her daughter at present. In 12 years' time, she will be twice the age of her daughter. What is the mother's age at present?

7. A man is 35 years old at present and his son is 7 years old. In how many years' time will he be twice the age of his son?

C. Other

8. The length of a rectangle is 5 metres longer than its breadth. Calculate the length and the breadth if the perimeter is equal to 30 m.

9. Mary sold 22 concert tickets at R5, R4, and R3 each. She sold twice as many R5 tickets as R4 tickets. If her total sales were R96, how many R5 tickets did she sell?

10. I have twice as many 20c coins as 10c coins, and twice as many 10c coins as 5c coins. If I have R3,15, calculate the number of each kind of coin.

D: Word problems with two unknowns

11. A grocer mixes x kilograms of tea at R6,50 per kg with y kilograms of tea at R8,20 per kg. The mixture is worth R261,20. He sells the mixture at R7,80 per kg and makes a profit of R19,60. How many kg of each brand of tea did he mix?

12. When a two digit number is added to twice the tens digit, the answer is 33. The new number formed by swapping the two digits is 63 greater than the original number. Determine the original number. (Let the tens digit be x and the unit digit be y.)

13. The area of a rectangle stays the same if the length is increased by 20 mm and the breadth is decreased by 10 mm. The area increases by 8 cm^2 if both the length and the breadth are increased by 10 mm. Determine the length and the breadth of the original rectangle.

14. The sum of four consecutive even numbers is 68. Determine the numbers.

E: Mixed problems

15. At present A is twice the age of B. Eight years ago A was 3 times the age of B. Calculate their present ages.

16. The sum of the ages of Sipho and Thabo is 72 years. In 12 years' time Sipho, the older person, will be six times the age that Thabo was 21 years ago.

    16.1 What are their present ages?

    16.2 What was the difference between their ages 15 years ago?

17. Bill is 10 years older than John. In 8 years' time, twice Bill's age will be equal to 3 times John's age. What are their present ages?

18. A farmer bought cattle and sheep for an amount of R53 800. He bought 100 animals in total. The sheep were R250 each and the cattle R730 each. How many sheep and how many cattle did he buy?

19. A man is allowed 25 shots at a target. For every bull's-eye (hitting the exact middle of the target) he receives R5,00, but for every miss he pays R2,50. How many bull's-eyes did he hit if he made a profit of R5,00?

20. A man can row at 5 km/h in stationary water. It takes him 4 hours to row upstream towards a bridge and back to his house. How far is the bridge from the house if the river flows downstream at 2 km/h?

21. Two towns A and B are 300 km apart. C is exactly halfway between A and B. Cyclist 1 departs from B to go to C at x km/h at 09:00, and one hour later Cyclist 2 departs from A to go to C, cycling 5 km/h faster than Cyclist 1.

    21.1 Calculate, in terms of x, the time each cyclist requires to reach C.

    21.2 Calculate at what time they will reach C.

Test A: Knowledge and routine procedures

1. Solve for x:

    1.1 $\frac{2x-3}{3} - 3x = \frac{2x}{6}$ (5)

    1.2 $4(3x+1) - 15 \geq 5(3x-1)$ (4)

    1.3 $tx - b = d - qx$ (3)

    1.4 $2x^2 - x = 1$ (4)

    1.5 $\frac{x-2}{x-1} = \frac{2x-1}{x+7}$ (6)

2. Solve for x and y: $x + y = -3$ and $4x + 3y = -8$. (5)

3. Solve for r in terms of V, π and h: $V = \pi r^2 h$. (3)

4. Rewrite the inequality $-2{,}5 < x \leq 5$ in interval notation and represent the inequality on a number line for $x \in \mathbb{R}$. (4)

5. Solve for x and represent your solution on a number line for $x \in \mathbb{R}$:

    5.1 $2x + 7 \geq 4x - 10$ (5)

    5.2 $3 < -2x + 1 \leq 6$ (5)

6. Given $(3x - 2)(x - 2) = k$, solve for x if:

    6.1 $k = 0$ (2)

    6.2 $k = 20$ (4)

Total 50

Test B: Complex procedures and problem solving

1. Solve for x:

    1.1 $1\frac{1}{2} - \frac{5x}{3} > \frac{1}{4}(5 - 3x) + \frac{1}{3}(3 - 5x)$ (5)

    1.2 $\frac{1}{6x^2 + x - 15} - \frac{1}{9 - 4x^2} = \frac{1}{6x^2 + 19x + 15}$ (8)

2. Solve for $\frac{x}{y}$ if $4x^2 + 22xy = 12y^2$. (6)

3. Solve for x and y: $\frac{3}{4}(x - y) = \frac{x+y}{2} - 5$ and $\frac{x+y}{2} = 2(x - y)$. (7)

4. Solve for x: $\frac{mx+n}{n} - \frac{nx+m}{m} = m + n$. (8)

5. Solve for x and y: $y = -4x + 12$ and $y = 4x^2 - 8x - 3$. (6)

6. The sum of the squares of two consecutive natural numbers is 85. Determine the numbers. (4)

7. Two towns A and B are 640 km apart. A motor cyclist leaves A at 08:00 and drives in the direction of B. A motor car driver leaves B for A at 10:30 and drives at twice the speed of the motor cyclist. They meet exactly halfway between A and B. Calculate the speed of each and the time at which they will meet. (6)

Total 50

Test C: Content and breakdown as for exam

1. Write down the value(s) of x that will satisfy the equations:
   1.1 $3(2 - x) - 2x = 6 - 5x$ (3)
   1.2 $(2x - 7)(y + 2) = 0$ (2)
   1.3 $k - \frac{m}{x} = t$ (4)

2. 2.1 Solve for x: $-1 \leq 2 - 3x < 8$. (5)
   2.2 Rewrite your answer using interval notation. (2)
   2.3 Illustrate your answer on a number line for $x \in \mathbb{R}$. (2)

3. A maths test has multiple choice questions worth 2 marks each and short questions worth 3 marks each. The test is out of 50 marks and there are 22 questions in total.
   3.1 Let the number of multiple choice questions be x and the number of short questions be y. Use all the given information to set up two linear equations. (2)
   3.2 Solve the two equations simultaneously to determine the number of multiple choice questions on this test. (3)

4. Solve for x:
   4.1 $\frac{3 - x}{1 - x^2} + \frac{2x + 4}{(x + 1)} = \frac{2x}{x - 1}$ (7)
   4.2 $6x^2 = 13x + 5$ (5)
   4.3 $px = p^2 - t^2 + tx$ (4)

5. Solve for x and y: $x + 2y = 1$ and $\frac{x}{3} + \frac{y}{2} = 1$. (6)

6. A man misses the train by 2 minutes if he walks from home to the station at an average speed of 4 km/h. However, if he jogs at an average speed of 8 km/h, he arrives at the station $5\frac{1}{2}$ minutes before the time of departure. What is the distance from the man's home to the station? (5)

Total 50

TOPIC 6: Trigonometric ratios and right-angled triangles

Trigonometry is a very old branch of Mathematics. Today it is used in the form of triangle measurement in surveying, navigation, engineering and astronomy. Trigonometric expressions and calculations are also used in magnetism, electricity and electronics.

In Grade 10 we work with trigonometry in the first quadrant, and solve right-angled triangles.

Solving trigonometric problems in all four quadrants is addressed in Topic 12 of this study guide. Both topics are examinable.

Knowledge and skills required for this topic

If you struggle with any of the work listed below, revise it before continuing with this Topic:

- the theorem of Pythagoras
- ratio and similarity
- a basic understanding of the Cartesian plane and point plotting
- adding, subtracting, multiplying and dividing fractions
- basic linear algebraic equations
- simplifying, adding, subtracting, multiplying and dividing surds
- basic triangle geometry.

Content of final exam

- Define the trigonometric ratios sin θ, cos θ and tan θ, using right-angled triangles.
- Define the reciprocals of the trigonometric ratios cosec θ, sec θ and cot θ, using right-angled triangles.
- Derive values of the trigonometric ratios for the special cases θ ∈ {0°; 30°; 45°; 60°; 90°} (without using a calculator).
- Use diagrams to determine the numerical values of the ratios for angles 0° to 90°.
- Use a scientific calculator for trigonometric calculations.
- Solve right-angled triangles using trigonometry.

Naming the sides of a right-angled triangle

Triangles are usually named using capital letters for the angles and small letters for the sides.

In this triangle:

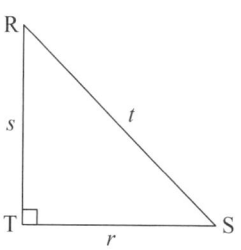

- the letter *t* is used for the side RS, because it is the side opposite the angle T
- the letter *r* is used for the side TS, because it is the side opposite the angle R
- the letter *s* is used for the side RT, because it is the side opposite the angle S.

However, any small letter can be used to represent the length of a side of a right-angled triangle.

Naming the angles of a right-angled triangle

- Any capital letters may be used, as in your Grade 9 geometry, e.g. $\hat{B}$.
- Letters of the Greek alphabet, such as α (alpha), β (beta), and most frequently, θ (theta), are used to represent the size of an angle.
- An angle may also be represented by small x or y.

In any right-angled triangle, if an angle θ is given, we refer to:

- the side next to θ as the **adjacent** side
- the side opposite θ as the **opposite** side
- the side opposite the right angle as the **hypotenuse**.

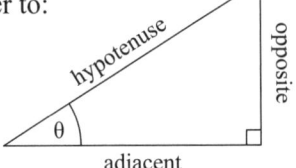

Definitions for the trigonometric ratios

The trigonometric ratios, (sine, cosine, tangent) and their reciprocal ratios (cosecant, secant, and cotangent) are defined as shown in the table below.

Trigonometric ratios		Reciprocal ratios	
sine (sin)	$\sin \theta = \dfrac{\text{opposite}}{\text{hypotenuse}} = \dfrac{O}{H}$	cosecant (cosec)	$\operatorname{cosec} \theta = \dfrac{\text{hypotenuse}}{\text{opposite}} = \dfrac{H}{O}$
cosine (cos)	$\cos \theta = \dfrac{\text{adjacent}}{\text{hypotenuse}} = \dfrac{A}{H}$	secant (sec)	$\sec \theta = \dfrac{\text{hypotenuse}}{\text{adjacent}} = \dfrac{H}{A}$
tangent (tan)	$\tan \theta = \dfrac{\text{opposite}}{\text{adjacent}} = \dfrac{O}{A}$	cotangent (cot)	$\cot \theta = \dfrac{\text{adjacent}}{\text{opposite}} = \dfrac{A}{O}$

These trigonometric ratios must be committed to memory.

SOHCAHTOA, (pronounced 'soccer tour') is a convenient memory device for defining sin θ, cos θ and tan θ.

SOH: $\sin \theta = \dfrac{O}{H}$ **CAH:** $\cos \theta = \dfrac{A}{H}$ **TOA:** $\tan \theta = \dfrac{O}{A}$

It is useful to remember that the reciprocal ratios can be written as:

$\operatorname{cosec} \theta = \dfrac{1}{\sin \theta}$ $\sec \theta = \dfrac{1}{\cos \theta}$ $\cot \theta = \dfrac{1}{\tan \theta}$

Using CAPITAL or small letters

Worked example

Use the diagram to complete the table.

> In trigonometry notation, when the name of a ratio is written in front of a capital letter that represents the angle, the angle symbol is not used, for example sin A, and not sin Â.

Remember

Upper case also means capital letters, and lower case means small letters.

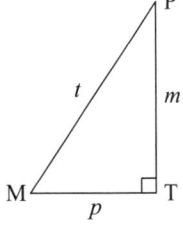

Ratio	Ratio (using capital letters)	Ratio (using small letters)	Ratio as a fraction: if $m = 4$ cm, $p = 3$ cm, and $t = 5$ cm
sin M			
cosec M			
cos M			
sec M			
tan M			
cot M			

Ratio	Ratio (using capital letters)	Ratio (using small letters)	Ratio as a fraction: if m = 4 cm, p = 3 cm, and t = 5 cm
sin P			
cosec P			
cos P			
sec P			
tan P			
cot P			

Solution

Ratio	Ratio (using capital letters)	Ratio (using small letters)	Ratio as a fraction: if m = 4 cm, p = 3 cm, and t = 5 cm
For $\hat{\text{M}}$, m is the opposite side, p is the adjacent side and t is the hypotenuse.	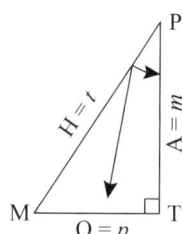		
sin M	$\sin \text{M} = \dfrac{\text{O}}{\text{H}} = \dfrac{\text{TP}}{\text{MP}}$	$\sin \text{M} = \dfrac{\text{O}}{\text{H}} = \dfrac{m}{t}$	$\sin \text{M} = \dfrac{\text{O}}{\text{H}} = \dfrac{4}{5}$
cosec M	$\operatorname{cosec} \text{M} = \dfrac{\text{H}}{\text{O}} = \dfrac{\text{MP}}{\text{TP}}$	$\operatorname{cosec} \text{M} = \dfrac{\text{H}}{\text{O}} = \dfrac{t}{m}$	$\operatorname{cosec} \text{M} = \dfrac{\text{H}}{\text{O}} = \dfrac{5}{4}$
cos M	$\cos \text{M} = \dfrac{\text{A}}{\text{H}} = \dfrac{\text{MT}}{\text{MP}}$	$\cos \text{M} = \dfrac{\text{A}}{\text{H}} = \dfrac{p}{t}$	$\cos \text{M} = \dfrac{\text{A}}{\text{H}} = \dfrac{3}{5}$
sec M	$\sec \text{M} = \dfrac{\text{H}}{\text{A}} = \dfrac{\text{MP}}{\text{TM}}$	$\sec \text{M} = \dfrac{\text{H}}{\text{A}} = \dfrac{t}{p}$	$\sec \text{M} = \dfrac{\text{H}}{\text{A}} = \dfrac{5}{3}$
tan M	$\tan \text{M} = \dfrac{\text{O}}{\text{A}} = \dfrac{\text{PT}}{\text{TM}}$	$\tan \text{M} = \dfrac{\text{O}}{\text{A}} = \dfrac{m}{p}$	$\tan \text{M} = \dfrac{\text{O}}{\text{A}} = \dfrac{4}{3}$
cot M	$\cot \text{M} = \dfrac{\text{A}}{\text{O}} = \dfrac{\text{MT}}{\text{TP}}$	$\cot \text{M} = \dfrac{\text{A}}{\text{O}} = \dfrac{p}{m}$	$\cot \text{M} = \dfrac{\text{A}}{\text{O}} = \dfrac{3}{4}$
Remember: Trig is about **position**. The adjacent and the opposite sides swop around for the other angle.			
sin P	$\sin \text{P} = \dfrac{\text{O}}{\text{H}} = \dfrac{\text{TM}}{\text{MP}}$	$\sin \text{P} = \dfrac{\text{O}}{\text{H}} = \dfrac{p}{t}$	$\sin \text{P} = \dfrac{\text{O}}{\text{H}} = \dfrac{3}{5}$
cosec P	$\operatorname{cosec} \text{P} = \dfrac{\text{H}}{\text{O}} = \dfrac{\text{PM}}{\text{MT}}$	$\operatorname{cosec} \text{P} = \dfrac{\text{H}}{\text{O}} = \dfrac{t}{p}$	$\operatorname{cosec} \text{P} = \dfrac{\text{H}}{\text{O}} = \dfrac{5}{3}$
cos P	$\cos \text{P} = \dfrac{\text{A}}{\text{H}} = \dfrac{\text{TP}}{\text{MP}}$	$\cos \text{P} = \dfrac{\text{A}}{\text{H}} = \dfrac{m}{t}$	$\cos \text{P} = \dfrac{\text{A}}{\text{H}} = \dfrac{4}{5}$
sec P	$\sec \text{P} = \dfrac{\text{H}}{\text{A}} = \dfrac{\text{MP}}{\text{TP}}$	$\sec \text{P} = \dfrac{\text{H}}{\text{A}} = \dfrac{t}{m}$	$\sec \text{P} = \dfrac{\text{H}}{\text{A}} = \dfrac{5}{4}$
tan P	$\tan \text{P} = \dfrac{\text{O}}{\text{A}} = \dfrac{\text{MT}}{\text{TP}}$	$\tan \text{P} = \dfrac{\text{O}}{\text{A}} = \dfrac{p}{m}$	$\tan \text{P} = \dfrac{\text{O}}{\text{A}} = \dfrac{3}{4}$
cot P	$\cot \text{P} = \dfrac{\text{A}}{\text{O}} = \dfrac{\text{PT}}{\text{TM}}$	$\cot \text{P} = \dfrac{\text{A}}{\text{O}} = \dfrac{m}{p}$	$\cot \text{P} = \dfrac{\text{A}}{\text{O}} = \dfrac{4}{3}$

Exercise 1

1. Use the diagram and the trig definitions to:

   1.1 write down the values of sin P, sec P and tan P as numerical fractions

   1.2 write down the values of cos T, cot T and cosec T as numerical fractions.

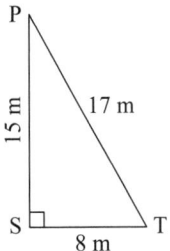

2. Using the diagram alongside:

   2.1 use capital letters to write down two ratios for sin B, cot B and sec B

   2.2 using △BAD, write down the ratios for tan B, cos D and cosec B using small letters.

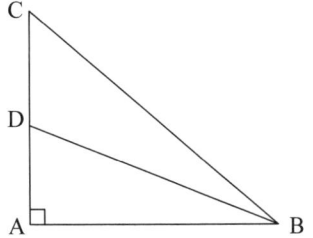

Trigonometry in the Cartesian plane: using ratios

In this section, trigonometric definitions are used to solve simple problems without the use of a calculator.

If we start with a line on the positive x-axis and rotate it anti-clockwise until it reaches a point P(x; y), we get a positive angle θ. If we rotate this line same line segment in a clockwise direction, we get a negative angle, θ.

In this section we will only work with positive angles in the first quadrant, in other words with $0° \leq θ \leq 90°$.

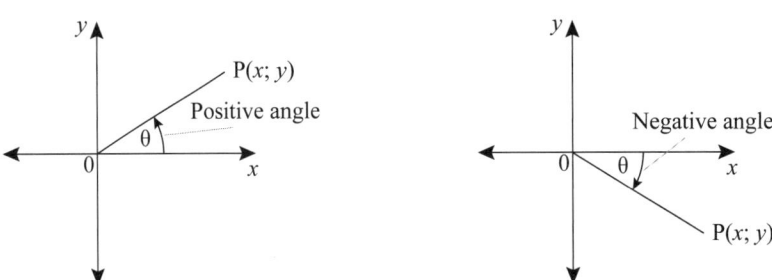

We can now define all the trigonometric ratios in terms of x, y and r. O is the y value, A is the x value, and H is the radius of the circle, r.

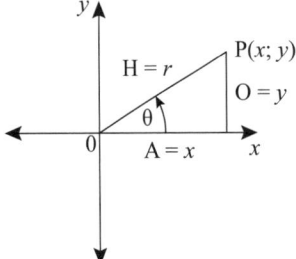

$$\sin θ = \tfrac{y}{r} \qquad \cos θ = \tfrac{x}{r} \qquad \tan θ = \tfrac{y}{x}$$

$$\text{cosec } θ = \tfrac{r}{y} \qquad \sec θ = \tfrac{r}{x} \qquad \cot θ = \tfrac{x}{y}$$

We can use the definitions given in terms of x, y and r to solve simple problems and to manipulate trigonometric ratios.

Two distinct types of problems fall under this section:

- **Type 1:** problems in which a specific point is given
- **Type 2:** problems in which an equation is given.

Type 1: A specific point is given

Worked examples

Consider a point P(3; 4) which lies in the first quadrant, forming the angle θ.

1. Find the values for sin θ, cos θ and cot θ.

2. Use your results to show that $\cot \theta = \frac{\cos \theta}{\sin \theta}$.

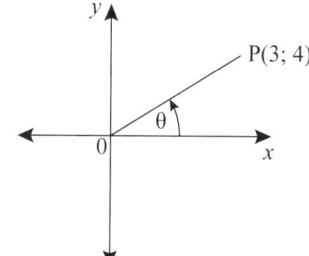

Solutions

1. The point given, P, has $x = 3$ and $y = 4$.

   Using Pythagoras: $r^2 = x^2 + y^2$

   $\therefore r = \sqrt{3^2 + 4^2} = 5$

   $\therefore \sin \theta = \frac{y}{r} = \frac{4}{5}$, $\cos \theta = \frac{x}{r} = \frac{3}{5}$ and $\cot \theta = \frac{x}{y} = \frac{3}{4}$

2. $\frac{\cos \theta}{\sin \theta} = \frac{3}{5} \div \frac{4}{5} = \frac{3}{5} \times \frac{5}{4} = \frac{3}{4}$

   $\cot \theta = \frac{3}{4} \therefore \cot \theta = \frac{\cos \theta}{\sin \theta}$

Type 2: An equation is given

Worked examples

Given: $13 \cos \theta = 12$ and $0° \le \theta \le 90°$.

1. Explain what the restriction $0° \le \theta \le 90°$ means.

2. Determine, without using a calculator, the value of:

   2.1 $\cos \theta$ **2.2** $\sin \theta$ **2.3** $\tan \theta$

3. Use your results to show that $\tan \theta = \frac{\sin \theta}{\cos \theta}$.

Solutions

1. The restriction $0° \le \theta \le 90°$ means that the angle θ lies in the first quadrant.

2. To solve this problem, first rearrange the equation so that you only have cos θ on the left hand side of the equation:

   $13 \cos \theta = 12$

   $\therefore \cos \theta = \frac{12}{13}$

   Use the trigonometric definitions to write down the given information in terms of x, y and r.

   $\cos \theta = \frac{x}{r} = \frac{12}{13} \therefore x = 12$ and $r = 13$

   Using Pythagoras: $y = \sqrt{r^2 - x^2} = \sqrt{13^2 - 12^2} = 5$

Use this information to sketch a triangle in quad 1:

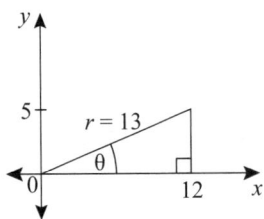

Use the trigonometric definitions and your sketch to answer the questions.

2.1 $\cos \theta = \frac{12}{13}$ **2.2** $\sin \theta = \frac{5}{13}$ **2.3** $\tan \theta = \frac{5}{12}$

3. To show that $\tan \theta = \frac{\sin \theta}{\cos \theta}$, separate the problem into two sections and deal with each separately.

LHS: $\tan \theta = \frac{5}{12}$

RHS: $\frac{\sin \theta}{\cos \theta} = \frac{\frac{5}{13}}{\frac{12}{13}} = \frac{5}{13} \times \frac{13}{12} = \frac{5}{12}$

$\therefore$ LHS = RHS $\therefore$ $\tan \theta = \frac{\sin \theta}{\cos \theta}$

Finding the numerical value of trigonometric expression

You will often be asked to find the numerical value of a trigonmetric expression, using the values of x, y and r given on your diagram.

Note the following:

* $3 \sin \theta$ means three times the fraction obtained by writing the ratio value of $\frac{y}{r}$.
* $\frac{\cos \theta}{3}$ means dividing the fraction from $\frac{x}{3}$ by 3.
* $\sin^2 \theta$ is the mathematical way of writing $(\sin \theta)^2$, so you will need to square the fraction obtained from $\frac{y}{r}$.
* $\cos \theta + \tan \theta$ requires the use of a LCD, because the denominators will be different.

Worked examples

If $6 \sin \theta - 3 = 0$ and $0° \leq \theta \leq 90°$, determine, without the use of a calculator, the value of:

1. $\sin^2 \theta + \cos^2 \theta$ 2. $\tan^2 \theta + \sec^2 \theta$.

Solutions

Use algebra to make $\sin \theta$ the subject of the formula:

$6 \sin \theta = 3$

$\sin \theta = \frac{3}{6} = \frac{1}{2}$

Use the fact that $\sin \theta = \frac{y}{r}$ to write down the values of y and r:

$\therefore y = 1$ and $r = 2$

Using the theorem of Pythagoras:

$x = \sqrt{r^2 - y^2} = \sqrt{2^2 - 1^2} = \sqrt{3}$

Use this information to set up a triangle in quad 1, as indicated by the restriction:

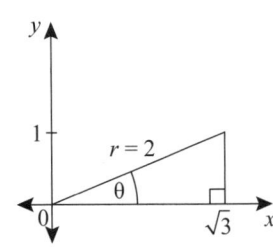

1. Use your sketch and the trigonometric definitions to complete the solution:

$$\therefore \sin^2 \theta + \cos^2 \theta = \left(\tfrac{1}{2}\right)^2 + \left(\tfrac{\sqrt{3}}{2}\right)^2$$
$$= \tfrac{1}{4} + \tfrac{3}{4} = 1$$

2. $\tan^2 \theta + \sec^2 \theta$
$$= \left(\tfrac{1}{\sqrt{3}}\right)^2 + \left(\tfrac{2}{\sqrt{3}}\right)^2$$
$$= \tfrac{1}{3} + \tfrac{4}{3}$$
$$= \tfrac{5}{3}$$

Exercise 2

1. Use the diagram and trigonometric definitions to determine:

    1.1 $\sin \beta$

    1.2 $\cot \beta$

    1.3 $(\sin \beta + \cos \beta)(\tan \beta + \cot \beta)$

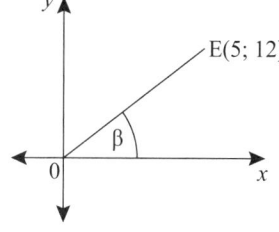

2. If $\sin \theta = \tfrac{3}{5}$ and $\theta \in [0°; 90°]$, find:

    2.1 $\cos \theta$ **2.2** $\tan \theta$ **2.3** $\sec^2 \theta - 1$

3. If $3 \tan \theta - 4 = 0$ and $0° \leq \theta \leq 90°$, find:

    3.1 $\sin \theta$ **3.2** $\cos \theta$ **3.3** $\sin^2 \theta + \cos^2 \theta$

4. Leaving your answers in surd form if necessary, use the diagram to determine:

    4.1 OD

    4.2 $\sin \theta \ \mathrm{cosec}\ \theta$

    4.3 $\cos \theta \sec \theta$

    4.4 $\sec^2 \theta - \tan^2 \theta$

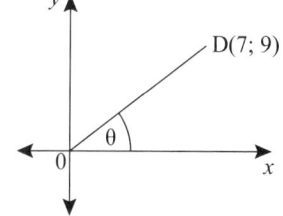

5. If $3 \sec \theta = 5$ and $0° \leq \theta \leq 90°$, find the value of $\tan^2 \theta + 1$ without using a calculator.

6. If $8 \cos \theta - 6 = 0$ and $\theta \in [0°; 90°]$, find:

    6.1 $\mathrm{cosec}\ \theta$

    6.2 $1 + \cot^2 \theta$

7. Leaving your answers in surd form if necessary, use the diagram to determine:

    7.1 $\sin \theta \cot \theta$

    7.2 $\cos \theta \tan \theta$

    7.3 $\mathrm{cosec}^2 \theta - \cot^2 \theta$

    7.4 Show that $\cos^2 \theta = 1 - \sin^2 \theta$.

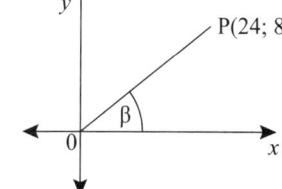

8. If $2\ \mathrm{cosec}\ \theta = 10$ and $0° \leq \theta \leq 90°$, find:

    8.1 $\sin \theta$ **8.2** $\cos \theta$ **8.3** $2 \sin \theta \cos \theta$

9. If $\frac{\cot \theta}{4} = \frac{3}{8}$ and $\theta \in [0°; 90°]$, find:

   9.1 $\tan \theta$ **9.2** $\sec \theta$

   9.3 Use your results to show that $\sec^2 \theta - 1 = \tan^2 \theta$.

Using special angles

You will be expected to work with and solve problems involving the special angles, without the use of a calculator. To do this you need to memorise all the special angle values or use triangles and the unit circle to work out the values for each of the special angles.

- The ratios for the special angles 30°, 45° and 60° can be determined using the triangles given in the table below, and the definitions in terms of O, A and H.
- The ratios for the angles 0° and 90° can be determined using a circle with a radius of 1 unit and the definitions in terms of y, x and r.

Learning the special angles by heart will be very useful.

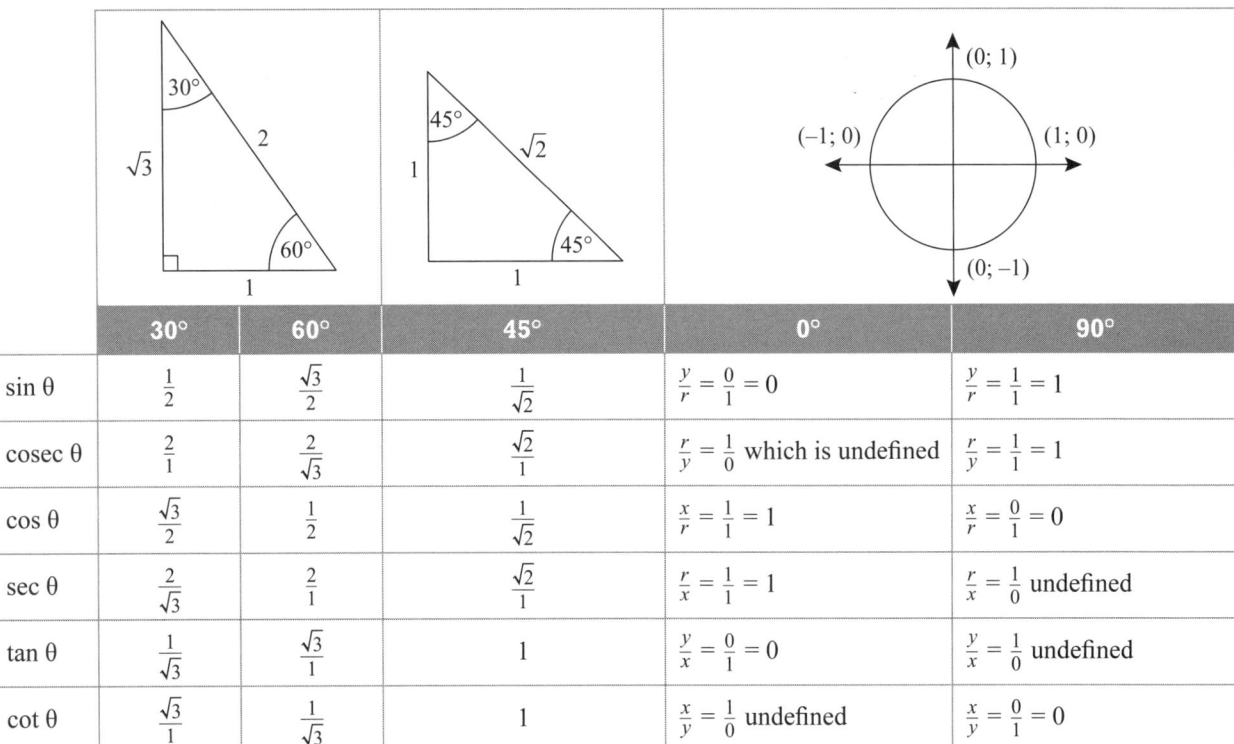

	30°	60°	45°	0°	90°
$\sin \theta$	$\frac{1}{2}$	$\frac{\sqrt{3}}{2}$	$\frac{1}{\sqrt{2}}$	$\frac{y}{r} = \frac{0}{1} = 0$	$\frac{y}{r} = \frac{1}{1} = 1$
$\operatorname{cosec} \theta$	$\frac{2}{1}$	$\frac{2}{\sqrt{3}}$	$\frac{\sqrt{2}}{1}$	$\frac{r}{y} = \frac{1}{0}$ which is undefined	$\frac{r}{y} = \frac{1}{1} = 1$
$\cos \theta$	$\frac{\sqrt{3}}{2}$	$\frac{1}{2}$	$\frac{1}{\sqrt{2}}$	$\frac{x}{r} = \frac{1}{1} = 1$	$\frac{x}{r} = \frac{0}{1} = 0$
$\sec \theta$	$\frac{2}{\sqrt{3}}$	$\frac{2}{1}$	$\frac{\sqrt{2}}{1}$	$\frac{r}{x} = \frac{1}{1} = 1$	$\frac{r}{x} = \frac{1}{0}$ undefined
$\tan \theta$	$\frac{1}{\sqrt{3}}$	$\frac{\sqrt{3}}{1}$	1	$\frac{y}{x} = \frac{0}{1} = 0$	$\frac{y}{x} = \frac{1}{0}$ undefined
$\cot \theta$	$\frac{\sqrt{3}}{1}$	$\frac{1}{\sqrt{3}}$	1	$\frac{x}{y} = \frac{1}{0}$ undefined	$\frac{x}{y} = \frac{0}{1} = 0$

Evaluating expressions using special angles

Worked example

Without using a calculator, determine the value of
$2 \cos^2 45° \cos 0° - 2 \sin 30° \cos 60° + \tan^2 60° - \sin 90°$.

Solution

$2 \cos^2 45° \cos 0° - 2 \sin 30° \cos 60° + \tan^2 60° - \sin 90°$

$= 2\left(\frac{1}{\sqrt{2}}\right)^2 \cdot (1) - 2\left(\frac{1}{2}\right) \cdot \left(\frac{1}{2}\right) + \left(\frac{\sqrt{3}}{1}\right)^2 - (1)$ Remember: $(\sqrt{x})^2 = x \therefore (\sqrt{3})^2 = 3$ and $\left(\frac{1}{\sqrt{2}}\right)^2 = \frac{1}{2}$

$= 2\left(\frac{1}{2}\right) \cdot (1) - 2\left(\frac{1}{2}\right) \cdot \left(\frac{1}{2}\right) + \left(\frac{3}{1}\right) - (1)$ Remember: $\frac{1}{2} \times \frac{1}{2} = \frac{1}{4}$

$= 2\left(\frac{1}{2}\right) \cdot (1) - 2\left(\frac{1}{4}\right) + (3) - (1)$

$= 1 - \frac{1}{2} + 2 = 2\frac{1}{2}$

Finding angles given a trigonometric equation

When finding the size of an angle, your equation will not usually be as simple as 'solve for θ if $\sin \theta = \frac{1}{2}$'. The Worked examples below gives some guidance on the mathematics of finding angle sizes.

Worked examples

1. $2 \sin \theta = \sqrt{3}$

$\quad\quad \sin \theta = \frac{\sqrt{3}}{2}$ Divide by 2.

$\quad \therefore \therefore \theta = 60°$

2. $\sin 2\theta = \frac{1}{2}$

$\quad\quad 2\theta = 30°$ You CANNOT divide by 2. First find the angle from $\sin 2\theta = \frac{1}{2}$.

$\quad \therefore \theta = 15°$ Only then divide the angle size by $\frac{1}{2}$.

3. $\sin (\theta + 10°) = \frac{\sqrt{3}}{2}$

$\quad\quad \theta + 10° = 60°$ Find the angle from $\sin \theta = \frac{\sqrt{3}}{2}$.

$\quad\quad \theta + 10° = 60° - 10°$ Subtract 10° in this step.

$\quad \therefore \theta = 50°$

Worked examples

In Grade 10, trigonometric equations are restricted to the first quadrant, so the restriction $\theta \in [0°; 90°]$ or $0° \leq \theta \leq 90°$ will be given with each question.

1. Solve for θ if $2 \cos \frac{\theta}{2} = \sqrt{3}$, for $\theta \in [0°; 90°]$.

2. Solve for θ if $4 \operatorname{cosec} 3\theta = 8$, for $0° \leq \theta \leq 90°$.

Solutions

1. $2 \cos \left(\frac{\theta}{2}\right) = \sqrt{3}$ for $\theta \in [0°; 90°]$ Use special triangles to determine the size of θ.

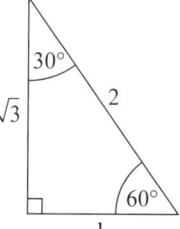

$\quad\quad \cos \left(\frac{\theta}{2}\right) = \frac{\sqrt{3}}{2}$

$\quad\quad\quad \frac{\theta}{2} = 30°$

$\quad \therefore \theta = 60°$

2. $4 \operatorname{cosec} 3\theta = 8$ for $\theta \in [0°; 90°]$

$\quad\quad \operatorname{cosec} 3\theta = \frac{8}{4} = \frac{2}{1}$

$\quad\quad\quad 3\theta = 30°$

$\quad \therefore \theta = 10°$

Exercise 3

1. Without a calculator, determine the value of each of the expressions given below:

    1.1 $\sin^2 30° - \cos^2 30°$

    1.2 $\sin 90° - (\tan 30° + \operatorname{cosec} 60°)^2$

    1.3 $\tan 0° + \tan 30° + \cot 60° - \operatorname{cosec} 60°$

    1.4 $\frac{\sin 45°}{\cos 45°} - 5 \operatorname{cosec} 90° + 3 \tan^2 30°$

1.5 $\sin^2 30° + \cos^2 30°$

1.6 $\frac{\sin 30°}{\cos 60°} - 2 \sin 90°(\tan 45° + \sin 30°)$

1.7 $\frac{2 \sin 45° \cos 30° \tan 60° \operatorname{cosec} 90°}{\sin 30° \cos 0° \cos 45°}$

2. Determine the acute angles θ in each of the equations given below, without the use of a calculator:

   2.1 $\sin θ = \frac{1}{2}$ for $θ \in [0°; 90°]$ **2.2** $2 \cos θ = 1$ if $0° \le θ \le 90°$

   2.3 $2 \tan 3θ - 2 = 0$ for $θ \in [0°; 90°]$

   2.4 $2 \cos (θ + 12°) - \sqrt{3} = 0$ if $0° \le θ \le 90°$

   2.5 $10 \operatorname{cosec} θ(θ - 21°) - 20 = 0$ if $0° \le θ \le 90°$

Exercise 4: revision

1. Using the diagram, write down three ratios for sin S, using capital letters.

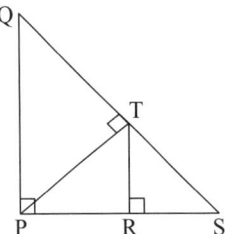

2. Now use these special angle values to simplify the expressions given below. Do not use a calculator and show all working.

   2.1 $\cos 60° \cot 45° - \cos^2 45°$ **2.2** $\sin (60° + 30°)$

   2.3 $\frac{\cos 45° \operatorname{cosec} 45° \tan^2 60°}{\sin 30°}$

3. Use special angles to solve for *x* in each equation:

   3.1 $x \cos 30° = \sin 30°$ **3.2** $x + \operatorname{cosec}^2 45° = \cos^2 45° - \cos 90°$

4. In each equation solve for θ for $θ \in [0°; 90°]$:

   4.1 $4 \tan θ - 4 = 0$ **4.2** $2 \sin θ = \sqrt{3}$

   4.3 $\tan 3θ - \sqrt{3} = 0$ **4.4** $\sqrt{3} \cot 3θ - 1 = 0$

   4.5 $2 \cos (2θ - 20°) = 1$ **4.6** $\frac{\sec (θ - 10°)}{2} = 1$

5. Point P(12; 5) is given. Use the trigonometric definitions and a sketch to prove that $\cos^2 θ - 1 = -\sin^2 θ$.

6. In the diagram, point Q(6; 3) is given. Without using a calculator, determine the values of:

   6.1 OQ   **6.2** $\sin θ$

   6.3 $\sqrt{5} \cos θ$ **6.4** $\sec^2 θ - \tan θ$

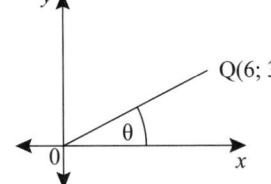

7. If $\sin θ = \frac{5}{8}$ and $0° < θ < 90°$, calculate the value of:

   7.1 $\cos θ$ **7.2** $\tan θ$

   7.3 Use your results to prove that $\cot θ = \frac{\cos θ}{\sin θ}$.

8. If $13 \sin β = 12$ and $β \in [0°; 90°]$, calculate the value of $\cos β + \cot β$.

Trigonometry using a calculator

Before using the definitions to solve problems involving triangles you need to know how to use your scientific calculator. Refresh your memory with these exercises before continuing.

Degrees, grades and radians are different units for measuring angles. For high school purposes, angles are always measured in degrees. **Make sure that your calculator is in degree mode.**

Finding the value of trig ratios given the angles

To evaluate trigonometric expressions where the angles are given:
- key the information in exactly as it is given, using your fraction button and closing all brackets
- press the **=** button and write down your answer, rounding off appropriately.

Worked examples

1. Use your calculator to find, correct to three decimal places:

   1.1 $\tan 36°$ **1.2** $5 \sin 75,6°$ **1.3** $\cos\left(\frac{35°}{2}\right)$ **1.4** $\frac{\cos 35°}{2}$

   1.5 $\sqrt{\sin 38°}$ **1.6** $\sin^2 38°$ **1.7** $\sec 55°$ **1.8** $\frac{2}{3}\cot 54,4°$

   1.9 $\tan(28° + 95°) - \sec^2 43°$ **1.10** $\frac{\tan^3 48,4°}{\sqrt{\sin 23°}} - 2\sec 78,3°$

Solutions

1. **1.1** $\tan 36° = 0,72654\ldots = 0,727$

   > **Note:**
   >
   > Your screen display reads **TAN(36** after you have keyed in the number. It is advisable to close the bracket. Some calculators will give you an error message if you don't close this bracket after keying in the value of the angle.

   1.2 $5 \sin 75,6° = 4,8429\ldots = 4,843$

   1.3 $\cos\left(\frac{35°}{2}\right) = 0,95371\ldots = 0,954$

   1.4 $\frac{\cos 35°}{2} = 0,40957\ldots = 0,410$

   1.5 $\sqrt{\sin 38°} = 0,784640\ldots = 0,785$

   > Failure to close a bracket in this calculation may result in an error message. If this happens, retype the expression, closing all brackets.

   1.6 $\sin^2 38° = 0,37903\ldots = 0,379$

   > Remember that $(\sin 38°)^2$ is written as $\sin^2 38°$. To avoid errors in reasoning, and to ensure that you get the right answer, type $\sin^2 38°$ into your calculator as $(\sin(38°))^2$.

   1.7 $\sec 55° = \frac{1}{\cos 55°} = 1,74344\ldots = 1,743$

> **Note**
>
> Notice that the expressions in 1.3 and 1.4 are different. Use your fraction buttons correctly to evaluate these expressions.

> **Note**
>
> To evaluate any of the reciprocal ratios, rewrite them in terms of sin, cos or tan:
>
> $\operatorname{cosec} \theta = \frac{1}{\sin \theta}$
>
> $\sec \theta = \frac{1}{\cos \theta}$
>
> $\cot \theta = \frac{1}{\tan \theta}$

1.8 $\frac{2}{3}\cot 54{,}4° = \frac{2}{3\tan 54{,}4°} = 0{,}47728\ldots = 0{,}477$

1.9 $\tan(28° + 95°) - \sec^2 43°$

$$= \tan(28° + 95°) - \frac{1}{(\cos(43°))^2}$$

$$= -3{,}40944 = -3{,}409$$

> Modern scientific calculators are programmed to work with complicated expressions as long as you key them in correctly. Note that answers can contain negative numbers.

1.10 $\dfrac{\tan^3 48{,}4°}{\sqrt{\sin 23°}} - 2\sec(78{,}3°) = \dfrac{(\tan(48{,}4°))^3}{\sqrt{\sin(23°)}} - \dfrac{2}{\cos(78{,}3°)} = -7{,}577$

Using a calculator to find angles

The **2nd F** or **SHIFT** key is used to find angles.

Using one of these keys and then keying in a trig ratio will display the size of the angle that produces that ratio. The size of angles is usually rounded off to one decimal place. Do not forget to write the degree symbol (°) with your answer, you can lose marks for leaving it out.

Worked examples

In each case, find the value of θ correct to one decimal place.

1. $\sin\theta = \frac{3}{5}$ 2. $\tan\theta = 6{,}78$

3. $\sec\theta = 3{,}7$ 4. $\operatorname{cosec}\theta = \frac{7}{4}$

Solutions

	1. $\sin\theta = \frac{3}{5}$	2. $\tan\theta = 6{,}78$	3. $\sec\theta = 3{,}7$	4. $\operatorname{cosec}\theta = \frac{7}{4}$
For reciprocals, rewrite in terms of sin, cos or tan.			$\frac{1}{\cos\theta} = \frac{3{,}7}{1}$ $\therefore \cos\theta = \frac{1}{3{,}7}$	$\frac{1}{\sin\theta} = \frac{7}{4}$ $\therefore \sin\theta = \frac{4}{7}$
Screen display				
Select Shift/2nd F and the required ratio	$\sin^{-1}($	$\tan^{-1}($	$\cos^{-1}($	$\sin^{-1}($
Key in the necessary figures and close any open brackets	$\sin^{-1}\left(\frac{3}{5}\right)$	$\tan^{-1}(6{,}78)$	$\cos^{-1}\left(\frac{1}{3{,}7}\right)$	$\sin^{-1}\left(\frac{4}{7}\right)$
Press **=**	$= 36{,}869\ldots$	$= 81{,}609\ldots$	$= 74{,}319\ldots$	$34{,}8499\ldots$
Write your answer in degrees, rounding off appropriately.	$\theta = 36{,}9°$	$\theta = 81{,}6°$	$\theta = 74{,}3°$	$\theta = 34{,}9°$

Solving trigonometric equations using a calculator

Worked examples

1.   Solve for θ if $2\sin\theta = 0{,}56$ for $\theta \in [0°; 90°]$.

2.   Solve for θ if $3\cos(2\theta - 10°) = 2$ for $\theta \in [0°; 90°]$.

Solutions

1. $2 \sin \theta = 0{,}56$

    $\sin \theta = \frac{0{,}56}{2}$

    Keys: Shift sin $\left(\frac{0{,}56}{2}\right) =$

    $\theta = 16{,}26° = 16{,}3°$

2. $3 \cos (2\theta - 10°) = 2$

    $\cos (2\theta - 10°) = \frac{2}{3}$

    Keys: Shift cos $\left(\frac{2}{3}\right) =$

    $2\theta - 10° = 48{,}189\ldots°$

    $2\theta = 48{,}189\ldots + 10°$

    $2\theta = 58{,}189\ldots°$

    $\therefore \theta = 29{,}0945\ldots° = 29{,}1°$

Exercise 5

1. If $A = 31°$ and $B = 78°$, evaluate the expressions given below, correct to three decimal places.

    1.1 $\frac{\sin 2A}{2}$

    1.2 $\frac{2 \sin 4A}{3}$

    1.3 $\cos^2 (A + B)$

    1.4 $\sin^2 A + \cos^2 B$

    1.5 $\sec 2A$

    1.6 $\operatorname{cosec} (B - A)$

    1.7 $\sqrt{\cot B - \tan B}$

    1.8 $3\sqrt{\tan A - \operatorname{cosec}^2 B}$

2. Solve for β, correct to one decimal place, if:

    2.1 $\sin \beta = 0{,}45$ and $\theta \in [0; 90°]$

    2.2 $3 \cos \beta = 1{,}3$ and $0° \leq \beta \leq 90°$

    2.3 $2 \tan 3\beta = 10$ and $0° \leq \beta \leq 90°$

    2.4 $4 \sin \beta - 1 = 3$ and $\theta \in [0°; 90°]$

    2.5 $\operatorname{cosec} 2\beta = \frac{7}{4}$ and $\theta \in [0°; 90°]$

    2.6 $21 \cot (\beta - 5°) - 23 = 0$ and $\theta \in [0°; 90°]$

Solving right-angled triangles

Trigonometry is used to solve problems involving triangles. In Grade 10 we will work only with right-angled triangles where the trigonometric definitions are used to find the size of missing sides and angles.

Finding the length of a side of a right-angled triangle, given an angle and a side of the triangle

Worked examples

1. Calculate the value of x in the following triangle. Give your answers correct to two decimal places.

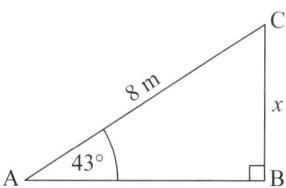

2. Hence, or otherwise, find the length of the other side.

Solutions

You are given an angle and a side. You need to find a side.

Steps for finding the length of a side, given an angle and a side of the triangle	
1. Label the triangle, using O, A and H.	1.
2. Set up a ratio in terms of what you 'want' over what you've 'got'. Use the trig definitions to identify the ratio you need to use.	$\frac{\text{want}}{\text{got}} = \frac{O}{H} = \sin\theta$ (don't need to show this step) $\sin\theta = \frac{CB}{AC}$ (do need to show this step)
3. Fill in the known values. Use your calculator to find the value of the unknown side. Round off appropriately.	$\frac{x}{8} = \sin 43°$ $x = 8\sin 43°$ $x = 5,46698\ldots = 5,46$
4. Use your answer and the theorem of Pythagoras to find the value of the remaining side.	2. $AB = \sqrt{AC^2 - BC^2}$ Pythagoras $AB = \sqrt{8^2 - 5,46^2} = 5,847 = 5,85$ m

Using reciprocal ratios to find a side

Worked examples

1. Calculate the value of side PQ in the triangle provided.

2. Hence or otherwise, find the value of side PR.

Solutions

Steps for using reciprocal ratios to find a side	
1. Label the triangle, using O, A and H.	1.
2. Set up a ratio in terms of what you 'want' over what you've 'got'. Use the trig definitions to identify the ratio you need to use.	$\frac{\text{want}}{\text{got}} = \frac{A}{O} = \cot\theta$ (don't need to show this step) $\cot\theta = \frac{PQ}{QR}$ (do need to show this step)
3. Fill in the known values. Use your calculator to find the value of the unknown side. Round off appropriately.	$\frac{PQ}{100} = \cot 56°$ $PQ = 100\cot 56°$ There is no 'cot' button on your calculator. $PQ = \frac{100}{\tan 56°}$ Use the reciprocal ratio $\cot\theta = \frac{1}{\tan\theta}$. $PQ = 67,45$ m
4. Use your answer and the theorem of Pythagoras to find the value of the remaining side.	2. $PR = \sqrt{PQ^2 + RQ^2}$ Pythagoras $PR = \sqrt{67,45^2 + 100^2} = 120,6218\ldots = 120,62$ m

Finding an angle given a right-angled triangle and two sides

Worked examples

1. Find the size of Â in the triangle given below. Round your answer off to one decimal place.

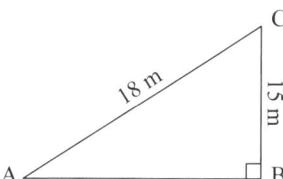

2. Now find the length of AB and the size of Ĉ.

Solutions

Steps for finding an angle in a right-angled triangle given two sides	
1. Label the triangle using O, A and H.	1. *(triangle diagram: C at top right, A at bottom left, B at bottom right; H = 18 m along AC, O = 15 m along CB, angle ? at A)*
2. Set up a ratio in terms of the two sides you are given. Use the trig definitions to identify the ratio you need to use.	$\frac{\text{got}}{\text{got}} = \frac{O}{H} = \sin\theta$ (don't need to show this step) $\sin A = \frac{CB}{CA}$ (do need to show this step)
3. Fill in the known values. Use your calculator to find the value of the unknown angle. Round off appropriately.	$\sin A = \frac{15}{18}$ $\hat{A} = 56,4426\ldots = 56,4°$
4. Use your answer and Pythagoras to find the value of the remaining side.	2. $AB = \sqrt{AC^2 - CB^2}$ Pythagoras $AB = \sqrt{18^2 - 15^2} = 3\sqrt{11} = 9,949\ldots = 9,9 \text{ m}$
5. Use your geometry facts to find the required angle.	$\hat{C} = 90° - 56,4° = 33,6°$ sum ∠s of △ = 180°

Steps for solving right-angled triangles: summary

In general when solving right-angled triangles:

1. Use the theorem of Pythagoras if you want to **find a side** and you are **given two sides**.

2. If you want to **find a side**, and you are **given one side and an angle**, decide on which trig ratio to use from $\frac{\text{unknown}}{\text{known}}$ or $\frac{\text{want}}{\text{got}}$.

3. If you want to **find an angle** and you are **given two sides**:
   - use θ to mark the angle you need to find
   - label your sides with O, A or H, according to the angle θ
   - decide which trig ratio to use from $\frac{\text{given side}}{\text{given side}}$ or $\frac{\text{got}}{\text{got}}$.

Exercise 6

1. In each of the triangles given below, use the small letters given to write down the ratio of the sides corresponding to the named trig ratios:

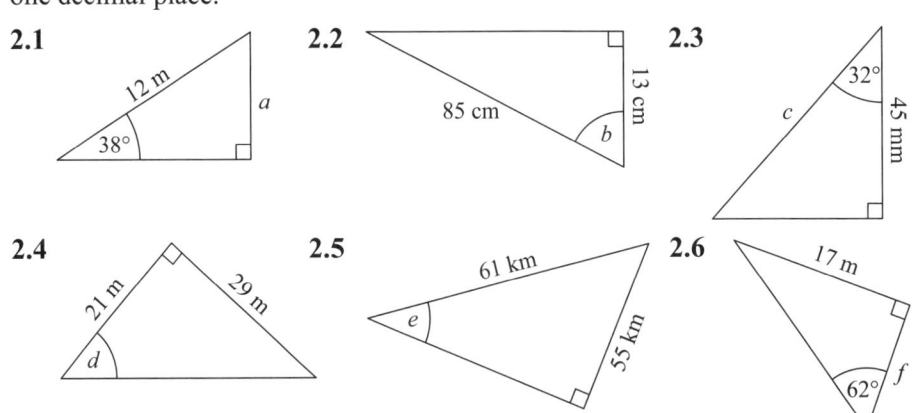

1.1

sin 44°   cot 44°

cosec 44°   tan 46°

sec 46°   cos 46°

1.2

cos θ   cot θ

cosec θ   sin (90° − θ)

sec (90° − θ) tan (90° − θ)

2. In the triangles below, find the values of the variables *a*, *b*, *c*, *d*, *e* and *f*. Give your answers for sides correct to two decimal places and for angles correct to one decimal place.

2.1   **2.2**   **2.3**

2.4   **2.5**   **2.6**

Right-angled triangles: complex procedures

More complex problems involve working in two dimensions, moving from one from one triangle into another.

Worked example

△PQR has P̂ = 90°, PQ = 28 m, QR = 34 m and SR = 24 m. Use this information to find values for *x* and *y*, correct to one decimal place.

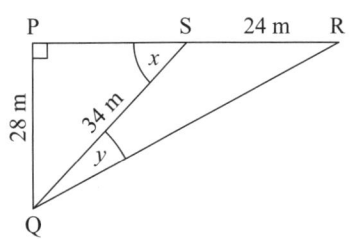

Solution

In △PQS: $\sin x = \dfrac{PQ}{QS} = \dfrac{28}{34}$

        $x = 55{,}43\ldots = 55{,}4°$

Remember: Use the shift/2nd button to find angles.

The second part of this problem is more involved and cannot be solved in one simple step. It helps to set up a strategy before you begin.

To find *y*: $y = P\hat{Q}R - P\hat{Q}S$

Begin by calculating the size of $P\hat{Q}S$ and $P\hat{Q}R$.

In $\triangle PQS$: $P\hat{Q}S = 90° - x = 90° - 55{,}4° = 34{,}6°$ sum of $\angle$s of $\triangle = 180°$

In $\triangle PQR$: $\tan P\hat{Q}R = \dfrac{PR}{PQ} = \dfrac{PS + SR}{PQ}$

To use side PR, calculate the size of PS:

In $\triangle PQS$: $PS = \sqrt{QS^2 - PQ^2} = \sqrt{34^2 + 28^2} = 2\sqrt{93}$ Pythagoras

$\tan P\hat{Q}R = \dfrac{PR}{PQ} = \dfrac{PS + SR}{PQ} = \dfrac{2\sqrt{93} + 24}{28}$

$\therefore P\hat{Q}R = 57{,}1°$

$\therefore y = P\hat{Q}R - P\hat{Q}S = 57{,}1° - 34{,}6° = 22{,}5°$

Note

Keeping your answer in surd form at this stage will result in a more accurate final answer.

Exercise 7: complex procedures

1. Find the values of x and y in each of the triangles. Round off the values of all angles to one decimal place and all sides to two decimal places.

  1.1
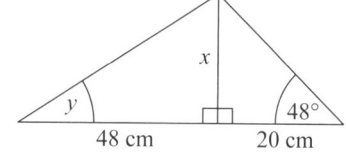

  1.2
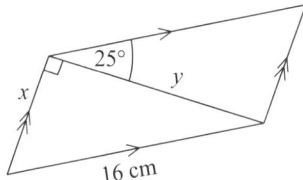

  1.3
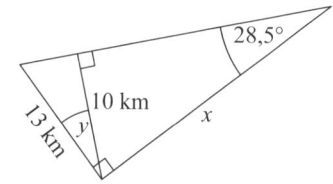

  1.4
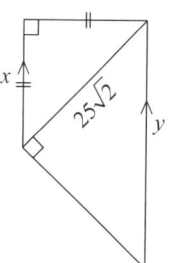

2. Given: $\triangle PSQ$ with $\hat{P} = 90°$, $\hat{Q} = 53°$, $PR \perp SQ$, and $SR = 65$ m.

  2.1 Find PR and RQ correct to two decimal places.

  2.2 Find the area of $\triangle PSQ$ correct to the nearest square metre.

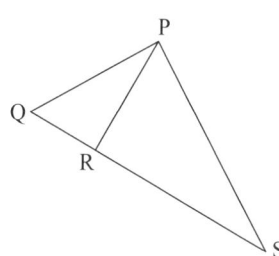

Exercise 8

1. Find the values of x and y in the triangle given below.

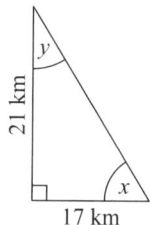

2. Given that $C\hat{B}D = D\hat{B}A = 20°$ and $AB = 50$ m, find the lengths of:

  2.1 DA, correct to two decimal places

  2.2 DC, correct to two decimal places.

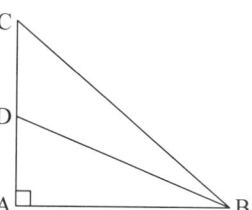

3. Find the values of x and y in the triangle given below.

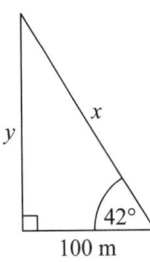

4. If D = 56,7° and E = 67,5°, calculate the value of each of the expressions given below, correct to two decimal places.

    4.1 $\tan (D + E)$ **4.2** $2 \sin E$ **4.3** $\cos 3D$

    4.4 $\frac{\sec E}{3}$ **4.5** $\cot (D + 23,6°)$ **4.6** $\sin^2 E - \cos^2 D$

    4.7 $3 \cos (3E - 30°) - \text{cosec}^2 D$ **4.8** $\frac{1 - \tan^2 E}{3 \cos D}$

    4.9 $\frac{2}{5} \tan D - \cot E + 2 \sin^2(D + E)$ **4.10** $\sqrt{\tan^3 E - \sin D}$

5. Solve for θ in each equation, given that $\theta \in [0°; 90°]$. Give your answers correct to one decimal place.

    5.1 $\tan \theta = 4,357$ **5.2** $3 \cos \theta = 1,657$

    5.3 $\sin 3\theta = 0,677$ **5.4** $\sec \theta = 3,37$

    5.5 $5 \text{ cosec } \theta = 2,578$ **5.6** $\cot 3\theta = 0,677$

    5.7 $5 \tan \theta - 4,357 = 0$ **5.8** $\cos (3\theta - 10°) = 0,6887$

    5.9 $4 \sin^2 \theta = 1,234$ **5.10** $\frac{2}{3} \sec \theta = 2,3887$

    5.11 $\frac{\text{cosec } \theta}{5} = \sin 34,5°$ **5.12** $5 - \cot 2\theta = 6,677$

Test A: Knowledge and routine procedures

1. Use the diagram given below to answer these questions:

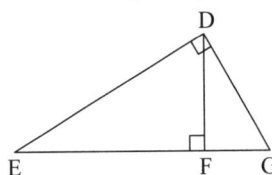

    1.1 Write down two ratios for sin E using capital letters.   (2)

    1.2 For △DEF, write down the trigonometric ratios for cos D and cot E using small letters.   (2)

2. Use a calculator to determine the value of each expression. Give your answers correct to three decimal places.

    2.1 $\frac{\sin^2 35,6°}{2}$ (1)

    2.2 $5 \sec 75°$ (2)

3. Use the information in the triangles given below to find the values of a, b, c, d and e. Give your answers correct to one decimal place.   (13)

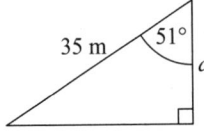

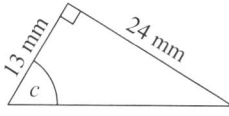

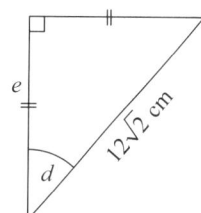

4. You may not use a calculator to answer this question. Use the diagram and trigonometric definitions to determine:

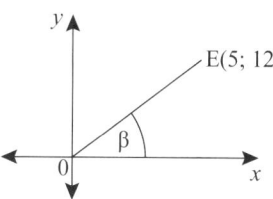

4.1 sin β (2)

4.2 tan β + sec β (3)

5. Given: 7 tan β = 9 and β ∈ [0°; 90°].

   5.1 Use a calculator to find the size of β correct to one decimal place. (2)

   5.2 Without using a calculator, and giving your answer in surd form, find the values of:

   5.2.1 O, A and H, given that 7 tan β = 9, and β ∈ [0°; 90°] (3)

   5.2.2 sin β cosec β (3)

   5.2.3 $\sec^2 β - \tan^2 β$ (4)

6. Use special angles to find the value of:

   6.1 sin 30° tan 45° cos 60° (4)

   6.2 sin 0° cos 45° − tan 60° sin 45° (5)

7. In the diagram, PT ⊥ ES, Ê = 63°, ET = 17 cm and TS = 38 cm. Use the trigonometric ratios and the information given to find the values of *x* and *y*, correct to one decimal place.

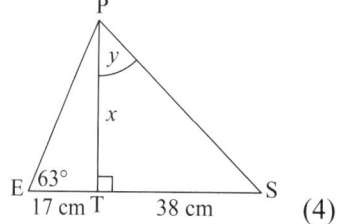

(4)

Total 50

Test B: Complex procedures and problem solving

1. Use the sketch given below to prove that sin θ = cos (90° − θ). (3)

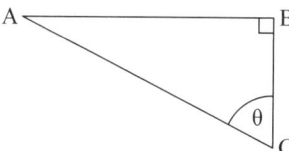

2. **2.1** Determine the value of $\frac{\sec 20° + \csc 20°}{\tan^2 20°}$ correct to one decimal place. (3)

2.2 Determine the value of x, correct to one decimal place, if $\frac{3}{5} \csc (3x - 100°) = 10$ and $(3x - 100°) < 90°)$. (4)

2.3 Determine the value of $\cot^2 (2\beta + 18°)$, correct to the nearest whole number, if $\sec \left(\frac{\beta}{2} - 15°\right) = 5 \sin 75{,}9°$ and $\frac{\beta}{2} - 15° < 90°$. (7)

3. Do not use a calculator for this question.

3.1 Find the value of $\left(\frac{2 \sin 45° \csc 90°}{\cos 60°}\right)^2$. Show all your working. (5)

3.2 Find the value of θ if $2 \sin 2\theta - \sec 0° = \sin 0°$ and $\theta \in [0°; 90°]$. (6)

4. If $\csc \beta = \frac{p}{2}$, determine the value of $\cos^2 \beta$ in terms of p. (5)

5. In the diagram, $CA \perp BC$, $AB = 100$ mm, $\hat{B} = 70°$, $CD = 85$ mm and $\hat{D} = (3x + 12°)$.

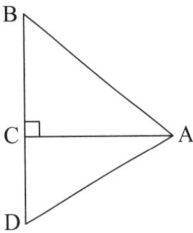

5.1 Solve for x, correct to one decimal place. (7)

5.2 Hence, or otherwise, prove that $\triangle BAD$ is not a right-angled triangle. (3)

6. In the diagram, $PQ \parallel VR \parallel TS$, $PT \perp US$, $\hat{U} = 60°$ and $UT = RS = WR = 10$ cm.

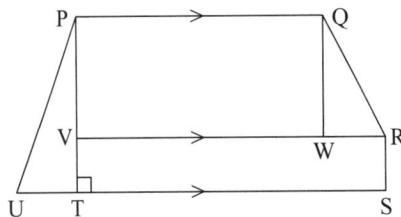

6.1 Find the length of WQ. Leave your answer in surd form. (4)

6.2 Hence, or otherwise, find the magnitude of $Q\hat{R}W$, correct to one decimal place. (3)

Total 50

Test C: Content and breakdown as for final exam

1. Use the diagram to answer these questions without using a calculator.

1.1 Determine the length of OP. (1)

1.2 Write down the ratios for $\cos \theta$ and $\cot \theta$. (2)

1.3 Use the information given to show that $\sec^2 \theta - 1 = \tan^2 \theta$. (4)

2. In △FAR, $\hat{A} = 90°$, $\hat{R} = \beta$ and AM ⊥ FR.

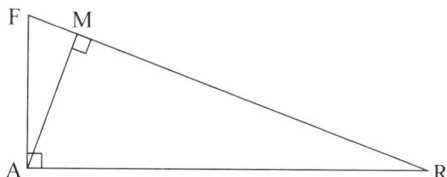

    2.1 Name one other angle in the diagram that is equal to β.   (1)

    2.2 Use capital letters to write down three ratios for sin β.   (3)

    2.3 If it is now given that AR = 10 cm and AM = 2,5 cm find the size of $\hat{R}$ correct to two decimal places.   (3)

3. Use special angles to answer these questions. You may not use a calculator and you must show all working.

    3.1 Determine the value of x if $x = \cos^2 30° + \tan 45° - \sin 90°$.   (4)

    3.2 Solve for θ if $2 \cos 2\theta - \sqrt{3} = 0$ and $\theta \in [0°; 90°]$.   (3)

4. Make sure that your calculator is in degree mode when answering this question. Use your calculator to calculate the value of x, correct to three decimal places, if:

    4.1 $x = \sin^2 34,5° + \sqrt{\cos 65,43°}$ (2)

    4.2 $x = \frac{\sec 34,5°}{3}$ (2)

5. Use your calculator to solve for θ in each of the equations given below. Give your answers correct to one decimal place.

    5.1 cosec $\theta = 3,456$ and $\theta \in [0°; 90°]$ (3)

    5.2 $10 \tan (\theta + 15°) = 234$ and $\theta \in [0°; 90°]$ (3)

6. Calculate the length of DR if AT ⊥ RD; $T\hat{A}R = 48°$; $A\hat{D}T = 35°$ and AD = 45 km. NB: △DAR is not a right-angled triangle.   (10)

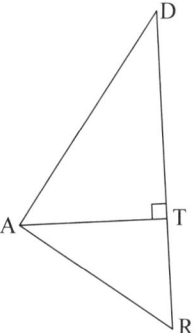

7. In the diagram, PQ = 80 cm and RS = 20 cm. $Q\hat{R}P = 40°$ and $P\hat{Q}S = 90°$. Find the size of $R\hat{P}S$ correct to one decimal place.   (9)

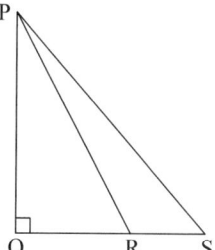

Total 50

TOPIC 7: Functions

In Grade 10 you will study the relationships between variables using tables, formulae and graphs. You will learn to work with functions and understand function notation.

You will learn how to draw the graphs of linear, quadratic, exponential, hyperbolic and trigonometric functions. You will then learn how to determine the equation of each of these functions given a graph.

You will also learn about the effect of different variables that result in a vertical shift, a vertical stretch, or a reflection about the axes. You will develop problem-solving techniques together with the graph work.

Knowledge and skills for this topic

If you struggle with any of the work listed below, revise it before continuing with this Topic:
- what is meant by the coordinates of a point in the Cartesian plane
- reading and interpreting tables and graphs
- a basic understanding of exponents
- good calculator knowledge and how to use the table function on the calculator
- concepts related to straight line graphs
- how to determine the gradient between any two points on a graph
- solving linear, quadratic, exponential and trigonometric equations to find intercepts with the axes.

Content of final exam
- Determine when a relationship is a function.
- Sketch graphs and use them to make deductions about linear, quadratic, hyperbolic, exponential and trigonometric functions.
- Determine defining equations of various functions from information given in a sketch.
- Work with function notation to determine the value of given x values and also to determine x-values for given y values.
- Determine the effect of the parameters a and q on the graphs of $y = a \cdot f(x) + q$, where $f(x)$ represents the standard form of a linear, quadratic, hyperbolic, exponential or trigonometric function.

Functions

A function can be described as a special type of relationship between two variables. The domain, x values or independent variable can be called the input values, and the range, y values or dependent variable can be called the output values. In a function, for every input value there is **one and only one output value**.

Function notation

- $f(x)$ is used for the function value (y), which is dependent on x.
- $f(-2) = 1$ indicates that the y value of the function f is 1 when $x = -2$.
- This gives you the coordinates (or ordered pair) $(-2; 1)$.

Worked example

Given $f(x) = 3x - 2$, determine the value/s of the following:

1. $f(2)$ **2.** $f(-4)$

3. x if $f(x) = 0$ **4.** x if $f(x) = 7$

Solutions

1. $f(2) = 3(2) - 2 = 4$

2. $f(-4) = 3(-4) - 2 = -14$

3. x if $f(x) = 3x - 2 = 0$ $\therefore 3x = 2 \therefore x = \frac{2}{3}$

4. x if $f(x) = 3x - 2 = 7$ $\therefore 3x = 9 \therefore x = 3$

Exercise 1

1. If $g(x) = -2x + 1$, determine the value/s of the following:

    1.1 $g(3)$ **1.2** $g(-2)$

    1.3 x if $g(x) = 0$ **1.4** x if $g(x) = 5$

2. If $h(x) = x^2 - 1$, determine the value/s of the following:

    2.1 $h(2)$ **2.2** $h(-3)$

    2.3 x if $h(x) = 0$ **2.4** x if $h(x) = 3$

3. If $f(x) = 2^x$, determine the value/s of the following:

    3.1 $f(0)$ **3.2** $f(1)$

    3.3 x if $f(x) = 0$ **3.4** x if $f(x) = 8$

4. if $p(x) = \frac{4}{x}$, determine the value/s of the following:

    4.1 $p(1)$ **4.2** $p(0)$

    4.3 x if $p(x) = 16$ **4.4** x if $p(x) = 1$

Restrictions

When finding the domain and range of a function, there may be restrictions you need to bear in mind. There are two main kinds that occur when working with fractions and/or exponential functions. Restrictions on fractions were covered in Topic 5. This is a quick recap, showing how these ideas will be applied to graphs.

Restrictions on fractions: division by zero is undefined

Worked example

Determine the restriction on the domain and range if $f(x) = \frac{4}{x-2} + 1$.

Solutions

Restrictions on domain: These are the permissible x values of $f(x)$.

The denominator of the fraction contains the variable x, so $x - 2$ cannot be equal to 0, $\therefore x - 2 \neq 0 \therefore x \neq 2$.

$\therefore$ Domain: $x \in \mathbb{R}, x \neq 2$ (in set builder notation)

or you can write

$\therefore$ Domain: $x \in (-\infty; 2) \cup (2; \infty)$ (in interval notation)

Restrictions on range: These are the permissible y values of $f(x)$.

In $f(x)$, the fraction $\frac{4}{x-2}$ will never be equal to zero, because the numerator of a hyberbola cannot be zero, therefore $y \neq 0$.

$\therefore$ Range: $y \in \mathbb{R}, y \neq 1$ (in set builder notation)

or you can write

$\therefore$ Range: $y \in (-\infty; 0) \cup (0; \infty)$ (in interval notation)

Restrictions on exponential functions

Worked example

Determine the restriction on the domain and range if $f(x) = 2^x - 1$.

Solutions

Restrictions on domain: Ask the following questions:

1. Can x be **negative**, or a **fraction**? Yes. Observe that as $x \to -\infty$, y becomes a fraction and $y \to 0$.

2. Can x be **zero**? Yes, $2^x = 1$.

3. Can x be **positive**? Yes. Observe that as $x \to \infty$, y also tends towards infinity, i.e. $y \to \infty$.

> **Note**
>
> In mathematics, this arrow $\to$ means 'tends towards'.

    $\therefore$ Domain: $x \in \mathbb{R}$ (set builder notation)

    Or $\therefore$ Domain: $x \in (-\infty; \infty)$ (interval notation)

Restrictions on range: While finding the permissible x values for this exponential function, most of the permissible y values have been determined.

If $x \to -\infty$, and $y \to 0$ for $f(x) = 2^x$, then $y \to -1$ for $f(x) = 2^x - 1$

$\therefore$ Range: $y \in \mathbb{R}, y > -1$ (set builder notation)

Or $\therefore$ Range: $y \in (-1; \infty)$ (interval notation)

Interval notation and set builder notation

We use notation when finding the domain, range or restrictions. We use two different types of notation in Grade 10:

1. **Interval notation (IN):** can only be used when the permissible values of the input and output variables is continuous, i.e. $x \in \mathbb{R}$ and $y \in \mathbb{R}$.

2. **Set builder notation (SBN):** this notation can be used for discrete (separate) input and output values. i.e. $x \in \mathbb{N}$ or $x \in \mathbb{N}_0$ or $x \in \mathbb{Z}$ or $x \in \mathbb{R}$.

Worked examples

Given $f(x) = \frac{1}{x-2}$, write down the following using both interval notation and set builder notation:

1. the domain of $f(x)$
2. the range of $f(x)$.

Solutions

1. Domain IN: $x \in (-\infty; 2) \cup (2; \infty)$; SBN: $x \in \mathbb{R}, x \neq 2$

2. Range IN: $y \in (-\infty; 0) \cup (0; \infty)$; SBN: $y \in \mathbb{R}, y \neq 0$

Exercise 2

For each of the following functions determine the domain and the range, using interval notation and set builder notation. Make sure the restrictions are clearly stated.

1. $f(x) = x^2 - 4$
2. $f(x) = -x^2 + 4$
3. $f(x) = 2x^2 - 2$

4. $g(x) = \frac{4}{x} - 1$
5. $g(x) = \frac{4}{x-1} - 1$
6. $g(x) = \frac{-4}{x+1} - 1$

7. $h(x) = 3^x + 1$
8. $h(x) = 2 \cdot 3^x + 1$
9. $h(x) = -3^x + 1$

10. $f(x) = \frac{9}{x-3} - 3$
11. $g(x) = 3 \cdot 2^x - 3$
12. $h(x) = -3x^2 + 12$

Zero points, roots or x-intercepts

The zero points of a function are the x values for which the function value (y) is zero, which are the values of x where the graph intersects the x-axis. These points can also be called the roots or the x-intercepts of the function.

Worked examples

Determine the zero points of the following functions:

1. $y = 3x - 2$
2. $y = 4 - x^2$
3. $y = 3^x - 27$

Solutions

To find the zero points of a function, let $y = 0$, because you are finding the x-intercept/s.

1. $3x - 2 = 0$

   $3x = 2$

   $\therefore x = \frac{2}{3}$

2. $4 - x^2 = 0$

   $(2 - x)(2 + x) = 0$

   $\therefore x = 2$ or $x = -2$

3. $3^x - 27 = 0$

   $3^x = 27$

   $3^x = 3^3$

   As this is an exponential equation, make your bases the same.

   $\therefore x = 3$

   Then you can equate the exponents.

Exercise 3

1. Determine the zero points of the following functions:

   1.1 $y = 1 + 2x$
   1.2 $y = x^2 - 9$
   1.3 $y = 2^x - 8$

   1.4 $y = -x^2 + 16$
   1.5 $y = 2 \cdot 3^x - 18$

2. Given $f(x) = 2x - 3$, determine:

   2.1 $f(-2)$ **2.2** $f(5)$ **2.3** $f(0)$

   2.4 $3 . f(-2)$ **2.5** x if $f(x) = 0$

   2.6 Write down the domain and range of $f(x)$ using interval notation.

3. Given $f(x) = x^2 - 1$, determine:

   3.1 $f(2)$ **3.2** $f(-2)$ **3.3** $f(0)$

   3.4 $3 . f(2)$ **3.5** x if $f(x) = 0$

   3.6 Write down the domain and range of $f(x)$ using set builder notation.

4. Given $f(x) = 5^x - 1$, determine:

   4.1 $f(1)$ **4.2** $f(-1)$ **4.3** $f(0)$

   4.4 $3 . f(1)$ **4.5** x if $f(x) = 0$

   4.6 Write down the domain and range of $f(x)$.

Show relationships between two variables using the equation and a table

One way of showing the relationships between two variables is by using a table. You will be given an equation (formula) and you will need to draw up a table of values by choosing an input value (x) and calculating the output value (y). A calculator can be used in order to draw up a table of values. This table of values can be used to plot points that you can use to create a graph.

It can also be used to find intercepts with axes, the equation of the axis of symmetry, the domain and the range, and whether the relationship is a function or not.

An **axis of symmetry** is a line through a graph so that each side is a mirror image of the other side.

Worked examples

Draw up a table of values for $y = x^2 - 4$, and determine the following for this relationship:

1. the shape by drawing a graph
2. coordinates of any intercepts with the axes
3. the equation of the axis of symmetry
4. the permissible domain and range
5. whether the relationship represents a function or not.

Solutions

Draw up a table with two rows. In the first row, write down four negative x values, zero and four positive x values. In the second row write down the corresponding y values, using algebra or the table mode on a calculator.

Steps for using a calculator in table mode to determine y values

The calculator steps for using the table mode are as follows
(using a CASIO $fx - 82ES$):

1. Press the mode button and select [3. Table]. You will then be prompted to enter $f(x)$.

2. Now type in the right hand side of the equation, which is $x^2 - 2$. To do this, type ALPHA x, x^2, $-$, 2 and then ▣. You will be prompted with the question [START?], which you should select.

3. Type the smallest **negative** x value, e.g. –4, and then ▣. Now select [END?]

4. Type the **largest** x value, e.g. 4, and then ▣. Now select [STEP?].

5. Select a step of 1 initially. You can always change it if you need to do so.

6. You will now have a table of x values and the corresponding y values on your calculator. Write down these values in your table.

x	–4	–3	–2	–1	0	1	2	3	4
y	12	5	0	–3	–4	–3	0	5	12

1. Shape: The shape of the graph is a curve called a parabola.

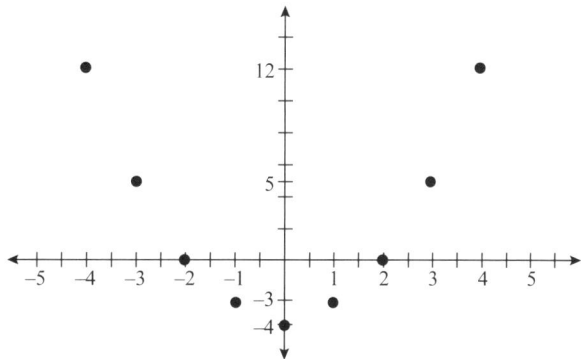

The points should then be joined by drawing a line through them all , because $x \in \mathbb{R}$, unless otherwise stated.

2. Cuts: coordinates of the intercepts with axes:

y-intercept: Let $x = 0$ ∴ $y = -4$

x-intercept: Let $y = 0$:

$$x^2 - 4 = 0$$
$$(x + 2)(x - 2) = 0$$
$$∴ x = -2 \text{ and } x = 2$$

3. The axis of symmetry is the y-axis, so the equation is $x = 0$.

4. Domain: $x \in \mathbb{R}$, Range: $y \in [-4; \infty)$

5. Using the table: Each x value has a unique y value, so this is a function.

   Using the graph, perform a vertical line test. This line will only cut the graph once for any given x value, so the relationship is a function, because every x value has only one y value.

Show relationships between two variables using the equation and a graph

The other way of showing the relationships between two variables is by using a graph. The sections that follow form a substantial part of functions work in Grade 10. You will work with the graph and equation (formula) for each type of function, and you will carry out the following:

- The equation will be given and you will need to draw the graph. You will then determine the effect a and q have on the graphs defined by $y = a \cdot f(x) + q$, where $f(x) = x$, $f(x) = x^2$, $f(x) = \frac{1}{x}$ and $f(x) = b^x$, $b > 0$, $b \neq 1$
- The graph of the function will be given in a sketch, and you will need to find the equation of the graph.
- You will answer questions about equations and inequalities related to their graphs.

Linear function

You can recognise a linear function by an equation in which the highest power of both x and y is 1 (i.e. just x and y). So if both variables are given on the same side of the equals sign, you need to add or subtract y from x.

Draw graph of $y = af(x) + q$ if $f(x) = x$, using the effect of a and q

In the standard form $f(x) = x$, the graph:

- passes through the origin $(0; 0)$
- has a gradient of 1, given by $a = \frac{\text{change in } y}{\text{change in } x}$
- has the point $P(1; 1)$.

The sign of a gives us information about the shape of the graph, because the gradient (slope) of the graph is equal to the value of a.

The **shapes** of standard linear functions are as follows:

- $f(x) = x$ for $a > 0$: The line slopes **up to the right**, and the gradient is 1 (the coefficient of x) and is **positive**.

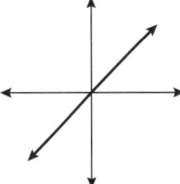

- $f(x) = -x$ for $a < 0$: The line slopes **down to the right**, and the gradient is -1 and is **negative**.

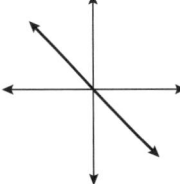

Effect of the value of a on the graph of $y = ax$

- If $a \neq 1$ and $q = 0$, the graph passes through the origin (0; 0).
- The value of a gives us the factor of the **vertical stretch**.
- The y values of the basic graph $f(x) = x$ will be multiplied by the value of a, which makes the gradient of the graph steeper, and the point P(1; 1) will become P′(1; a).
- The gradient is a $= \frac{y_2 - y_1}{x_2 - x_1}$.
- The value of q gives us information about the intersection, in other words the cut, through the y-axis, and the **vertical shift** of the graph is equal to q. This shift is vertically **upwards** if $q > 0$ and vertically **downwards** if $q < 0$.

Effect of the value of q on the graph of $y = x + q$, i.e. for $a = 1$

The **vertical shift** of standard linear functions is as follows:

- For $y = x + q$ if $\boldsymbol{q > 0}$, the line shifts q units **upwards**.
  - The cut on the y-axis is positive (i.e. > 0) and equal to q, as shown in the graph.
  - You would draw the basic graph, $f(x)$, as a broken line, so that you can always tell which is the basic graph.

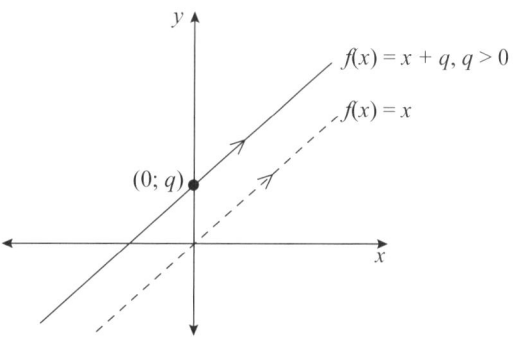

- For $y = x + q$ if $\boldsymbol{q < 0}$, the line shifts q units **downwards**.
  - The cut on the y-axis is **negative** (i.e. < 0) and equal to q, as shown in the graph.

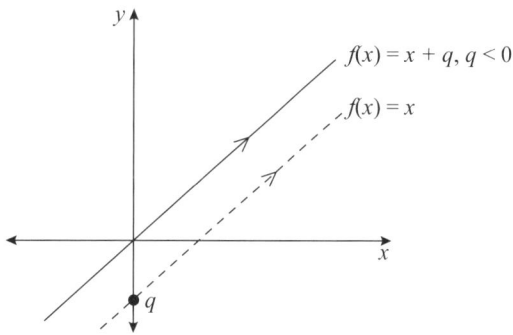

- All the y values of the basic graph $f(x) = x$, will have the value of q added to obtain the formula $y = x + q$.
- The graphs will be parallel to one another and the point P(1; 1) will become P′(1; 1 + q).
- Increasing the y values in this way can be described as a vertical shift by a factor of q.

Worked example

Draw the graph of $f(x) = 2x - 3$ using the effect of a and q. Indicate clearly all points of intersection with the axes.

Solution

$f(x) = 2x - 3$, $a = 2$, $q = -3$

• Start with the basic graph $y = x$:

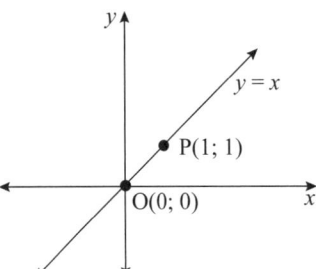

Notice that this graph passes through the origin $(0; 0)$.

• Indicate a point on the graph, let $x = 1$ ∴ P(1; 1)

• Deal with the effect of a first:
   $a = 2$ gives you a vertical stretch by a factor of 2

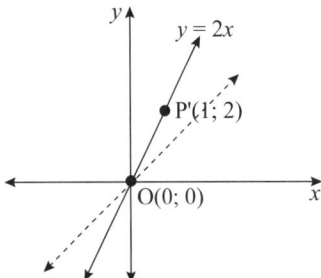

• Now deal with the effect of q:
   $q = -3$ gives you a vertical shift of 3 units downwards

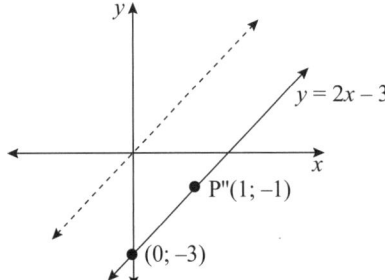

Now that you have drawn the graph you can see that it has an x-intercept. Calculate the x-intercept by letting $y = 0$:

$2x - 3 = 0$

∴ $x = \frac{3}{2}$

∴ x-intercept is $\left(\frac{3}{2}; 0\right)$

Quadratic function

A quadratic function is recognised by an equation in which the highest power of y is 1 and the highest power of x is 2 (i.e. we have x^2).

On the standard form $f(x) = x^2$ we examine the effect of a and q to obtain the graph of $y = a \cdot x^2 + q$.

The **sign of a** gives us information about the **shape** of the graph.

The **shape** of standard quadratic functions are as follows:

- For $y = a . x^2$ if $a > 0$: the graph is a smiley graph and has a **minimum function value**.

- For $y = a . x^2$ if $a < 0$: the graph is a frowny graph and has a **maximum function value**.

Effect of the value of a on the graph $y = ax^2$, $q = 0$

See Figure 1:

- If $a \neq 1$ and $q = 0$, the graph passes through the origin $(0; 0)$.
- The value of a gives us the factor of the **vertical stretch**.
- The y values of the basic graph $f(x) = x^2$ will be multiplied by the value of a, so for the graph of $y = ax^2$ the point P(1; 1) will become P'(1; a).
- If $a > 1$ the graph becomes thinner because the y values of $y = x^2$ become larger each time the y value is multiplied by a.

Note

The term 'vertical stretch' can be confusing, as parabolas become wider or narrower. If a is a whole number, the arms move closer to the y-axis. If a is a fraction, the arms move away from the y-axis.

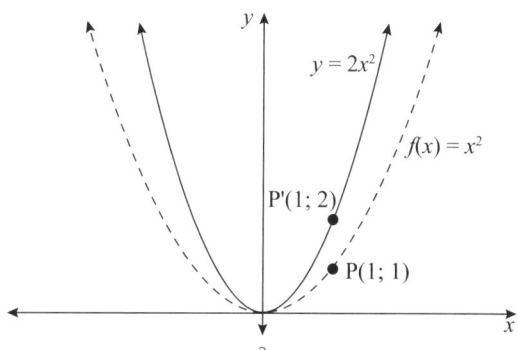

Figure 1: Graph of $y = 2x^2$

See Figure 2:

- If $0 < a < 1$, the graph becomes wider. The y values of $f(x) = x^2$ become smaller, because each y value is multiplied by the value of a, which is a fraction that is smaller than 1. This gives the graph of $y = \frac{1}{2}x^2$.

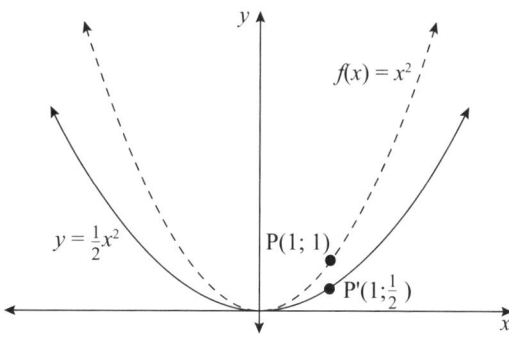

Figure 2: Graph of $y = \frac{1}{2}x^2$

Effect of the value of q on the graph $y = ax^2$

The value of q gives us information about the **vertical shift** of the graph, which is equal to q. This shift is vertically **upwards** if $q > 0$ and vertically **downwards** if $q < 0$.

Because we are working with quadratic equations in the form $y = ax^2 + bx + q$, $b = 0$, the value of q will also become the cut on the y-axis.

The **vertical shift** of a standard quadratic function is as follows:

See Figure 3:

- For $y = x^2 + q$ if $q > 0$, the graph shifts **upwards** by q units.
- The cut on the y-axis is **positive** and equal to q. As a result there are no x-intercepts.

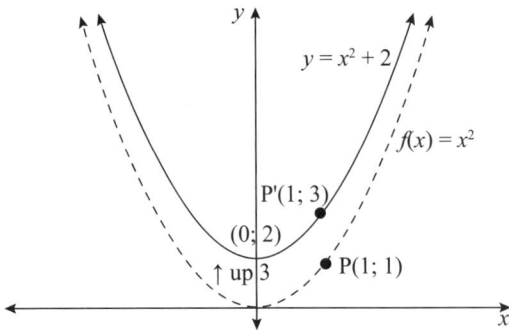

Figure 3: Graph of $y = x^2 + 2$

See Figure 4:

- For $y = x^2 + q$ if $q < 0$, the graph shifts **downwards** by q units.
- The cut on the y-axis is **negative** and equal to q. From the graph it is clear that now there are x-intercepts.

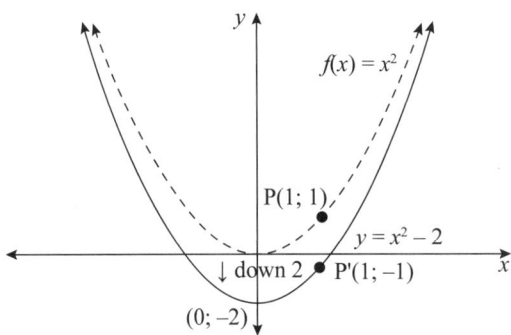

Figure 4: Graph of $y = x^2 - 2$

Worked example

Sketch the graph of $y = 2x^2 - 2$, using the effect of a and q on $f(x) = x^2$. Indicate clearly all points of intersection with the axes.

Solution

$y = 2x^2 - 2$ is in the form $y = a \cdot f(x) + q$.

S Shape: $a > 0$, smiley graph:

For $a > 1$, the graph will become thinner than the standard graph $f(x) = x^2$. This is because the value $a = 2$ indicates a vertical stretch by a factor of 2.

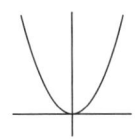

C Cut on the y-axis (y-intercept): the value q gives the value of the y-intercept and also indicates a **vertical shift** of 2 units downwards.

$\therefore$ y-intercept: $(0; -2)$

From the graph it is clear that the x-intercepts need to be found.

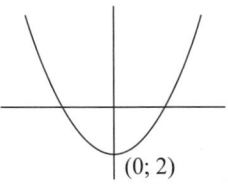

R Roots (x-intercept/s): Let $y = 0$:
$$2x^2 - 2 = 0$$
$$2(x^2 - 1) = 0$$
$$2(x - 1)(x + 1) = 0$$
$$\therefore x = 1 \text{ or } x = -1$$

A quadratic function has another two important properties, the axis of symmetry and the minimum value:

A Axis of symmetry: this is a vertical line drawn halfway between the two zero points: $x = 0$.

M Minimum value: this value is always found on the axis of symmetry, which passes through the turning point: $f(0) = -2$.

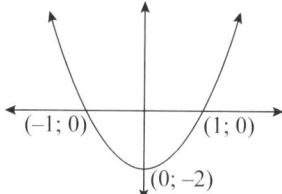

Exercise 4

1. The standard form of a linear function is $f(x) = x$. State the effect of a and q on each of the following graphs defined by $y = a \cdot f(x) + q$. Also sketch a graph for each function.

   1.1 $y = -2x + 4$ **1.2** $y = 3x + 2$ **1.3** $y = -3x - 6$

   1.4 $y = x + 3$ **1.5** $y = -\frac{1}{2}x + 2$ **1.6** $2y = 4x - 2$

2. The standard form of a quadratic function is $f(x) = x^2$. State the effect of a and q on each of the following graphs defined by $y = a \cdot f(x) + q$. Also sketch a graph for each function.

   2.1 $y = 3x^2 - 3$ **2.2** $y = \frac{1}{2}x^2 - 8$ **2.3** $y = -3x^2 + 1$

   2.4 $y = x^2 - 3$ **2.5** $y = 3 - 4x^2$ **2.6** $y = -2x^2 - 1$

Function of the hyperbola

You can recognise the function of a hyperbola (called the hyperbolic function) by an equation where the highest power of y is 1 and the highest power of x is 1. If both variables are on the same side of the equals sign, you **multiply** one variable by the other. This means they will have a constant product.

On the standard form $f(x) = \frac{1}{x}$, we examine the effect of a and q to obtain the graph of $y = a \cdot f(x) + q = \frac{a}{x} + q$

Properties of the hyperbola

1. Domain: $x \in \mathbb{R}, x \neq 0$, indicating a vertical asymptote.

   Range: $y \in \mathbb{R}, y \neq 0$ indicating a horizontal asymptote.

2. The **sign and value of a** gives information about the **shape** of the graph.

   S The shapes expected from the standard equation of a hyperbola are as follows:

The **sign** of a indicates where the graph will be drawn:

$a > 0$: The graph will be in the 1st and 3rd quadrants, because the product of x (the domain) and y (the range) is positive.	$a < 0$: the graph is in the 2nd and 4th quadrants, because the product of the domain and range is **negative**.

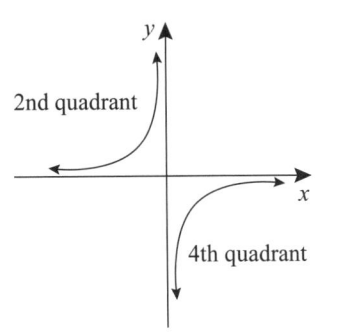

3. Asymptote and axes of symmetry:

An asymptote is a line that a curve gets continueally closer to, but never touches or intersects.

   A In both graphs above, the x-axis ($y = 0$) is the horizontal asymptote, and the y-axis ($x = 0$) is the vertical asymptote.

      In Grade 10 we will shift the horizontal asymptote. You will learn how to shift the vertical asymptote in Grade 11.

   A Axes of symmetry:

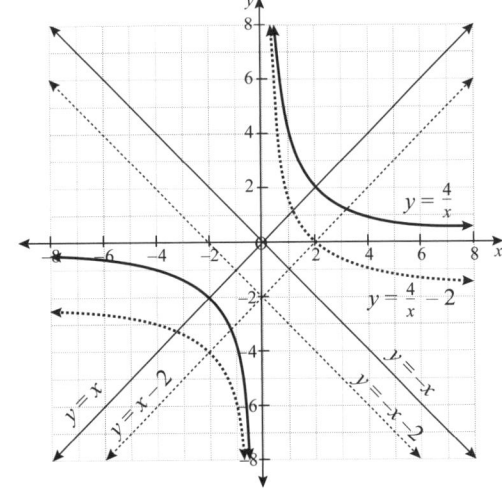

- If $y = \frac{1}{x}$, the axes of symmetry are $y = x$ and $y = -x$.

- If $y = a\left(\frac{1}{x}\right) = \frac{a}{x}$, the axes of symmetry are $y = x$ and $y = -x$.

- If $y = \frac{a}{x} + q$, the axes of symmetry are: $y = x + q$ and $y = -x + q$.

Effect of the value of a on graph of hyperbolic function

- The value *of a* gives us the factor of the **vertical stretch**.
- The y values of the basic graph $f(x) = \frac{1}{x}$ will be multiplied by the value of a, so the point P(1; 1) will become P′(1; a).
- If $a > 1$, the new graph will lie above the graph of $f(x) = \frac{1}{x}$, as you will see in the worked example below.

Worked example

Draw the graph of $y = \frac{4}{x}$ using the effect of a and q on $f(x) = \frac{1}{x}$. Indicate clearly all points of intersection with the axes, one other point on the graph, and the equations of all asymptotes.

Solutions

$y = \frac{4}{x}$ is in the form $y = a \cdot f(x)$, so:

S Shape: $a > 0$, which indicates that the graph is in the 1st and 3rd quadrants.

* 4 indicates that the graph has undergone a **vertical stretch** by a factor of 4.
* There are no intercepts with the axes.
* Make sure you plot at least one point on each graph. An easy point to use will be the point where $x = 1$.

A Asymptote/s: horizontal asymptote is the x-axis, vertical asymptote is the y-axis.

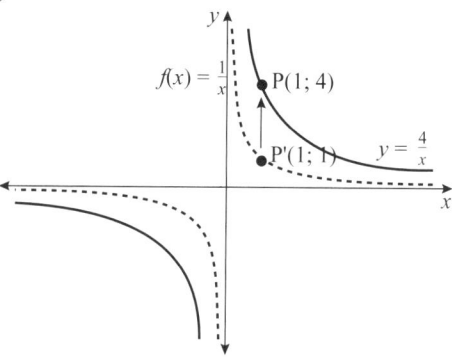

If $0 < a < 1$, the new graph will lie below the graph of $f(x) = \frac{1}{x}$, as shown in the graph below.

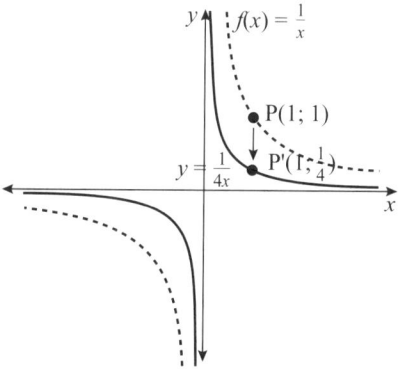

Effect of the value of q on graph of hyperbolic function

* The value **of q** gives us information about the **vertical shift of** the graph, which is equal to q. This shift is vertically **upwards** if $q > 0$ and vertically **downwards** if $q < 0$.
* Because we are working with hyperbolic functions in the form $y = \frac{a}{x} + q$:
  * the value of q will become the **horizontal asymptote** of the graph
  * the **vertical asymptote** remains the y-axis, i.e. $x = 0$.
* The **vertical shift** of standard hyperbolic function $y = x^2 + q$ for $q > 0$ is that the graph shifts **upwards** by q units.
* The horizontal asymptote is above the x-axis, so its equation is $y = q$ (as shown in the worked example below) and the axes of symmetry become $y = x + q$ and $y = -x + q$.

Worked example

Sketch the graph of $y = \frac{1}{x} + 2$, using the effect of a and q on $f(x) = \frac{1}{x}$. Indicate clearly all points of intersection with the axes, one other point on the graph, and the equation/s of all asymptote/s.

Solution

$y = \frac{1}{x} + 2$ is in the form $y = a \cdot f(x) + q$, so:

S Shape: $a > 0$, which indicates that the graph is in the 1st and 3rd quadrants.

C Cut/y-intercept: let $x = 0$. This is a restriction for this graph and indicates that $x = 0$ is a vertical asymptote, so there is no y-intercept.

R Roots/x-intercepts: let $y = 0$:

$\frac{1}{x} + 2 = 0$

$\therefore x = -\frac{1}{2}$

A Asymptote: horizontal asymptote indicated at $y = 2$ (vertical shift of 2 units upwards).

The horizontal asymptote is above the x-axis, so the equation is $y = 2$.

A Axes of symmetry: $y = x + 2$ and $y = -x + 2$.

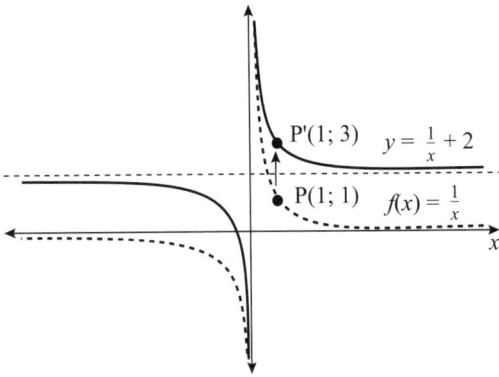

Worked example

Sketch the graph of $y = \frac{1}{x} - 2$, using the effect of a and q on $f(x) = \frac{1}{x}$. Indicate clearly all points of intersection with the axes, one other point on the graph, and the equations of all asymptotes.

Solution

$y = \frac{1}{x} - 2$ is in the form $y = a \cdot f(x) + q$, so:

S Shape: $a > 0$, which indicates that the graph is in the 1st and 3rd quadrants.

C Cut/y-intercept, let $x = 0$. This is a restriction for this graph and indicates that $x = 0$ is a vertical asymptote. So there is no y-intercept.

R Roots/x-intercepts, let $y = 0$:

$\frac{1}{x} - 2 = 0$

$\therefore x = \frac{1}{2}$

A Asymptote, horizontal: at $y = -2$ (vertical shift of 2 units downwards).

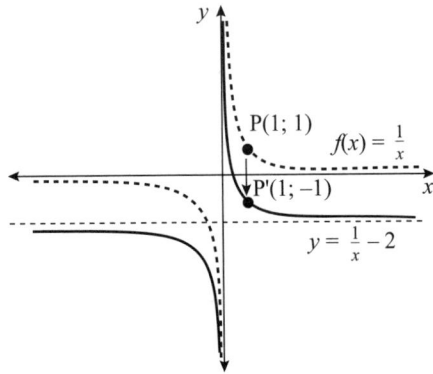

Worked example

Sketch the graph of $y = \frac{-4}{x} + 2$, using the effect of a and q on $f(x) = \frac{1}{x}$. Indicate clearly all points of intersection with the axes, one other point on the graph, and the equations of all asymptotes and the axes of symmetry.

Solution

$y = -\frac{4}{x} + 2$ is in the form $y = a \cdot f(x) + q$, so:

S Shape: $a < 0$, which indicates that the graph is in the 2nd and 4th quadrants.

4 indicates that the graph has undergone a vertical stretch by a factor of 4.

C Cut/y-intercept: let $x = 0$: This is a restriction for this graph and indicates that the vertical asymptote is at $x = 0$. So there is no y-intercept.

R Roots/x-intercepts, let $y = 0$:

$-\frac{4}{x} + 2 = 0$

$\therefore x = 2$

A Asymptote, horizontal: indicated at $y = 2$ (vertical shift of 2 units upwards).

Axes of symmetry: $y = x + 2$ and $y = -x + 2$.

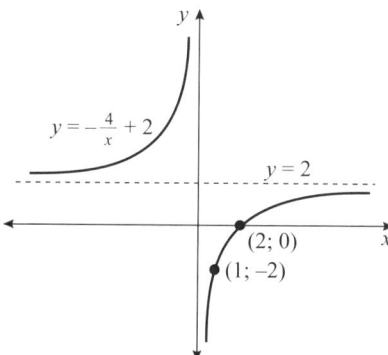

Exercise 5

1. If the standard form of a hyperbolic function is $f(x) = \frac{1}{x}$, state the effect of a and q on each of the following graphs defined by $y = a \cdot f(x) + q$. Also sketch each graph.

    1.1 $y = \frac{3}{x} - 1$ **1.2** $y = \frac{-3}{x} - 1$ **1.3** $y = \frac{3}{x} + 1$ **1.4** $y = \frac{-3}{x} + 1$

    1.5 $y = \frac{9}{x} - 3$ **1.6** $y = \frac{9}{x} - 1$ **1.7** $y = \frac{-9}{x} + 3$ **1.8** $y = \frac{5}{x} - 2$

    1.9 $y = \frac{5}{x} + 2$ **1.10** $y = \frac{16}{x} - 4$

2. Given that $f(x) = \frac{-4}{x} + 8$, determine the following:

    2.1 $f(1)$

    2.2 x if $f(x) = 0$

    2.3 the equation of the horizontal asymptote.

Exponential function

You can recognise an exponential function by an equation in which x is an exponent.

On the standard form $f(x) = b^x$, $b > 0$, $b \neq 1$, we examine the effect of a and q to obtain the graph of $y = a \cdot f(x) + q = a \cdot b^x + q$.

Domain: $x \in \mathbb{R}$

Range: $y > q$, $y \in \mathbb{R}$

The **sign and value of *a* and *b*** give us information about the **shape** of the graph:

S   Shape expected from the standard equation of an exponential function is as follows:

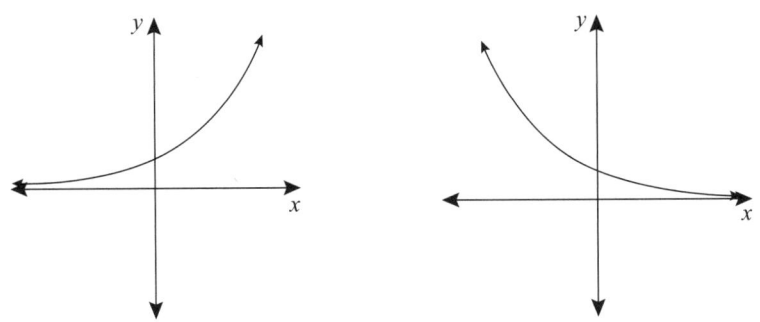

Type of exponential function	Values of *b*	Sketch
Increasing exponential function	$b > 1$	
Decreasing exponential function	$0 < b < 1$	

Note that exponential graphs in the form $y = ab^x$ have a horizontal asymptote of $x = 0$.

In Grade 10 we will shift the graph vertically. You will learn how to shift the graph horizontally in Grade 11.

*   The value of *a* gives us the factor of the vertical stretch.
*   The *y* values of the basic graph $f(x) = b^x$ will be multiplied by the value of *a*, i.e. $y = a \cdot b^x$, so the point P(0; 1) will become P'(0; *a*).

Effect of *a* on exponential function if *a* > 0

If *a* > 1, the new graph will lie **above** the graph of $f(x) = b^x$, as shown in the worked example below.

Worked example

Draw the graph of $y = 2 \cdot 3^x$ by comparing it to the graph of $f(x) = 3^x$. Indicate any points of intersection with the axes and one other point on the curve. Label any asymptotes with their equations.

Solution

S   Shape $a = 3$ for $a > 1$, so the graph is an increasing exponential graph.

C   *y*-intercept: let $x = 0$ ∴ $y = 2$

R   Roots: there is no *x*-intercept.

A   Asymptotes: the *x*-axis is the horizontal asymptote because $q = 0$.

Find another point on the graph, e.g. if $x = 1$ then $y = 6$.

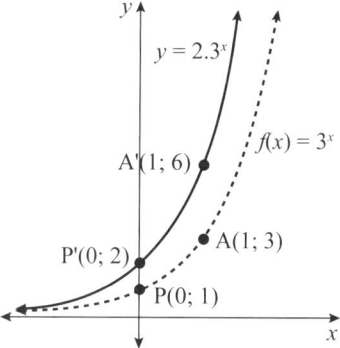

If $0 < a < 1$, the new graph will lie below the graph of $f(x) = b^x$.

Effect of a on exponential function if $a < 0$

If $a < 0$, the new graph will be a reflection of the graph of $f(x) = b^x$ in the x-axis, multiplied by the positive value of a.

Worked example

Draw the graph of $y = -2 \cdot 3^x$ by comparing it to the graph of $f(x) = 3^x$. Indicate clearly the point of intersection with the y-axis and another point on the graph.

Solution

First draw the graph of $f(x) = 3^x$:

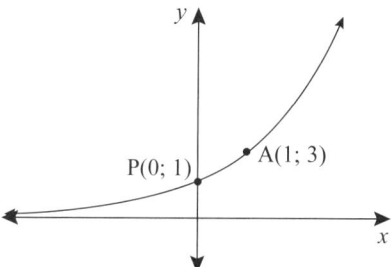

Then draw the graph of $y = 2 \cdot 3^x$:

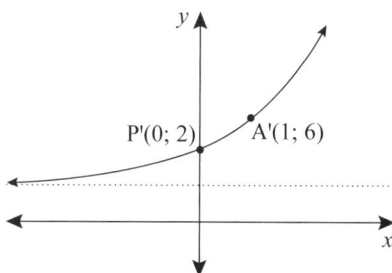

Now draw the graph of $y = -2 \cdot 3^x$ (all the y values must now change sign, indicating a reflection in the x–axis):

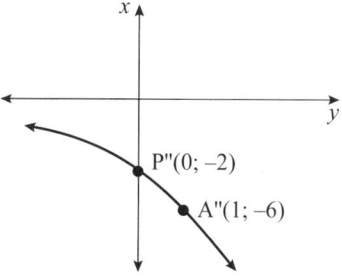

Effect of q on exponential function

- The **value of q** gives us information about the **vertical shift** of the graph, which is equal to q. This shift is vertically **upwards** if $q > 0$ and vertically **downwards** if $q < 0$.
- Because we are working with exponential functions in the form: $y = a \cdot b^x + q$, the value of q will become the **horizontal asymptote** of the graph.
- The **vertical shift** of a standard exponential function is as follows:
  - $y = b^x + q$ for $q > 0$: the graph shifts **upwards** by q units.
  - The horizontal asymptote is above the x-axis, and the equation is $y = q$.

Worked example

Draw the graph of $y = 2^x + 1$ by comparing it to the graph of $f(x) = 2^x$. Indicate clearly the point of intersection with the y-axis and another point on the graph. Label any asymptotes with their equations.

Solution

Draw the graph of $f(x) = 2^x$ for $q = 1$, indicating a vertical shift of 1 unit upwards. So add 1 to every y value and draw the new graph.

The horizontal asymptote is $y = 1$.

Notice that for $y = b^x + q$, $q < 0$, the graph shifts **downwards** by q units. The horizontal asymptote has the equation is $y = q$.

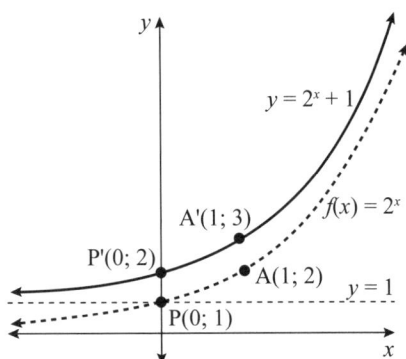

Exercise 6

1. Using the standard form of an exponential function $f(x) = b^x$, state the effect of a and q on each of the following graphs defined by $y = a \cdot f(x) + q$. Also sketch a graph for each function.

   1.1 $y = 3^x - 1$ **1.2** $y = -3^x - 1$ **1.3** $y = 3^x + 1$

   1.4 $y = -3^x + 1$ **1.5** $y = 3 \cdot 2^x - 3$ **1.6** $y = \frac{1}{2} \cdot 5^x - 1$

Find the equation if given a sketch

In this section we will find the equation of the given graph and answer questions applying mathematical facts you've already learned.

Linear function

Worked examples

Determine the equation in the form $y = a \cdot f(x) + q$ of each of the graphs provided.

1.

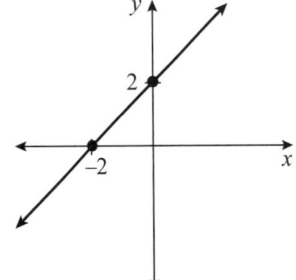

2.

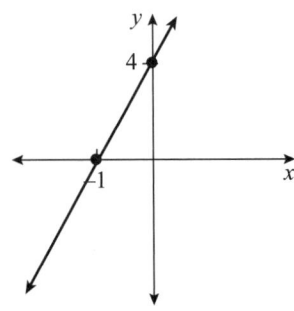

3.

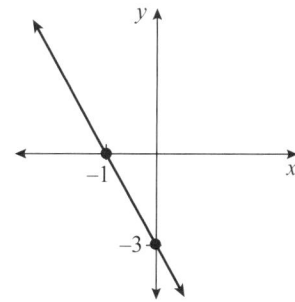

Solutions

1. $y = a \cdot f(x) + q$

In a linear function $f(x) = x$:

$\therefore y = ax + 2$

$0 = -2a + 2$ Substitute $(-2; 0)$ for $(x; y)$.

$a = 1$

$\therefore y = x + 2$

2. $y = a \cdot f(x) + q$

$f(x) = x$

$y = ax + 4$

$0 = -a + 4$ Substitute $(-1; 0)$ for $(x; y)$.

$a = 4$

$\therefore y = 4x + 4$

3. $y = a \cdot f(x) + q$

$f(x) = x$

$y = ax - 3$

$0 = -a - 3$ Substitute $(-1; 0)$ for $(x; y)$.

$a = -3$

$\therefore y = -3x - 3$

Find the points of intersection of two linear functions

To determine the coordinates of the point where two graphs intersect:

1. Make the y values of the two functions equal to each other.

2. Solve the two equations simultaneously.

Worked examples

1. Determine the coordinates of the point of intersection of $y = 2x - 3$ and $y = -2x - 7$.

2. Use the accompanying graph and determine:

   2.1 the equations of f and g

   2.2 the coordinates of P.

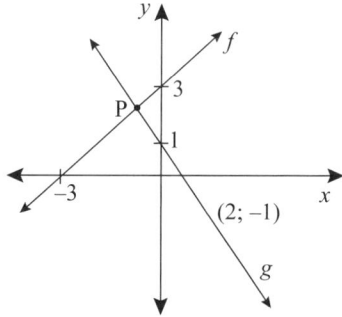

Solutions

1. Graph optional.

$2x - 3 = -2x - 7$

$4x = -4$

$x = -1$

Substitute -1 for x in $y = 2x - 3$:

$\therefore y = -5$

$\therefore$ Point of intersection $(-1; -5)$

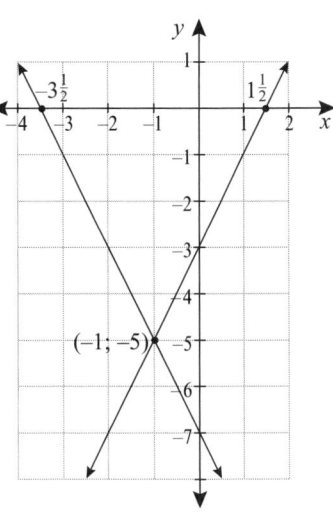

2. **2.1** $f: y = ax + q, q = 3$

$y = ax + 3$

$0 = -3a + 3$ Substitute $(-3; 0)$ for $(x; y)$.

$3a = 3$

$a = 1$

$\therefore y = x + 3$

$g: y = ax + q, q = 1$

$y = ax + 1$

$-1 = 2a + 1$ Substitute $(2; -1)$ for $(x; y)$.

$-2 = 2a$

$a = -1$

$y = -x + 1$

2.2 At P, $f(x) = g(x)$

$x + 3 = -x + 1$

$2x = -2$

$x = -1$

$y = x + 3$

$\therefore y = -1 + 3 = 2$

$\therefore$ P$(-1; 2)$

Worked examples: complex procedures

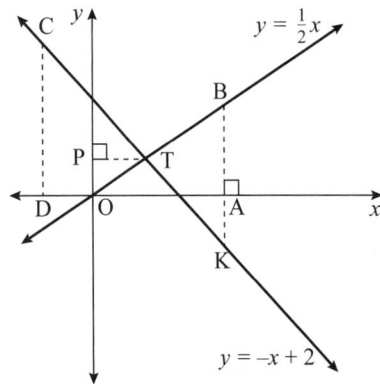

Use the graph and determine the length of:

1. CD if OD $= 1$

2. BK if OA $= 3$

3. PT

Solutions

1. Determine the y coordinate of C:

   $OD = 1$ ∴ at D $x = -1$

   Substitute -1 for x in $y = -x + 2$:

   $y = -(-1) + 2 = 3$

   ∴ $C(-1; 3)$

   ∴ $CD = 3$

2. BK = difference between y
   coordinates of B and K:

   $OA = 3$ ∴ at B and K $x = 3$

   B: Substitute 3 for x in $y = \frac{1}{2}x$:

   $y = \frac{1}{2}(3) = 1\frac{1}{2}$

   ∴ $B\left(3; 1\frac{1}{2}\right)$

   K: Substitute 3 for x in $y = -x + 2$:

   $y = -3 + 2 = -1$

   ∴ $K(3; -1)$

   ∴ $BK = 2\frac{1}{2}$ units

3. PT = x coordinate of T, the point of intesection of the two graphs:

   $\frac{1}{2}x = -x + 2$

   $x = -2x + 4$

   $3x = 4$

   ∴ $x = \frac{4}{3}$

   ∴ $PT = \frac{4}{3}$

Exercise 7

1. Determine the equation for each straight line graph.

   1.1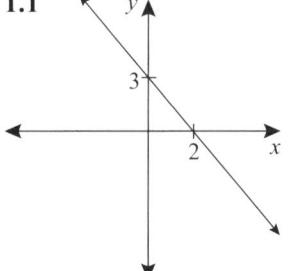

   1.2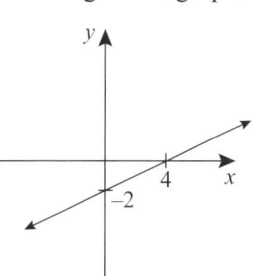

   1.3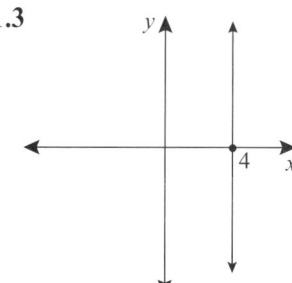

   1.4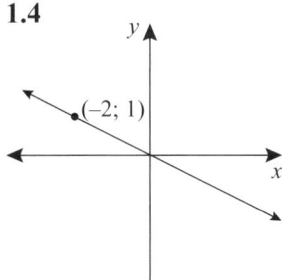

   1.5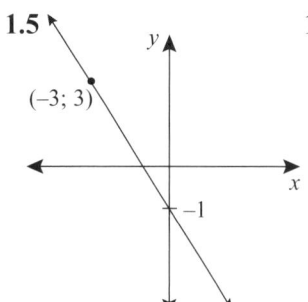

   1.6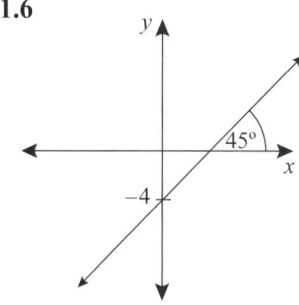

   1.7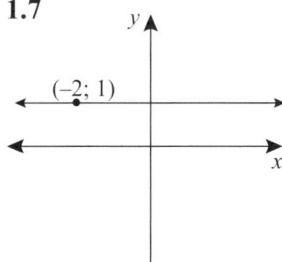

   1.8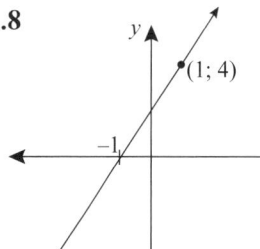

   1.9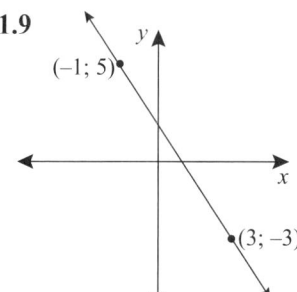

2. Determine the equation of g in each graph.

2.1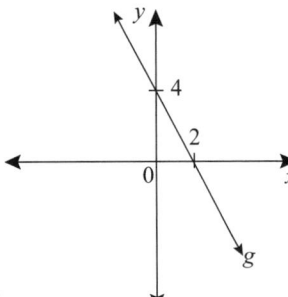

2.2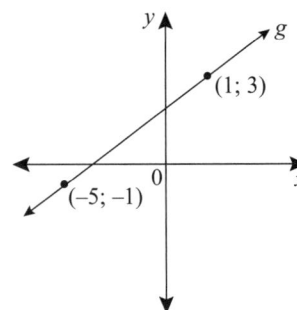

2.3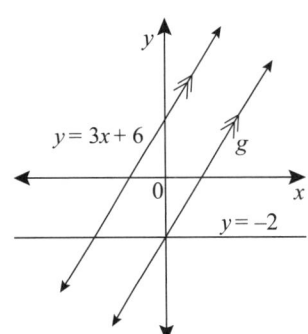

2.4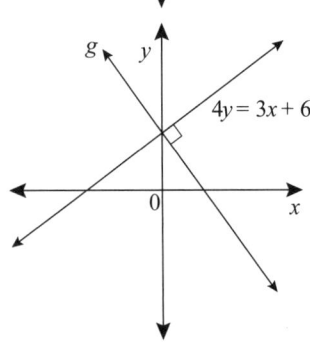

3. Given: $y = -3x + 6$ and $y = 2x + 1$.

    3.1 Determine the coordinates of the point of intersection of the two graphs.

    3.2 Sketch graphs of both equations on the same set of axes. Indicate clearly the coordinates of the point of intersection determined in 3.1.

4. **4.1** Determine the equation of AC.

    4.2 Determine the lengths of AB and CD.

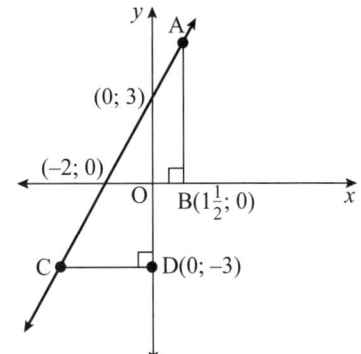

5. **5.1** Calculate the length of :

    5.1.1 PT

    5.1.2 AB

    5.1.3 CD

    5.1.4 DE

    5.2 Use the graph to determine the values of x if :

    5.2.1 $5x - 6 > 20 - 8x$

    5.2.2 $f(x) \cdot g(x) < 0$.

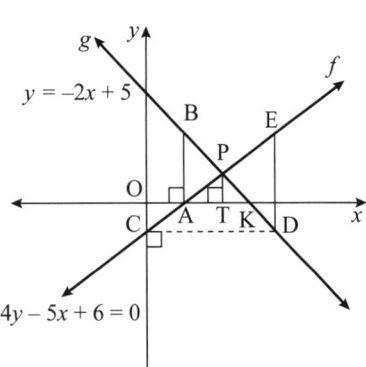

Quadratic function

In this section you will use a sketch to help you to find the equation of the function in the form $y = a \cdot f(x) + q$, where $f(x) = x^2$.

Type 1: Find equation given the x-intercepts and the y-intercept

To find the equation of the function if you are given two roots (the x-intercepts) and the y-intercept, use: $y = a(x - \text{one root})(x - \text{other root})$.

Worked example

Determine the equation of the graph in the sketch.

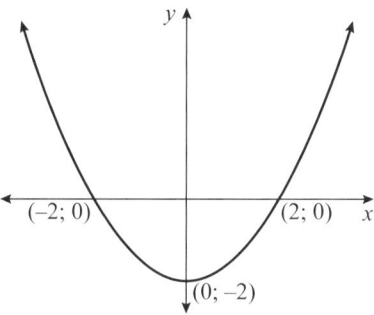

Solution

$$y = a(x - 2)(x - (-2))$$
$$y = a(x^2 - 4)$$
$$-2 = a(0 - 4) \quad \text{Substitute } (0; -2) \text{ for } (x; y).$$
$$a = \tfrac{1}{2}$$
$$\therefore y = \tfrac{1}{2}(x^2 - 4)$$
$$\therefore y = \tfrac{1}{2}x^2 - 2$$

OR $y = ax^2 + q$, where $q = -2$

$\qquad\qquad$ y value at the turning point

$$y = ax^2 - 2$$
$$0 = 4a - 2 \qquad \text{Substitute } (2; 0) \text{ for } (x; y).$$
$$a = \tfrac{1}{2}$$
$$\therefore y = \tfrac{1}{2}x^2 - 2$$

Type 2: Find equation given x-intercepts and another point on the graph

To find the equation if you are given two roots (the x-intercepts) and another point on the graph, use $y = a(x - \text{one root})(x - \text{other root})$.

Worked example

Determine the equation of the graph in the sketch.

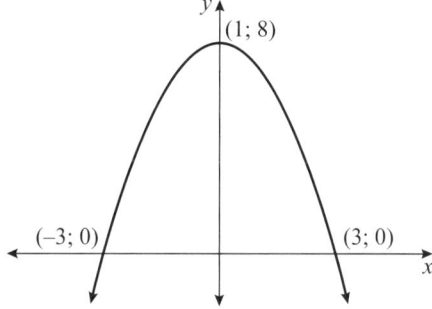

Solutions

$$y = a(x - 3)(x + 3)$$
$$y = a(x^2 - 9)$$
$$8 = a(1 - 9) \qquad\qquad \text{Substitute } (1; 8) \text{ for } (x; y).$$
$$a = -1$$
$$\therefore y = -1(x^2 - 9)$$
$$\therefore y = -x^2 + 9$$

Type 3: Find equation given the *y*-intercept and another point on the graph

To find the equation when you are given the *y*-intercept and another point on the graph, use $y = ax^2 + q$, where q is the *y*-intercept (the *y* value at the turning point).

Worked example

Determine the equation of the graph in the sketch.

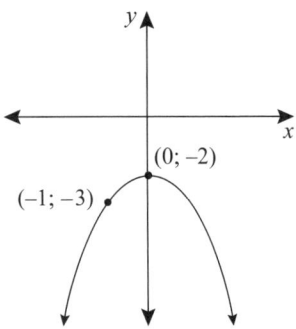

Solution

$y = ax^2 - 2$

$-3 = a(-1)^2 - 2$ Substitute (–1; –3) for (*x*; *y*).

$a = -1$

$\therefore y = -x^2 - 2$

Type 4: Find equation given two points on the graph: complex procedures

To find the equation when you are given two points on the graph, you use $y = ax^2 + q$, substitute the coordinates of both points into the equation, and set up two simultaneous equations to solve.

Worked example

Determine the equation of the graph in the sketch.

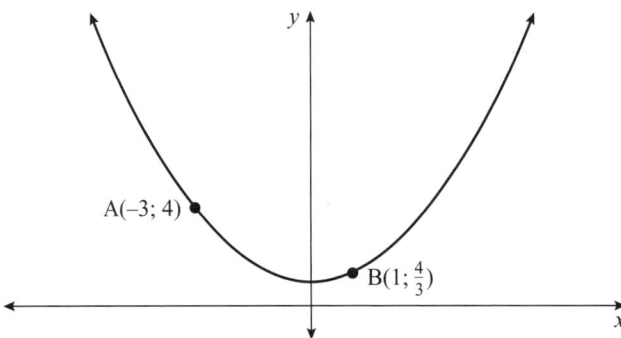

Solution

The general form of the equation is $y = ax^2 + q$.

Substitute (–3; 4) and $\left(1; \frac{4}{3}\right)$ into this equation.

$4 = 9a + q$

$\therefore q = 4 - 9a$ ①

$\frac{4}{3} = a + q$ × 3

$4 = 3a + 3q$ ②

Substitute equation ① into equation ②:

$$4 = 3a + 3(4 - 9a)$$

$$4 = 3a + 12 - 27a$$

$$24a = 8$$

$$a = \tfrac{1}{3}$$

$$\therefore y = \tfrac{1}{3}x^2 + q \qquad \text{Substitute } (-3; 4) \text{ to find } q.$$

$$4 = \tfrac{1}{3}(9) + q$$

$$q = 1$$

$$\therefore y = \tfrac{1}{3}x^2 + 1$$

Worked examples: revision

1. Sketch the graphs of $f(x) = x^2 - 4$ and $g(x) = 2x - 4$ on the same set of axes.

2. Use your graphs to determine the coordinates of the points of intersection of the two graphs.

3. Determine the range of f.

4. Determine a if $(3; a)$ is a point on the graph of f.

5. For which values of x will both graphs increase when x increases?

6. Determine b if $(2b; 3)$ is a point on graph g.

7. For which values of x will each of the following be correct?

   7.1 $x^2 - 4 > 0$ **7.2** $x^2 - 4 > 2x - 4$

   7.3 $x^2 - 4 \leq 2x - 4$ **7.4** $g(x) = 0$

Solutions

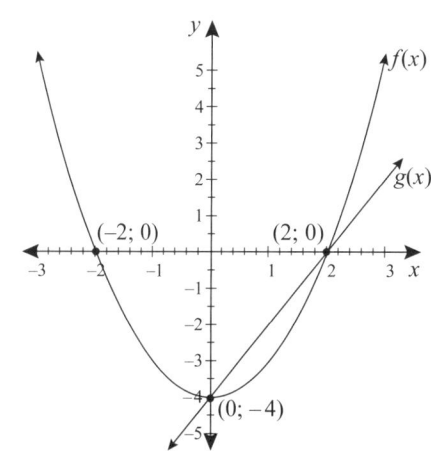

1. The graph of $f(x) = x^2 - 4$ is a parabola.

   y-intercept: -4

   x-intercepts: $x^2 - 4 = 0$

   $\therefore x = \pm 2$

   The graph of $g(x) = 2x - 4$ is a straight line.

   y-intercept: -4

   x-intercept: $2x - 4 = 0$

   $\therefore 2x = 4$

   $\therefore x = 2$

2. Points of intersection: $(0; -4)$ and $(2; 0)$

3. Range of f: $\{y : y \geq -4\}$

4. $f(x) = x^2 - 4$

   $a = 3^2 - 4 \qquad \text{Substitute } (3; a) \text{ into the equation.}$

   $\therefore a = 5$

5. Both graphs increase for $x > 0$ (because both graphs have a positive gradient for $x > 0$). Note that g is an increasing function and that f is an increasing function to the right of the turning point.

6. $g(x) = 2x - 4$

$3 = (2b - 2)2$ ⟵ Substitute $(2b; 3)$ into the equation.

$3 = 4b - 4$

$\therefore b = \frac{7}{4}$

7. 7.1 To find where $x^2 - 4 > 0$ we use $f(x)$, since $f(x) = x^2 - 4$.

For $x^2 - 4 > 0$, we are looking for the x-values for which $f(x)$ (or y) is positive.

From the graph:

$\therefore x < -2$ or $x > 2$

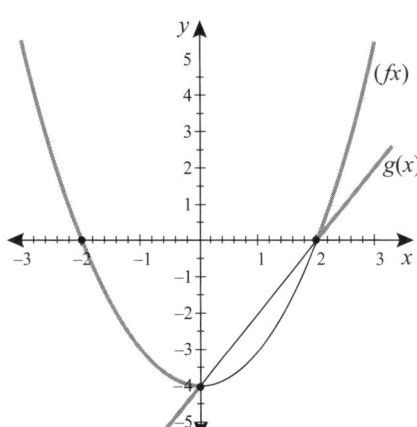

7.2 To find the x-values where $x^2 - 4 > x - 2$, we are looking for the region in which $f(x) > g(x)$, in other words, where the parabola is greater/higher than the straight line. From the graph:

$\therefore x < 0$ or $x > 2$

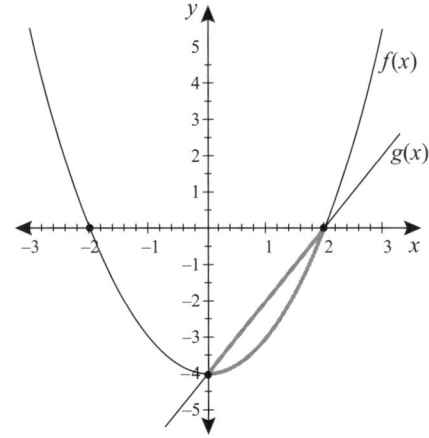

7.3 To find the x-values where $x^2 - 4 \leq 2x - 4$, we look for the region in which $f(x) \leq g(x)$, in other words, where the parabola is less/lower than or the same as the straight line.

From the graph:

$\therefore 0 \leq x \leq 2$

7.4 From the graph, $g(x) = 0$ is the x-intercept of g.

$\therefore x = 2$

Exercise 8

1. Determine the equation for each graph.

1.1

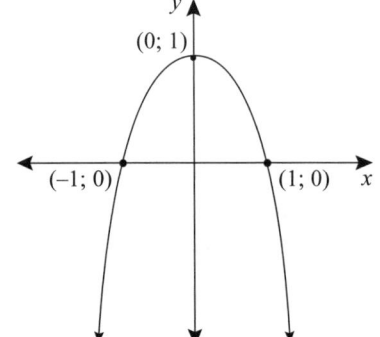

1.2

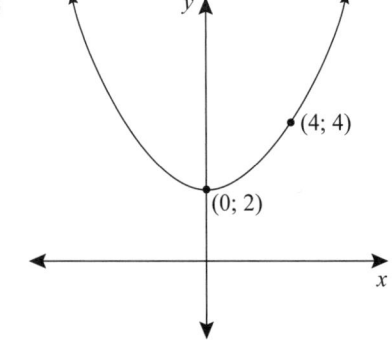

1.3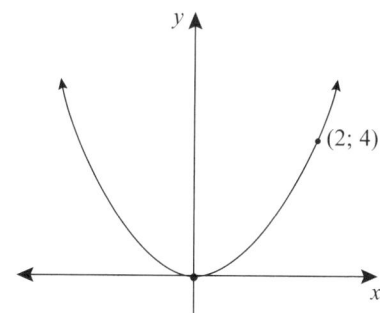

1.4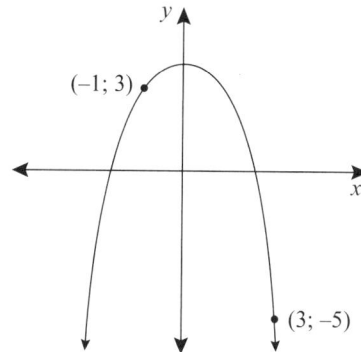

2. Determine the equation of the parabola in the form $y = ax^2 + c$ for each graph.

2.1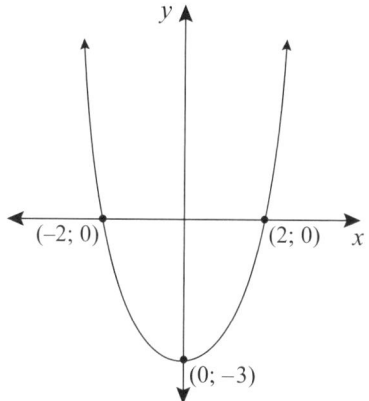

2.2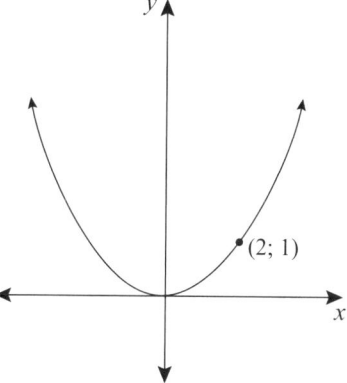

2.3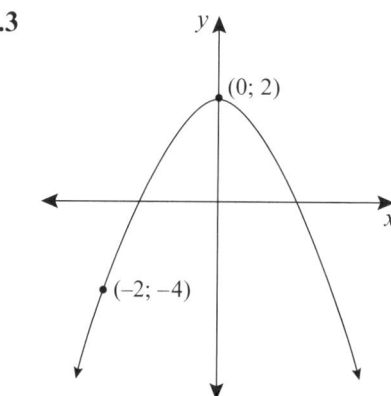

2.4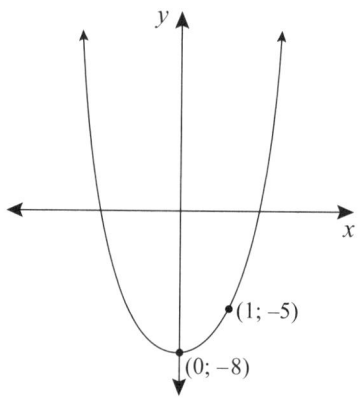

2.5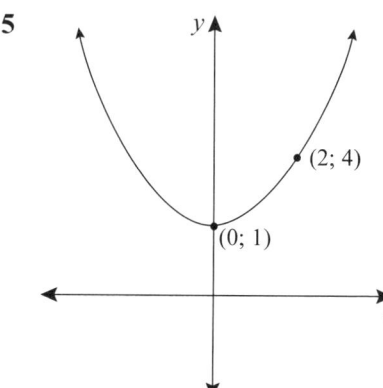

2.6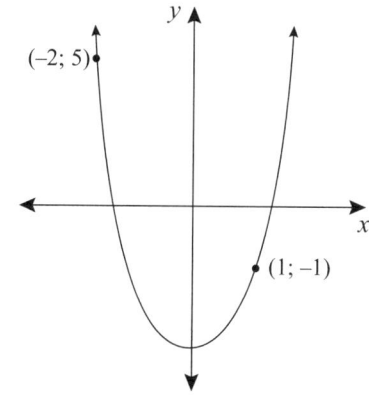

3. Given: $f(x) = -3x^2 + 12$.

   3.1 Sketch the graph of f. Indicate clearly the coordinates of the intercepts with the axes and the turning point.

   3.2 Does f have a maximum value? Determine this value.

3.3 Determine the value of k if:

3.3.1 $f(3) = k$ **3.3.2** $f(k) = 3$

3.4 What are the domain and range of f?

4. Use the graphs of f and g on the right to determine:

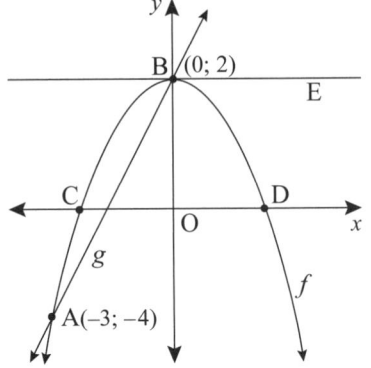

    4.1 the equation of f

    4.2 the coordinates of C

    4.3 the length of CD

    4.4 the range of f

    4.5 the equation of line AB

    4.6 the equation of BE

    4.7 the equation of the straight line perpendicular to AB that passes through the origin.

5. Sketch graphs for $f(x) = -x^2 + 4$ and $g(x) = 2x + 4$ on the same set of axes.

    5.1 Give the range of f.

    5.2 For which values of x will $-x^2 + 4 = 2x + 4$?

    5.3 For which values of x will $-x^2 + 4 > 2x + 4$?

    5.4 For which values of x will $-x^2 + 4 \geq 0$?

    5.5 If $(2a; -5)$ is a point on the graph of $-x^2 + 4$, determine a.

    5.6 Determine $f(1) - g(0)$.

    5.7 Determine a if:

        5.7.1 $f(-3) = a$ **5.7.2** $f(a) = 2$

    5.8 For which values of x will both graphs increase as x increases?

6. Given: $f : x \rightarrow x^2 - 1$

    6.1 Determine the following lengths.

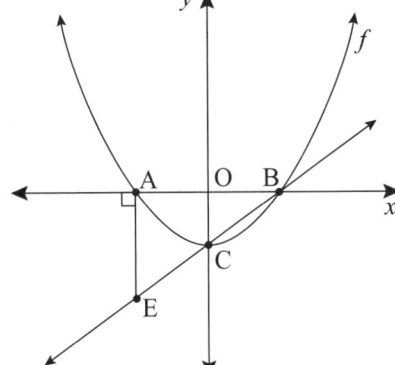

        6.1.1 OA

        6.1.2 OB

        6.1.3 OC

    6.2 What is the gradient of line CB?

    6.3 Determine the equation of the straight line CB.

    6.4 Determine the length of AE.

7. In the figure $f(x) = x^2 - 9$ and g is a straight line.

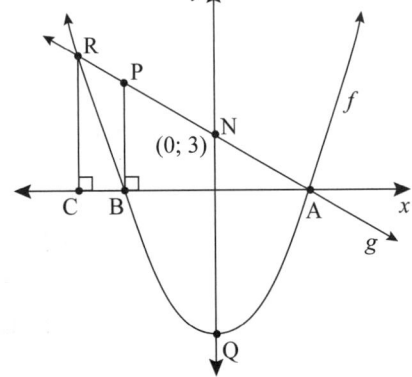

    7.1 Determine the coordinates of Q.

    7.2 Determine the length of OA.

    7.3 Determine the equation of g.

    7.4 Determine the following lengths.

        7.4.1 NQ

        7.4.2 BP

        7.4.3 AP

7.5 Give the range of f.

7.6 For which values of x will both f and g decrease as x increases?

7.7 Determine the equation of CR.

Hyperbolic function

Type 1: Find equation where the x-axis is the horizontal asymptote, i.e. $q = 0$

If $q = 0$, then $y = a\left(\frac{1}{x}\right) = \frac{a}{x}$.

- If the graph is in the 1st and 3rd quadrant then $a > 0$.
- If the graph is in the 2nd and 4th quadrants the $a < 0$.

Worked example

Determine the equation of this graph and its axes of symmetry.

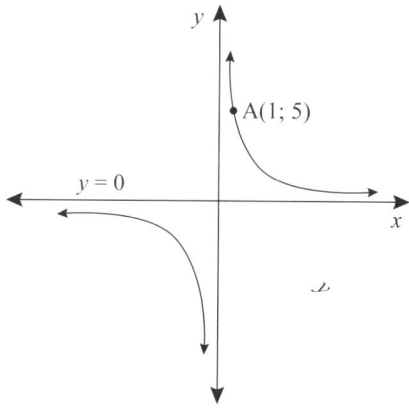

Solution

$y = \frac{a}{x}, q = 0$

$5 = \frac{a}{1}$ Substitute (1; 5) for $(x; y)$.

$a = 5$

$\therefore y = \frac{5}{x}$

Axes of symmetry: $y = x$ and $y = -x$

Note

Notice that in this case, the value of a does not change the axes of symmetry. They are the same as the axes of symmetry of the basic graph.

Type 2: Find equation where horizontal asymptote is $y = q$ and y-intercept is given

Worked example

Determine the equation of the graph and the axes of symmetry from the sketch of the graph.

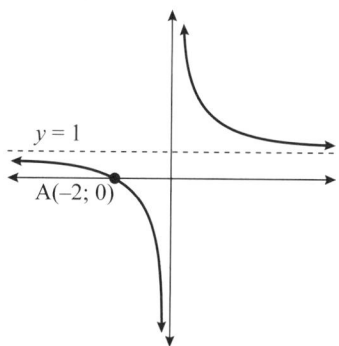

Solution

Note that the value of q indicates a vertical shift and gives the value of the horizontal asymptote.

$\therefore y = a\left(\frac{1}{x}\right) + q = \frac{a}{x} + q$

$q = 1 \therefore y = \frac{a}{x} + 1$

$0 = \frac{a}{-2} + 1$ Substitute (–2; 0) for $(x; y)$.

$a = 2$

$\therefore y = \frac{2}{x} + 1$

Axes of symmetry: $y = x + 1$ and $y = -x - 1$

Exponential function

From a sketch you will find the equation of the exponential function in the form $f(x) = a \cdot b^x + q$.

Type 1: Find equation given the y-intercept and another point on the graph

The form of the equation that we need to find is $y = a(b^x)$.

Worked examples

1. Determine the equation of the graph in the form $(x) = a \cdot b^x + q$.

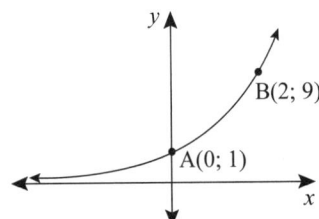

2. Now determine:

    2.1 $f(3)$.   **2.2** x if $f(x) = \frac{1}{3}$

    2.3 the average gradient of $f(x)$ between points A and B.

Solutions

1. The equation is in the form $f(x) = a \cdot b^x + q$, **and** is an increasing function.

    $q = 0$ because the x-axis is the horizontal asymptote.

    $a = 1$ because the y-intercept A is (0; 1).

    $y = 1 \cdot b^x + 0$ Substitute B(2; 9) to calculate b.

    $9 = b^2, b > 0$ This is an increasing function.

    $b = 3$

    $\therefore f(x) = 3^x$

2.   **2.1** $f(3) = 3^3 = 27$

    2.2 $3^x = \frac{1}{3} = 3^{-1}$

        $\therefore x = -1$

    2.3 Average gradient $= \frac{y_B - y_A}{x_B - x_A} = 4$

Type 2: Find equation given y-intercept, another point on graph, and horizontal asymptote

$y = a(b^x) + q$ is the form of the equation we want.

Worked examples

1. Determine the equation of the graph in the form $(x) = a \cdot b^x + q$.

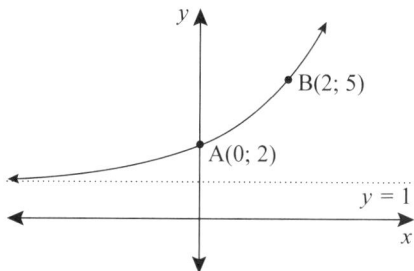

2. Determine:

   2.1 $f(-1)$ **2.2** x if $f(x) = 17$

   2.3 the average gradient of $f(x)$ between points A and B.

Solutions

1. The equation is in the form: $f(x) = a \cdot b^x + q$

$q = 1$	horizontal asymptote
$2 = a \cdot b^0 + 1$	Substitute A(0; 2) for $(x; y)$.
$a = 1$	
$y = 1 \cdot b^x + 1$	
$5 = b^2 + 1$	
$4 = b^2, b > 1$	From the shape of the graph, this is an increasing function.
$b = 2$	
$\therefore f(x) = 2^x + 1$	

2. **2.1** $f(-1) = 2^{-1} + 1 = \frac{1}{2} + 1 = \frac{3}{2}$

   2.2 $2^x + 1 = 17$

       $2^x = 16 = 2^4$ Make the bases the same.

       $\therefore x = 4$

   2.3 Average gradient $= \frac{y_B - y_A}{x_B - x_A} = \frac{5-2}{2-0} = \frac{3}{2}$

Exercise 9

In this exercise we will combine hyperbolic and exponential functions.

1. Write down the equation for each graph in the form $f(x) = a \cdot bx + q$ or $g(x) = \frac{a}{x} + q$.

   1.1 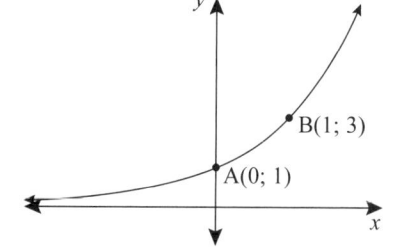   **1.2**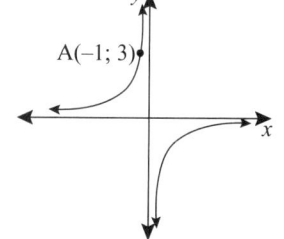

1.3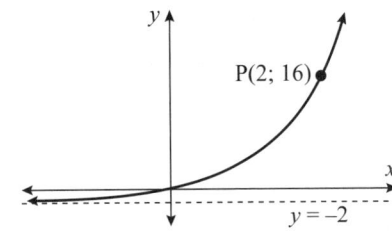

1.4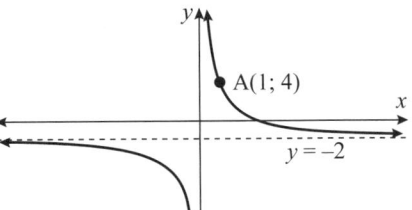

1.5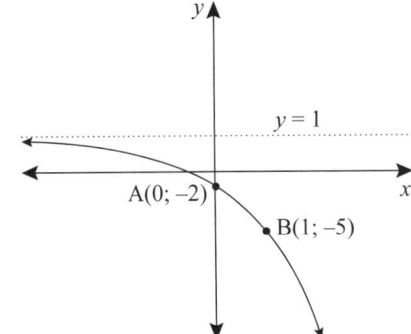

1.6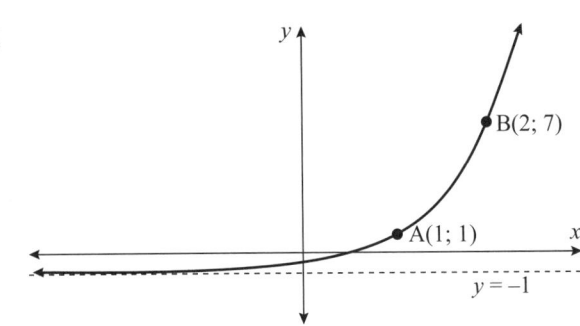

2. **2.1** Using the equation you have found for 1.1, determine:

    2.1.1 the average gradient between A and B

    2.1.2 the y value when $x = -2$.

  2.2 Using the equation found in 1.2, determine the following:

    2.2.1 the value of x for which $y = 6$, and the coordinates of point B

    2.2.2 the average gradient between A and B

    2.2.3 whether the function is increasing or decreasing between $x = -4$ and $x = -2$.

  2.3 Using the equation found in 1.3, determine:

    2.3.1 the average gradient between O and P

    2.3.2 the x-value of the function when $y = 4$

    2.3.3 the equations of the axes of symmetry.

  2.4 Using the equation found in 1.4, determine the following:

    2.4.1 the coordinates of point B, which is the x-intercept

    2.4.2 the average gradient between A and B

    2.4.3 the domain and range

    2.4.4 the values of x for which the function values are negative

    2.4.5 the equations of the axes of symmetry.

  2.5 Using the equation found in 1.5, determine:

    2.5.1 the average gradient between A and B

    2.5.2 the coordinates of the x-intercept

2.5.3 the y value of the function when $x = 3$

2.5.4 the x-value of the function when $y = -\frac{1}{2}$

2.6 Using the equation found in 1.6, determine:

2.6.1 the coordinates of the x and y-intercepts

2.6.2 the average gradient between A and B

2.6.3 the value of x for which $y = -\frac{7}{8}$.

3. The exponential function $h(x) = a \cdot 2^x + q$ has a horizontal asymptote at $y = 1$ and passes through P(0; –2). Determine the values of a and q.

4. The graphs of $f(x) = 2x^2 - 8$ and $g(x) = 2^x - 8$ are shown in the sketch.

   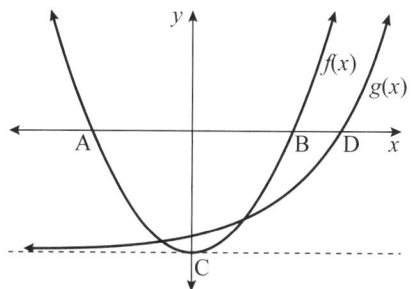

   4.1 Determine the coordinates of the points A, B, C and D.

   4.2 Determine the equation of the asymptote of $g(x)$.

   4.3 If $g(x)$ is reflected in the x-axis to obtain $h(x)$, write down the equation of $h(x)$.

Trigonometric functions

Sketching basic trigonometric graphs

To plot the points of basic trigonometric graphs, choose the table mode on a calculator, then:

- Type $f(x) =$ and your equation.

- Select [Start?], type 0°, select [End?], type 360°, select [Step?], type 90°.

Now you can copy down the table and then plot the coordinates of the points in the table.

Graph of $y = \sin x$

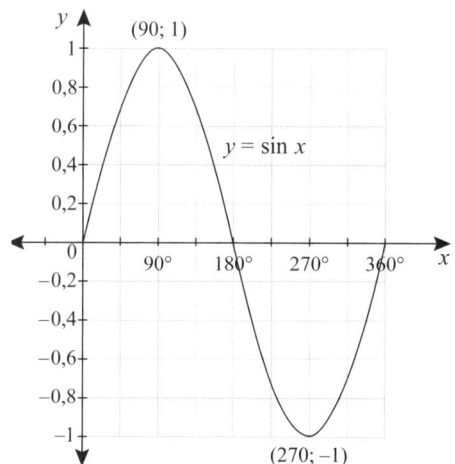

The properties of $y = \sin x, x \in [0°; 360°]$ are:

- Period: length of one pattern or wave: 360°
- Amplitude: half of the distance between the maximum and minimum points of the graph: 1 unit
- Range: permissible y values: $-1 \leq y \leq 1, y \in \mathbb{R}$
- Zero points: $x = 0°$ or 180° or 360°

Graph of $y = \cos x$

The properties of $y = \cos x$, $x \in [0°; 360°]$ are:

- Period: length of one pattern or wave: $360°$
- Amplitude: distance of max/min points from the axis: 1 unit.
- Range: permissible y values: $-1 \le y \le 1$, $y \in \mathbb{R}$
- Zero points: $x = 90°$ or $270°$

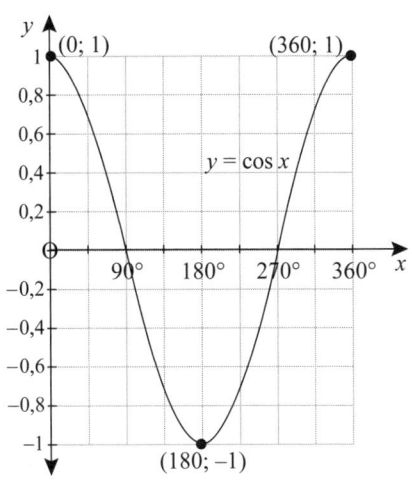

Graph of $y = \tan x$

The properties of $y = \tan x$, $x \in [0°; 360°]$ are:

- Period: length of one pattern or wave: $180°$
- Asymptotes: $x = 90°$ and $x = 270°$
- Domain: permissible x values: $x \in [0°; 90°) \cup (90°; 270°) \cup (270°; 360°]$
- Range: permissible y values: $y \in \mathbb{R}$
- Zero points: $x = 0°$ or $180°$ or $360°$

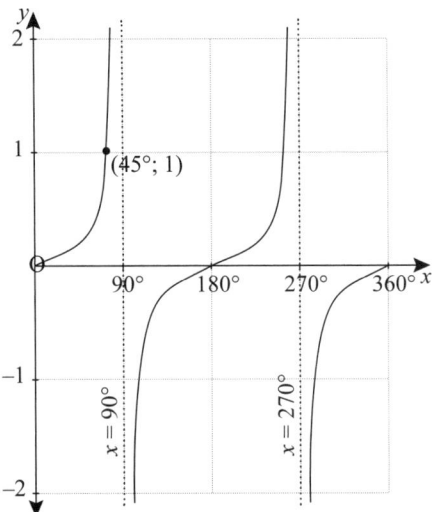

Effect of a and q on graphs of $y = a \sin x + q$, $y = a \cos x + q$ and $y = a \tan x + q$ for $x \in [0°; 360°]$

The **effect of a** is a **vertical stretch** of the basic graphs, in the following ways:

1. a changes the amplitude of the graph to the positive value of a.
2. The period remains the same.
3. The asymptotes remain the same.
4. The zero points remain the same.
5. If a is **negative**, the graph is **reflected in the x-axis**.

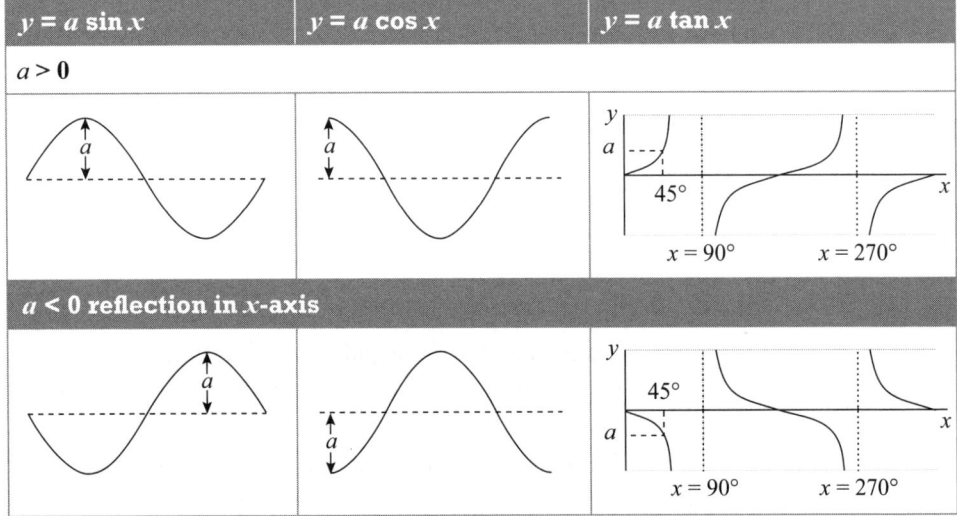

The **effect of q** is a **vertical shift** (up or down) of the basic graph, in the following ways:

1. If $q > 0$, the graph shifts upwards q units.

2. If $q < 0$, the graph shifts downwards q units.

3. The period remains the same.

4. The amplitude remains the same.

5. The asymptotes remain the same.

6. The zero points will change and need to be calculated by making $y = 0$.

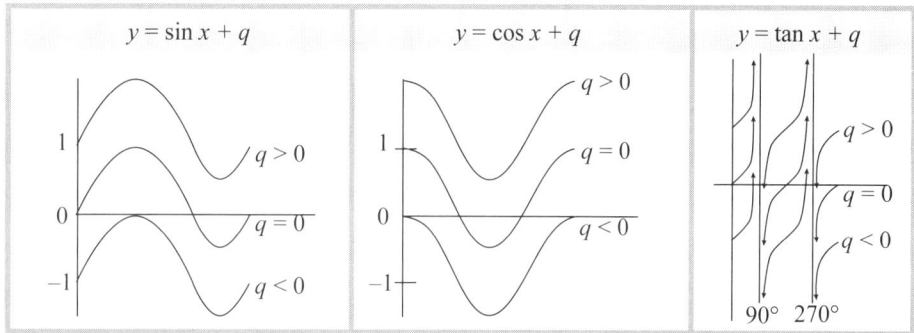

Exercise 10

1. Draw each **set** of graphs on its own set of axes.

Set 1	Set 2	Set 3
$y = \sin x$	$y = \cos x$	$y = \tan x$
$y = 3 \sin x$	$y = 3 \cos x$	$y = 3 \tan x$
$y = \frac{1}{2} \sin x$	$y = \frac{1}{2} \cos x$	$y = \frac{1}{2} \tan x$
$y = -2 \sin x$	$y = -2 \cos x$	$y = -2 \tan x$

2. Draw each set of graphs on its own set of axes.

Set 1	Set 2	Set 3
$y = \sin x$	$y = \cos x$	$y = \tan x$
$y = \sin x + 1$	$y = \cos x + 1$	$y = \tan x + 1$
$y = \sin x - 2$	$y = \cos x - 2$	$y = \tan x - 2$
$y = -\sin x + 1$	$y = -\cos x + 1$	$y = -\tan x - 1$

Finding the equation of the function from a sketch

Worked examples

1. The figure alongside represents the graph of $y = a \sin x$, $x \in [0°; 360°]$ with A(90°; 2).

   Determine the equation, amplitude and period of the graph.

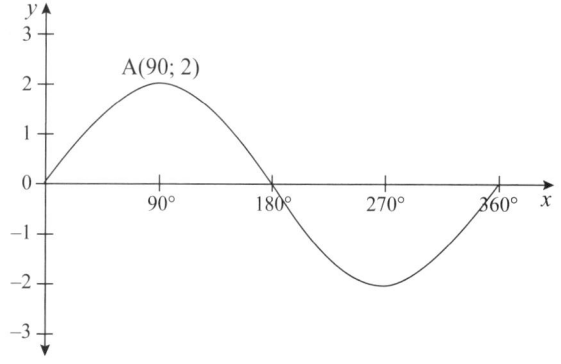

2. The graph below represents the function of $y = a \tan x$, $x \in [0°; 360°]$ with P(45°; 0,5). Determine the equation of the graph, the amplitude and period of the graph.

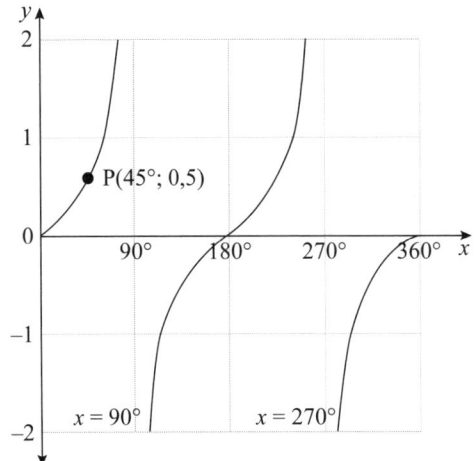

Solutions

1. The value of a is the amplitude of the graph, so the amplitude is 2.

   Equation: $y = 2 \sin x$

   Period: 360°

2. The value of a is the y value when $x = 45°$ on the graph of $y = a \tan x$, so the value of $a = 0,5$.

   Equation: $y = 0,5 \tan x$

   Period = 180°

Exercise 11

1. The graph of $y = 3 \sin x$, $x \in [0°; 360°]$ is shown. Use the graph to write down the following:

   1.1 coordinates of A, B, C and D

   1.2 the period

   1.3 the amplitude

   1.4 the range

   1.5 the turning points

   1.6 the zero points of the function

   1.7 the value of x for which $3 \sin x > 0$.

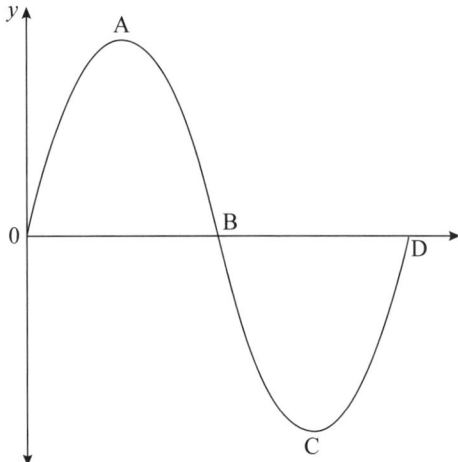

2. The graph shows the functions $y = \cos x$ and $y = a \sin x$ where $x \in [0°; 360°]$.

2.1 Write down the value of a.

2.2 Determine the coordinates of A, B, C, D and E.

2.3 What is the amplitude of the function $y = a \sin x$?

2.4 Write down the zero points of $y = a \sin x$.

2.5 For which values of x is $\cos x$ negative?

2.6 Use the graph to determine how many solutions the equation $\cos x = a \sin x$ has in the interval $x \in [0°; 360°]$.

3. On the same system of axes sketch graphs of $y = 2 \cos x$ and $y = \tan x$ for the interval $[0°; 360°]$. Indicate the x- and y-intercepts of both graphs with the axes as well as the asymptotes of $y = \tan x$. Using your sketch graphs, answer the following questions.

3.1 For which value of x will $2 \cos x$ increase in value as x increases?

3.2 What is the minimum value of $2 \cos x$?

3.3 What are the equations of the asymptotes of $y = \tan x$?

3.4 What is the period of $y = \tan x$?

3.5 Indicate on the x-axis of your graph where to find the solutions of the equation $\tan x = 2 \cos x$. (Use A, B, C, etc.)

3.6 Determine any two of the solutions to $2 \cos x - \tan x = 2$.

4. The graphs of $f(x) = \tan x$ and $g(x) = a \cos x$, $x \in [0°; 360°]$ are shown in the diagram.

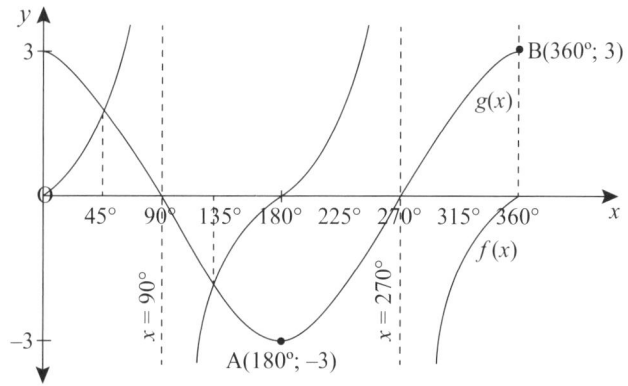

4.1 Write down the value of a in $g(x) = a \cos x$.

4.2 Determine each of the following:

4.2.1 amplitude of $g(x)$

4.2.2 period of $f(x)$

4.2.3 maximum value of $g(x)$

4.2.4 number of solutions to the equation $\tan x = a \cos x$, $x \in [0°; 360°]$.

4.3 Use the graph to write down one value of x for which $g(x) - f(x) = 3$.

4.4 Use the graph to determine the value/s of x in the indicated interval, for which:

4.4.1 $g(x) = -3$

4.4.2 $f(x) = \tan x$ is not defined

4.4.3 $f(x)$ increases as $g(x)$ decreases

4.4.4 $f(x) = 0$, i.e. the zero points of $f(x)$.

Test A: Knowledge and routine procedures

1. **1.1** Change the equation $3x - 2y = 5$ to the standard form $y = ax + q$. (1)

    1.2 Determine the x and y intercepts of this function. (2)

2. Given: $4x - 3y - 12 = 0$. Determine:

    2.1 y-intercept (2)

    2.2 x-intercept (2)

    2.3 gradient of the line. (2)

3. Draw each graph on its own sets of axes. Label important properties of each function.

    3.1 $y = -3x + 1$ (2)

    3.2 $y = \frac{2}{x} + 1$ (3)

    3.3 $x = 1$ (2)

    3.4 $y = 2x^2 - 2$ (3)

4. Determine the equations of the following graphs:

    4.1

    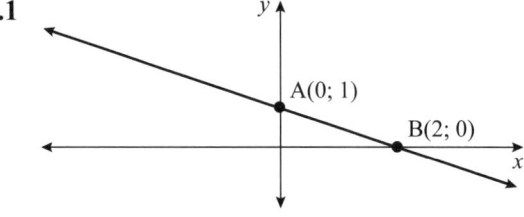

    (2)

    4.2

    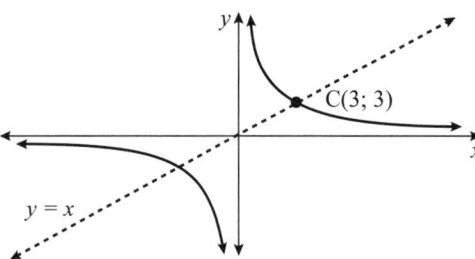

    (2)

    4.3

    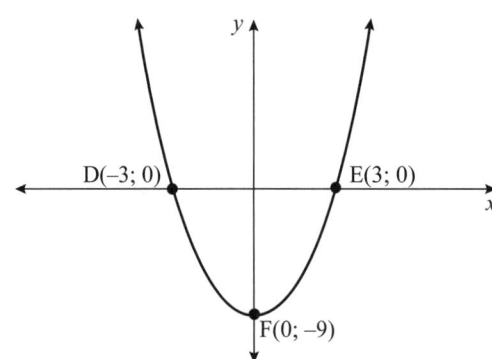

    (4)

    4.4

    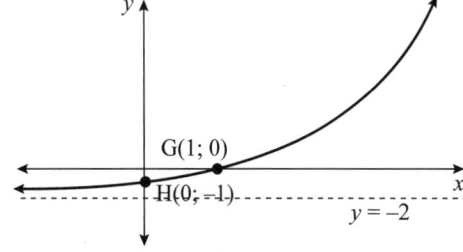

    (6)

5. Determine the gradient of the line passing through the points $(-3; 5)$ and $(1; -3)$. (2)

6. Given: $f(x) = -3x + 6$ and $g(x) = x + 2$.

   6.1 Determine the points of intersection of the two graphs, i.e. solve for x if $f(x) = g(x)$. (3)

   6.2 Draw the graphs of f and g indicating the points of intersection with the axes. (6)

   6.3 Determine the equation of the line parallel to $f(x)$ passing through the origin. (2)

   6.4 Determine the equation of the line perpendicular to $g(x)$ passing through the origin. (2)

   6.5 Determine $f(5)$. (2)

   Total 50

Test B: Complex procedures and problem solving

1. Given $f(x) = -x + 2$ and $y = \frac{-3}{x}$, determine:

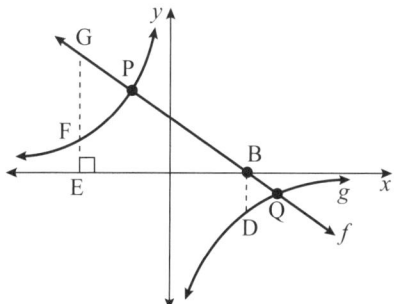

   1.1 the coordinates of P and Q (6)

   1.2 the coordinates of B (2)

   1.3 the distance BD (2)

   1.4 the length of GF if E $(-6;0)$ (3)

2. Given: $f(x) = \frac{4}{x} - 1$, and $g(x)$, which is the axis of symmetry of $f(x)$ with a positive gradient.

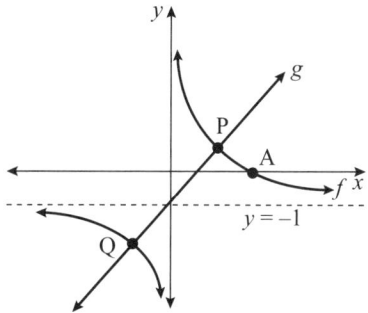

   2.1 Determine the coordinates of A. (2)

   2.2 Determine the equation of $g(x)$. (2)

   2.3 Calculate the coordinates of P and Q, the points of intersection of $f(x)$ and $g(x)$. (5)

   2.4 Write down the equation of $h(x)$ if $h(x)$ is a reflection of $f(x)$ in the x-axis. (2)

3. The graphs of $y = -x + 4$ and $2y - x - 2 = 0$ are shown in the diagram. Determine the lengths of:

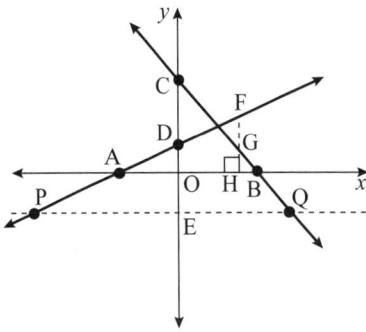

3.1 AB   (2)

3.2 CD   (2)

3.3 PQ if OE = 1 unit   (4)

3.4 FG if OH = 3 units.   (4)

4. The graphs of f, g, h and m in the sketch represent the graphs of a hyperbola, straight line, semi-circle and a parabola respectively.

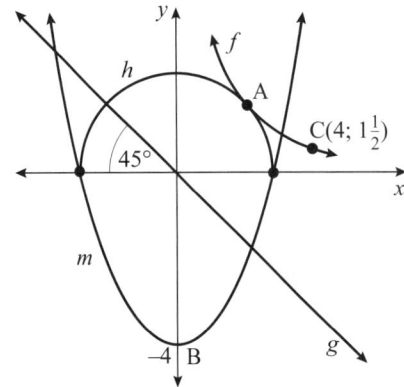

4.1 Determine the equations of f, g h and m.   (10)

4.2 Give the coordinates of A, the point of intersection with the axis of symmetry.   (2)

4.3 Determine the equation of the line joining B and C.   (3)

Total 50

Test C: Content and breakdown as for exam

1. The following are given: $f(x) = 3x + 9$ and $g(x) = -x^2 + 9$.

   1.1 Sketch graphs of f and g on the same set of axes. Indicate the intercepts with the axes.   (6)

   1.2 Determine the equation of h, the straight line that is perpendicular to the graph of f and passes through the point $(0; -6)$.   (2)

   1.3 Determine algebraically the coordinates of the intercept of the graphs of f and h.   (5)

   1.4 Give the range of g.   (2)

   1.5 For which values of x is g an increasing function?   (2)

   1.6 Use the graphs and determine the value(s) of x for which the following is true: $g(x) \geq f(x)$.   (2)

2. The graphs of $f(x) = 2x + 3$ and $g(x) = -x + 6$ are given below.

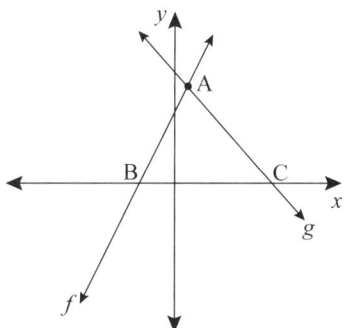

2.1 Determine the coordinate of A. (4)

2.2 Determine the values of x for which the following are true.

    2.2.1 $f(x) > 0$ (2)

    2.2.2 $g(x) \leq 0$ (2)

2.3 Sketch the graph of $y = 2 \cdot 3^x - 6$. Indicate clearly any points of intersection with the axes and the equation of the asymptote. (4)

3. Below are the graphs of parabolas f and g. The graph of g represents the graph of f shifted down $4\frac{1}{4}$ units. A is the point $(4; 6)$.

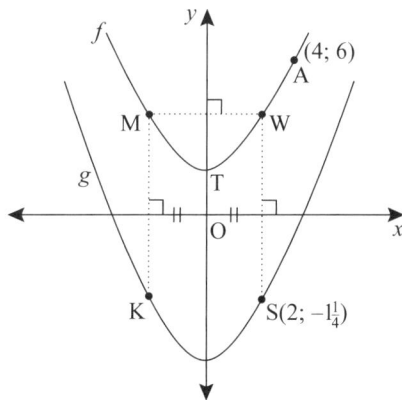

Determine:

3.1 the equation of f (7)

3.2 the equation of g (4)

3.3 the coordinates of point M (2)

3.4 the distance MW (2)

3.5 the distance OT (1)

3.6 the value of $f(4) - g(4)$ (1)

3.7 the range of g. (2)

Total 50

TOPIC 8: Euclidean geometry

Note that for use in schools, there are enough exercises and tests within this Topic to be used both for the three weeks in Term 2 and for the one week in Term 3.

In Grade 10 you will revise the geometry you studied in Grades 8 and 9 regarding lines, angles and triangles in order to use these concepts and ideas in quadrilaterals. In Grade 10 you will study special quadrilaterals by defining each shape and learning about and proving the different properties for each of the shapes. The quadrilaterals you will learn about are the kite, parallelogram, rectangle, rhombus, square and trapezium. You will learn how to work with line segments joining the midpoint of two sides of a triangle. You will then use this information to solve geometry riders (questions involving shapes, angles and lines).

Knowledge and skills for this topic

If you struggle with any of the work listed below, revise it before continuing with this Topic:

*   validating geometry concepts with reasons
*   working with angles related to lines and triangles
*   recognising and classifying different types of triangles, and indicating their properties from a given sketch
*   using the theorem of Pythagoras
*   solving problems related to shapes using congruence and similarity
*   setting up basic equations and solving for the variable.

Content of final exam

*   Given a diagram of triangles or quadrilaterals and triangles containing some measurements, find the size of angles or lengths of sides using:
    *   geometry axioms and theorems on lines and triangles
    *   definitions and theorems of quadrilaterals.
*   Solve problems and prove geometry riders using the properties of parallel lines, triangles and quadrilaterals.

Angles and lines: revision

In geometry we work with shapes. Every shape consists of angles and lines. A good tip which works 99% of the time is as follows:

*   When **proving facts related to angles**, use **algebra** by introducing a variable.
*   When **proving facts related to lines** use **congruence or similarity**.

In Grade 8 you were taught how to work with angles within lines and different shapes. In a sketch some angles were given and others contained variables, which you solved for, and then used the answer to find other angles.

In Grade 9 you were taught how to work with line segments within triangles and quadrilaterals, using congruence and similarity to prove them equal or in proportion.

Lines and triangles: revision

You must know all the theorems given below and be able to use them. For each theorem we supply a sketch, make a statement and give a reason for the statement. The theorems you need to learn are given as full proofs in this section.

When giving a reason, always make sure you can answer two questions within your given reason:

1. Which theorem am I using?

2. Where is it being used?

Theorems on lines

Here are a few of the basic theorems you may remember from previous work.

- The angles on a straight line are supplementary. Their sum is 180°.

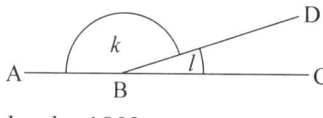

$k + l = 180°$ straight line ABC

- The angles around a point add up to 360°.

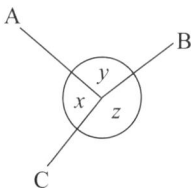

$x + y + z = 360°$ angle of revolution

- Vertically opposite angles are equal.

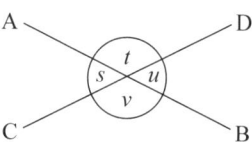

$s = u$ vertically opposite angles

$t = v$ vertically opposite angles

Parallel lines

Parallel lines have a number of specific properties.

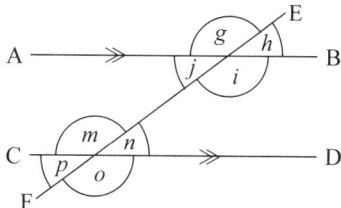

- Corresponding angles on parallel lines are equal.

   For example:

   $o = i$ corresponding angles, AB ∥ CD

   $m = g$ corresponding angles, AB ∥ CD

- Alternate angles between parallel lines are equal.

  For example:

  $j = n$ alternate angles, AB ∥ CD

  $i = m$ alternate angles, AB ∥ CD

- Co-interior angles between parallel lines are supplementary.

  $m + j = 180°$ co-interior angles, AB ∥ CD

  $i + n = 180°$ co-interior angles, AB ∥ CD

How to remember parallel line theorems

You can use the following diagram to help you remember the parallel line theorems.

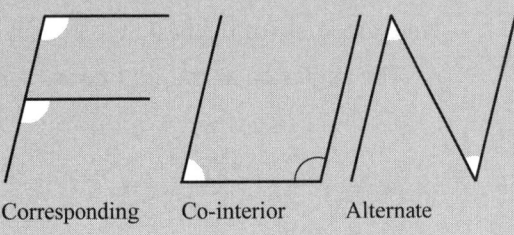

Corresponding   Co-interior   Alternate

The converses of these theorems are also true.

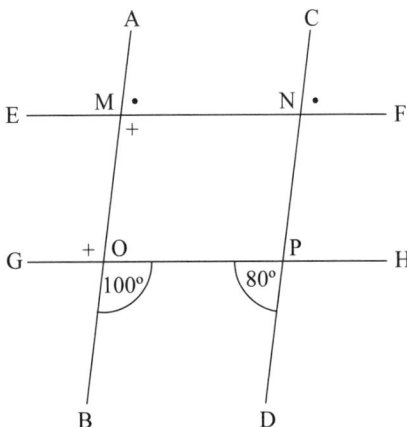

- If corresponding angles are equal, then they are on parallel lines.

  For example:

  AB ∥ CD   AM̂N equals corresponding CN̂F

- If alternate angles are equal, then they are on parallel lines.

  EF ∥ GH   GÔM equals alternate OM̂N

- If co-interior angles are supplementary, then they are between parallel lines.

  AB ∥ CD   BÔP + co-interior OP̂D are supplementary

Theorems on triangles

The five most important triangle theorems are explained briefly below.

- The sum of the interior angles of a triangle is 180°.

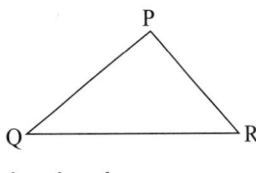

  $\hat{P} + \hat{Q} + \hat{R} = 180°$ sum of angles of △PQR

- The exterior angle of a triangle is equal to the sum of the two interior opposite angles.

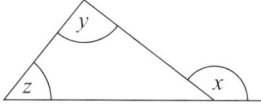

$x + y = z$ exterior angle of a triangle equals sum of interior opposite angles

- The angles opposite the equal sides of an isosceles triangle are equal.

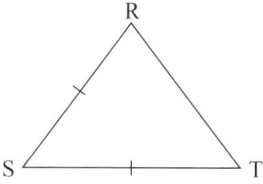

$\hat{R} = \hat{T}$ angles opposite equal sides, isosceles triangle

The converse is also true: if two angles of a triangle are equal, then it is an isosceles triangle.

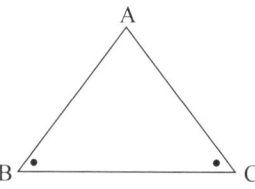

$AB = AC$ sides opposite equal angles

- The angles of an equilateral triangle are all 60°.

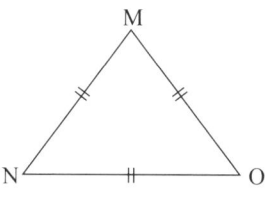

$\hat{M} = \hat{N} = \hat{O} = 60°$ angles in an equilateral triangle

- The theorem of Pythagoras.

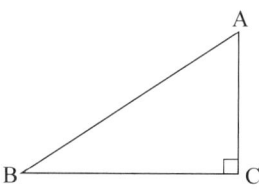

$AB^2 = AC^2 + BC^2$ theorem of Pythagoras

The converse of the theorem of Pythagoras is also true: if the theorem of Pythagoras holds true, the triangle is a right-angled triangle.

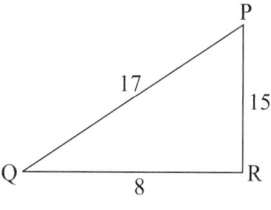

$\hat{R} = 90°$

$$PQ^2 = 17^2 \qquad PR^2 + QR^2 = 15^2 + 8^2$$
$$= 289 \qquad\qquad = 225 + 64$$
$$= 289$$

converse of Pythagoras: $PQ^2 = PR^2 + QR^2$

Worked examples

1. ABC ‖ EFG and BF and CE intersect at D.
   Calculate angles a, b, c, d, e and f.

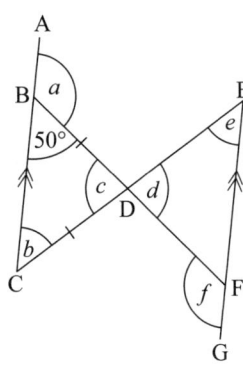

2. In △ABC, $\hat{B} = 15°$.
   AD bisects BÂC and AD = AC.
   Calculate:

   2.1 BÂD

   2.2 $\hat{C}$

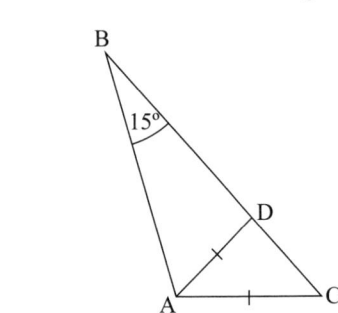

3. In △ABC on the right, $\hat{A} = 90°$
   and AB = AC. XY is a line that
   passes through A, and BH and
   CK are perpendicular to XY.
   Prove that AH = CK.

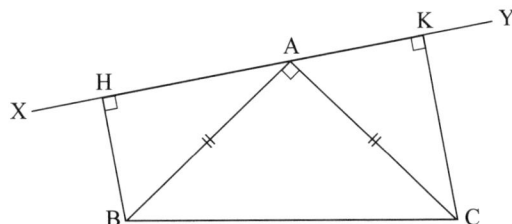

4. In the figure, ST = 150 mm,
   MT = 120 mm and SM = 90 mm.

   4.1 Prove that $\hat{M} = 90°$.

   4.2 P is a point on TM such that $\hat{T} = P\hat{S}T$.
   Calculate the length of PS, correct
   to two decimal places.

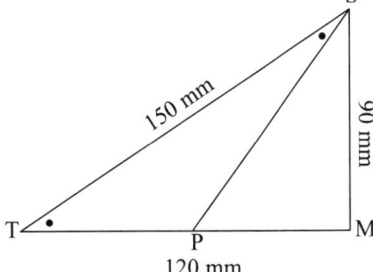

Solutions

1. $a + 50° = 180°$ straight line ABC

   $\therefore a = 130°$

   $b = 50°$ angles opposite equal sides, BD = DC

   $c + b + 50° = 180°$ sum of interior angles of △BCD

   $\therefore c = 80°$

   $\therefore d = 80°$ vertically opposite angles

   $e = b = 50°$ alternate angles, BC ‖ EF

   $f = d + e$ exterior angle of △DEF equals sum of two
   interior opposite angles

   $\therefore f = 130°$

2. **2.1** Let $D\hat{A}C = D\hat{A}B = x$ DA bisects $B\hat{A}C$, given

$A\hat{D}C = \hat{B} + B\hat{A}D$ exterior angle of $\triangle BAD$ equals sum of interior opposite angles $= 15° + x$

$\hat{C} = A\hat{D}C$ angles opposite equal sides, AD = AC

$\quad = 15° + x$

$15° + x + x + 15° + x = 180°$ sum of interior angles of $\triangle ABC$

$\therefore 3x = 150°$

$\therefore x = 50°$

$\therefore B\hat{A}D = x$

$\qquad = 50°$

2.2 $\hat{C} = 15° + x$

$\quad = 65°$

3. In $\triangle HBA$ and $\triangle KAC$:

$AB = CA$ given

$\hat{H} = \hat{K} = 90°$ given

Let $H\hat{A}B = x$

$\therefore x + 90° + H\hat{B}A = 180°$ sum of angles of $\triangle HBA$

$\therefore H\hat{B}A = 90° - x$

$x + 90° + C\hat{A}K = 180°$ straight line HAK

$\therefore C\hat{A}K = 90° - x$

$\therefore H\hat{B}A = C\hat{A}K$ sum of interior angles of $\triangle AKC$

$\therefore \triangle HBA \equiv \triangle KAC$ AAS

$\therefore AH = CK$ $\triangle HBA \equiv \triangle KAC$

4. **4.1** $TS^2 = 150^2$

$\qquad = 22\ 500$

$TM^2 + SM^2 = 120^2 + 90^2$

$\qquad\qquad = 14\ 400 + 8\ 100$

$\qquad\qquad = 22\ 500$

$\therefore \hat{M} = 90°$ converse of Pythagoras: $TS^2 = TM^2 + SM^2$

4.2 $SP = PT$ sides opposite equal angles, $\hat{T} = T\hat{S}P$

$PM = TM - TP$

$\qquad = 120 - PS$ TP = PS

$PS^2 = SM^2 + PM^2$ Pythagoras

$PS^2 = 90^2 + (120 - PS)^2$

$PS^2 = 8\ 100 + 14\ 400 - 240PS + PS^2$

$\therefore 240PS = 22\ 500$

$\therefore PS = 93{,}75$ mm

Exercise 1

Determine the magnitude (size) of the angles marked with letters. Only write down the answers.

1.

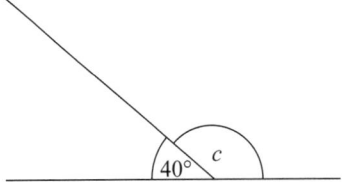

2.

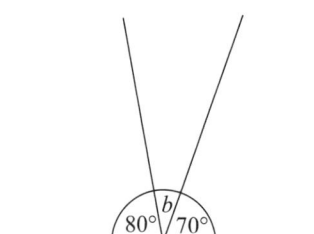

3.

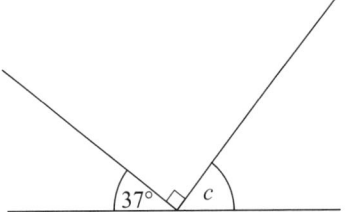

4.

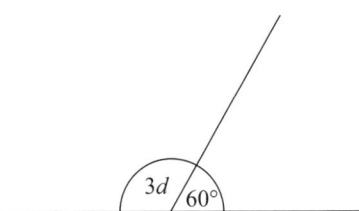

5.

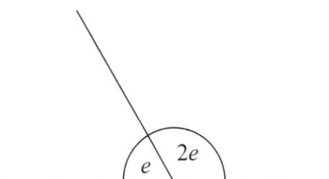

6.

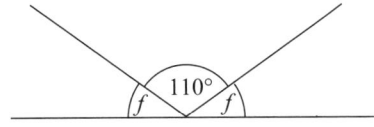

7.

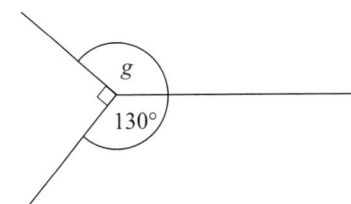

8.

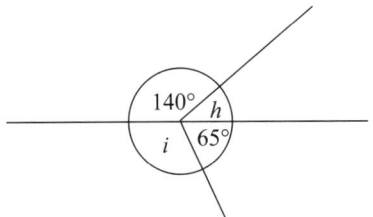

9.

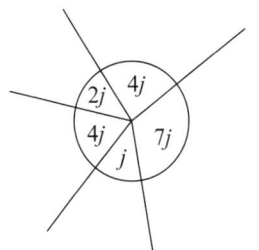

10.

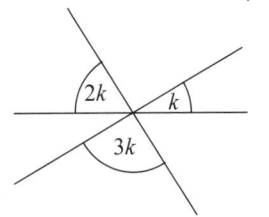

11.

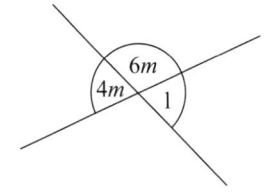

12.

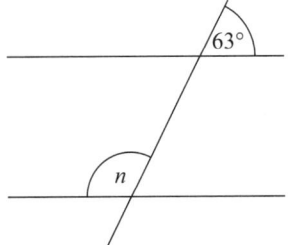

13.

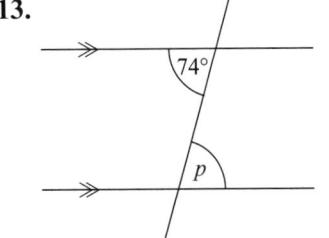

14.

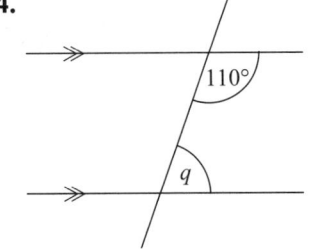

15.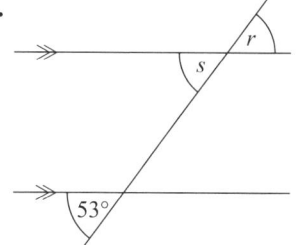

16.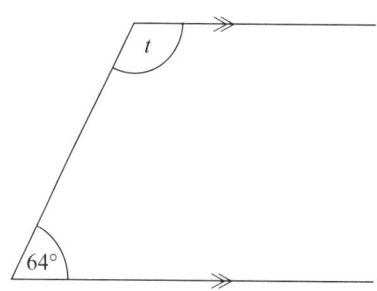

17.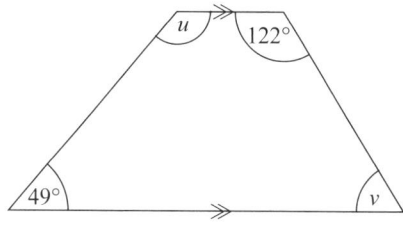

18.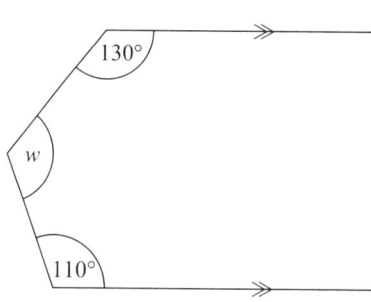

19.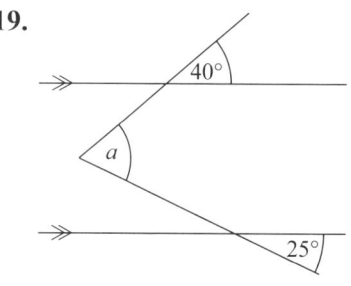

20.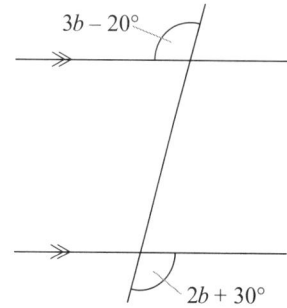

21.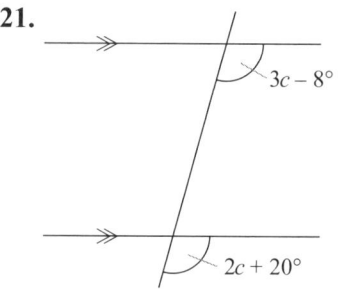

22.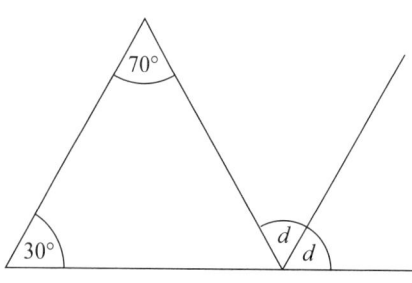

23.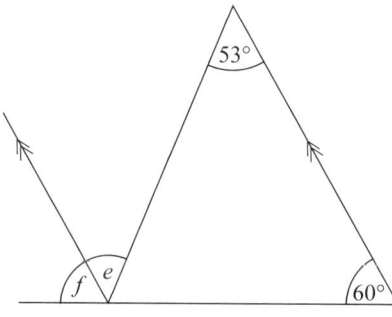

24.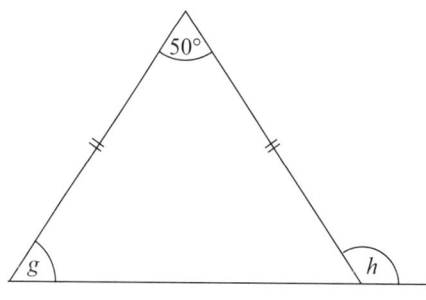

25.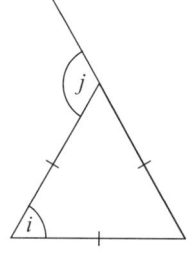

26.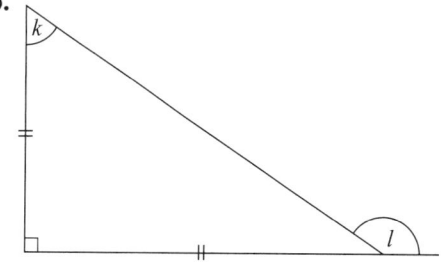

27.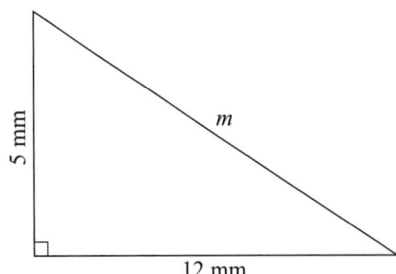

28.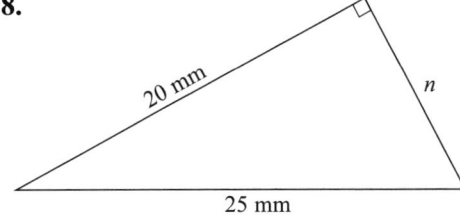

29.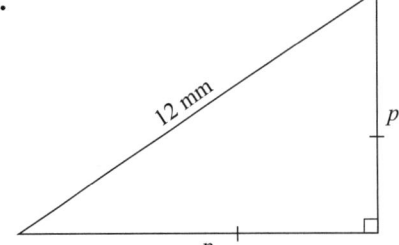

30.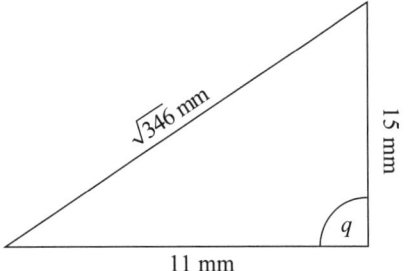

Exercise 2

1. Determine the value of x in each diagram. Give complete statements and reasons.

   1.1 $BC = AC = AD$ and $BD = DC$.

   1.2 $AB = AC$, $DC = BC$ and $D\hat{C}A = 12°$.

   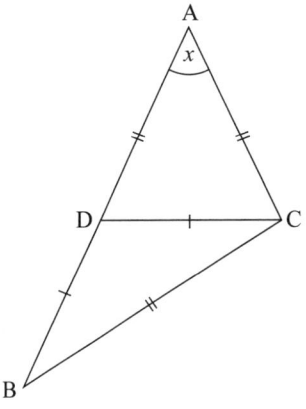

   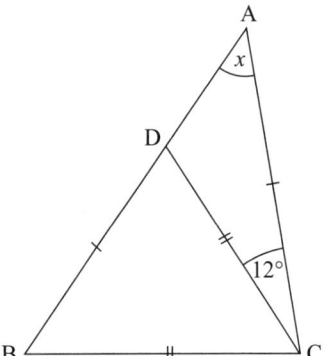

   1.3 $D\hat{B}A = x + 80°$, $\hat{A} = 70°$ and $A\hat{C}E = 2x + 20°$.

   1.4 $AD \parallel BC$, AK bisects $B\hat{A}D$ and BK bisects $A\hat{B}C$.

   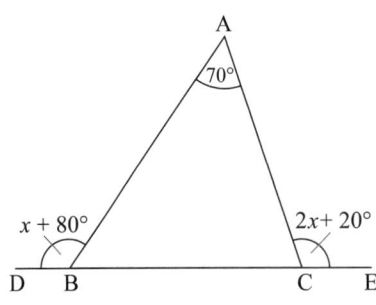

   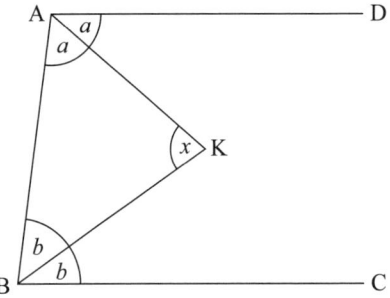

   1.5 $CA \parallel ED$,
   $\hat{B} = 55°$,
   $C\hat{A}D = x + 15°$
   and $C\hat{E}D = 110° - x$.

   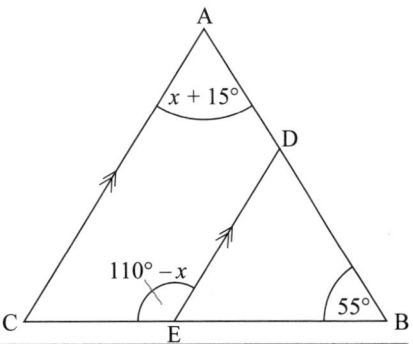

1.6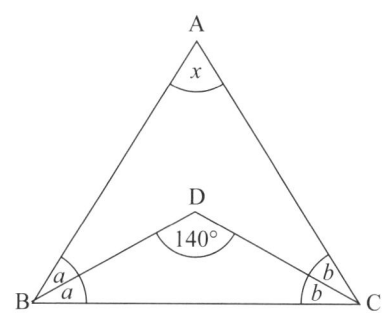

1.7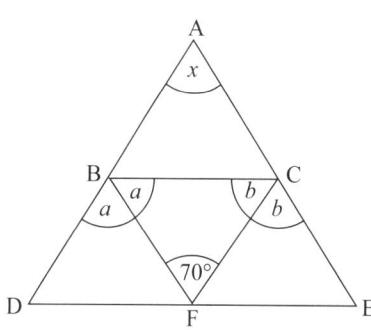

2. In △ABC, BÂC = 70° and AB̂C = AĈB.
   BC is produced to D.
   CE bisects AĈD. AE ∥ BD.
   Calculate:

   2.1 AĈE

   2.2 AÊC

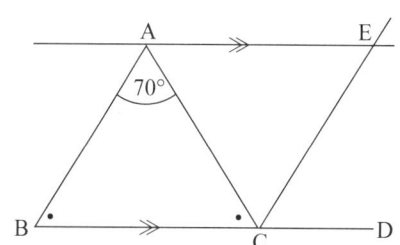

3. In △ABC, Â = 2B̂ = 2Ĉ. Calculate Â, B̂ and Ĉ.

4. In △PQR below, QR is produced to S. If PR̂S = 2P̂, prove P̂ = Q̂.

   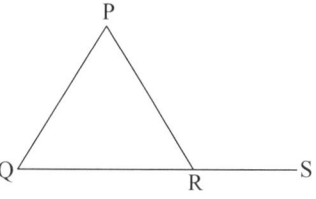

5. In △ABC, BA is produced to D. If AB = AC and AE bisects DÂC, prove AE ∥ BC.

   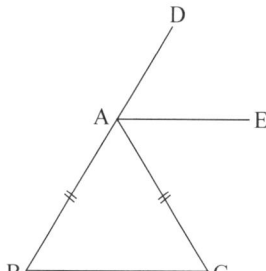

6. D is a point on BC of △ABC such that BD = DC and AE ⊥ BC.

   6.1 Write AB² in terms of BD, AD and DE.

   6.2 Write AC² in terms of AE, CD and DE.

   6.3 Hence, prove that
   AB² + AC² = 2AD² + 2BD².

   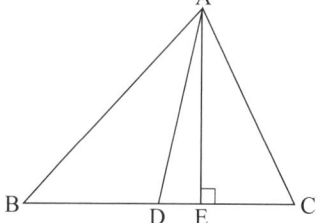

Congruence: revision

Two shapes are congruent when they have exactly the same shape and size – in other words they are identical. There are four ways of proving that two triangles are congruent. You can prove that:

1. the three sides are equal (SSS)

2. two sides and the included angle are equal (SAS)

3. an angle of 90°, the hypotenuse and one other side are equal (RHS)

4. two angles and one side are equal (AAS).

Two sides and a non-included angle (SSA) is *not* a case for congruency.
Three angles (AAA) is also not a case for congruency.

> **Abbreviations commonly used in proofs:**
>
> S – side
> A – angle
> R – right angle
> H – hypotenuse
>
> The symbol ≡ is used to show congruency.

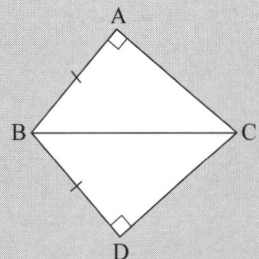

How to prove two triangles are congruent

The method you use to prove the congruency of triangles depends on the case of congruency you decide to use. Note the layout used to prove that triangles ABC and DBC alongside are congruent.

In △ABC and △DBC:

AB = DB	given sides are equal
BÂC = BD̂C = 90°	given; right angles
BC = BC	common side; hypotenuse
∴ △ABC ≡ △DBC	RHS

Because we have proved that the triangles are congruent, we can now say AC = DC; AB̂C = CB̂D and AĈB = DĈB.

Rules to follow when proving that triangles are congruent

1. State which triangles are being used.

2. You can write the numbers 1 to 3 along the left hand side below your statement because three things need to be proved equal to prove congruency.

3. Name the triangles, with the letters in the correct order, that you have proved congruent.

4. Always state the case of congruency you have used, e.g. AAS or SSS.

Worked example

Use the diagram provided to prove that △ADE ≡ △BEC.

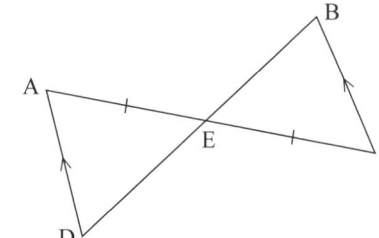

Solution

Proof: In △ADE and △BEC:

1.	Â = Ĉ	alternate ∠s, AD ∥ BC
2.	D̂ = B̂	alternate ∠s, AD ∥ BC
3.	AE = EC	given
	∴ △ADE ≡ △CBE	AAS

We will cover more on proving congruent triangles in the section of this Topic on quadrilaterals.

Quadrilaterals

Basic theorems of quadrilaterals

The following two basic theorems apply to all quadrilaterals.

• The sum of the interior angles of a quadrilateral is 360°.

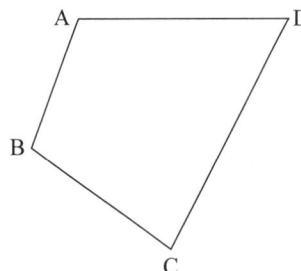

$$Â + B̂ + Ĉ + D̂ = 360°$$ sum of angles of quadrilateral ABCD

- The sum of the exterior angles of a quadrilateral is 360°.

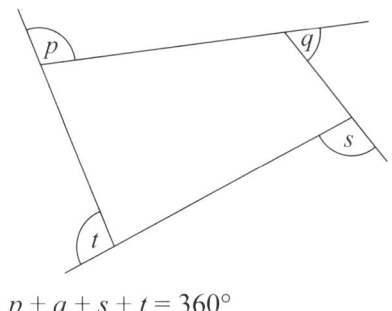

$p + q + s + t = 360°$ sum of exterior angles of quadrilateral PQST

Properties of special quadrilaterals

The table gives the definitions and properties of a few special quadrilaterals.

Quadrilateral	Definition	Properties (alternative definition)
Square	A square is a rectangle with two adjacent sides equal.	As for a rectangle: • both pairs of opposite sides are parallel • diagonals bisect each other • diagonals bisect the area. Also: • All angles are 90°. • Diagonals are equal. (PR = QS) • All the sides are equal. • Diagonals bisect each other at right angles. • Diagonals bisect the angles at 45°.
Rhombus	A rhombus is a parallelogram with two adjacent sides equal.	As for a parallelogram: • both pairs of opposite sides are parallel • both pairs of opposite angles are equal • diagonals bisect each other • diagonals bisect the area. Also: • All sides are equal. • Diagonals bisect each other at right angles. • Diagonals bisect the angles.
Rectangle	A rectangle is a parallelogram with angles of 90°.	As for a parallelogram: • both pairs of opposite sides are equal • both pairs of opposite sides are parallel • diagonals bisect each other • diagonals bisect the area. Also: • All angles are 90°. • Diagonals are equal (PR = QS).
Parallelogram	A parallelogram is a quadrilateral with both pairs of opposite sides parallel.	• Both pairs of opposite sides are equal. • Both pairs of opposite angles are equal. • Diagonals bisect each other (PO = OR and QO = OS). • Diagonals bisect the area.

Quadrilateral	Definition	Properties (alternative definition)
Trapezium ![trapezium ABCD]	A trapezium is a quadrilateral with one pair of opposite sides parallel.	• Two pairs of angles are supplementary. $\hat{A} + \hat{B} = 180°$ and $\hat{D} + \hat{C} = 180°$ Co-interior angles, AD ∥ BC.
Kite ![kite diagrams]	A kite is a quadrilateral with two pairs of adjacent sides equal.	• Diagonals intersect at right angles. • One diagonal bisects the other. • One diagonal bisects the angles.

Proving two theorems on parallelograms

Theorem 1: The opposite sides and angles of a parallelogram are equal

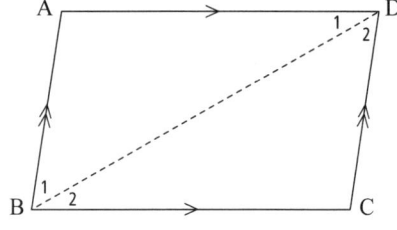

Given: Parallelogram ABCD.
From definition AD ∥ BC and AB ∥ DC.

To prove: AB = DC, AD = BC, $\hat{A} = \hat{C}$ and $A\hat{B}C = A\hat{D}C$

Construction: Diagonal BD

Proof:

In △ABD and △CDB:

(a) BD = DB   common
(b) $\hat{D}_1 = \hat{B}_2$ alt. ∠s, AD ∥ BC
(c) $\hat{B}_1 = \hat{D}_2$ alt. ∠s, AB ∥ DC

 ∴ △ABC ≡ △CDB   AAS

 ∴ AD = BC, AB = DC and $\hat{A} = \hat{C}$

 $\hat{B}_1 + \hat{B}_2 = \hat{D}_1 + \hat{D}_2$ proved

 ∴ $A\hat{B}C = A\hat{D}C$

Theorem 2: The diagonals of a parallelogram bisect one another

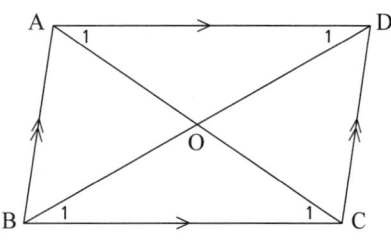

Given: Parallelogram ABCD.
From definition AB ∥ DC and AD ∥ BC.

From theorem 1:
AB = DC and AD = BC

To prove: AO = OC and BO = OD

Proof:

In $\triangle AOD$ and $\triangle COB$:

(a) $\hat{A}_1 = \hat{C}_1$ alt. $\angle$s, AD $\parallel$ BC

(b) $\hat{B}_1 = \hat{D}_1$ alt. $\angle$s, AD $\parallel$ BC

(c) $AD = BC$ opp. sides parm are equal

      $\therefore \triangle AOD \equiv \triangle COB$ AAS

      $\therefore AO = OC$ and $BO = OD$

Remember

You can abbreviate 'parallelogram' as 'parm'.

Sufficient conditions

When you are working with proofs in geometry, it is important to know what the sufficient conditions are for each shape. These are the conditions that give you enough information to prove that a shape is a square, trapezium, etc.

For example, a shape that has a pair of parallel sides is not necessarily a parallelogram. You would need to prove that the other two sides are also parallel, or that the parallel sides are equal in length.

The table below gives the sufficient conditions for the quadrilaterals.

Parallelogram Quadrilateral ABCD is a parallelogram if we can prove that: • both pairs of opposite sides are parallel, or • both pairs of opposite sides are equal, or • both pairs of opposite angles are equal, or • the diagonals bisect one another, or • one pair of opposite sides is equal and parallel.	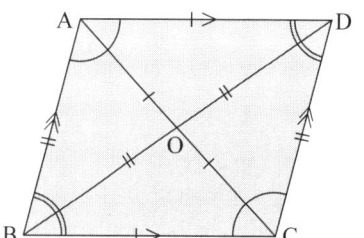
Rectangle Quadrilateral ABCD is a rectangle if we can prove that: • it is a parallelogram with an interior angle of 90°, or • it is a quadrilateral with four right angles.	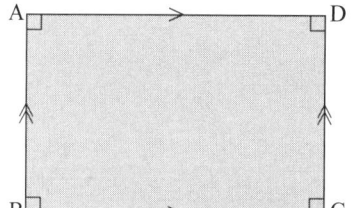
Rhombus Quadrilateral ABCD is a rhombus if we can prove that: • it is a parallelogram with one pair of adjacent sides equal, or • it is a quadrilateral with all four sides equal, or • the diagonals of the parallelogram bisect at 90°.	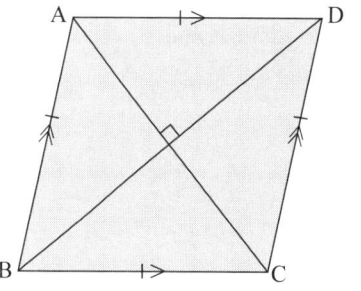
Square Quadrilateral ABCD is a square if we can prove that: • it is a rectangle (a parallelogram with a right angle) and one pair of adjacent sides equal, or • it is a rhombus (a parallelogram with one pair of adjacent sides equal) and an interior angle of 90°, or • it is a quadrilateral with all four sides equal and four right angles.	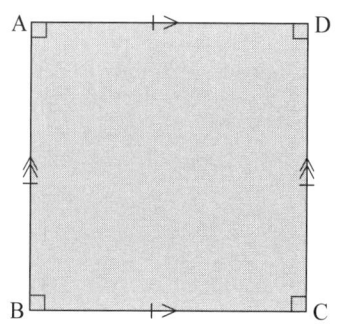

Kite Quadrilateral ABCD is a kite if we can prove that: • two pairs of adjacent sides are equal, or • one diagonal bisects the other diagonal, or • one diagonal bisects the angles.	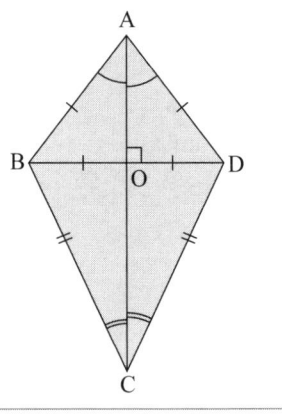
Trapezium Quadrilateral ABCD is a trapezium if we can prove that: • one pair of opposite sides is parallel.	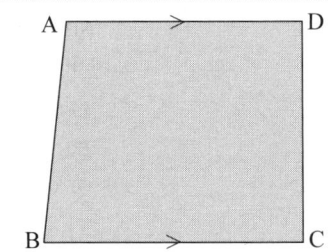

Worked examples

> **Note**
>
> In this example we work with angles and use algebra.

1. In the diagram ABCD is a parallelogram.
   CD is produced to E so that $\hat{BEC} = 56°$.
   EB is produced to G so that BG = DC.
   If $\hat{GAB} = x$, calculate the magnitude of x.

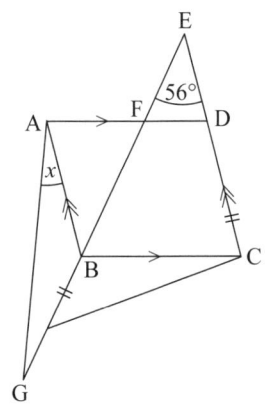

> **Note**
>
> In this example we work with lines and use Pythagoras.

2. EBCF is a rhombus. FD = 50 mm,
   OC = 30 mm and $\hat{FCD} = \hat{D}$.
   Calculate the length of BF.

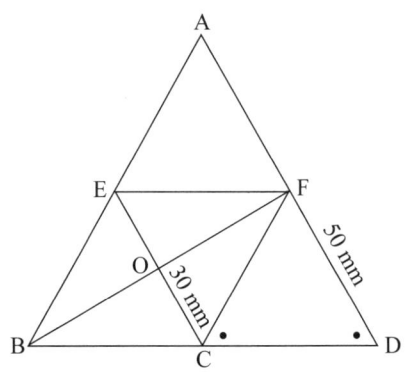

3. PQRS is a parallelogram.
   A and B are points on SR so
   that SA = BR.
   PB bisects $\hat{QPS}$.
   Prove that:
   3.1 SP = SB
   3.2 QA bisects $\hat{PQR}$.

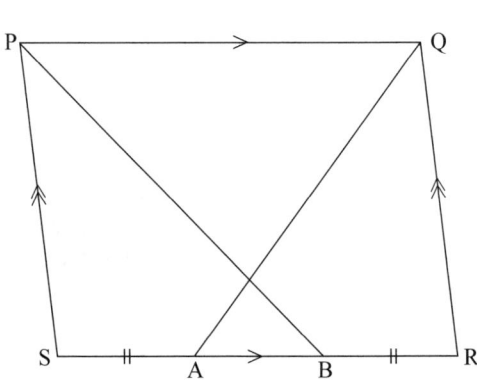

Solutions

1. $\hat{ABF} = \hat{E} = 56°$ alternate angles, EC ∥ AB

      $\hat{ABF} = x + \hat{G}$ exterior angle of $\triangle AGB$

      $AB = DC$ opposite sides of parallelogram are equal

      $DC = BG$ given

      $\therefore AB = BG$

      $\hat{G} = x$ isosceles triangle

      $\hat{ABF} = 2x$

      $\therefore x = 28°$

2. $FC = FD = 50$ mm   sides opposite equal angles, $\hat{D} = \hat{FCD}$

      $\hat{FOC} = 90°$ diagonals of a rhombus bisect at right angles

      $OF^2 = FC^2 - OC^2$ Pythagoras

      $\therefore OF = 40$ mm

      $\therefore BF = 80$ mm   diagonals of a rhombus bisect

3. **3.1** $\hat{QPB} = \hat{SPB}$ PB bisects $\hat{QPS}$, given

         $\hat{SBP} = \hat{QPB}$ alternate angles, PQ ∥ SR

         $\therefore SP = SB$ sides opposite equal angles

  3.2 $SA = BR$ given

         $\therefore SA + AB = BR + AB$

         $\therefore SB = AR$

         $\therefore AR = SP$

         $SP = RQ$ opposite sides of parallelogram are equal

         $\therefore AR = RQ$

         $\therefore \hat{BAQ} = \hat{AQR}$ AR = RQ

         $\hat{BAQ} = \hat{PQA}$ alternate angles, PQ ∥ SR

         $\therefore \hat{PQA} = \hat{AQR}$

         $\therefore$ QA bisects $\hat{PQR}$

Exercise 3

1. A rectangle is a parallelogram where the angles are 90°. Name one property of the diagonals of a rectangle that is not applicable to all parallelograms.

2. Name three properties of a rhombus that are not applicable to all parallelograms.

3. Choose from the list the quadrilaterals that have the properties given below. (More than one answer may be correct.)

> parallelogram   rectangle   rhombus   square   kite

    3.1 Diagonals are always equal.

    3.2 Diagonals always intersect at right angles.

    3.3 Diagonals bisect the area.

    3.4 Diagonals bisect the angles.

    3.5 Diagonals bisect each other.

4. In each figure, determine the size of each angle indicated by a letter. Only write down the solutions.

4.1
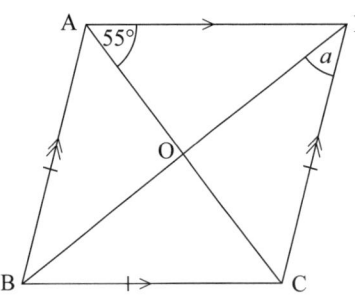

4.2
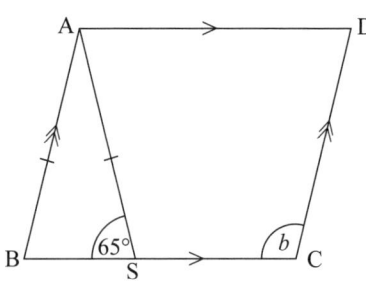

4.3
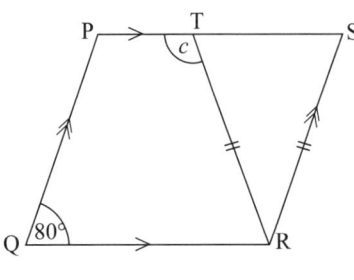

4.4
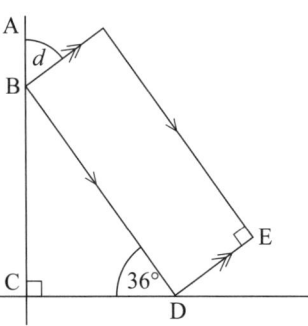

4.5
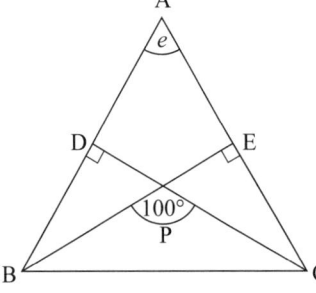

4.6
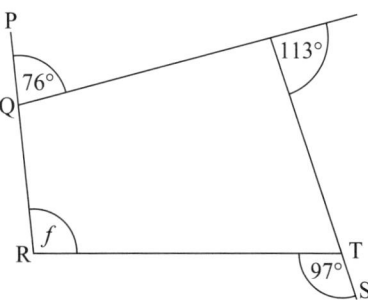

4.7
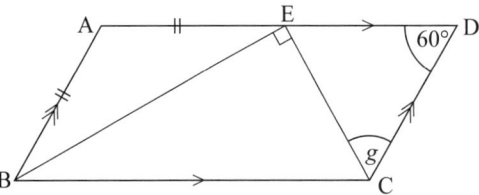

4.8
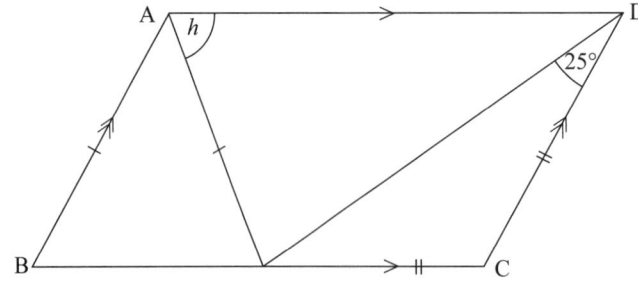

4.9
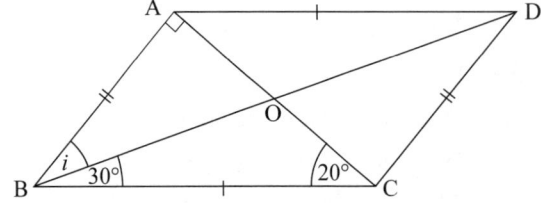

4.10

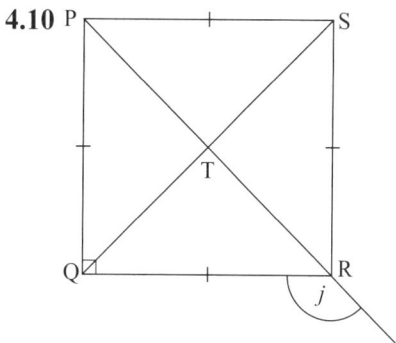

Exercise 4

Show all working and give reasons.

1. The diagonals of parallelogram ABCD intersect at O.

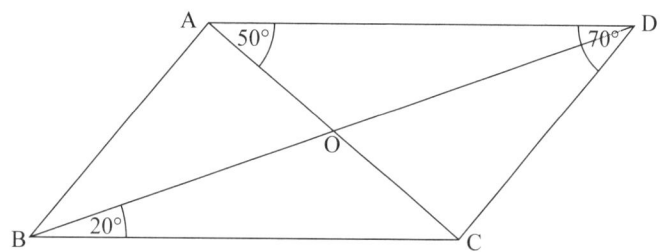

   If CÂD = 50°, AD̂C = 70° and DB̂C = 20°, calculate the following.

   1.1 AB̂O   **1.2** BÔC   **1.3** OÂB

2. Show, by calculation, that quadrilateral ABCD is a parallelogram if AE = AB, FD = FC, AÊB = 62°, ED̂F = 68° and FD̂C = 56°.

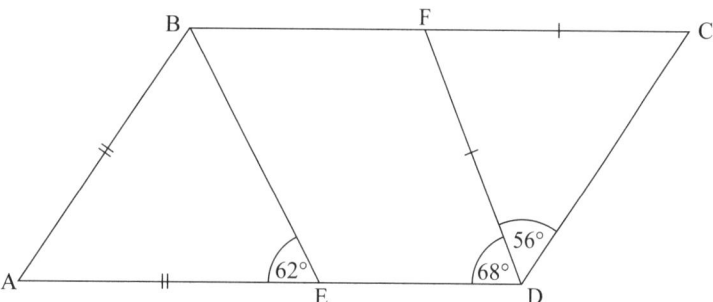

3. Quadrilateral ABCD is a rhombus with CB̂D = 36°.

   Calculate Ĉ.

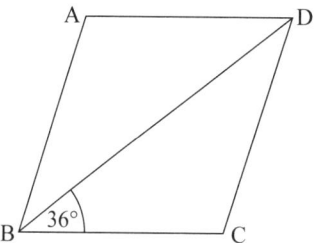

4. In parallelogram ABCD, AB = 50 mm, BC = 80 mm and BÂD = 110°.

   E is a point on AD so that AE = AB and BÊC = 75°.

   Calculate the following.

   4.1 CÊD

   4.2 the lengths of the sides of △CED

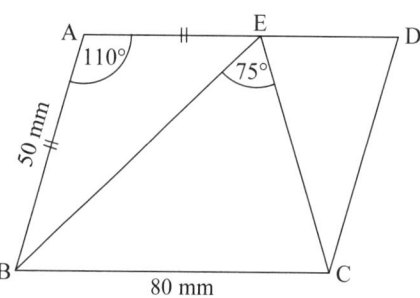

---

5. Diagonals PR and QS in rectangle PQRS intersect at O. PR = 80 mm and QR = 60 mm. If PŜQ = 20°, calculate the following.

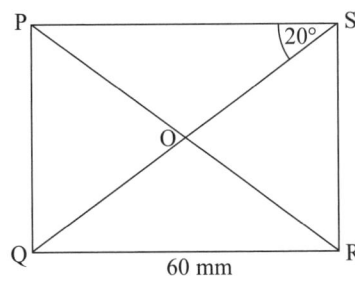

 5.1 PÔQ   **5.2** PR̂Q

 5.3 PS   **5.4** OP

 5.5 OQ

6. KMNO is a square and A is a point on KO such that KÂM = 50°. MA and NK intersect at P. Calculate MP̂N.

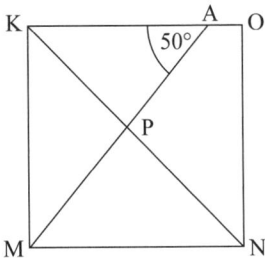

7. In parallelogram ABCD, CB̂D = 30°, BÂC = 75° and CÔD = 48°.

 Calculate the following.

 7.1 BD̂E

 7.2 BĈF

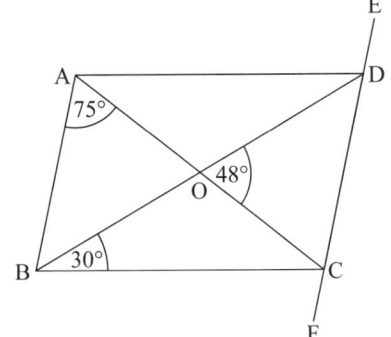

8. In parallelogram ABCD, E is a point on AD so that ED = DC = EB. EĈB = 35° and AB̂E = x. Calculate the value of x.

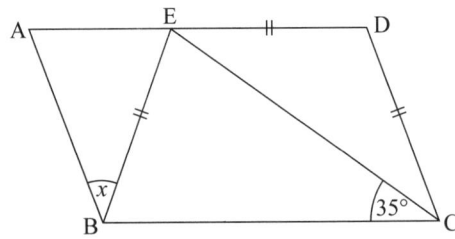

9. In parallelogram PQRS, T is a point on QR so that SR = TR = 60 mm, ST̂R = 30° and PQ = PT. Determine the length of QT.

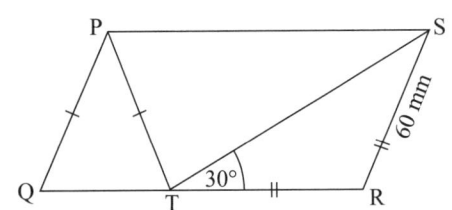

10. In rectangle RSTU, V is the midpoint of ST. VT = 12 mm, UW = 7 mm and RS = 16 mm. Prove that RV̂W = 90°.

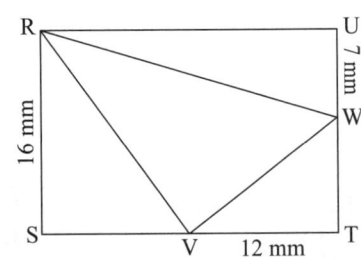

11. In rhombus ABCD, AB = 26 mm and
    diagonal BD = 48 mm.
    Calculate the length of diagonal AC.

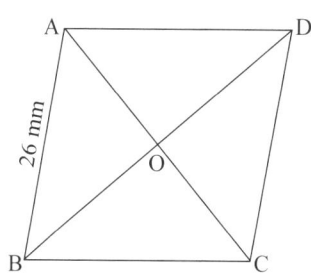

12. In quadrilateral KMNR,
    KR ∥ MN. P is a point on
    MN so that KM = KP and
    PN = NR. If MK̂P = 56°
    and KP̂R = 87°, prove that
    KMNR is a parallelogram.

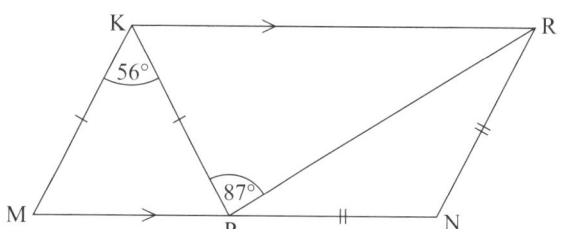

13. PQRS is a parallelogram with
    QP̂R = 50° and QŜR = 40°.
    Prove that PQRS is a rhombus.

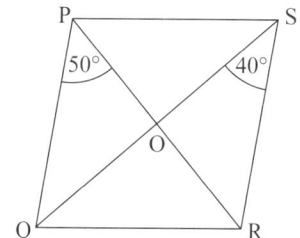

14. In the figure, BC = EF, AD ∥ EF
    and AB = AE, Â = 48°
    and CF̂D = 20°.
    Calculate the magnitude of D̂.

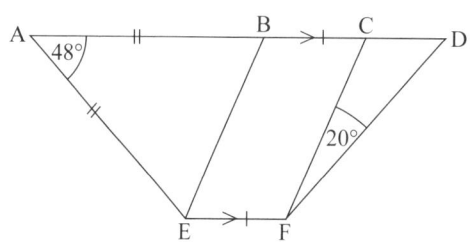

15. ABCD is a parallelogram.
    BE and CE bisect B̂ and Ĉ
    respectively, BE = √210 mm,
    CE = √46 mm and AB = 8 mm.
    Calculate the following lengths.
    15.1 BC
    15.2 ED

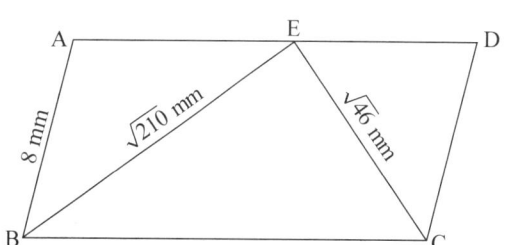

16. Diagonals PR and QS of
    parallelogram PQRS intersect at T.
    If PT = PQ and PT̂S = 120°,
    prove that PQRS is a rectangle.

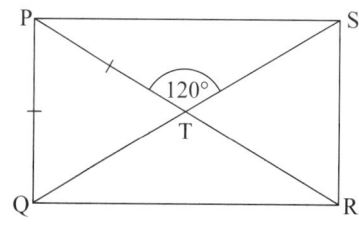

17. In parallelogram ABCD,
    F is a point on BC so that
    AF̂B = B̂ and AF = FC.
    If AF̂D = 72°, calculate Ĉ.

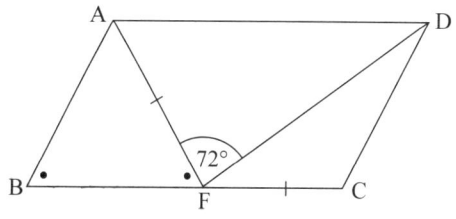

18. In the diagram AQ ∥ BP, AB ∥ DC, QP ∥ AC and AB = BC.

AD̂C = 112° and QP̂A = 20°.

Determine the magnitude of AÔD.

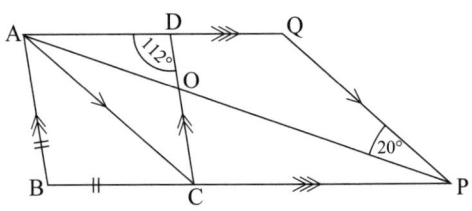

Exercise 5

1. KPMN is a parallelogram.

ON bisects KN̂M and OM bisects NM̂P.

Prove the following.

   1.1 NÔM = 90°

   1.2 △KNO is an isosceles triangle

   1.3 KP = 2KN

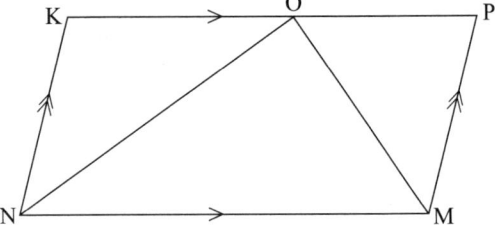

2. In quadrilateral ABCD:

Â = x, B̂ = 5x, Ĉ = 4x and D̂ = 2x.

Prove that ABCD is a trapezium.

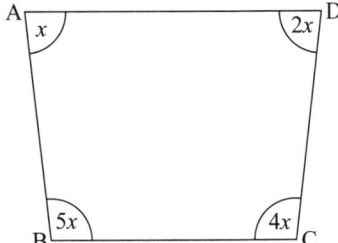

3. N and M are points on the diagonal AC of parallelogram ABCD such that AN = MC.

Prove the following.

   3.1 △ABN ≡ △CDM

   3.2 NBMD is a parallelogram

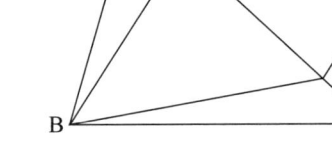

4. The diagonals of rhombus ABCD intersect at O.

AD̂B = x and AĈD = 2x + 30°.

Calculate the value of x.

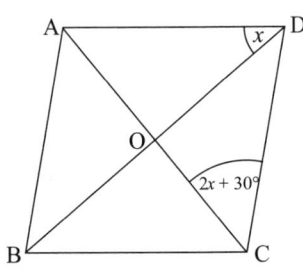

5. In quadrilateral PQRS, PQ ∥ SR.

The diagonals intersect at O and QO = OS.

Prove that PQRS is a parallelogram.

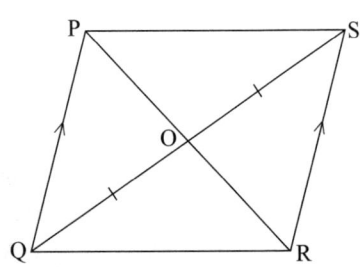

6. In parallelogram ABCD, AE bisects BÂD
   and DE bisects AD̂C.

   AF ∥ ED and AE ∥ FD.

   Prove that AEDF is a rectangle.

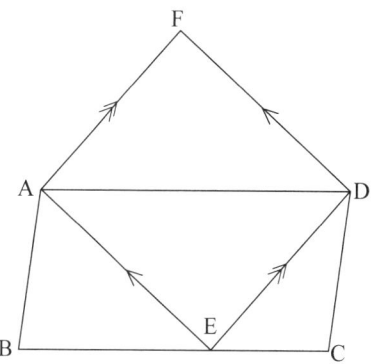

7. △PAQ and △BPS are both equilateral,
   PS ∥ QR and PQ ∥ SR.

   7.1 Prove that AQ̂R = RŜB.

   7.2 Prove that △AQR ≡ △RSB.

   7.3 If the perimeter of PQRS is 26 cm
       and PS = 8 cm, determine the
       perimeter of △AQP.

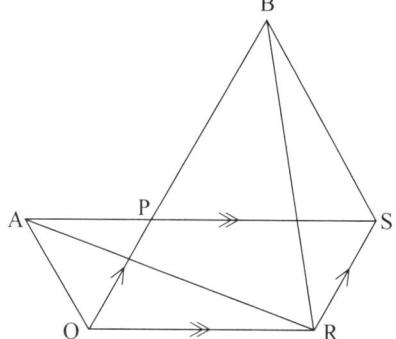

8. In parallelogram ABCD, P and Q are
   points on AB and CD respectively.

   PQ passes through O, the midpoint of AC.

   Prove that PBQD is a parallelogram.

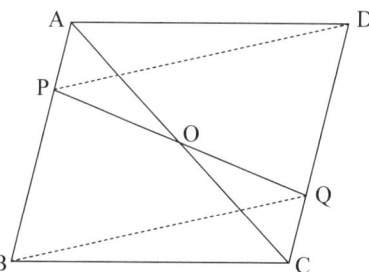

9. M and N are points on AB so that AM = BN.

   MK ∥ BC and NH ∥ AC.

   Prove the following.

   9.1 △AMK ≡ △NBH

   9.2 MBHK is a parallelogram

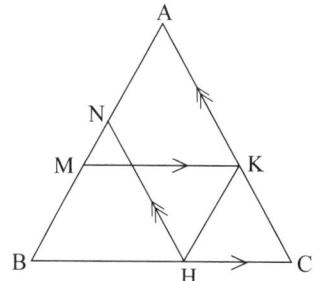

10. ABCD and ABOE are parallelograms.

    Prove that EAOD is also a
    parallelogram.

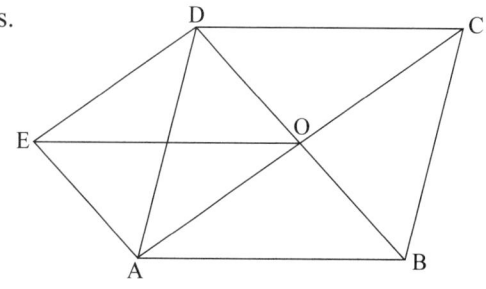

11. ABCD is a rhombus. OF = FC and OE = EC.
Prove that OECF is a rhombus.

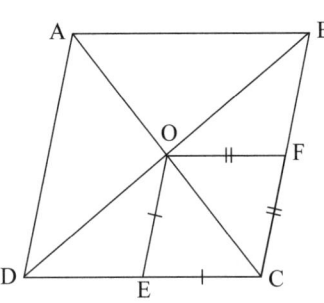

12. PTUQ and RTUS are parallelograms.
Prove that PQSR is also a
parallelogram.

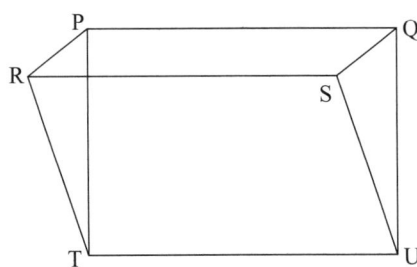

13. ABCD is a parallelogram. BE
and CE bisect $\hat{B}$ and $\hat{C}$ respectively,
with E on AD.
Show that BC = AB + CD.

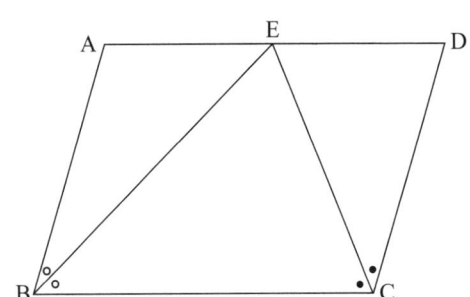

14. PR ∥ AB, AC ∥ BR and QB = RC.
Prove that $Q\hat{P}R = 2A\hat{Q}P$.

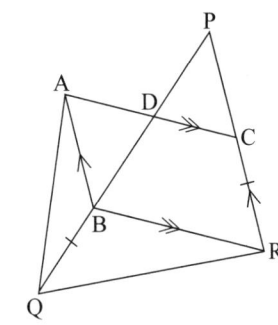

15. ABCD is a parallelogram.
E is a point on AD so that AE = AB,
CE = CD and BE = BC.
Calculate the magnitude of $B\hat{E}C$.

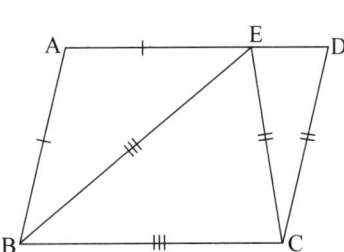

16. BCDE and AODE are parallelograms.
Prove the following.
16.1 CO = 2EF
16.2 ABOE is a parallelogram

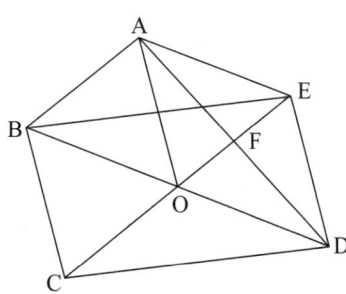

17. ABCD is a rhombus. DÂF = x and AF̂D = $3x$.

Prove that AF bisects DÂC.

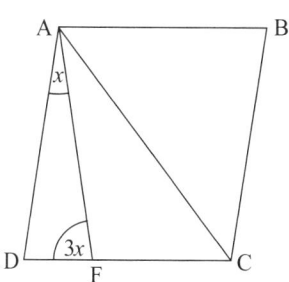

18. In the figure, AB ∥ DC, AF ∥ BE,
DÂE = BÂE = x and AB̂F = EB̂F = y.

Prove that AB = FE.

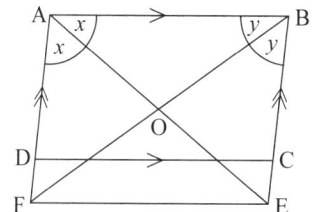

19. STNM is a parallelogram.

PMNQ is a straight line so that
PM = MS and NQ = NT.

PS is produced to meet the extension
of QT at R.

Prove that R̂ = 90°.

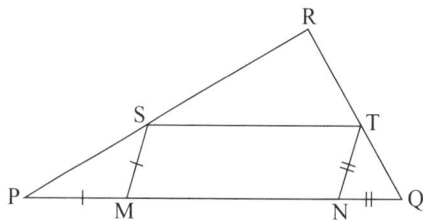

20. ABCD is a square.

E and F are the midpoints of AD and DC
respectively.

Prove that BE ⊥ AF.

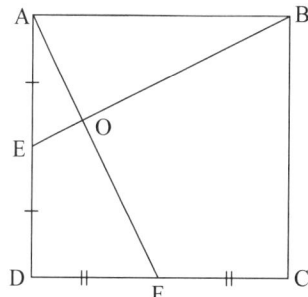

Line segments joining the midpoints of two sides of a triangle

You will need to know how to use these two theorems in solving geometry problems.

Theorem 1: Midpoint theorem

The line segment joining the midpoints of two sides of
a triangle is parallel to the third side and equal to half
its length.

In △QPR: PS = SQ and PT = TR

∴ ST ∥ QR and ST = $\frac{1}{2}$ QR

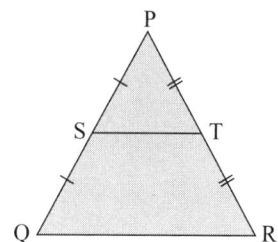

Theorem 2: Converse of midpoint theorem

The line passing through the midpoint of one side of
a triangle, and parallel to a second side, bisects the
third side.

In △ABC: AD = DB and DE ∥ BC

∴ AE = EC

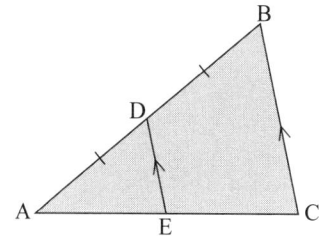

Worked examples

P, Q, R and S are the midpoints of AB, BC, CD and DA respectively. These are the sides of quadrilateral ABCD.

Prove that:

1. PQRS is a parallelogram
2. the perimeter of PQRS = AC + BD

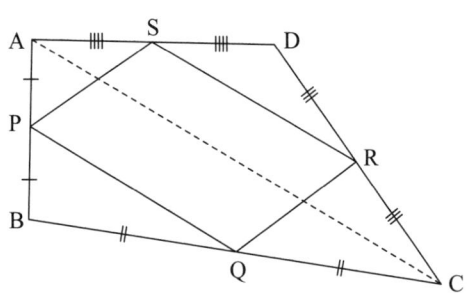

Solutions

In order to use one of the midpoint theorems, you will need a triangle, so you will have to construct one at AC or BD.

1. Construction: Draw AC.

   PQ ∥ AC and PQ = $\frac{1}{2}$AC   in △BAC, AP = PB and BQ = QC

   SR ∥ AC and SR = $\frac{1}{2}$AC   in △DAC, DS = SA and DR = RC

   ∴ SR ∥ PQ ∥ AC and SR = $\frac{1}{2}$AC = PQ

   ∴ PQRS is a parallelogram   one pair of opposite sides is equal and parallel

2. Similarly by joining BD we can prove that PS = QR = $\frac{1}{2}$BD.

   ∴ Perimeter PQRS = PS + QR + SR + PQ

   $$= \tfrac{1}{2}BD + \tfrac{1}{2}BD + \tfrac{1}{2}AC + \tfrac{1}{2}AC$$

   $$= BD + AC$$

Exercise 6

1. In the diagram, AF = FB, AG = GD, GH ∥ AE and EI = OD.

   Prove that FG ∥ HI if BCD is a straight line.

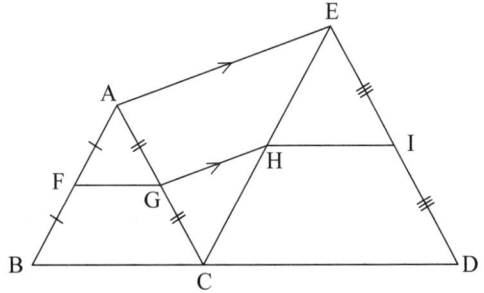

2. QR = 10 mm, PM = MQ, PN = NR, PQ ∥ SR and SR = 4 mm.

   Calculate:

   2.1 MN

   2.2 PQ.

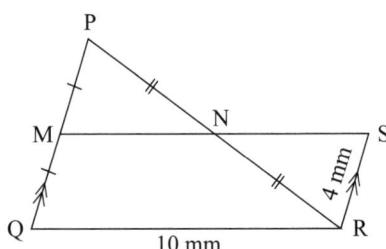

3. AD = BD, AE = EC and GC ∥ BE.

   3.1 Prove that AO = OG.

   3.2 Prove that OBGC is a parallelogram.

   3.3 If OF = 15 cm, calculate the length of AF.

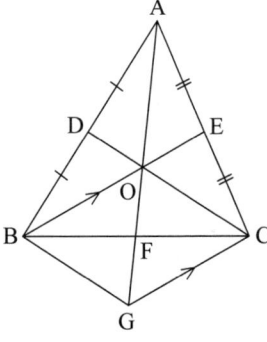

4. ABCD is a rhombus with E the midpoint of DC. If BD = 80 mm and AC = 60 mm, determine the magnitude (size) of:

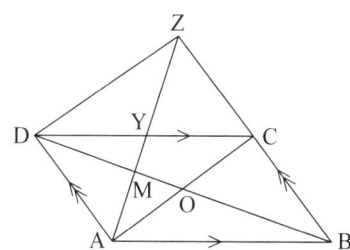

   4.1 BÔC

   4.2 BC

   4.3 OE

5. In the diagram, ZY = YA, CD ∥ AB and AD ∥ CB. Prove that:

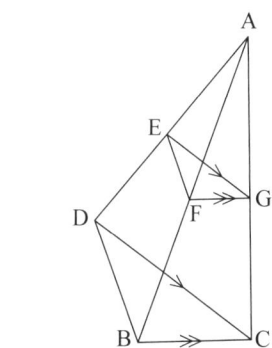

   5.1 DZ ∥ AC

   5.2 area △ZAB = area quad. ABCD.

6. AF = FB, FG ∥ BC and EG ∥ DC. Prove that EF ∥ DB.

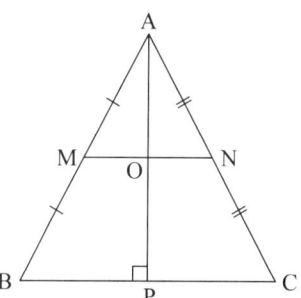

7. In △ABC, AM = MB and AN = NC. AP ⊥ BC, and AP and MN intersect at O. Prove that area △AMN = $\frac{1}{4}$ area △ABC.

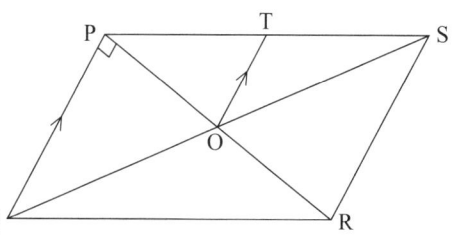

8. In parallelogram PQRS, PQ = 6 cm, PS = 10 cm and QP̂R = 90°. OT ∥ PQ. Calculate, with reasons:

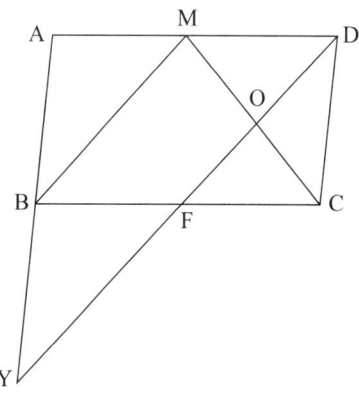

   8.1 PR   **8.2** TS

   8.3 OT

   8.4 the area of parallelogram PQRS.

9. In parallelogram ABCD: AD = 2 AB, and M is the midpoint of AD.
AB is produced to Y such that YB = BA.
BM, MC and YD are joined. YD intersects BC and MC at F and O respectively.

Prove that:

   9.1 BM ∥ YD

   9.2 BM bisects AB̂C

   9.3 O is the midpoint of MC.

10. AD ∥ EG ∥ BC and EF ∥ AC.

If BF = FC, prove that CG = GD.

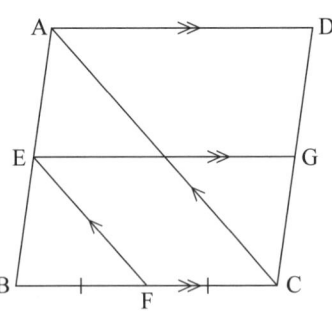

11. O is the centre of the semi-circle with diameter AC.

D is any point on the semi-circle.

AD is produced to B such that AD = DB.

Prove that BC = AC.

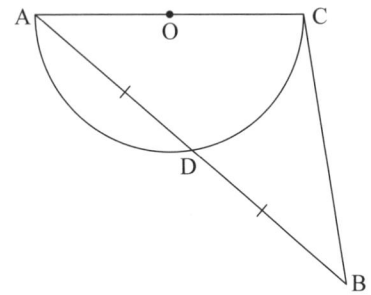

12. D and E are midpoints of AB and BC respectively.

If AC = $2x + 2$, calculate ED in terms of x.

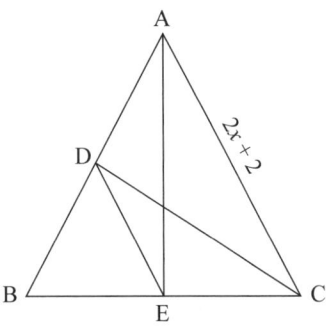

Test A: Knowledge and routine procedures

1. In △ABC, AB = AC.

Through O, a point on BC, OY is drawn perpendicular to BC.

Y is a point on BA produced, and YO and AC intersect at P.

If OP̂C = 40°, calculate:

1.1 the interior angles of △APY

1.2 complete: AY = …

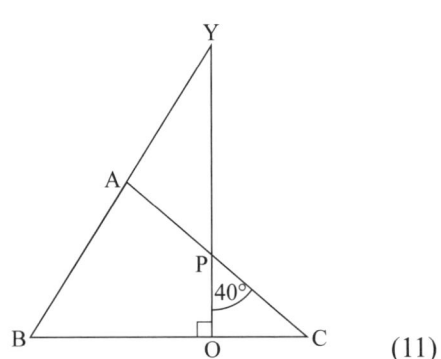

(11)

2. In the diagram, AC = CE, BC = CD and AE ∥ BD.

Prove that △ABC ≡ △EDC.

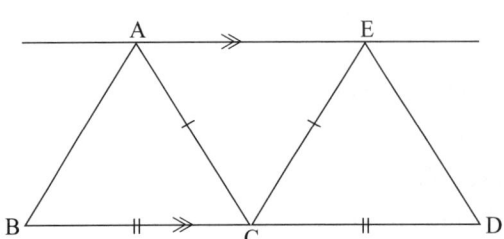

(9)

3. Quadrilateral ABCD is a parallelogram if AB̂C = AD̂C and BÂD = BĈD.

Use the diagram to list four more sets of conditions for which ABCD will be a parallelogram.

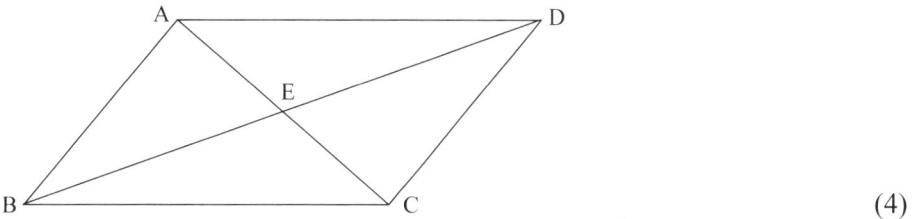

(4)

4. In the figure, ABCD is a parallelogram with BÂO = 4x, AD̂B = 2x, BD̂C = 3x and AÔB = 75°. Calculate the following.

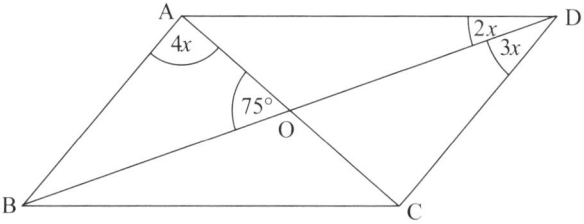

4.1 AB̂D in terms of x

4.2 the value of x

4.3 the magnitude of BĈD

(6)

5. In quadrilateral ABCD,
AO = OC and AD ∥ BC.

Prove the following.

5.1 △AOD ≡ △COB

5.2 ABCD is a parallelogram

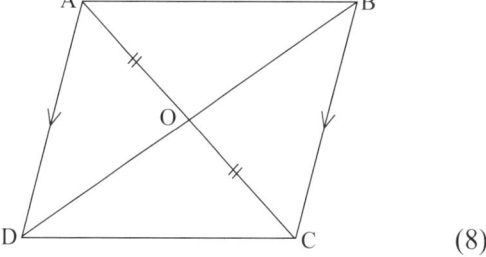

(8)

6. O is the midpoint of AB and OE ∥ BF.

BC = 38 mm, AD = 20 mm, DE = 12 mm and EF = 30 mm.

Give the values of:

6.1 DC   **6.2** AE

6.3 OD   **6.4** BF.

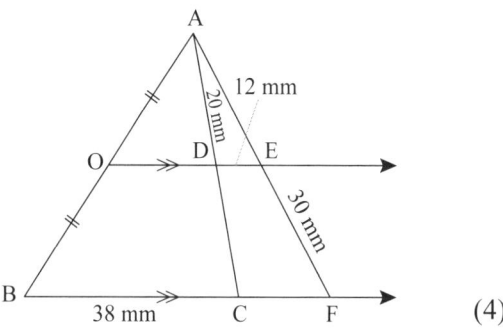

(4)

7. ABCD is a rhombus.

The diagonals intersect at O and OF ∥ BC.

AC = 16 cm and DB = 30 cm.

Calculate OF.

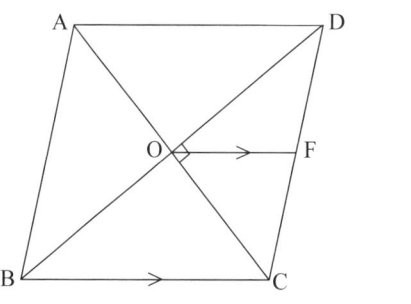

(8)

Total 50

Test B: Complex procedures and problem solving

1. In the diagram, AB ∥ CE, BF = BD, Â = 50° and BD̂E = 70°.

   Determine the magnitude of x if AB̂F = x.

   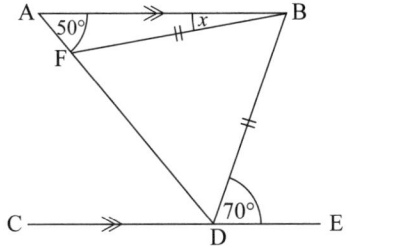

   (5)

2. In △PQR, PQ = PR.

   A is a point on PR such that QA = QP.

   Calculate the magnitude of R̂.

   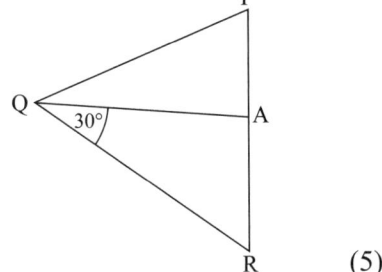

   (5)

3. ABCD is a parallelogram. E is a point on AD so that AE = AB and EC = CD. BÊC = 90°. Calculate EB̂C.

   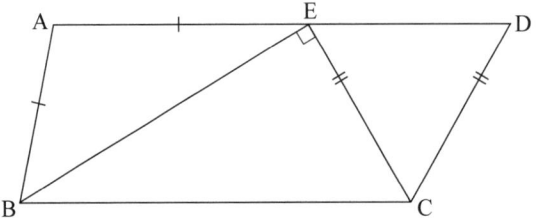

   (7)

4. ABCD is a parallelogram with E and F the midpoints of AB and DC respectively. ED and BF intersect AC at G and H respectively.

   Prove the following.

   4.1 EBFD is a parallelogram

   4.2 EG = HF

   4.3 GBHD is a parallelogram

   (14)

5. A and B are the midpoints of PQ and PR respectively.

   OD = DQ and OE = ER.

   Prove that ADEB is a parallelogram.

   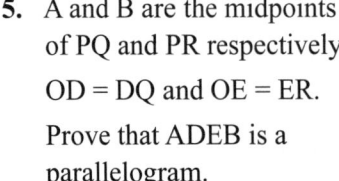
   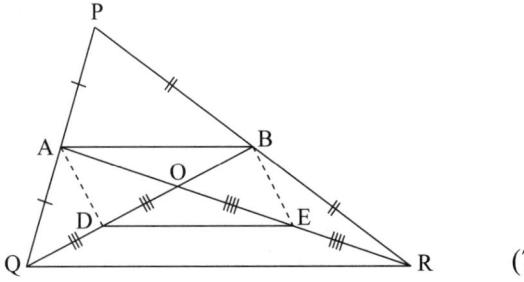

   (7)

6. In the diagram, $\hat{P}_1 = \hat{P}_2$.
   TC ⊥ PD and TC is
   produced to A.
   BC ∥ PT and CO ∥ TV.
   Prove that BO = $\frac{1}{2}$PV.

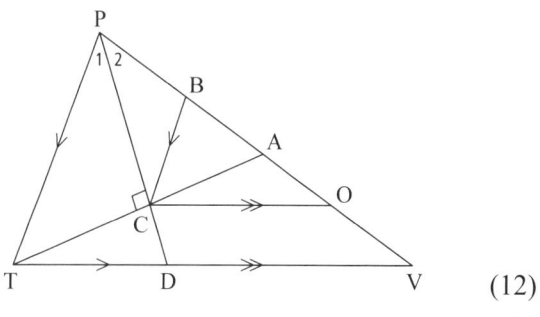

(12)

Total 50

Test C: Content and breakdown as for exam

1. In the figure, $\hat{B} = 68°$,
   $E\hat{D}C = 20°$, AD = DC and DE
   ∥ BC. Calculate, giving
   appropriate reasons:
   1.1 DĈB
   1.2 BD̂C
   1.3 AD̂C
   1.4 DĈE
   1.5 AÊD

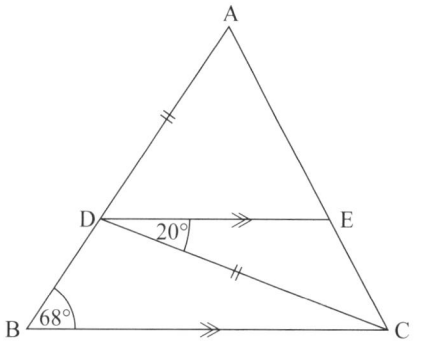

(11)

2. In the figure, $\hat{A} = x$,
   $A\hat{C}E = 50°$, $B\hat{E}C = 130°$
   and $E\hat{C}G = 2x - 110°$.
   2.1 Determine the value
   of x.
   2.2 Is AB ∥ FG?
   Motivate your answer.

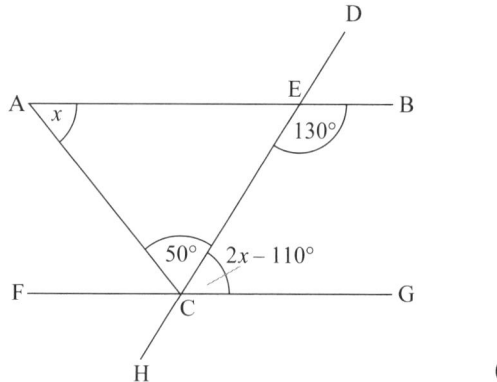

(4)

3. PQRS is a parallelogram.
   M is a point on PQ so that PS = MS
   and MQ = QR. $S\hat{M}R = 54°$
   and $M\hat{R}Q = x$.
   Determine the value of x.

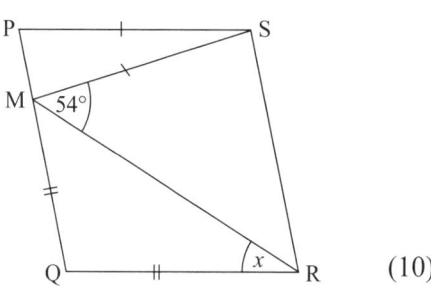

(10)

4. ABCD is a rhombus.
   The diagonals intersect at O and
   BC is produced to E. $O\hat{B}C = 40°$.
   Calculate the following.
   4.1 BÂO   **4.2** AD̂C
   4.3 DĈE

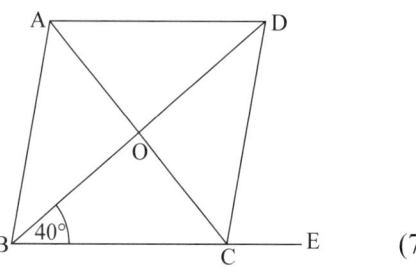

(7)

5. F and E are midpoints of AB and AC respectively.

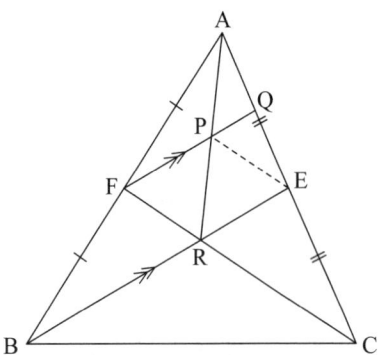

FPQ ∥ BRE. Prove that:

5.1 AQ = $\frac{1}{4}$AC

5.2 FPER is a parallelogram

5.3 area △AFR = area △BFR. (11)

6. ABCD is a trapezium with AD ∥ BC. OT is drawn parallel to AD, with O the midpoint of AB and T a point on DC.

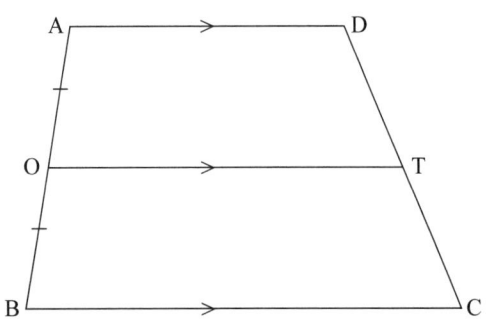

Prove that:

6.1 DT = TC

6.2 2OT = AD + BC (construct a line). (7)

Total 50

Analytical geometry uses the coordinate system to represent and work with geometric shapes. Geometric properties and theorems can be translated into algebraic definitions and equations giving another method of proving geometric properties.

In Grade 10, we work with work the algebraic properties relating to straight lines, triangles and quadrilaterals in two dimensions, represented in the Cartesian plane.

Knowledge and skills for this topic

If you struggle with any of the work listed below, revise it before continuing with this Topic:

- addition and simplification of surds
- basic algebraic manipulation and factorisation
- linear equations involving fractions
- quadratic equations
- representation and point plotting within the Cartesian plane
- equations of straight lines and gradient
- the theorem of Pythagoras to solve problems in right-angled triangles
- properties and classification of triangles
- properties and classification of quadrilaterals.

Content of final exam

- Represent geometric figures in a Cartesian coordinate system.
- Apply, for any two points, formulae for calculating:
  - the distance between two points
  - the gradient of the line segment joining two points
  - conditions for parallel and perpendicular lines
  - the coordinates of the midpoint of a line segment.

Formulae

$\text{Distance} = \sqrt{(x_2 - x_1)^2 + (y_2 - y_1)^2}$

$\text{Midpoint} = \left(\frac{x_1 + x_2}{2}; \frac{y_1 + y_2}{2} \right)$

$\text{Gradient } (m) = \frac{y_2 - y_1}{x_2 - x_1}$

Note: These formulae will be provided in all tests and examinations.

Straight lines: revision

In your work on functions, you learnt how to sketch, find the equation and interpret straight line graphs. Your knowledge of straight line graphs, gradients, parallel and perpendicular lines will be useful when solving problems using analytic geometry methods. Let's start with some important revision of these concepts that will also be useful in this section.

The general form for any straight line is given by:

$$y = mx + c$$

where m is the gradient (i.e. the slope of the graph) and c the y-intercept.

A line with a **positive gradient** slopes **up to the right**, while a line with a **negative gradient** slopes **down to the right**.

We can find the gradient of any line using the definition:

$$\text{Gradient} = \frac{\text{vertical rise}}{\text{horizontal run}}$$

Reading directly from the graph in the diagram alongside:

$$m = \frac{3}{1}$$
$$c = 3$$

The equation of this graph is given by:

$$y = 3x + 3$$

To find the equation of a straight line graph:

• Identify the y-intercept, c.
• Read off the gradient, m.
• Substitute for m and c into the general form $y = mx + c$.

Worked examples

Find the equations for the graphs.

1.

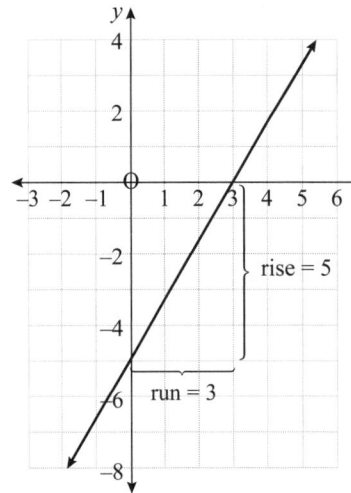

2.

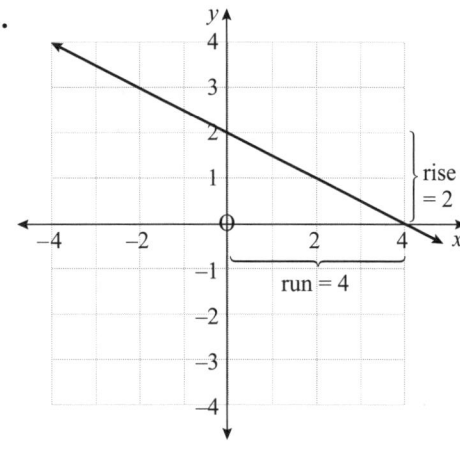

Solutions

1. $\text{Gradient} = \frac{\text{vertical rise}}{\text{horizontal run}}$

$$m = \frac{5}{3}$$
$$c = -5$$

Equation: $y = \frac{5}{3}x - 5$

2. This graph has a negative gradient (it slopes down to the right).

$$m = -\frac{2}{4} = -\frac{1}{2}$$
$$c = 2$$
$$y = -\frac{1}{2}x + 2$$

Using the gradient formula

> ## Gradient
>
> If $A(x_1; y_1)$ and $B(x_2; y_2)$ are any two points, the gradient of AB is given by:
>
> $$m_{AB} = \frac{y_2 - y_1}{x_2 - x_1}$$

Using the definition: gradient $= \frac{\text{vertical rise}}{\text{horizontal run}}$:

$$m = \frac{3}{1}$$

The formula $m = \frac{y_2 - y_1}{x_2 - x_1}$ allows us to find the gradient between any two points by substitution.

Let $A(0; 3)$ be $(x_1; y_1)$ and $B(-1; 0)$ be $(x_2; y_2)$.

Then, by substitution:

$$m = \frac{0 - 3}{-1 - 0} = \frac{-3}{-1} = \frac{3}{1}$$

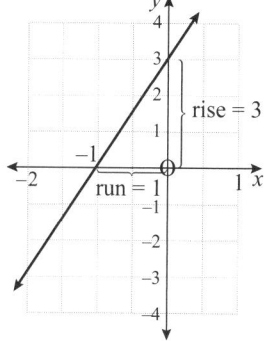

Worked example

Determine the gradient of the line that joins the points $P(-2; 3)$ and $Q(-3; -4)$.

Solution

$m_{PQ} = \frac{y_2 - y_1}{x_2 - x_1}$ ⟶ State the formula.

Let $P(-2; 3)$ be $(x_1; y_1)$ and $Q(-3; -4)$ be $(x_2; y_2)$.

$m_{PQ} = \frac{-4 - 3}{-3 - (-2)}$ ⟶ Substitute.

$\quad\;\; = \frac{-7}{-1} = \frac{7}{1}$ ⟶ Simplify fully.

Worked example

Determine the gradient of the line that joins the points $A\left(\frac{3}{2}; -3\right)$ and $B\left(-2; \frac{5}{2}\right)$.

Solution

$m_{AB} = \frac{y_2 - y_1}{x_2 - x_1}$ ⟶ State the formula.

Let $A\left(\frac{3}{2}; -3\right)$ be $(x_1; y_1)$ and $B\left(-2; \frac{5}{2}\right)$ be $(x_2; y_2)$.

$m_{AB} = \frac{\frac{5}{2} - (-3)}{-2 - \frac{3}{2}}$ ⟶ Substitute.

$\quad\;\; = \frac{\frac{11}{2}}{-\frac{7}{2}} = -\frac{11}{7}$ ⟶ Simplify fully.

Horizontal and vertical lines

Line AB is parallel to the x-axis:
(a horizontal line)

This line has the general solution $y = c$, where c is the y-intercept.

Line CD is parallel to the y-axis:
(a vertical line)

This line has the general solution $x = c$, where c is the x-intercept.

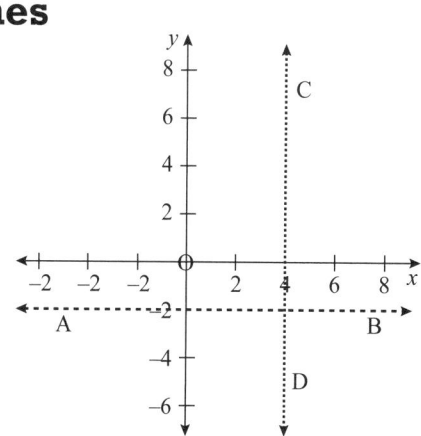

Equation for AB: $y = -2$ $\quad m_{AB} = 0$

Equation for CD: $x = 4$ $\quad m_{CD} = \frac{\infty}{0}$ or undefined

Relationship between parallel and perpendicular lines

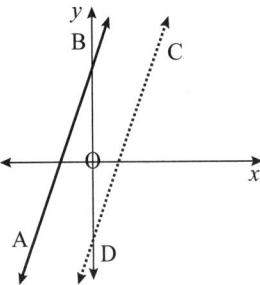

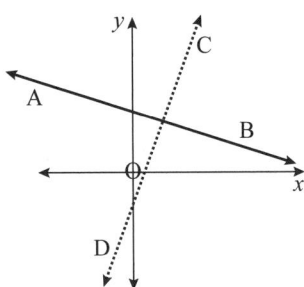

- Parallel lines have equal gradients.
  If AB || CD then $m_{AB} = m_{CD}$.

- The product of the gradients of perpendicular lines is –1.
  If AB $\perp$ CD then $m_{AB} \times m_{CD} = -1$.
  Conversely: If $m_{AB} \times m_{CD} = -1$ then AB $\perp$ CD.

Gradient and collinear points

Collinear points are points that lie in a straight line.

> **Collinear points**
>
> Points $A(x_A; y_A)$, $B(x_B; y_B)$ and $C(x_C; y_C)$ are collinear if:
>
> - $m_{AB} = m_{BC}$
>
> - $\frac{y_A - y_B}{x_A - x_B} = \frac{y_B - y_C}{x_B - x_C}$
>
> Note that a common point B is used in both gradients.

Worked examples

Consider the points A(2; 5), B(0; 2), C(–3; 4) and D(–2; –1).

1. Find the gradients for the line segments AB, BC and AD.

2. Which lines segments are parallel, and which are perpendicular?

3. Represent points A, B, C and D on a Cartesian plane. What can you conclude about points A, B and D?

Solutions

1. Let A(2; 5) be $(x_1; y_1)$ and B(0; 2) be $(x_2; y_2)$.

   Then $m_{AB} = \frac{2-5}{0-2} = \frac{-3}{-2} = \frac{3}{2}$

   Let B(0; 2) be $(x_1; y_1)$ and C(–3; 4) be $(x_2; y_2)$.

   Then $m_{BC} = \frac{4-2}{-3-0} = -\frac{2}{3}$

   Let A(2; 5) be $(x_1; y_1)$ and D(–2; –1) be $(x_2; y_2)$.

   Then $m_{AD} = \frac{-1-5}{-2-2} = \frac{-6}{-4} = \frac{3}{2}$

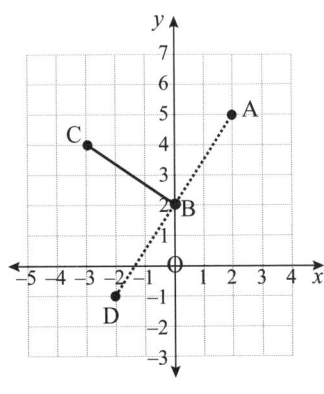

2. AB || AD $m_{AB} = m_{AD}$

   AB $\perp$ BC $m_{AB} \times m_{BC} = -1$

   AD $\perp$ BC $m_{AD} \times m_{BC} = -1$

3. Points A, B and D form a straight line.

   Points A, B and D are collinear.

   $m_{AB} = m_{AD}$ and the line segments share a common point B.

Vertical and horizontal distances

Worked examples

R(3; 5), S(3; –1) and T(–1; –1) are the vertices
of △RST.

1. Calculate the vertical distance RS.

2. Calculate the horizontal distance ST.

3. Use the theorem of Pythagoras to calculate the
   distance between points R and T. Leave your
   answer in simplified surd form.

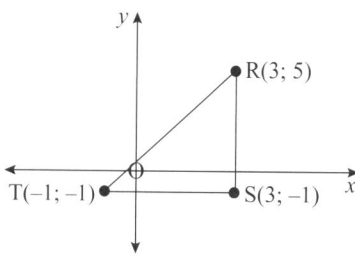

Solutions

1. The vertical distance RS = 5 – (–1) = 5 + 1 = 6 units.

2. The horizontal distance ST = 3 – (–1) = 3 + 1 = 4 units.

3. $TR^2 = RS^2 + TS^2$ Pythagoras.

   $TR^2 = 6^2 + 4^2$

   $TR^2 = 52$

   $\therefore TR = \sqrt{52} = 2\sqrt{13}$

> **Remember**
>
> **How to simplify
> a surd:**
>
> $\sqrt{52} = \sqrt{4 \times 13}$
>
> $= 2\sqrt{13}$

Using the distance formula

The distance formula can be used to find the distance between any two points.

Worked example

Calculate the distance between A(6; –8) and B(–2; 3), correct to one decimal place.

> **Note**
>
> The distance between
> any two points
> A(x_1; y_1) and
> B(x_2; y_2) is:
>
> $\sqrt{(x_2 - x_1)^2 + (y_2 - y_1)^2}$

Solution

Write down the formula: $AB = \sqrt{(x_2 - x_1)^2 + (y_2 - y_1)^2}$

Let A(6; –8) be (x_1; y_1) and B(–2; 3) be (x_2; y_2).

Substitute: $\sqrt{(-2 - 6)^2 + (3 - (-8))^2}$

Simplify: $\sqrt{(-8)^2 + (11)^2}$

        $= \sqrt{185} \approx 13{,}6$

Using the midpoint formula

The midpoint formula allows us to find the coordinates of the midpoint between
any two points.

> **The midpoint of a line segment**
>
> The coordinates of the midpoint between points A(x_1; y_1) and B(x_2; y_2) is:
>
> $M\left(\dfrac{x_1 + x_2}{2}; \dfrac{y_1 + y_2}{2}\right)$

Worked example

Determine the midpoint of the line joining $(-3; 2)$ and $(5; -6)$.

Solution

$$\text{Midpoint} = \left(\frac{x_1 + x_2}{2}; \frac{y_1 + y_2}{2}\right) \qquad \text{Write down the formula.}$$

$$= \left(\frac{-3+5}{2}; \frac{2-6}{2}\right) \qquad \text{Substitute values into the formula.}$$

$$= \left(\frac{2}{2}; \frac{-4}{2}\right) = (1; -2) \qquad \text{Simplify fully.}$$

Worked example

Determine the coordinates of M, the midpoint of AB if $A\left(-\frac{1}{2}; -3\right)$ and $B(2; -5)$.

Solution

$$M\left(\frac{x_1 + x_2}{2}; \frac{y_1 + y_2}{2}\right). \qquad \text{Write down the formula.}$$

Let $A\left(-\frac{1}{2}; -3\right)$ be $(x_1; y_1)$ and $B(2; -5)$ be $(x_2; y_2)$.

$$\therefore M\left(\frac{-\frac{1}{2}+2}{2}; \frac{-3+(-5)}{2}\right) = \left(\frac{3}{4}; -4\right) \qquad \text{Substitute and simplify.}$$

Exercise 1

1. Given: points A(2; 5) and B(14; 9).

    1.1 Find the distance between A and B, leaving your answer in simplified surd form.

    1.2 Calculate the gradient of the line segment AB.

    1.3 Find the coordinates for M, the midpoint of AB.

    1.4 Write down the gradient for DE if DE ∥ AB.

    1.5 Write down the gradient for RT if RT ⊥ AB.

2. Given: points P(–4; 7) and Q(1; –5).

    2.1 Find the distance between Q and P.

    2.2 Calculate the gradient of the line segment QP.

    2.3 Find the coordinates for M, the midpoint of QP.

    2.4 Write down the gradient for RT if RT ∥ QP.

    2.5 Write down the gradient for ME if ME ⊥ QP.

3. Given: points $R\left(3; \sqrt{2}\right)$ and $S\left(-1; -\sqrt{2}\right)$.

    3.1 Find the distance between R and S, leaving your answer in simplified surd form.

    3.2 Calculate the gradient of the line segment RS.

    3.3 Find the coordinates for M, the midpoint of RS.

    3.4 Write down the gradient for TZ if TZ ∥ RS.

    3.5 Write down the gradient for ZB if ZB ⊥ RS.

4. For each set of points:

    (a) determine m_{AB} and m_{CD}.

    (b) state whether the line segments AB and CD are parallel, perpendicular or neither.

4.1 A(1; 3), B(5; 5), C(3; 2), D(7; 4)

4.2 A(2; 0), B(8; 2), C(4; 2), D(2; –4)

4.3 A(4; 8), B(7; 2), C(–3; –4), D(0; 2)

4.4 A(5; 5), B(3; –7), C(–7; 3), D(5; 1)

5. 5.1 Calculate the distance from the origin to T(–4; –7), correct to two decimal places.

5.2 Find the gradient for the line from the origin to T(–4; –7).

5.3 Find the coordinates for the point halfway between the origin and T(–4; –7).

6. Which of the points A(3; 5) and B(–4; –4) is closer to the origin?

7. Consider the equations:

A. $x = 5$	**B.** $y = 2x - 3$	**C.** $y = -3$
D. $2y + 3x = 10$	**E.** $3y - 2x = 1$	**F.** $2y + 3x + 1 = 0$
G. $2y = 5$	**H.** $x + 4 = 0$	

7.1 Which of these equations are parallel to one another?

7.2 Which are perpendicular to one another?

8. For each set of points given below, find:

(a) the point of bisection of AB **(b)** the gradient of AB.

8.1 A($4p$; $-2n$) and B($2p$; $6n$)

8.2 A(a; $3b$) and B($-a$; $-3b$)

8.3 A($2n^2$; $-2p$) and B(n^2; $-p$)

9. Show that the points P(3; 6) and Q(–1; 4) are the same distance from S(2; 3).

10. Show that the points A(1; 2), B(4; 5) and C(7; 8) are collinear.

Working backwards

To solve problems in the Cartesian plane you may be required to 'work backwards', i.e. to find coordinate points when distance, midpoint or gradient are given. To 'work backwards' you have to use the given information and the relevant formulae to set up an equation, which you then need to solve.

In each case you will have to interpret the given information and use it to set up and to solve an equation.

Worked example

A(a; –1) and B(5; 3) are two points. If AB = 5, find the value(s) of a.

Solution

In this problem you will use a given **distance** to find coordinate points.

$$AB = \sqrt{(x_2 - x_1)^2 + (y_2 - y_1)^2}$$

Let A(a; –1) be (x_1; y_1) and B(5; 3) be (x_2; y_2).

$$\therefore 5 = \sqrt{(5 - a)^2 + (3 - (-1))^2}$$

Square both sides to get rid of the surd:

$$25 = (5 - a)^2 + (3 - (-1))^2$$

$$25 = 25 - 10a + a^2 + 16$$

$$0 = a^2 - 10a + 16$$

Use factors to solve the quadratic equation:

$$0 = (a - 8)(a - 2)$$

$$\therefore a = 8 \text{ or } a = 2$$

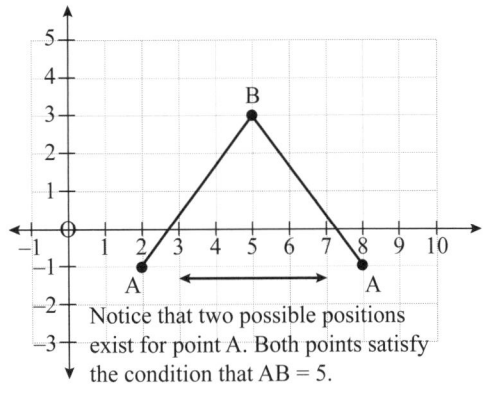

Notice that two possible positions exist for point A. Both points satisfy the condition that AB = 5.

Worked example

Point $M\left(\frac{1}{2}; -3\right)$ is the midpoint of AB with $B\left(-\frac{7}{2}; 8\right)$.
Determine the coordinates of A.

Solution

In this problem you will use a given **midpoint** to find coordinate points.

$$M = \left(\frac{x_1 + x_2}{2}; \frac{y_1 + y_2}{2}\right) = \left(\frac{1}{2}; -3\right)$$

Let $A(x; y)$ be $(x_1; y_1)$ and $B\left(-\frac{7}{2}; 8\right)$ be $(x_2; y_2)$.

$$M\left(\frac{x + \left(-\frac{7}{2}\right)}{2}; \frac{y + 8}{2}\right) = \left(\frac{1}{2}; -3\right) \qquad \text{By substitution.}$$

Work with the x- and y-coordinates separately.

Solve the equations for x and y separately.

$$\therefore \frac{x + \left(-\frac{7}{2}\right)}{2} = \frac{1}{2} \qquad \text{and} \qquad \frac{y + 8}{2} = -3$$

$$x - \frac{7}{2} = 1 \qquad\qquad\qquad y + 8 = -6$$

$$x = \frac{9}{2} \qquad\qquad\qquad y = -14$$

$$\therefore A\left(\frac{9}{2}; -14\right)$$

Worked example

Use a given **gradient** to find coordinate points.

$A(a; -3)$ and $B\left(6; -\frac{3}{2}\right)$ are two points. If the gradient of AB = 3, determine the value(s) of a.

Solution

$$m_{AB} = \frac{y_2 - y_1}{x_2 - x_1} = 3$$

Let $A(a; -3)$ be $(x_1; y_1)$ and $B\left(6; -\frac{3}{2}\right)$ be $(x_2; y_2)$.

$$\therefore \frac{-\frac{3}{2} - (-3)}{6 - a} = 3$$

Solve for a in the linear equation:

$$\frac{3}{2} = 3(6 - a)$$

$$\frac{3}{2} = 18 - 3a$$

$$3a = 18 - \frac{3}{2}$$

$$\therefore a = \frac{11}{2}$$

Finding the coordinates of the fourth vertex of a parallelogram

The fourth vertex of any parallelogram can be determined using careful inspection.

Worked example

Points P(–1; 2), Q(6; 4) and R(1; –3) are given. Determine the coordinates for point S if PQRS forms a parallelogram.

Solution

S(–6; –5).   Determined by inspection.

Method to determine the fourth vertex of a parallelogram

Note

While it is not necessary to show any working, a rough sketch will help you to find the fourth vertex without making unnecessary mistakes.

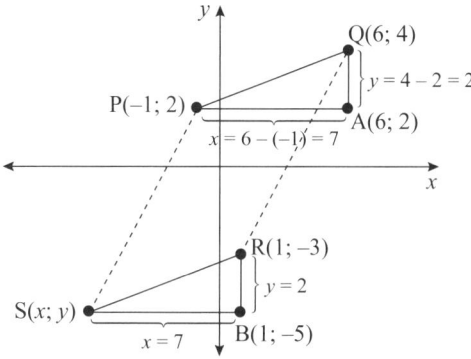

1. Set up a rough drawing of the required parallelogram.
2. Roughly construct $\triangle$QAP:
   - Determine the coordinates for A.
   - Use the change in the x and y values respectively to determine the coordinates for A.
3. Roughly construct $\triangle$RBS:

   $\triangle$QAP = $\triangle$RBS

   $\therefore$ QA = RB = 2

   $\therefore$ B(1; –3 –2) = B(1; –5)

   $\therefore$ PA = SB = 7

   $\therefore$ S(1 – 7; –5) = S(–6; –5)

Now let's put everything together in a typical exam question involving working backwards.

Worked examples

$P\left(-\frac{3}{2}; \frac{3}{2}\right)$, Q(7; 1), R($x$; 2) and S(1; y) are points in a Cartesian plane.

1. Calculate the value(s) for x if:
   1.1 the gradient of QR is –0,5
   1.2 P is the midpoint of QR
   1.3 QR = $\sqrt{2}$ units.

2. Calculate the value(s) for x and y if PRQS is a parallelogram.

Solutions

1. **1.1** $m_{QR} = \dfrac{1-2}{7-x} = -0,5$

$\therefore \dfrac{-1}{7-x} = -\dfrac{1}{2}$

$2 = 7 - x$

$\therefore x = 5$

1.2 $\left(-\dfrac{3}{2}; \dfrac{3}{2}\right) = \left(\dfrac{7+x}{2}; \dfrac{1+2}{2}\right)$

$\therefore -\dfrac{3}{2} = \dfrac{7+x}{2}$

$\therefore 7 + x = -3$

$\therefore x = -10$

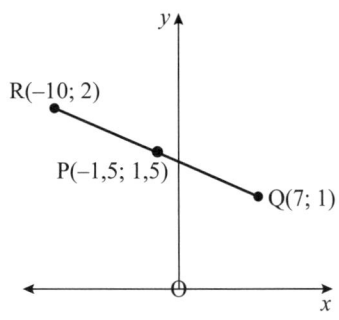

1.3 $QR = \sqrt{2}$

$\sqrt{2} = \sqrt{(7-x)^2 + (1-2)^2}$

$2 = 49 - 14x + x^2 + 1$

$0 = x^2 - 14x + 48$

$0 = (x-6)(x-8)$

$x = 6 \text{ or } x = 8$

2. Using inspection to find a fourth vertex:
    $R\left(\dfrac{1}{2}; 2\right)$ and $S\left(1; \dfrac{1}{2}\right)$

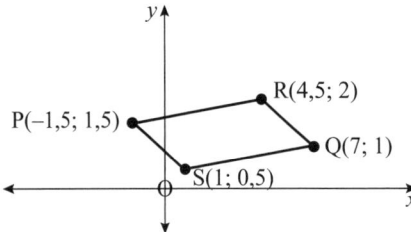

Exercise 2

1. Find t if $(1; -2,5)$ is the midpoint of the line segment from $(t; -4)$ to $(4; -1)$.

2. Points A(2; p) and B(–3; 0) are given.

    2.1 If $m_{AB} = \dfrac{1}{10}$, find the value(s) of p.

    2.2 If AB $\perp$ TM and $m_{TM} = -5$, find the value(s) of p.

3. C is the point $(1; -2)$. The point D lies in the second quadrant and has coordinates $(x; 5)$. If the length of CD is $\sqrt{53}$ units, calculate the value of x.

4. If M is the midpoint of AB, find the unknown coordinates p and t.

    4.1 M(3; 5), A(0; 4), B(6; p)   **4.2** M(1; t), A(–4; 6), B(p; 5)

5. AC = AB where A(2; 5), B(2; t) and C(5; 1).

    5.1 Use the information to calculate the value(s) of t.

    5.2 If t lies in on the y-axis, find the fourth vertex of the rhombus ABCD.

    5.3 If B lies in on the 1st quadrant, find the fourth vertex of the rhombus ABCD.

6. Points A(2; p) and B(3; –6) are given. Determine the value(s) of p and t in each case below:

    6.1 If AB has a gradient of $-\frac{2}{3}$ **6.2** If AB = $\sqrt{17}$

    6.3 If M(t; 10) is the midpoint of AB.

7. M(3; 2) is the midpoint of the line segment from P(–3; 4) to T. Find the coordinates of T.

8. Points P(–2; 3), Q(a; –7) and R(–8; b) are given. Find the value(s) of a and b if:

    8.1 PQ = $2\sqrt{41}$

    8.2 $m_{PQ} = 2$ and Q, P and R are collinear

    8.3 Q is the midpoint of PR.

9. If MA = MB , and A, B and M are collinear, find the value(s) for p and t in each case.

    9.1 M(2; 3), A(–3; 1), B(p; t)   **9.2** M(8; 5), A(p; t), B(10,2)

10. The diameter of a circle, centre C, stretches from A(1; 1) to B(5; 7).

    10.1 Find the coordinates of C.   **10.2** Find the radius of the circle.

    10.3 Show that D(5; 1) lies on the circle.

11. Find the coordinates of point A if points P(2; 7), Q(–3; 4) and A are collinear and PQ = QA.

12. If the points C(–1; 7), A(5; 1) and N(–4; y) are collinear, find the value of y.

13. The line segment joining the points A(–4; y) and B(–2; 5) is perpendicular to the line segment DE which has a gradient of $\frac{2}{3}$. Find the value(s) for y.

Layered questions and problem solving

The basic analytical geometry skills can be used to solve problems and to identify and classify geometric shapes. The following table summarises some of the common uses for the basic skills/formulae.

Formula	Some common uses for this formula
Distance formula: $\sqrt{(x_2 - x_1)^2 + (y_2 - y_1)^2}$	• Find the distance between two points. • Prove that a triangle is isosceles/equilateral/right-angled. • Prove that a quadrilateral is a parallelogram/square, etc. • Determine perimeter.
Gradient: $m = \frac{y_2 - y_1}{x_2 - x_1}$	• ∥ lines: $m_1 = m_2$ • ⊥ lines: $m_1 \times m_2 = -1$ • Show that an angle is right-angled. • Determine whether three points are collinear (i.e. form a straight line). • Determine whether a quadrilateral is a parallelogram.
Midpoint: $\left(\frac{x_1 + x_2}{2}, \frac{y_1 + y_2}{2}\right)$	• Find the midpoint of a line segment. • Find the length of a line segment, given one point and the midpoint.

Typical exam questions will require you to use a mixture of these tools, with the easier questions specifically instructing you to find a distance, a gradient or a midpoint. More difficult questions will require you to select the tool/tools necessary to answer the question.

Worked examples

Points A(–4; –1), B(2; 3) and C(6; –3) are the vertices of △ABC.

1. Find the midpoints P and Q of the sides AB and BC respectively.

2. Show that PQ ∥ AC.

3. Show that 2PQ = AC.

4. If ABCD are the vertices of a parallelogram, find the coordinates of point D.

Solutions

1. $P = \left(\frac{x_1 + x_2}{2}; \frac{y_1 + y_2}{2}\right) = \left(\frac{-4 + 2}{2}; \frac{-1 + 3}{2}\right) = P(-1; 1)$

   $Q = \left(\frac{x_1 + x_2}{2}; \frac{y_1 + y_2}{2}\right) = \left(\frac{2 + 6}{2}; \frac{3 - 3}{2}\right) = Q(4; 0)$

Note

Use **gradient** to show that lines are **parallel**:

$m = \frac{y_2 - y_1}{x_2 - x_1}$

2. $m_{PQ} = \frac{1 - 0}{-1 - 4} = -\frac{1}{5}$

   $m_{AC} = \frac{-1 + 3}{-4 - 6} = \frac{2}{-10} = -\frac{1}{5}$

   ∴ PQ ∥ AC

Note

Use **distance** to when working with lengths:

$\sqrt{(x_2 - x_1)^2 + (y_2 - y_1)^2}$

3. $PQ = \sqrt{(0 - 1)^2 + \left(4 - (-1)\right)^2} = \sqrt{(-1)^2 + (5)^2} = \sqrt{26}$

   $AC = \sqrt{\left(6 - (-4)\right)^2 + \left(-3 - (-1)\right)^2} = \sqrt{(10)^2 + (-2)^2} = \sqrt{104} = 2\sqrt{26}$

   ∴ AC = 2PQ

4. Set up a diagram to help you to find the fourth vertex: D(0; –7):

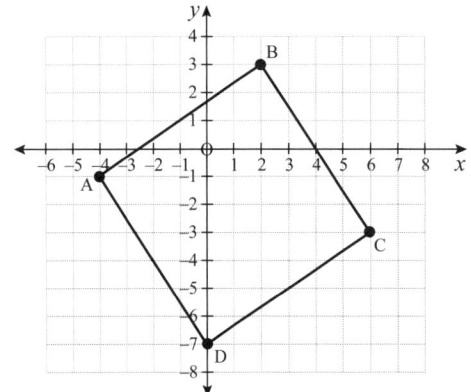

When working with analytical problems it often helps to set up a neat sketch, which will help you to visualise the problem and to assess the accuracy and validity of your solutions and/or conclusions.

A typical problem involves the use of formula in a layered fashion, i.e. a set of information is given and you have to use it to answer a few different questions. You will need to select the formula that you will then use to answer the question.

Worked examples

Given △ABC, with points A(3; 4), B(5; –3) and C(–2; 2):

1. determine whether △ABC is isosceles, scalene or equilateral

2. calculate the perimeter of △ABC, correct to two decimal places

3. show that the points A(3; 4), B(5; –3) and D(9; –17) are collinear.

Note

To find the **lengths of sides** we use the distance formula:

$$\sqrt{(x_2 - x_1)^2 + (y_2 - y_1)^2}$$

To determine if points are **collinear**, use gradient $= \frac{y_2 - y_1}{x_2 - x_1}$

Solutions

1. $AB = \sqrt{(5-3)^2 + (-3-4)^2} = \sqrt{53}$

   $BC = \sqrt{(-2-5)^2 + (2+3)^2} = \sqrt{74}$

   $AC = \sqrt{(-2-3)^2 + (2-4)^2} = \sqrt{29}$

   $\triangle ABC$ a scalene triangle   All sides have different lengths.

2. Perimeter $= \sqrt{53} + \sqrt{74} + \sqrt{29} = 21,2675\ldots = 21,27$ units

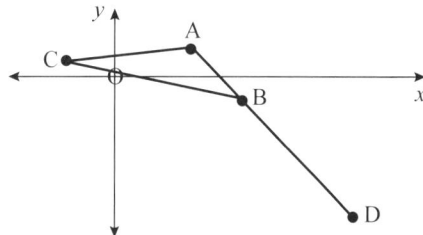

3. $m_{AB} = \frac{-3-4}{5-3} = -\frac{7}{2}$

   $m_{BD} = \frac{-3-(-17)}{5-9} = \frac{14}{-4} = -\frac{7}{2}$

   $m_{AB} = m_{BD} = -\frac{7}{2}$

   $\therefore$ A, B and D are collinear points.

Worked examples

Quad FGHI has coordinates F(–3; 2), G(3; 6), H(9; –2) and I(3; –6).

1. Show that quad FGHI is a parallelogram.

2. Show that the diagonals of quad FGHI bisect each other.

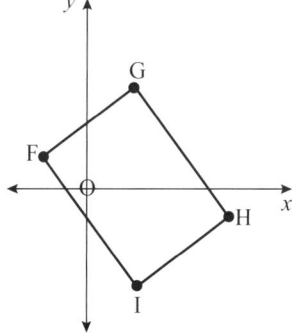

Solutions

1. $m_{FI} = \frac{-6-2}{3-(-3)} = \frac{-8}{6} = -\frac{4}{3}$

   $m_{GH} = \frac{-2-6}{9-3} = \frac{-8}{6} = -\frac{4}{3}$

   $\therefore$ FI $\parallel$ GH $m_{FI} = m_{GH}$

   $m_{FG} = \frac{6-2}{3-(-3)} = \frac{4}{6} = \frac{2}{3}$

   $m_{IH} = \frac{-2-(-6)}{9-3} = \frac{4}{6} = \frac{2}{3}$

   $\therefore$ FG $\parallel$ IH $m_{FG} = m_{IH}$

   FGHI is a parallelogram.   Both pairs of opposite sides are parallel.

2. Midpoint of FH: $= \left(\frac{x_1 + x_2}{2}; \frac{y_1 + y_2}{2}\right) = \left(\frac{-3 + 9}{2}; \frac{2 + (-2)}{2}\right) = (3; 0)$

    Midpoint of GI: $= \left(\frac{x_1 + x_2}{2}; \frac{y_1 + y_2}{2}\right) = \left(\frac{3 + 3}{2}; \frac{6 + (-6)}{2}\right) = (3; 0)$

    The diagonals meet in a common midpoint.

    ∴ The diagonals bisect each other.

It is always wise to look back at your diagram and check that your values are realistic. Ask yourself if the answers you calculated look correct. This will help you to spot silly errors.

Exercise 3

1. Use the information on the diagram to answer the following questions.

    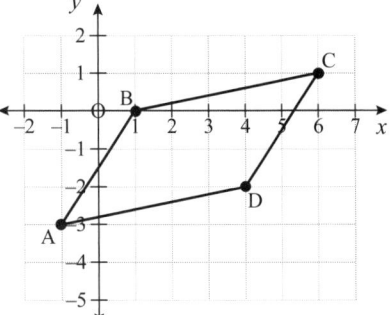

    1.1 Show that the diagonals of ABCD bisect each other.

    1.2 Show that one pair of opposite sides is both equal and parallel.

    1.3 What do these results show figure ABCD to be?

2. Points A(2; 4) B(6; 4) and C(6; 7) are given.

    2.1 Find the gradients for the line segments AB, AC and BC.

    2.2 What do the gradients tell us about these line segments?

    2.3 What do these gradients tell us about △ABC?

    2.4 Find the distances AB, AC and BC.

    2.5 What do these distances/lengths tell us about △ABC?

3. The points P(2; 7), E(–6; –9), and T(6; –5) in a Cartesian plane form an triangle. Use the information to determine:

    3.1 the gradient of PE

    3.2 the coordinates of M, the midpoint of PE

    3.3 the gradient of TM.

    3.4 What can you say about line segments TM and PE? Give a reason for your answer.

    3.5 Find the area of △PET, correct to two decimal places.

4. The points A(5; 3), B(1; –3) and C(–1; t) are given. Use these two points to answer the questions that follow.

    4.1 If the points A, B and C are collinear, find the value of t.

    4.2 If point D lies in the first quadrant and has the coordinates (p; 1) find the value(s) of p that will make △ABD a right-angled triangle (i.e. Â = 90°).

5. The coordinates of the corners of a shape are A(2; 4), B(4; 1), C(2; –2), D(–2; –2), E(–4; 1) and F(–2; 4).

    5.1 Make a neat drawing of this shape on a Cartesian plane.

    5.2 Find the length of the sides AB, BC and CD.

    5.3 Find the perimeter of this shape, correct to two decimal places.

    5.4 Write down the equations for the line segments FA and DC.

5.5 Show that the sides BC and EF are parallel.

5.6 Which other pairs of sides are parallel? Support you answer with analytical working.

5.7 What type of polygon is this?

6. P(–4; –1), A(2; 3) and T(6; –3) are the vertices of a triangle.

    6.1 Show that △PAT is right-angled.

    6.2 Show that $A\hat{P}T = 45°$. Support your working with full reasons.

    6.3 Calculate the area of △PAT.

Geometric shapes: complex procedures and problem solving

When solving more complicated problems, **take time to think before you start**. There is often more than one solution path to a problem. You need to decide on an appropriate method of solving the problem without wasting time and doing a lot of unnecessary working steps. But at the same time, you need to be sure that your problem is solved. It is easy to leave out essential steps, so take time to brainstorm.

Classifying triangles

Worked example

Prove that P(–2; 5), Q(1; –2) and R(3; 3) are the vertices of a right-angled isosceles triangle.

In this case:
1. To prove that the triangle is a right-angled triangle, we can:
   - use distance and the theorem of Pythagoras
   - OR use gradients to prove that we have a right angle.
2. To prove that the triangle is isosceles, we need to use distance to show that two sides are equal in length.

Solution

Note

In this problem, using the distance formula to find the lengths of all three sides is an efficient method because it can be used simultaneously both to prove that the triangle is right-angled and that it is isosceles.

$$PR = \sqrt{(3+2)^2 + (3-5)^2} = \sqrt{29}$$

$$RQ = \sqrt{(3-1)^2 + (3+2)^2} = \sqrt{29}$$

$$PQ = \sqrt{(1+2)^2 + (-2-5)^2} = \sqrt{58}$$

$$PR^2 + RQ^2 = 58$$

$$PQ^2 = 58$$

$$\therefore PR^2 + RQ^2 = PQ^2$$

$$\therefore \hat{R} = 90° \qquad \text{Pythagoras}$$

$$PR = RQ \text{ and } \hat{R} = 90°$$

$$\therefore △PQR \text{ is an isosceles right-angled triangle.}$$

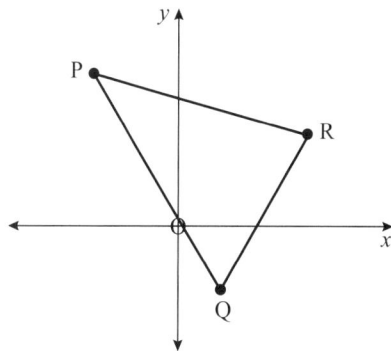

Classifying quadrilaterals

A thorough working knowledge of quadrilaterals and their properties is necessary for the successful completion of these types of problems.

The flow diagrams that follow summarise how we can use analytic methods to classify quadrilaterals.

Using gradients

Gradients can be used to show that opposite sides are parallel. Begin by proving that the quadrilateral is a parallelogram. Then look for right angles and or equal sides.

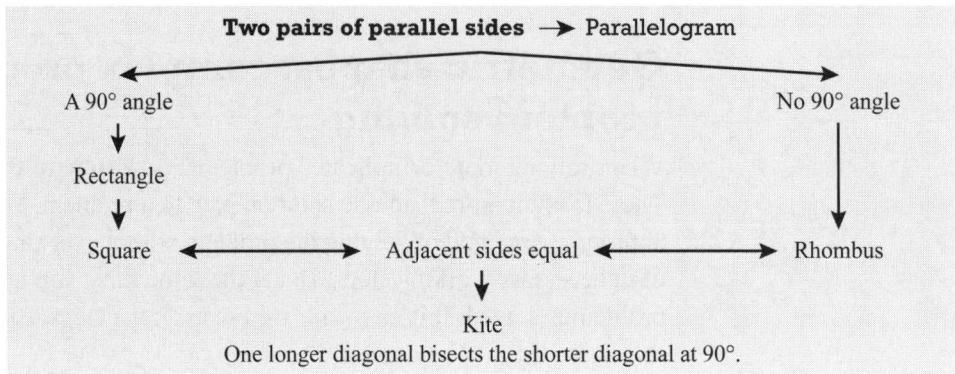

Using diagonals

The properties relating to the diagonals of the various quadrilaterals can be used to classify quadrilaterals. If the diagonals bisect each other, the quadrilateral is at least a parallelogram.

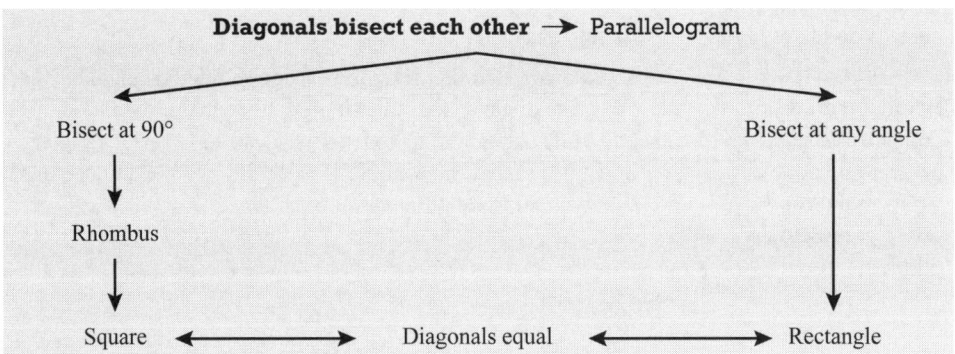

Worked example

Consider the points A(0; –4), B(–5; –2), C(–3; 3) and D(2; 1). Use analytical methods to classify this quadrilateral.

Solution

$m_{AB} = \frac{-2 - (-4)}{-5 - 0} = \frac{2}{-5} = -\frac{2}{5}$

$m_{BC} = \frac{-2 - 3}{-5 - (-3)} = \frac{-5}{-2} = \frac{5}{2}$

$m_{CD} = \frac{1 - 3}{2 - (-3)} = \frac{-2}{5} = -\frac{2}{5}$

$m_{AD} = \frac{1 - (-4)}{2 - 0} = \frac{5}{2}$

$\therefore$ AB $\parallel$ CD $m_{AB} = m_{CD} = -\frac{2}{5}$

$\therefore$ AD $\parallel$ CB $m_{BC} = m_{AD} = \frac{5}{2}$

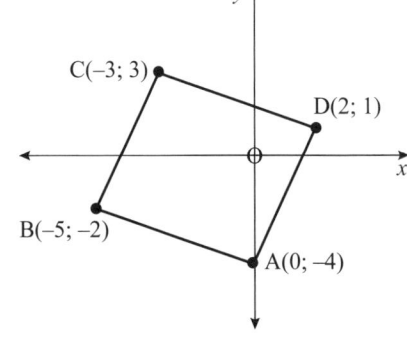

∴ ABCD is at least a parallelogram Both pairs of opposite sides are parallel.

$AD \perp DC$ $\quad m_{AD} \times m_{DC} = -1$

∴ ABCD is at least a rectangle A parallelogram with a right angle.

$AB = \sqrt{(-5-0)^2 + (-2+4)^2} = \sqrt{29}$

$BC = \sqrt{(-3+5)^2 + (3+2)^2} = \sqrt{29}$

∴ ABCD is a square A rectangle with adjacent sides equal.

Alternate solution

Using diagonals

Start by showing that the diagonals bisect each other: $\left(\dfrac{x_1 + x_2}{2}, \dfrac{y_1 + y_2}{2}\right)$

i.e. that they have a common midpoint.

Midpoint of CA: $\left(\dfrac{-3+0}{2}; \dfrac{3-4}{2}\right) = \left(-\dfrac{3}{2}; -\dfrac{1}{2}\right)$

Midpoint of BD: $\left(\dfrac{-5+2}{2}; \dfrac{-2+1}{2}\right) = \left(-\dfrac{3}{2}; -\dfrac{1}{2}\right)$

∴ ABCD is at least a parallelogram Diagonals bisect each other.

Use the **gradients** of the diagonals to show that they **bisect at right angles**:
$m = \dfrac{y_2 - y_1}{x_2 - x_1}$

$m_{CA} = \dfrac{-4-3}{0+3} = \dfrac{-7}{3}$

$m_{BD} = \dfrac{1+2}{2+5} = \dfrac{3}{7}$

∴ $CA \perp BD$ $\quad m_{CA} \times m_{BD} = -1$

∴ ABCD is at least a rhombus Diagonals bisect at right angles.

Use the **distance** to show that the diagonals are **equal** in length: $\sqrt{(x_2 - x_1)^2 + (y_2 - y_1)^2}$

$CA = \sqrt{(-3-0)^2 + (3+4)^2} = \sqrt{(-3)^2 + (7)^2} = \sqrt{58}$

$BD = \sqrt{(-5-2)^2 + (-2-1)^2} = \sqrt{(-7)^2 + (-3)^2} = \sqrt{58}$

∴ $CA = BD$

ABCD is a square Diagonals are of equal length and bisect at right angles.

Exercise 4

1. Quad PQRS has P(2; 2), Q(4; –4), R(10; –2) and S(8; 4).

    1.1 Prove that PQRS is a parallelogram.

    1.2 Now prove that PQRS is a rectangle.

    1.3 Now prove that PQRS is a square.

2. Determine whether the triangle with vertices P(4; –3), E(8; 5) and T(–2; 3), is equilateral, isosceles or scalene.

3. A quadrilateral has vertices J(1; 7), K(8; 8), L(2; 0), and M(–5; –1).

    3.1 Are the diagonals perpendicular? Explain your answer.

    3.2 What type of quadrilateral is JKLM? Give a reason for your answer.

4. Quadrilateral BAND has vertices B(–10; 0) , A(–3; –1), N(–2; –8) and D(–9; –7).

    4.1 Show that the diagonals bisect each other at 90°.

    4.2 Name two types of quadrilaterals that have diagonals that bisect each other at right angles.

    4.3 Show that BAND is not a square.

    4.4 If the diagonals bisect each other at point M, find the coordinates of point E such that ME = 2MA and points D, M, A and E are collinear, and E has $y > 0$.

    4.5 What type of quadrilateral is BAND?

    4.6 Calculate the area of BAND.

5. The points R(2; 5), U(–2; 1) and S(2; –3) are given.

    5.1 Write down the coordinates of T such that RUST forms a parallelogram.

    5.2 Prove that the parallelogram RUST is in fact a square.

6. Points P(–3; 7), Q(7; 2) , R(4; –4) and S(–7; –5) form a quadrilateral.

    6.1 Find the coordinates of A, B C and D, the midpoints of PQ, QR, RS and SP respectively.

    6.2 What kind of quadrilateral is ABCD?

7. A triangle has vertices A(–4; 0), B(0; 3) and C(0; –3).

    7.1 Show that △ABD is isosceles.

    7.2 Find the coordinates of point D so that the vertices ABCD are the vertices of a rhombus.

    7.3 Find the coordinates of point E so that the vertices ACBE are the vertices of a rhombus.

8. K(4; 9), I(5; 6), T(1; 3) and E(1; 8) form the vertices of a kite. Calculate the area of this kite.

Test A: Knowledge and routine procedures

1. Points R(2; 5) and T(–1; 3) are given.

    1.1 Find the length of RT.   (2)

    1.2 Find the gradient of RT.   (2)

    1.3 Find the midpoint of RT.   (2)

2. Points D(2; 5), E (–4; –4) , F(0; 2) and G (–4; 9) are given.

    2.1 Find m_{DE} and m_{EF}. What can you conclude about points D, E and F?   (6)

    2.2 Find m_{DG}. What can you conclude about the line segments DE and DG?   (4)

3. Points C(–4; 3), A(6; 1) and T(2; –3) are given.

    3.1 Set up a neat sketch of the points C, A and T in the Cartesian plane.   (1)

    3.2 Show that the points C, A and T are the vertices of a right-angled triangle.   (5)

    3.3 Find the fourth vertex, S, so that CATS forms a rectangle.   (2)

4. For Q(–3; –5), U(2; –3), A(3; 5) and D(–2; 3), show that QA and UD bisect each other. (4)

5. The line joining the points S(4; –2) and T(6; p) has a gradient of 1. Find the value of p. (3)

6. M(3; 2) is the midpoint of AB with A(4; 7) and B(a; b) Find the value(s) of a and b. (2)

7. The distance between (2; 3) and (6; y) is 5. Calculate the value(s) of y. (5)

8. The points P(–2; –2), E(1; –2) and T(1; 2) are given.

   Jonty says that △PET is an isosceles right-angled triangle.

   Paula says that △PET is a scalene right-angled triangle.

   Zandie says that △PET is scalene BUT NOT right-angled.

   Who is correct?

   8.1 Use coordinate methods and show all your working to support your answer fully. (10)

   8.2 Use some of your working from 8.1 to calculate the area of △PET. (2)

   Total 50

Test B: Complex procedures and problem solving

1. The points P(2; 7), Q(–3; 4) and R(3; –3) are given.

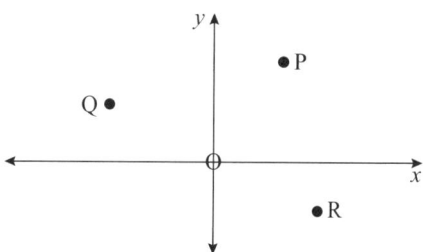

   1.1 Find the coordinates of point S, so that PQRS forms a parallelogram. (2)

   1.2 Find the coordinates of point T so that PQRT forms a trapezium with QP ∥ RT and RT = 2QP. (5)

2. Two equal circles with centres A(1; 1) and B(4; 5) touch a third circle with centre P as shown in the diagram.

   If A, P and B lie on a straight line, find:

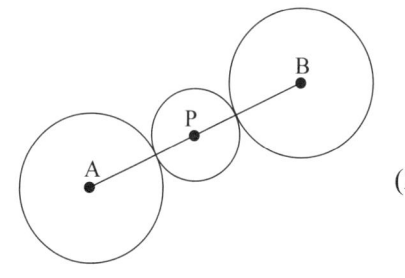

   2.1 the coordinates of P (2)

   2.2 the length of the diameter of the small inner circle if the two equal circles have a radius of 2 units each. (4)

   2.3 Label the point where the circle with centre A and the smaller circle with centre P, meet T. Find the coordinates of point T.
   (*Hint*: It may be useful to consider similarity.) (6)

3. The diagram shows the course for a long-distance mountain bike race.

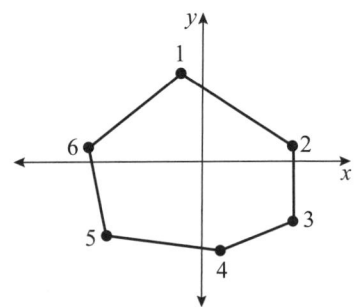

The start of the race is at checkpoint 1 with coordinates (–2; 12). Checkpoint 2 is at (10; 2). The coordinates for the respective checkpoints 3, 4, 5 and 6 are: 3 (a; b), 4 (2; –12), 5 (c; –10) and 6 (d; 2).

3.1 Find the distance between checkpoints 1 and 2 if each unit on the map equals 2 km on the mountain. Round your answer off to the nearest km. (4)

3.2 The race course from checkpoint 2 (10; 2) to check point 3 (a; b) is parallel to the y-axis. The distance between checkpoints 2 and 3 is 20 km. Find the values for a and b. (2)

3.3 The gradient between checkpoints 4 and 5 is $-\frac{1}{6}$. Find the value for c. (4)

3.4 The distance from checkpoint 6 to the finish at checkpoint 1 is $10\sqrt{2}$ units. Use this information to find the value for d. (5)

3.5 The course from the start to checkpoint 2 is exceptionally steep and the organisers have decided to set up **three** additional water points between checkpoints 1 and 2. If these water points are to be equally spaced, find the coordinates for these additional water points. Label them A, B and C. (6)

4. K(4; 9), I(5; 6), T(1; 3) and E(1; 8) form the vertices of a quad. Without using any distance formulae, prove that these points are the vertices of a kite. (10)

Total 50

Test C: Content and breakdown as for exam

1. $\triangle$ABC has vertices A(–3; 3), B(7; –3) and C (5; 5).
   1.1 Write down the coordinates of the midpoint, D of AB. (2)
   1.2 Use analytical methods to prove that AD $\perp$ DC. (5)
   1.3 Calculate the area of $\triangle$ABC. (6)

2. Points A(–5 ; 6), B(3; –2) and C(p; 6) are given.
   2.1 Find the value of p in point C if the gradient of BC is 2. (3)
   2.2 If it is given that AB = BC, find the value(s) of p in point C. (8)

3. In the diagram, D(–3; 3), E(3; –5), F(–1; k) are three points in the Cartesian plane.

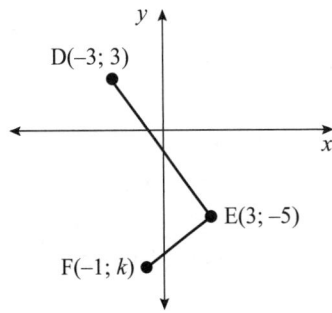

3.1 Calculate the length of DE. (2)

3.2 Calculate the gradient of DE. (2)

3.3 Determine the value of k if $D\hat{E}F = 90°$. (4)

3.4 If $k = -8$, determine the coordinates of M, the midpoint of DF. (2)

3.5 Determine the coordinates of point G such that quadrilateral GEFD is a rectangle. (2)

4. The plan for a wheelchair ramp has a gradient of $\frac{1}{7}$ from point A to C with BC $\perp$ AB: C(b; 8), B(12; 4) and A(a; 4).

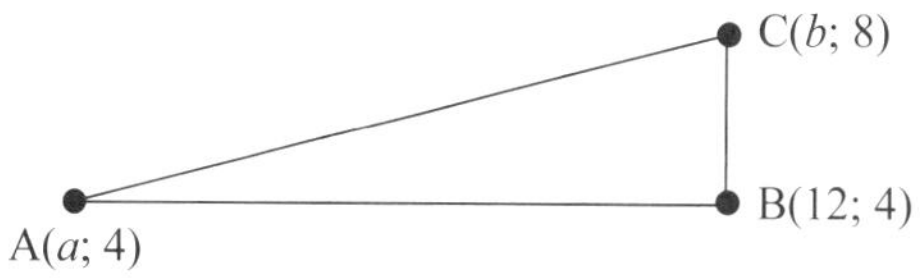

Use the information to find the values for a and b. (4)

5. Given: E(0; 0) and D(12; 15). Find the coordinates of the:

5.1 midpoint of ED (2)

5.2 points that divide ED into FOUR equal parts (4)

5.3 points that divide ED into THREE equal parts. (4)

Total 50

In Grade 10, you will study growth linked to simple and compound interest and the effects of fluctuating foreign exchange rates. You will solve problems related to hire purchase agreements, for which simple interest will be used. For problems that involve inflation and population growth, you will learn how to use compound interest.

Financial mathematics is a model of real-life situations and needs to be used as a tool for determining different variables related to situations. In Grade 10, these situations will take into account the growth/increase of an initial amount called the **present** value, using a situation of either constant increase or exponential increase.

Knowledge and skills for this topic

If you struggle with any of the work listed below, revise it before continuing with this Topic.

- ratio, percentage and decimals
- reading and interpreting tables and graphs
- a basic understanding of exponents
- good calculator knowledge and how to use the memory functions on the calculator.

Content of final exam

- Perform calculations and solve problems using the simple interest formula: $A = P(1 + in)$
- Perform calculations and solve problems using the compound interest formula: $A = P(1 + i)^n$
- Perform calculations using different exchange rates.

Calculator work essential for financial mathematics

It is critical to be able to change percentages to decimals and decimals to percentages.

Note

When doing calculations in financial mathematics, work to two decimal places. This is an automatic assumption for financial calculations; the examiner will not always specify this in the question.

Worked examples

Change each percentage into a decimal.

1. 20%   **2.** 8%   **3.** 7,5%   **4.** 4,7%

Solutions

Percentages are rate/ratios related to 100%.

1. $\text{Decimal} = \frac{\text{stated \%}}{100\%} = \frac{20}{100} = 0{,}2$ **2.** $\text{Decimal} = \frac{\text{stated \%}}{100\%} = \frac{8}{100} = 0{,}08$

3. $\text{Decimal} = \frac{\text{stated \%}}{100\%} = \frac{7{,}5}{100} = 0{,}075$ **4.** $\text{Decimal} = \frac{\text{stated \%}}{100\%} = \frac{4{,}7}{100} = 0{,}047$

Exercise 1

1. Change each percentage to a decimal.

   1.1 4,5% **1.2** 5,1% **1.3** 4,9% **1.4** 5,6%

Use a calculator to determine the value of the following variables, correct to two decimal places.

2. $P = 2\,000(1 + 0,055(3))$

3. $P = \dfrac{5\,000}{(1 + 0,045(3))}$

4. $n = \dfrac{\left(\frac{4\,500}{2\,650} - 1\right)}{0,04}$

5. $i = \dfrac{\left(\frac{4\,500}{2\,650} - 1\right)}{5}$

6. $P = 20\,000\left(1 + \frac{0,0485}{12}(4 \times 12)\right)$

7. $P = 20\,000(1 + 0,051)^5$

8. $P = \dfrac{7\,500}{(1 + 0,048)^3}$

9. $i = \sqrt[3]{\dfrac{10\,000}{7\,800}} - 1$

10. $P = 2\,000\left(1 + \frac{0,05}{4}(3 \times 4)\right)$ Could be simple interest with quarterly payments.

11. $P = 2\,000\left(1 + \frac{0,05}{12}(3 \times 12)\right)$ Could be simple interest with monthly payments.

12. $P = 20\,000\left(1 + \frac{0,051}{12}\right)^{5 \times 12}$ Could be interest per annum, compounded monthly.

13. $P = \dfrac{7\,500}{\left(1 + \frac{0,048}{12}\right)^{3 \times 12}}$ Could be interest per annum, compounded monthly.

Interest and interest rates

The interest earned on an investment or charged on a loan is the **actual amount of money**.

The **interest rate** earned or charged is the **percentage or rate of interest** applied over a stated time period.

There are two different types of interest rates:

- simple interest increase or growth
- compound interest increase or growth.

The interest earned is stated as a percentage (r = rate of interest), generally per annum.

When items or goods increase in value using the words **simple interest**, then **straight-line appreciation** can be used in the question. This means that the item increases by the same amount every year.

If items and goods increase in value using the words **compound interest**, then **exponential growth** can be used. The growth of the money becomes more for every time period.

The formulae and graphs representing simple and compound increase/growth are given in the following table.

Type of increase/ growth	Formula	Graph	Type of increase
Simple interest	$F = P(1 + in)$		Straight-line or 'constant rate'
Compound interest	$F = P(1 + i)^n$		Exponential or 'increases more and more with time'

In the formulae in the table, F represents the *Future value of money* and P represents the *Present value*, while *i* represents the interest rate per period (as a decimal), and *n* represents the number of interest periods.

Simple interest

Interest is calculated as a percentage of the **present value only**.

We generally associate hire purchase with simple interest, unless otherwise stated. Hire purchase occurs through a contract made up of regular payments. Sometimes money may be borrowed for a few months or over a period of days.

The formula is: F or A = P(1 + *in*) where:

F or A is the accumulated amount or **future value** of money.
P is the principal amount or **present value** of money invested/borrowed.
i is the interest rate, as a decimal, per period.
n is the number of interest periods.

Worked examples

1. Tim borrows R20 000 for 3 years at 8% p.a. simple interest. How much interest did he pay?

2. Bongani borrows R5 000 from a friend and is charged a simple interest rate of 9% per annum. How much interest will he pay at the end of 4 months?

3. John borrows R20 000 for 90 days at 8% p.a. simple interest (assume 365 days per year). How much interest did he pay?

Solutions

1. F = P(1 + *in*), where P = R20 000

   $i = \frac{8}{100} = 0{,}08$ *i* represents the interest rate per period as a decimal.

   and $n = 3$ *n* represents the number of interest periods.
   F = 20 000(1 + 0,08(3)) = R24 800
   Interest paid = R4 800

2. F = P(1 + *in*), where P = R5 000,

   $i = \frac{9}{100} \times \frac{1}{12} = 0{,}0075$ *i* represents the interest rate per period i.e. 1 month.

   and $n = 4$ *n* represents the number of interest periods i.e. 4 months.

   F = 5 000(1 + 0,0075(4)) = R5 150
   Interest paid = R150

3. F = P(1 + *in*), where P = R20 000,

   $i = \frac{8}{100} \times \frac{1}{365} = \frac{0{,}08}{365} = 0{,}000219 \ldots$ *i* represents the interest rate per day as a decimal.

   and $n = 90$ *n* represents the number of interest periods i.e. 90 days.

   $F = 20\,000\left(1 + \frac{0{,}08}{365} \times 90\right) = R20\,394{,}52$
   Interest paid = R394,52

Exercise 2

1. Thulani invests R10 000 for 5 years earning a simple interest rate of 9% per annum. How much interest does he earn?

2. Jane borrows R7 000 from a friend, who charges her simple interest at 12% per annum. She will repay this money over a period of 120 days.
   2.1 What is the interest rate per day?
   2.2 How much will Jane pay to her friend in total?
   2.3 How much interest will Jane pay?

3. Karabo invests R25 500 for 3 years and earns simple interest at a rate of 8% per annum.
   3.1 How much money will Karabo have in this account after 3 years?
   3.2 How much interest will Karabo earn over 3 years?
   3.3 If she had left the money in this account for double the time, i.e. 6 years, determine the following:
      3.3.1 The amount of money she will have after 6 years
      3.3.2 The amount of interest she will have after 6 years

4. Nazeem invests R15 000 into an account earning simple interest at a rate of 5,4% per annum. If he takes the money out after 4 months, how much money will he have?

Manipulation of the simple interest formula

At times, different variables, such as the original amount, the interest rate or the time period, need to be calculated.

To calculate	Manipulation of simple interest formula: $F = P(1 + in)$
P	$P = \dfrac{F}{(1 + in)}$
i	$1 + in = \dfrac{F}{P}$ $in = \dfrac{F}{P} - 1$ $i = \dfrac{\frac{F}{P} - 1}{n}$
n	$n = \dfrac{\frac{F}{P} - 1}{i}$

Worked examples

1. Peter paid R25 000 at the end of 5 years in settlement of a loan. He was charged simple interest at 5% per annum. How much did Peter borrow?

2. Determine the annual simple interest rate Sandra will earn if she invests R5 800 and is able to withdraw R8 250 after 4 years.

3. Gerrie invests R8 000 into an account at a simple interest rate of 9% per annum. How many years will it take to accumulate R15 000?

Solutions

1. $P = \dfrac{F}{(1+in)}$, F = R25 000, r = 9% p.a. so i = 0,05 and n = 5

   $P = \dfrac{25\,000}{(1+0,05(5))} = R20\,000$

2. $i = \dfrac{\frac{F}{P}-1}{n} = \dfrac{\frac{8\,250}{5\,800}-1}{4} = 0,1056$. The interest rate, r = 10,56% p.a.

3. $n = \dfrac{\frac{F}{P}-1}{i}$, F = R15 000, P = R8 000 and i = 0,09:

   $n = \dfrac{\frac{15\,000}{8\,000}-1}{0,09} = 9,72$ years, so it will take approximately 10 years.

Exercise 3

1. Joshua lends money to people, charging them simple interest. Determine the original amount for each loan based on the given information.

   1.1 Future Value (F) = R20 000, rate 5% per annum over 2 years

   1.2 F = R100 000, rate 7% p.a. over 120 days

   1.3 F = R55 000, rate 6% per annum over 5 years

   1.4 F = R5 000, rate 2% per month over 1 year

2. Molly lends R6 000 to Xolani on condition he pays her R8 000 after 2 years. Use the simple interest formula to find the rate of interest Molly charged him.

3. How long (correct to the nearest month) will it take to pay off each loans?

   3.1 A loan of R5 000 at 8% p.a. with a final amount of R8 000

   3.2 A loan of R65 000 at 12% p.a. with a final amount of R75 000

   3.3 A loan of R10 000 at 14% p.a. with a final amount of R15 000

4. Phemelo invests R36 255 into an account earning simple interest at a rate of 6% p.a. Determine the future value of the money after:

   4.1 1 year   **4.2** 2 years   **4.3** 3 years   **4.4** 4 years.

   Determine the interest she earned after:

   4.5 2 years   **4.6** 4 years.

   4.7 What do you observe about the answers to questions 4.1 to 4.4? Explain why this happens.

   4.8 What do you observe about the answers to questions 4.5 and 4.6?

Hire purchase

Hire purchase is a method of regular payments, using a simple interest rate to purchase expensive household items over a fixed period of time.

How to work with hire purchase:

Note

The interest may also be referred to as a 'constant rate'.

- If there is a deposit, calculate this first.
- Deduct the deposit from the item price to obtain the balance outstanding or Present Value (P).
- Now calculate the Future Value (F) using the simple interest formula.
- Divide F by the number of regular payments to obtain the value of each payment.
- To determine the interest paid, subtract the original price of the item from all payments that have been made, including the deposit.

Worked examples

1. Fred wants to buy a laptop advertised for R12 800. A 15% deposit is required; thereafter interest will be charged at 9% per annum on the balance. If he pays off the balance of the loan in 18 monthly instalments, determine:

   1.1 the value of the deposit

   1.2 the value of the monthly instalments

   1.3 the amount of interest paid.

2. Jana signed a hire purchase agreement for a washing machine and stove, the total selling price being R6 500. She pays a deposit of R1 625 and is required to pay 18 monthly instalments of R295,20 each.

   2.1 Determine the Present (initial) Value of the loan.

   2.2 Determine the total value of all payments made.

   2.3 How much interest did Jana pay?

   2.4 Calculate the interest rate charged per annum.

Solutions

1. **1.1** Deposit $= \frac{15}{100} \times \frac{12\,800}{1} = $ R1 920

   1.2 Balance = R12 800 − R1 920 = R10 880

   Total to be repaid: Future value $= P(1 + in)$
   $$= R10\,880\left(1 + \frac{0,09}{12}(18)\right)$$
   $$= R12\,348,80$$

   Monthly instalments = R12 348,80 ÷ 18 = R686,04

   1.3 Interest paid = money paid − cost price

   Interest paid = R686,04 × 18 + R1 920 − R12 800
   $$= R1\,468,80$$

2. **2.1** P = R6 500 − R1 625 = R4 875

   2.2 Total = R1 625 + R295,20 × 18
   $$= R6\,938,60$$

   2.3 Interest = total of all payments − selling price
   $$= R6\,938,60 - R6\,500,00$$
   $$= R438,60$$

   2.4 To calculate the interest rate:

   P = R4 875

   F = R295,20 × 18 = R5 313,60

   $i = \frac{\frac{F}{P} - 1}{n} = \frac{\frac{5\,313,60}{4\,875} - 1}{18} = 0,005$ per month i.e. interest rate is 6% per annum.

   Multiply by 12, because the value of n is in months, so i is per month.

Exercise 4

1. Angelina purchases a bicycle with a marked price of R7 500. She can buy it on hire purchase with a deposit of R1 200 and 9% p.a. interest rate on the balance outstanding. She must repay the outstanding amount over 20 months.

Determine:

1.1 the balance outstanding

1.2 the interest rate per period

1.3 the monthly payments

1.4 the interest paid.

2. The cash price of a cricket bat is R4 450. Tulani buys it on hire purchase by making a deposit of 20% of the cash price and paying 15 monthly instalments. If the interest rate is 12% per annum, determine:

   2.1 the amount of the deposit

   2.2 the balance outstanding after the deposit is paid

   2.3 the future value of the loan

   2.4 the monthly instalment

   2.5 the interest paid.

3. A fridge is advertised for R6 200 cash or a deposit of R1 200 and 36 monthly instalments of R174,31. Determine:

   3.1 the future value of the fridge after 36 months

   3.2 the interest rate per annum.

4. Derek buys furniture on hire purchase over 2 years from a furniture store. The total value of the furniture bought is R16 500 and the interest charged is 10% p.a. Calculate the monthly instalment Derek needs to pay.

Compound interest

Interest is calculated on the **present** value or **principal** amount for the first interest period. Thereafter it is calculated on the present value together with the interest. Compound interest is used on calculations involving inflation, population growth and many other real-life problems where exponential growth appears to occur.

Worked examples

1. Rebecca takes a short-term loan of R5 000 from a bank. She is charged interest at 8% p.a. compounded annually. How much interest will she pay if she settles the loan at the end of 2 years?

2. Terry borrows R20 000 for 3 years at 8% p.a. compounded half-yearly. How much interest did he pay?

3. Mavis borrows R8 000 for 9 months at 10% p.a. compounded monthly. How much interest did she pay?

Solutions

1. $F = P(1 + i)^n$, $P = $ R5 000, $i = \frac{8}{100} = 0,08$ *i* represents the interest rate per period as a decimal, and *n* represents the number of interest periods.

   and

   $n = 2$

   $F = 5\,000(1 + 0,08)^2 = $ R5 832

   Interest paid = R832

2. $F = P(1 + i)^n$, $P = R20\,000$, $i = \frac{8}{100} \times \frac{6}{12} = 0{,}04$

    and $n = 3 \times 2 = 6$

    $F = 20\,000(1 + 0{,}04)^6 = R25\,306{,}38$

    Interest paid $= R5\,306{,}38$

3. Interest is compounded monthly, so the periods of interest are monthly calculations.

    $F = P(1 + i)^n$, $P = R8\,000$, $i = \frac{10}{100} \times \frac{1}{12} = 0{,}0083$ *i* in this case is is a non-terminating decimal. Place this value into the memory of the calculator and use this value in your calculation.

    and $n = 9$

    $F = 8\,000(1 + 0{,}0083\ldots)^9 = R8\,620{,}39$

    Interest paid $= R620{,}39$

Exercise 5

1. Thomas invests R120 000 for 3 years earning compound interest at a rate of 6% per annum.

    1.1 How much money will Thomas have in this account after 3 years?

    1.2 How much interest does he earn?

2. Nadine deposits R4 500 into a fixed deposit account. She earns compound interest at 5,6% p.a. compounded monthly.

    2.1 Calculate the amount of money she will have in this account after 18 months.

    2.2 Determine the interest she earns.

3. Chwayita wins R250 000 in a competition. She decides to use 20% of her winnings on an overseas trip. She decides to invest the rest of the money and is faced with two options:

    Option A is an account earning 5,75% per annum compounded monthly.
    Option B is an account earning simple interest at 6,2% per annum.

    Determine which option is the better under the following conditions:

    3.1 a short-term investment of 2 years

    3.2 a long-term investment of 10 years.

4. Determine the value of money in each account:

    4.1 a deposit of R2 800 for 5 years earning 4,8% p.a. compounded quarterly

    4.2 a deposit of R2 800 for 5 years earning 4,8% p.a. compounded monthly

    4.3 a deposit of R2 800 for 10 years earning 4,8% p.a. compounded monthly

Note

The present value of each investment is the same.

5. Use your answers to question 4 to answer these questions. Give a reason for each answer.

    5.1 If the compounding of the interest rates change and the number of years remain the same, what happens to the future values of the investments?

    5.2 If the length of time of the investment changes, what happens to the future values?

5.3 Using the answers from questions 4.2 and 4.3, does the interest earned double when the number of years is doubled? Validate this answer with calculations.

Manipulation of the compound interest formula

At times, different variables, such as the original amount, the interest rate or the time period, need to be calculated.

To calculate	Manipulation of compound interest formula $F = P(1 + i)^n$
P (initial amount)	$P = \dfrac{F}{(1 + i)^n}$
i (interest rate per period)	$(1 + i)^n = \dfrac{F}{P}$ $i = \sqrt[n]{\dfrac{F}{P}} - 1$

Worked examples

1. Lamela paid R25 000 at the end of 5 years in settlement of a loan on which he was charged 5% compounded annually. How much did Lamela borrow?

2. On Monday Irfaan borrows R10 from a friend and promises to give him R12, on Friday (5 days later), as payment for this loan. Determine the daily interest rate he paid his friend.

Note

Extremely short-term loans carry a very high rate of interest. In Grade 11 you will learn how to convert this to an effective annual interest rate.

Solutions

1. Calculate the present value (original amount of the loan).
$$P = \frac{F}{(1 + i)^n} = \frac{25\,000}{(1 + 0,05)^5} = R19\,588,15$$

2. $i = \sqrt[n]{\dfrac{F}{P}} - 1 = \sqrt[5]{\dfrac{12}{10}} - 1 = 0,03713728934$
The daily interest rate is 3,7%.

Exercise 6

1. Masekhane lends money to people, charging them compound interest. They will repay the loan at the end of the time specified as a lump sum. For each loan, (i) determine the original amount (P) and (ii) the interest paid (i).

    1.1 Final amount (F) = R16 585, rate 4,8% per annum over 3 years

    1.2 F = R10 000, rate 16,5% p.a. compounded daily, over 120 days

    1.3 F = R5 800, rate 6% per annum compounded quarterly, over 5 years

    1.4 F = R1 000, rate 2% per month over 1 year

    1.5 F = R1 000, rate 2% p.a. compounded monthly for 1 year

2. Rohelna lends R7 500 to Xolani on condition he pays her R10 000 after two years. Using the compound interest formula, what annual rate of interest did Rohelna charge him?

3. Determine the annual interest rate for each loan.

    3.1 A loan of R5 000 taken out for 3 years with a future value (final amount) of R8 000

    3.2 A loan of R65 000 taken out for 4 years with a future value of R75 000

    3.3 A loan of R10 000 taken out for 2 years with a future value of R15 000

4. Determine the total amount (F) after 10 years if an amount of R15 000 (present value) is invested as follows:

   4.1 9% per annum simple interest

   4.2 % p.a. compound interest.

5. Maya invests R35 000 at 9% p.a. compounded semi-annually. Determine the future value of her investment after 5 years.

6. Raphi is able to draw R68 575 after investing a lump sum of money 3 years ago into an account for which she earned 5,5% p.a. compounded weekly. How much money did she invest?

7. Zazalathi is able to withdraw R25 768,45 after an initial investment of R15 000 in an account earning 5,8% p.a. simple interest. Determine the number of years of his investment.

8. Geena deposits R350 000 into a savings account, leaving the money in the account for 5 years. Determine how much money will be in the account if interest is calculated as follows:

   8.1 5,6% p.a. simple interest

   8.2 5,3% p.a. compounded monthly

   8.3 Which investment is better? Validate your answer.

Inflation

Inflation is a sustained increase in the national average price level. Under inflation, the buying power of your money gets less year by year. If inflation is at 10% per year, at the end of the year goods and services will cost 10% more than they did at the beginning of the year. This table summarises the inflation history of South Africa between 1994 and 2013.

Historic inflation in South Africa (CPI) – by year			
Annual inflation (Dec vs Dec)	Inflation %	Annual inflation (Dec vs Dec)	Inflation %
2013	5,30	2003	−1,63
2012	5,71	2002	13,51
2011	6,41	2001	4,59
2010	3,37	2000	6,99
2009	6,04	1999	2,24
2008	9,35	1998	8,95
2007	7,57	1997	6,17
2006	4,82	1996	9,31
2005	2,02	1995	6,93
2004	2,20	1994	9,83

Source: http://www.inflation.eu/inflation-rates/south-africa/historic-inflation/cpi-inflation-south-africa.aspx, accessed 14 December 2014

Inflation and buying power

People will only have more buying power if wages and salaries go up more than inflation. So if someone's wages increase by 8% per year, but inflation is 5% per year, real buying power increases by only 3%.

In manufacturing, prices rise to cover wage increases and other price increases. This results in an **upward inflationary spiral**.

Inflation can sometimes spiral completely out of control, called **hyperinflation**. This happened in Germany after World War I; in 1923 the Price Index was $1,02 \times 10^{10}$, a monthly inflation rate of 322%, which meant prices quadrupled each month.

Worked examples

A 125 g packet of chips costs R9,49. The average Price Index for 2013 for South Africa of 5,30% per annum is maintained. Determine the price of a 125 g packet of chips in:

1. 5 years' time

2. 10 years' time

3. Determine the price increase over the first 5 years and the second 5 years.

4. Although the average rate of increase (percentage) is constant, what is happening with the actual price increases?

Solutions

1. $F = P(1 + i)^n = 9,49(1 + 0,053)^5 = R12,29$

2. $F = P(1 + i)^n = 9,49(1 + 0,053)^{10} = R15,91$

3. Price increase 1st 5 years = R2,80
   Price increase 2nd 5 years = R3,62

4. Prices are increasing more rapidly with time.

Exercise 7

1. Graph A shows the annual inflation rate in South Africa (a developing country) from 1958 to 2013. Graph B shows the annual inflation rate in Great Britain (a developed country) from 1958 to 2013.

Graph A: Historic CPI inflation South Africa (yearly basis)

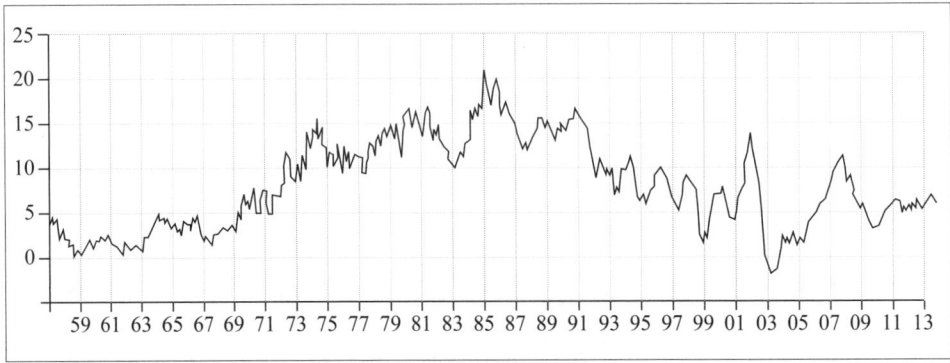

Source: http://www.inflation.eu/inflation-rates/south-africa/historic-inflation/cpi-inflation-south-africa.aspx, accessed 14 December 2014

Graph B: Historic CPI inflation Great Britain (yearly basis)

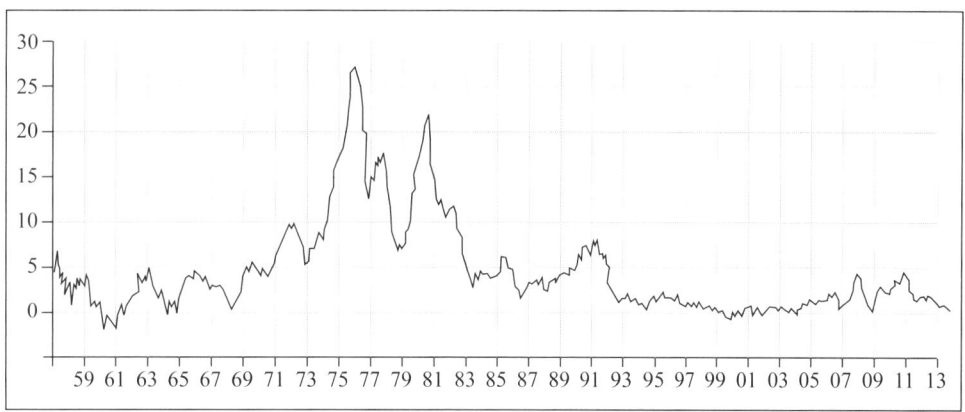

Source: http://www.inflation.eu/inflation-rates/great-britain/historic-inflation/cpi-inflation-great-britain.aspx, accessed 14 December 2014

Note

In tests and exams you could be asked to make predictions based on the information given in graphs

1.1 Use the graphs to write a paragraph explaining how South African inflation rates compare with the rates of Great Britain over the period of time shown.

1.2 Write a paragraph on what you think may happen with the South African inflation rate.

2. Assume a consistent inflation rate of 5,3% for each cost.

2.1 What is the expected cost of a new car in 3 years' time that costs R168 000 today?

2.2 What would it cost to replace a washing machine in 5 years' time that has a present selling price of R3 599?

2.3 What will be the cost of a holiday in Zanzibar in 2 years' time, that now costs R12 000?

2.4 Workers earning R36 000 per annum get a wage increase to match inflation. What can they expect to earn in 3 years' time?

Exchange rates

The exchange rate is the price of one currency (money unit) in terms of another, for example:

R10,66 = 1 US dollar (US$1)
R17,88 = 1 British pound (£1)
R14,30 = 1 Euro (€1)
R0,09 = 1 Jamaican dollar (J$1)
R0,33 = 1 Thai baht (฿1)

These rates change daily and are determined on the foreign exchange market.

Worked examples

1. Lamela is travelling to America. He buys US dollars from the bank at R10,66 per US dollar. How many dollars will he get for R25 000?

2. Bianca is going on an awesome island holiday to Thailand. She buys Thai bahts from the bank and receives ฿5 000. How much rand must she pay the bank?

Solutions

1. R10,66 = US$1

R25 000 = $\frac{25\,000}{10,66}$ = US$2 345,22

2. ฿1 = R0,33

฿5 000 = R0,33 × 5 000 = R1 650

Exercise 8

1. Use the exchange rates given on the previous page to convert these amounts to ZAR (South African rand):

   1.1 US$27 589 **1.2** ฿4 375 **1.3** £19 000

   1.4 J$19 000 **1.5** €55 725

2. Convert the following ZAR values to the currency in brackets.

   2.1 R20 000 (British pound) **2.2** R10 000 (Thai baht)

   2.3 R15 000 (US dollar) **2.4** R5 000 (Jamaican dollar)

   2.5 R15 000 (euro)

Test A: Knowledge and routine procedures

1. Colleen purchased a new motor bike for R32 000. The annual inflation rate of motor bikes is 12%. Determine how much she will pay for the same type of bike after 5 years under the following conditions:

   1.1 straight-line appreciation (3)

   1.2 exponential appreciation. (3)

2. Dean buys furniture on hire purchase over 3 years. The total value of the furniture bought is R26 800. He is required to pay a deposit of 12% and the interest charged is 9,5% p.a. Calculate:

   2.1 the present value of the loan (4)

   2.2 the future value of the loan (3)

   2.3 the monthly instalment (2)

   2.4 the total interest he will pay. (3)

3. A car is purchased for R168 000. The inflation is expected to be 7% per annum for the next 5 years.

   3.1 Calculate the price of a new car, of the same type, in 5 years' time. (3)

   3.2 A deposit of 20% is required. Determine the present value of the loan. (3)

4. A financial institution offers two investment options:
   Option A: 8,5% per annum simple interest
   Option B: 8,2% per annum compounded monthly

   4.1 Pat has R100 000 to invest for 1 year. Which option would be the best for Pat? Show all workings. (5)

   4.2 Would Pat's choice be different if she was investing the money for 6 months? Show all your working. (5)

5. Convert each amount to ZAR (South African rand).

   5.1 US$12 000 if R10,66 = US$1 (2)

   5.2 ฿6 000 if R0,33 = ฿1 (2)

   5.3 £7 500 if R17,88 = £1 (2)

   5.4 J$7 500 if R0,09 = J$1 (2)

6. Convert the following rand values to the currency in brackets.

   6.1 R20 000 (British pounds) if R17,88 = £1 (2)

6.2 R10 000 (Thai baht) if R0,33 = ฿1 (2)

6.3 R15 000 (Euro) if R14,30 = €1 (2)

6.4 R5 000 (Jamaican dollar) if R0,09 = J$1 (2)

Total 50

Test B: Complex procedures and problem solving

1. R9 500 was invested in a fund, which after 2 years had a value of R11 238,50. Calculate the annual interest rate under the following conditions:

   1.1 The fund earned simple interest. (4)

   1.2 The interest is compounded annually. (4)

2. Thandi and Xolani wish to buy their first home. The bank will allow their repayments to be no greater than 30% of their combined net salaries. Thandi's gross salary is R13 650 per month with deductions of 20% from her salary. Xolani's gross salary is R12 890 per month with monthly deductions of 22% from his salary.

   2.1 Determine the value of each net salary (take home pay after deductions). (4)

   2.2 Determine the maximum value of the bond repayments they can afford. (5)

   2.3 They apply for a bond of R650 000. The bank charges interest at 8% p.a. compounded monthly on bonds.

       2.3.1 Determine the interest that must be paid after the first month on a bond of this nature. (4)

       2.3.2 Show all workings to determine whether they can afford this bond. (4)

3. A newly married couple buy a washing machine for R3 299 and a dishwasher for R3 699, on hire purchase, for their new home. They pay a deposit of 20% of the purchase price, after which they make regular monthly payments for 36 months. Interest is charged at 18% per annum.

   3.1 Calculate the value of the deposit. (3)

   3.2 Determine the value of the money they still owe. (3)

   3.3 Determine the monthly instalments. (4)

   3.4 Calculate how much interest they paid. (3)

4. **4.1** Due to inflation, the purchase price of a photocopier valued at R64 000 increases at a rate of 7,5% per annum. Determine the cost of a new machine in 8 years' time. (4)

   4.2 A car that costs R165 000 is advertised in the following way: 'No deposit necessary and the first payment due 5 months after date of purchase'. The interest rate quoted is 9,8% per annum compounded monthly.

       4.2.1 Determine the amount owing 5 months after the purchase date. (3)

       4.2.2 Calculate the interest paid over this period of 5 months. (2)

       4.2.3 If a deposit of R20 000 is paid on the date of purchase, determine the amount owing 5 months after the purchase date. (3)

   Total 50

Test C: Content and breakdown as for exam

1. **1.1** Determine the annual percentage interest rate, compounded annually, if a lump sum value is invested in order for it to double in 6 years. (4)

   1.2 R20 400 is invested at 5,2% p.a. for 4 years. Determine the value of the investment if:

   1.2.1 the investment earns simple interest (3)

   1.2.2 the investment earns compound interest (3)

   1.2.3 calculate the interest earned in question 1.2.1 (2)

   1.2.4 calculate the interest earned in question 1.2.2. (2)

2. Two brothers each inherit an amount of R42 500 and decide to invest the money for a period of 7 years. They invest the money as follows:

   Ben: 6,5% per annum simple interest. At the end of 7 years, Ben will receive a bonus of exactly 4% of the present value of the investment.

   Josh: 6% per annum compounded monthly.

   Who will receive the larger amount after 7 years? Validate your answer with appropriate calculations. (10)

3. Fahieda wants to purchase a house that costs R250 000. She is required to pay a 15% deposit. She will borrow the balance from the bank. Calculate:

   3.1 the value of the deposit (2)

   3.2 the amount Fahieda will borrow from the bank (2)

   3.3 The bank charges interest at a rate of 8% per annum, compounded monthly on the loan amount. Determine the amount of interest Fahieda will pay at the end of the first month of the loan. (4)

4. **4.1** A loan of R85 000 is taken out to start a small business. The loan must be repaid in total at the end of 6 years. Determine:

   4.1.1 how much must be repaid if the interest is calculated at 9% per annum simple interest (3)

   4.1.2 the interest paid. (2)

   4.2 R5 000 is invested into a fund paying i% compound interest per annum. After 3 years the value of the fund is R6 850. Calculate the value of i and hence the annual interest rate. (3)

5. **5.1** Convert R320 000 to British pounds if R17,88 = £1. (2)

   5.2 Convert €58 450 to ZAR (South African Rand) if R14,30 = €1. (2)

   5.3 Gershwin takes a gap year and on arrival in the United States buys a small car for US$1 325. Calculate the cost of this car in ZAR if R10,66 = US$1. (2)

   5.4 Lionel is travelling to Jamaica. He buys Jamaican dollars from the bank at R0,09 per Jamaican dollar. How many Jamaican dollars will he get for R15 000? (2)

   5.5 Penny wants to purchase an item from Brazil. She buys the item for 225 Brazilian Real (R$). If R4,66=R$1, determine the equivalent in ZAR she paid for the item. (2)

   Total 50

In this topic, we work with representation and interpretation of data.

Knowledge and skills for this topic

If you struggle with any of the statistics studied in Grades 8 and 9, revise it before continuing with this Topic.

Content of final exam

- Apply measures of central tendency for ungrouped data:
  - mean, median and mode.
- Apply measures of central tendency for grouped data:
  - calculate **mean estimate** of grouped and ungrouped data
  - identify **modal interval**
  - identify **interval** in which the mode lies.
- Find the range as a measure of dispersion extended to include:
  - percentiles
  - quartiles
  - interquartile range
  - semi-interquartile range
- Determine a five-point summary (maximum, minimum and quartiles) and a box-and-whisker diagram.
- Use the statistical summaries (measures of central tendency and dispersion) and graphs to analyse and make meaningful comments on the context associated with the given data.

Frequency tables

Frequency is a measure of how often something happens, and a frequency table shows the number of times a score or event takes place, or happens.

Worked example

Below are the marks out of 10 obtained by a group of 30 Grade 10 learners for a short Mathematics test.

5	6	4	8	5	1	2	9	5	5
2	3	3	4	8	7	9	2	3	1
5	4	6	10	9	6	7	6	5	6

Draw a frequency table for the above data and answer the following questions.

1. Which test mark appears most frequently?
2. Which test mark appears least frequently?
3. How many learners scored 9 out of 10?
4. How many learners scored 50% or more?

Solution

Mark	Tally	Frequency
1				2
2					3
3					3
4					3
5								6
6							5
7				2
8				2
9					3
10			1
Total		**30**

1. 5

2. 10

3. 3 learners

4. $6 + 5 + 2 + 2 + 3 + 1 = 19$ learners

Discrete and continuous data

Only whole numbers are used in the data in the previous example. This means that they are discrete data and can only take on certain values. (Discrete means separate.) A similar type of data, which is called nominal data, refers to data that can only belong to certain categories, such as colour, name, type or class. Discrete and nominal data can easily be summarised in a frequency table.

Continuous data can have many possible values, such as height, time, mass and temperature. It is difficult to organise such data in a frequency table because there will be very few bits of information that are the same. Instead, the data is first grouped into suitable intervals.

How to group data

In order to group data effectively you need to choose the start of the first interval, which must be below the lowest score, and the end of the last interval, which must be above the highest score. You should size the intervals so that you have between 5 and 10 groups. Intervals do not have to be the same size – particularly if there are outliers (scores that fall far below or above the rest of the data).

You should be careful when indicating the size of an interval you have chosen. For example, if you choose an interval of 10 and the first two groups are 0–10 and 10–20, where will you place a score of 10? Will it be in the first or second group? To avoid this problem, you can indicate the intervals in the following ways:

* $0 \leq x < 10$ and $10 \leq x < 20$
* $0–9^+$ and $10–19^+$

This means that the first interval is from 0 to 10, but does not include a score of 10.

Measures of central tendency

Measures of central tendency are all measures of the average of a set of data.

*   The mean is the closest measure of what we usually call the average. Calculate it as follows:

    mean = $\frac{\text{sum of all scores}}{\text{number of scores}} \bar{x} = \frac{\sum x}{n}$

*   The median score is the middle score. If n is the number of scores:

    median = $\left(\frac{n+1}{2}\right)$th score

    Arrange the scores in order from lowest to highest. If there is an even number of scores and you have to find the 4,5th score, for example, then you must find the mean of the 4th and 5th scores.

    If there is an odd number of scores, we simply use the score found by using the formula. For example if there are 13 scores, the median is score number 7, i.e. $\frac{(13+1)}{2} = 7$.

*   The mode is the score that occurs most often.

Averages of grouped and table data

Sometimes you may be given grouped data or data in a frequency table and not the raw scores. It is possible to work out the averages in these situations, but they are often approximations.

Worked examples

1. Find the mean, median and mode of the data in the following frequency table.

Score	Frequency
5	3
6	4
7	2
8	3
Total	**12**

2. The following table shows the examination marks and symbol distribution for a class of 40 learners.

Symbol	Interval %	Frequency
G	20–29	2
H	30–39	6
E	40–49	4
D	50–59	12
C	60–69	10
B	70–79	3
A	80–89	2
A$^+$	90–100	1
Total		**40**

2.1 Determine an estimate for the mean, correct to two decimal places.

2.2 Determine the median interval.

2.3 Determine the modal interval.

Solutions

1. To find the mean, we need to find the total of the scores. Do this by multiplying the frequency of each mark by the mark itself and write this value in another column of the table.

Score	Frequency	Frequency × score
5	3	$3 \times 5 = 15$
6	4	$4 \times 6 = 24$
7	2	$2 \times 7 = 14$
8	3	$3 \times 8 = 24$
Total	12	77

$$\text{Mean} = \frac{77}{12}$$

$$= 6{,}42$$

$$\text{Median} = \text{middle score}$$

$$= \left(\frac{(n+1)}{2}\right)\text{th mark}$$

$$= \left(\frac{13}{2}\right)\text{th mark}$$

$$= 6{,}5\text{th mark}$$

$$= 6 \qquad \text{The 6th mark and the 7th mark are both 6; from the table.}$$

$$\text{Mode} = 6 \qquad \text{This is the highest frequency in the table.}$$

2.1 To estimate the mean for grouped data, we need to find the total of all the marks. Since we don't know what each mark is, we have to estimate the mean by using the mid-interval mark as an estimate. We then multiply each mid-interval mark by the frequency.

> **Finding the mid-interval score**
> Multiply the lower and upper limit for the interval and divide the answer by 2.

Symbol	Interval %	Frequency (f)	Mid-interval score (x)	Frequency × mid-interval ($f \times x$)	Cumulative frequency (Cf)
G	20–29	2	24,5	$2 \times 24{,}5 = 49$	2
H	30–39	6	34,5	$6 \times 34{,}5 = 207$	$2 + 6 = 8$
E	40–49	4	44,5	$4 \times 44{,}5 = 178$	$8 + 4 = 12$
D	50–59	12	55,5	$12 \times 55{,}5 = 666$	$12 + 12 = 24$
C	60–69	10	65,5	$10 \times 65{,}5 = 655$	$24 + 10 = 34$
B	70–79	3	75,5	$3 \times 75{,}5 = 226{,}5$	$34 + 3 = 37$
A	80–89	2	85,5	$2 \times 85{,}5 = 171$	$37 + 2 = 39$
A$^+$	90–100	1	95	$1 \times 95 = 95$	$39 + 1 = 40$
Total		40		$\sum(fx) = 2\,247{,}5$	

$$\text{Mean (approx)} = \frac{\sum(fx)}{n} = \frac{2\,247{,}5}{40} = 56{,}19\%$$

2.2 Determining **cumulative frequency** is useful when determining the median interval. The median in this case lies between the 20th and the 21st scores, both of which lie in the interval 50–59. So the median interval is: 50–59 or a D symbol.

2.3 The mode is the most common mark. Since we do not know what the marks are, we cannot tell which mark is the most common. We can, however, give the modal interval, which is 50–59 or a D symbol.

Measures of dispersion

Knowing the average of a set of scores is interesting, but it does not tell us much about the data itself. It is often useful to know how the information is spread around the average. To do this we can measure the dispersion of the data. There are several ways to do this.

Range

The range of a set of data is the simplest measure of dispersion as it gives the difference between the highest and the lowest values.

Range = highest value – lowest value

The interquartile and semi-interquartile range

The interquartile range gives a better idea of the spread of data as it is not affected by outliers (scores that are extremely high or low), unlike the range.

In order to calculate the interquartile and semi-interquartile range you need to be able to calculate percentiles and quartiles.

- **Percentiles:** A set of ordered data may be divided into percentiles by dividing it into 100 equal groups. Each group of values represents a percentile. There are various ways to calculate quartiles and percentiles.
  We suggest you use the following formula. If there are n ordered values, the kth value that marks the pth percentile is given by:

  $k = \frac{p(n+1)}{100}$

  It should be obvious that percentiles are only really a useful measure for large sets of data. There should be at least 100 values – can you see why?

- **Quartiles:** These divide a set of data into four equal groups. The first quartile (Q_1) is the value below which 25% of the scores lie, and it corresponds to the 25th percentile. The second quartile (Q_2) divides the values in half and is the same as the median. It corresponds to the 50th percentile. Likewise, the third quartile (Q_3) corresponds to the 75th percentile.

  Quartiles are only really useful for groups of 12 or more values. Can you say why?

- **Interquartile range (IQR):** The interquartile range is the difference between the first and third quartiles. It contains the middle half of the values.
  $IQR = Q_3 - Q_1$

- **Semi-interquartile range:** This range is half of the interquartile range.
  Semi-interquartile range $= \frac{IQR}{2}$
  $= \frac{Q_3 - Q_1}{2}$

Worked examples

1. The heights of 11 plants in a nursery (in centimetres) are given below.

102	104	105	107	108	109
110	112	115	118	118	

Determine the following for the data.

1.1 mean **1.2** mode **1.3** range

1.4 median **1.5** interquartile range

2. The masses (in grams) of eight bars of the same chocolate chosen at random off a supermarket shelf are given below.

109,8	109,0	110,2	110,6	109,7	110,3	111,4	108,5

Determine the interquartile range.

3. The ages of 750 learners at Matebaseke High School are given in the frequency table.

Calculate the following:

3.1 interquartile range

3.2 age of a child who falls in the 23rd percentile

3.3 semi-interquartile range

3.4 number of learners in the fourth quartile.

Age	Frequency
12	9
13	197
14	169
15	155
16	103
17	99
18	14
19	4

Solutions

1. **1.1** Mean $= \frac{102 + 104 + 105 + 107 + 108 + 109 + 110 + 112 + 115 + 118 + 118}{11}$

$= \frac{1\,208}{11}$

$= 109{,}82$ cm

1.2 Mode $= 118$ cm

1.3 Range $= 118 - 102$

$= 16$ cm

1.4 $\text{Median} = \left(\frac{(n+1)}{2}\right)\text{th value}$

$\qquad\qquad = \left(\frac{12}{2}\right)\text{th value}$

$\qquad\qquad = 6\text{th value}$

$\qquad\qquad = 109 \text{ cm}$

Note

When the data set is small we don't use the formula to find the quartiles, we simply find the middle of the middle.

1.5

102	104	105	107	108	109	110	112	115	118	118
		Q_1			Q_2			Q_3		

$\text{IQR} = Q_3 - Q_1 = 115 - 105 = 10$

2.

108,5	109,0	109,7	109,8	110,2	110,3	110,6	111,4
	Q_1					Q_3	

$$Q_2 = \frac{109,8 + 110,2}{2} = 110$$

$\text{IQR} = Q_3 - Q_1 = 110,6 - 109,0 = 1,6$

3. For larger data sets we use the relevant formulae to find quartiles and percentiles.

Note

Recall that working out the cumulative frequency will help you to identify quartiles and percentiles.

Age	Frequency	Cumulative frequency
12	9	9
13	197	206
14	169	375
15	155	530
16	103	633
17	99	732
18	14	746
19	4	750

3.1 $Q_1 = \frac{1}{4}(n+1) = \frac{1}{4}(750+1) = 187,75$

$\therefore$ Read the first quartile as data value number 188.

$\therefore Q_1 \approx 13$

$Q_3 = \frac{3}{4}(n+1)$

$\qquad = \frac{3}{4}(750+1) = 563,25$

$\therefore$ Read the third quartile as data value number 563.

$\therefore Q_3 \approx 16$

$\text{IQR} = 16 - 13 = 3$

3.2 The percentile formula is: $k = \frac{P}{100}(n+1)$

The 23rd percentile: $k = \frac{23}{100}(750+1) = 172,73$

$\therefore$ Read the 23rd percentile as data value number 173.

$\therefore P_{23} \approx 13$

3.3 The semi-interquartile range $= \frac{Q_3 - Q_1}{2} = \frac{3}{2} = 1,5$

3.4 The number of learners in the fourth quartile will be 25% of 750.

$\frac{25}{100} \times 750 = 187,5 \approx 188$

$\therefore$ 188 learners lie in the fourth quartile.

Five-number summaries and box-and-whisker diagrams

Five number summaries and box-and-whisker diagrams are used to represent and to **analyse the spread of data about the median**.

A **five-number summary** uses the following measures of dispersion:

- Minimum: the smallest value in the data set
- Lower quartile: the first quartile/median of the lower half of the data values
- Median: the second quartile/value that divides the data values in half
- Upper quartile: the third quartile/median of the upper half of the values
- Maximum: the largest value in the data set

A **five-number summary** is used to draw a box-and-whisker diagram.

A **box-and-whisker diagram** is a graphic representation of the five-point summary.

- 25% of the data values lie within each of the quarters. The size of the quarters speaks to the dispersion of the data values within each quarter.
- 50% of all the data values lie below the median and 50% lie above the median.

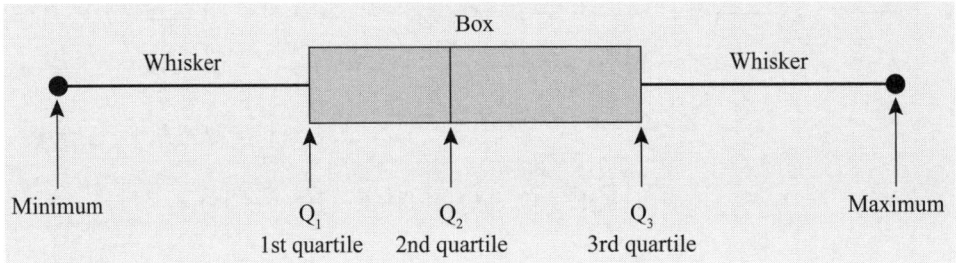

Worked examples

The test results for Ms Banks's French class are given in the table below.

77	45	67	46	67	89	34	56	55	73	77
78	79	90	35	44	57	89	65	87	66	

1. Use an ordered stem-and-leaf plot to arrange the test marks from lowest to highest.

2. Give the five-number summary for this data.

3. Give the IQR and the semi-IQR for this data.

4. Draw a box-and-whisker diagram to represent the five-number summary.

Solutions

1.
```
3 | 4  5
4 | 4  5  6
5 | 5  6  7
6 | 5  6  7  7
7 | 3  7  7  8  9
8 | 7  9  9
9 | 0                    Key: 3|4 = 34
```

2. Lowest score: 34

Q_1: $Q_1 = \frac{46 + 55}{2} = 50{,}5$ The median of the lower half of the scores lies between score numbers 5 and 6.

Median: 67   Middle-most score in this case is score number 11.

Q_3: $Q_3 = \frac{78 + 79}{2} = 78{,}5$ The median of the upper half of the scores lies between score numbers 16 and 17.

Highest score: 90

3. $IQR = Q_3 - Q_1 = 78{,}5 - 50{,}5 = 28$

$Semi\text{-}IQR = \dfrac{Q_3 - Q_1}{2} = \dfrac{28}{2} = 14$

4.

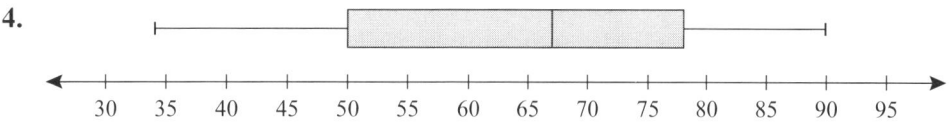

Box-and-whisker diagrams are used to compare two or more sets of data.

Worked example

Mr Gerhardt teaches German. The final exam results for his Grade 10 learners are summarised in the box-and-whisker diagram below.

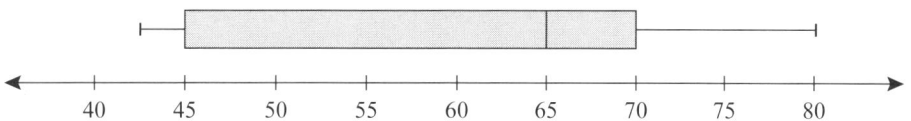

Use the box-and-whisker diagram to answer the following:

1. What was the highest score?   **2.** What is the range of scores?

3. What is the median score?   **4.** What is the upper quartile?

5. What percentage of learners scored above 65%?

6. What percentage of learners scored below 45%?

7. If the pass mark for German studies is 40%, what percentage of this class passed their German studies?

8. What percentage of learners scored between 45% and 70%?

Solution

1. Highest score: 80%   **2.** Range = 80–42 = 38%

3. Median = 65%   **4.** Upper quartile = 70%

5. Median = 65% $\therefore$ 50% of the learners scored above 65%.

6. $Q_1 = 45\%$ $\therefore$ 25% of the learners scored below 45%.

7. Lowest score obtained = 42%, $\therefore$ all the learners passed.

8. $Q_1 = 45\%$ and $Q_3 = 70\%$ $\therefore$ 50% of the learners scored between 45% and 70%.

> **Note**
> We cannot use a box-and-whisker diagram to determine how many learners were in the group nor can it be used to calculate the mean.

Box-and-whisker diagrams are useful to compare data.

The French and German exam results are shown below.

The differences and similarities between the two sets of results can be seen at a glance.

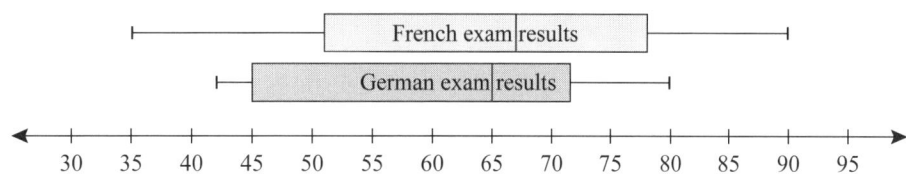

The list below includes some of the differences and similarities between these two sets of results.

- The medians for both groups are almost equal: 67% for French and 65% for German.
- The French results are less consistent with a lower minimum and a higher maximum.
- The French results have a greater range from 34% to 90% as opposed to the German results, which range from 42% to 80%.
- The IQR and the semi-IQR for both subjects are fairly similar:
  - The IQR for the French results is 28%, from 50,5% to 78,5% and the semi-IQR 14%.
  - The IQR for the German results is 25%, from 45% to 70% and the semi-IQR 12,5%.

Exercise 1

1. The table gives the masses of 20 Grade 10 learners.

| 42 | 51 | 56 | 55 | 51 | 54 | 51 | 60 | 66 | 45 |
| 62 | 54 | 51 | 56 | 61 | 52 | 69 | 64 | 58 | 48 |

   1.1 Set up an ordered stem-and-leaf plot for this data.

   1.2 What is the range for the masses of these learners?

   1.3 What is the median mass?

   1.4 What is the upper quartile?

   1.5 Determine the IQR and the semi-IQR for this data set.

2. The stem-and-leaf plot gives the test scores for Ms van Huysteen's Maths class.

$$
\begin{array}{c|ccccccccc}
6 & 0 & 1 & 2 & 3 & 5 & 5 & 5 & 6 & 7 \\
7 & 3 & 4 & 4 & 4 & 8 & 9 \\
8 & 0 & 0 & 3 & 3 & 4 & 9 \\
9 & 2 & 3 & 7 & 8 \\
\end{array}
$$
   Key: $\frac{6}{0} = 60$

   2.1 How many learners are in the class?

   2.2 What are the lowest and the highest scores? Write down the range for this data set.

   2.3 What is the median score?

   2.4 Find Q_1 to Q_3 and then write down the IQR and the semi-IQR for this data set.

   2.5 Set up an accurate box-and-whisker diagram to summarise this data set.

3. The box-and-whisker diagram below represents the Maths scores on a maths test.

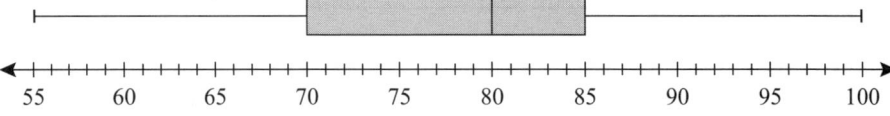

   Answer the following questions if it is possible to do so. (If it's impossible you must say it is impossible. Do not leave a blank answer.)

   3.1 What is the highest score?

   3.2 What is the range of the scores?

3.3 What is the median score?

3.4 What is the mean score?

3.5 How many learners wrote the test?

3.6 What is the upper quartile?

3.7 What percentage of the learners scored above 80?

3.8 What percentage of the learners scored below 70?

3.9 What percentage of the learners scored below 55?

3.10 What percentage of the learners scored between 70 and 85?

4. Pete, the owner of Pete's Pizza Den, wanted to compare the number of pizzas he sold between 5.00 p.m. and 7.00 p.m. with the number he sold between 7.00 p.m. and 9.00 p.m. He collected data for 28 days.

5.00 p.m.–7.00 p.m.													
23	37	64	56	60	50	48	47	47	51	66	38	23	56
43	52	58	62	44	58	50	49	55	57	48	34	45	38

7.00 p.m.–9.00 p.m.													
64	78	80	58	68	48	50	59	67	78	66	54	49	76
58	61	55	60	63	75	74	64	50	67	59	73	67	74

4.1 Set up an ordered back-to-back stem-and-leaf plot to represent this data.

4.2 Find the range of the pizzas sold between:

    4.2.1 5.00 p.m.–7.00 p.m.   **4.2.2** 7.00 p.m.–9.00 p.m.

4.3 Find the median number of pizzas sold between:

    4.3.1 5.00 p.m.–7.00 p.m.   **4.3.2** 7.00 p.m.–9.00 p.m.

4.4 Find the upper quartile and lower quartile of the number of pizzas sold between:

    4.4.1 5.00 p.m.–7.00 p.m.   **4.4.2** 7.00 p.m.–9.00 p.m.

4.5 Set up two box-and-whisker diagrams, using a common scale line to represent the dispersion of pizzas sold between:

    4.5.1 5.00 p.m.–7.00 p.m.   **4.5.2** 7.00 p.m.–9.00 p.m.

Representing data

Displaying data on graphs and charts can make it easier to interpret information.

Pie charts

A pie chart is a circle that is divided into sectors. Each sector represents a frequency of the data.

Converting data for a pie chart

To calculate the size of a sector: $\frac{score}{total} \times 360°$

To calculate the score that a sector represents: $\frac{angle\ size}{360°} \times total$

To calculate the percentage that a sector represents: $\frac{angle\ size}{360°} \times 100$

Worked example

In a class of 30 learners, 12 have blue eyes, 9 have brown eyes, 4 have dark brown eyes and 5 have green eyes. Draw a pie chart to represent this information. Label each sector with the percentage it represents.

Solution

Eye colour	Frequency	Angle	Percentage
Blue	12	$\frac{12}{30} \times 360° = 144°$	$\frac{12}{30} \times 100 = 40\%$
Brown	9	$\frac{9}{30} \times 360° = 108°$	$\frac{9}{30} \times 100 = 30\%$
Dark brown	4	$\frac{4}{30} \times 360° = 48°$	$\frac{4}{30} \times 100 = 13\frac{1}{3}\%$
Green	5	$\frac{5}{30} \times 360° = 60°$	$\frac{5}{30} \times 100 = 16\frac{2}{3}\%$
Total	**30**	**360°**	**100%**

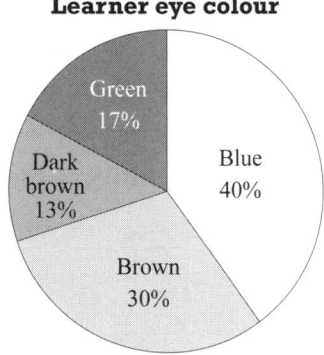

Learner eye colour

Bar graphs

A bar graph uses the height of bars to indicate the frequency of each category or class. Bar graphs are used for discrete (or nominal) data. This type of data consists of only whole values or descriptions, such as colour, type, number of people or objects, grade (or even age and shoe size).

Worked example

The table shows the frequency of each flavour of sweet found in a packet. Draw a bar chart to represent the data in the table.

Flavour	Frequency
Grape	18
Cherry	12
Strawberry	13
Orange	9
Lime	16

Solution

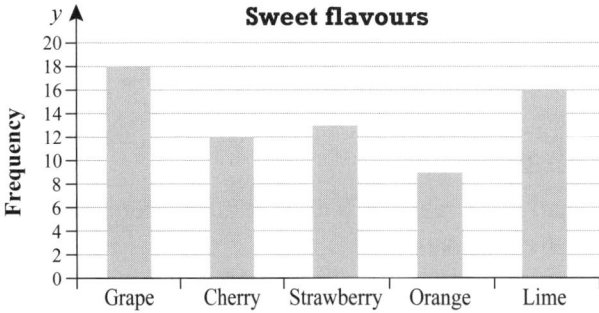

Histograms

A histogram is more complicated than a bar graph. It has the following properties:

- There are no spaces between the bars.
- It can only be used for continuous data.
- The y-axis measures the frequency, divided into interval size.

Since the data for a histogram is continuous it must be grouped first.

Worked example

The ages of 60 adults attending an HIV-testing clinic are given below.

23	20	24	26	29	34	35	37	39	40
25	24	26	28	30	34	39	45	54	58
60	65	64	63	62	69	20	23	21	24
27	28	39	31	32	35	40	42	43	45
52	57	58	46	64	67	48	51	56	59
42	45	27	56	53	52	45	46	32	33

Group the data appropriately and construct a histogram to represent it.

Solution

Group the data in 10-year intervals. Answers will vary depending on the size of
the intervals you choose, as well as the intervals you choose (such as 20 to 30
or 15 to 25).

Interval (years)	Tally	Frequency	Cumulative frequency
$15 \le x < 25$	ЖЖ |||	8	$\frac{8}{10} = 0,8$
$25 \le x < 35$	ЖЖ ЖЖ ЖЖ	15	$\frac{15}{10} = 1,5$
$35 \le x < 45$	ЖЖ ЖЖ		11	$\frac{11}{10} = 1,1$
$45 \le x < 55$	ЖЖ ЖЖ ||	12	$\frac{12}{10} = 1,2$
$55 \le x < 65$	ЖЖ ЖЖ		11	$\frac{11}{10} = 1,1$
$65 \le x < 75$	|||	3	$\frac{3}{10} = 0,3$

Do you see how the area of the bars gives the frequency for each group? The
width of first bar, for example, is 10 years and its height is 0,8. So, the frequency is
$10 \times 0,8 = 8$ people.

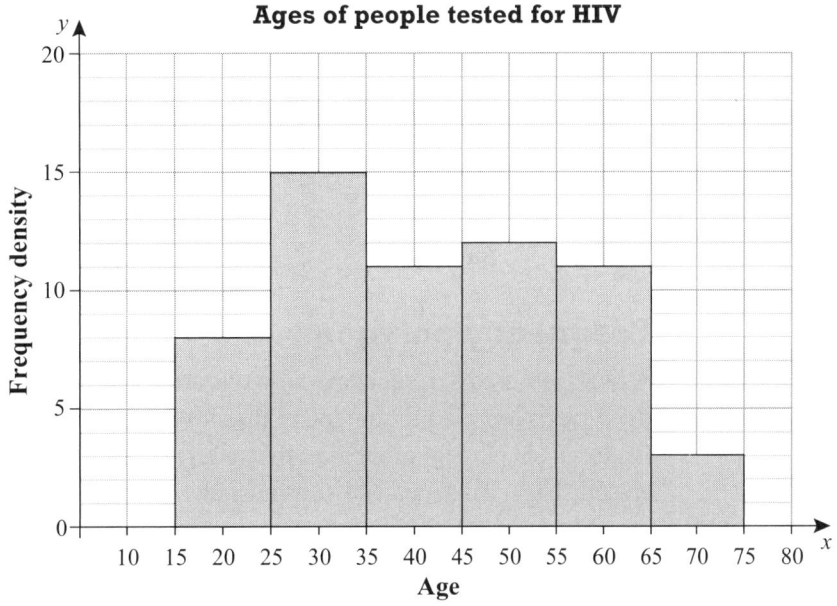

Ages of people tested for HIV

Broken line graphs

A broken-line graph is formed by joining each plotted score with a line segment. They are commonly used in the Life Sciences, for example, to plot changes in temperature over a period of time. Broken-line graphs are used for discrete/nominal/categorical data only.

Worked example

The table shows the temperature (in °C) as measured at different times on one day in Parys.

Time of day	07:00	08:00	11:00	12:00	13:00	15:00	17:00	19:00
Temperature (°C)	8	12	16	20	22	18	10	6

Draw a broken-line graph and use it to estimate the following:

1. the temperature at 10:00

2. the temperature at 14:00

3. the time when the temperature first reached 13 °C

4. the time(s) when the temperature was 17 °C.

Solution

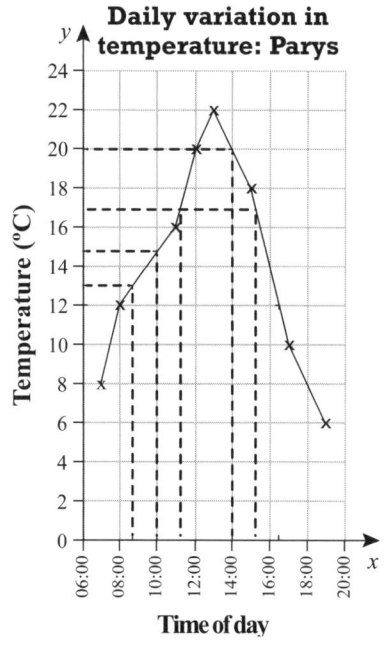

1. 14,6 °C

2. 20 °C

3. 09:00 (approximately)

4. 11:15 and 15:15 (approximately)

Frequency polygons

As for histograms, frequency polygons are used for continuous data, but broken-line graphs are used to join the midpoints of the intervals. Because it is a polygon (a closed shape with more than two sides), a frequency polygon must always be a closed shape. This means that it must always begin and end on the x-axis. Unlike on a histogram, the y-axis of a frequency polygon measures the frequency and not the frequency density.

Worked example

The following information shows the time taken (in seconds) by a sample group of 50 university students to complete a task in a psychology experiment.

7,0	16,0	10,1	20,0	16,4	14,7	12,6	21,4	20,2	19,0
9,8	20,8	18,4	15,2	11,8	7,2	8,4	12,8	9,3	9,6
22,9	14,1	8,9	7,0	7,1	13,0	15,5	16,1	16,6	16,7
9,6	17,6	14,2	12,6	7,9	16,7	19,6	19,9	15,3	11,8
21,1	18,6	22,3	21,9	15,6	9,2	14,4	10,7	20,8	10,3

Group the data using intervals of 2. Start at 6 and draw a frequency polygon to represent the data.

Solution

Use a frequency table first to tally the data and determine the frequency for each group.

The frequency of each interval is plotted at the midpoint of the interval. So, for the first interval, $6 \leq x < 8$, the frequency (5) is plotted at 7 on the x-axis. The first and last points are connected to the x-axis at the midpoints of the previous and following interval.

Interval (seconds)	Tally	Frequency
$6 \leq x < 8$	$\cancel{||||}$	5
$8 \leq x < 10$	$\cancel{||||}\ ||$	7
$10 \leq x < 12$	$\cancel{||||}$	5
$12 \leq x < 14$	$||||$	4
$14 \leq x < 16$	$\cancel{||||}\ |||$	8
$16 \leq x < 18$	$\cancel{||||}\ ||$	7
$18 \leq x < 20$	$\cancel{||||}$	5
$20 \leq x < 22$	$\cancel{||||}\ ||$	7
$22 \leq x < 24$	$||$	2

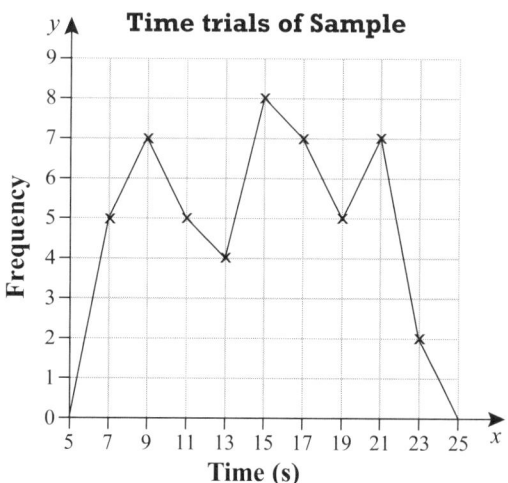

Exercise 2

1. In a Mathematics exam, 40 learners scored the following marks (in per cent).

25	40	61	43	29	87	94	66	63	59
29	28	63	54	39	86	89	95	96	54
48	49	53	54	63	67	69	84	83	85
74	73	83	95	74	72	83	84	85	63

1.1 Make a copy of the table and complete it.

Intervals (%)	Tally	Frequency	Cumulative frequency
20–29			
30–39			
40–49			
50–59			
60–69			
70–79			
80–89			
90–99			

1.2 How many learners obtained more than 90%?

1.3 If 40% is a pass, how many learners failed?

1.4 What is the median interval?

1.5 What is the modal interval?

1.6 Work out an estimate of the mean for this grouped data.

1.7 Draw a histogram for this set of data.

2. Draw a broken-line graph to represent the following information.

Year	Population
2003	1 200
2004	1 500
2005	2 000
2006	2 200
2007	2 500

3. Draw a bar graph to represent the following favourite subjects of a Grade 10 class.

Subject	Sesotho	Mathematics	Physical Sciences	Life Orientation	Geography
Frequency	5	10	8	3	4

4. The compound bar graph to the right shows the number of learners in each grade at a particular school in the Eastern Cape.

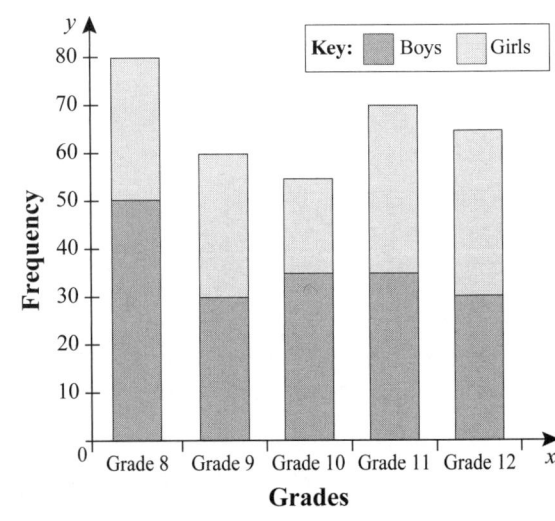

4.1 How many learners are there in Grade 10?

4.2 How many boys are there in Grade 12?

4.3 How many more boys than girls are there in Grade 8?

4.4 Which grade has an equal number of boys and girls?

4.5 How many learners are there in the school in total?

4.6 Which grades have the same number of girls?

5. The table shows a learner's informal assessment marks for Mathematics.

   5.1 Draw a broken-line graph for the information in the table.

   5.2 What is the learner's mean mark for the year?

   5.3 In which assessment(s) did she obtain a mark higher than his average?

Section of work	Assessment number	Mark obtained (%)
Types of number	1	52
Exponents	2	67
Quadrilaterals	3	74
Factorising	4	60
Trigonometry	5	78
Graphs	6	85
Statistics	7	94

6. In a survey of the popularity of different flavours of chips, customers were asked to indicate their favourite flavour from the following list.

   The customers' responses are shown below.

R	R	R	H	N	N	N	H	S	H
O	C	H	R	R	N	O	S	R	O
R	C	H	R	S	R	N	R	C	R
R	S	C	O	N	S				

   R Radical relish
   C Cheesy corn
   H Hot stuff
   S Salted mix
   N Nibble mania
   O Original flip

   6.1 Create and complete a frequency table for the above information.

   6.2 Draw a pie chart to represent the customers' choices.

   6.3 Which flavour is the most popular?

7. Determine the following for each set of numbers below.

   (a) mean   **(b)** mode   **(c)** median   **(d)** range

7.1	5	10	12	5	5	14	6	10	12	5
7.2	3	6	1	8	9	1	4			
7.3	5	6	8	9	6	5	6	9	5	6	8	9

8. Determine the mean, median and mode of the data in each frequency table.

   8.1

Value	Frequency
5	3
7	8
9	2
10	3
12	4

   8.2

Value	Frequency
10	3
15	5
20	2
25	4
30	1
Total	**15**

9. Calculate an estimate of the mean for the grouped data given in the table below.

Interval	Frequency
10–19	2
20–29	3
30–39	7
40–49	1
50–59	2

10. The heights (in centimetres) of 40 learners are given below.

142	170	162	131	145	146	147	160	159	150
141	132	169	173	139	146	152	154	140	145
161	163	156	157	171	168	166	151	152	132
142	150	161	138	170	132	149	150	138	152

10.1 Construct a frequency table for the data. Use six equal intervals.

10.2 Use the table to do the following.

    10.2.1 Construct a frequency polygon.

    10.2.2 Approximate the mean height.

10.3 Calculate the actual mean. By how much does your answer differ from your approximation for 10.2.2?

10.4 Set up a stem-and-leaf plot for this data.

10.5 Use your stem-and-leaf plot to determine the median and the mode for this set of data.

10.6 How many learners lie above the 20th percentile?

10.7 How many learners lie between the 20th and the 60th percentiles?

Test A: Knowledge and routine procedures

1. Determine the mean, median and mode for each set of numbers.

    1.1   1   3   5   8   8   11   13   (3)

    1.2   385   391   392   392   392   395   396   400   402   402   405   406   (3)

    1.3   18   20   12   25   17   32   26   15   (4)

2. **2.1** Determine the range, interquartile range and semi-interquartile range for the following data.

9	1	9	3	9	3	3	3	3	9
9	9	9	7	1	2	2	4	5	2
9	7	6	10	6	8	4			

                                                                              (6)

    2.2 Write down the five-number summary for the data set given in 2.1 above. Use this to set up a neat scaled box-and-whisker diagram to represent the spread of this data.   (5)

3. The 40 learners of a Grade 10 class indicated their favourite sports as follows.

S	soccer	☐
C	cricket	☐
H	hockey	☐
N	netball	☐
R	rugby	☐
B	basketball	☐

Their choices are given below.

H	C	R	C	B	N	C	S	H	B
B	R	N	S	S	S	H	S	R	B
R	N	S	H	S	R	B	S	H	R
S	C	C	H	C	R	N	B	C	H

3.1 Draw a frequency table for the above data. (3)

3.2 Use the frequency table to draw a bar chart. (5)

3.3 Write down the mode. (1)

3.4 Why is it not possible to find the mean of the data? (2)

4. The table below shows the sizes of each province and the total size of South Africa.

Province	Size (km^2)
Eastern Cape	169 580
Free State	129 480
Gauteng	17 010
KwaZulu-Natal	92 100
Limpopo	123 910
Mpumalanga	x
Northern Cape	361 830
North West	116 320
Western Cape	129 370
South Africa	1 219 090

Adapted from Census 2001: Census in Brief, Statistics South Africa

4.1 Find the value of x. (2)

4.2 What is the median province size? (2)

4.3 What is the mean province size (correct to the nearest km^2)? (2)

4.4 Compare the median and the mean. Which value do you think is a more useful measure of central tendency? Briefly explain why. (3)

4.5 Draw a line graph to represent the data. (5)

4.6 Answer the questions below if the information were to be represented as a pie chart.

    4.6.1 What angle size would the sector that represents the Western Cape be? (2)

    4.6.2 Which province would be represented by the sector that is 27,2° in size? (2)

Total 50

Test B: Complex procedures and problem solving

1. **1.1** Determine the range, interquartile range and semi-interquartile range for the following set of numbers.

44	62	35	33	54	60	64	53	43	41
37	55	54	52	43	56	47	43	65	40
56	43	58	48	54	58	40	54	46	61
37	27	54	46	62	52	38	46	55	64
47	47	59	60	50	50	44	37	74	28

(5)

  1.2 Why is the interquartile range a more reliable measure of dispersion than the range? (2)

2. In order to make sure that heavy vehicles are not overloaded, there are weigh stations situated at various places on South Africa's roads. The frequency table shows the statistics for a particular weigh station for one month.

Gross vehicle mass (GVM) (kg)	Frequency
$3\,000 \leq x < 5\,000$	49
$5\,000 \leq x < 7\,000$	72
$7\,000 \leq x < 9\,000$	123
$9\,000 \leq x < 11\,000$	135
$11\,000 \leq x < 13\,000$	109
$13\,000 \leq x < 15\,000$	86

  2.1 Estimate the mean of the data (to the nearest kg). (3)

  2.2 Draw a frequency polygon to represent the information in the table. (5)

  2.3 Why can the modal score not be found for the data? (2)

  2.4 In which interval does the 25th percentile lie? (3)

3. The following figures show the daily income (in rand) for a small business over a period of 30 working days.

436	520	980	684	798	842	346	472	591	257
596	745	892	385	427	763	445	676	848	907
605	412	306	584	527	634	701	549	1 002	812

  3.1 Calculate the mean daily income. (3)

  3.2 Group the data into five equal intervals and hence construct and complete a frequency table. (4)

  3.3 Use the frequency table to construct a histogram. (5)

  3.4 The owner of the business decides that if he is to keep the business, he needs a daily income of R600 or more for at least 60% of the time. By suitable calculation, determine what income the owner is earning 60% of the time and hence suggest whether he should keep the business. (6)

4. Softball is a popular sports code in many schools. A selector for the schools provincial team wants to determine which of two leagues in his area, league A or league B, scored more home runs. The stats for the season's home runs scored by the teams in leagues A and B are summarised in the back-to-back stem-and-leaf plot shown below.

League A					Stem	League B				
			7	1	**14**	2	5			
					13	0	4			
8	8	4	3	0	**12**	2	6	7	7	9
					11	6	7			
				0	**10**	1	8			
			7	5	**9**	4				
				9	**8**					
				3	**7**					

Key: $\frac{10}{1} = 101$ home runs.

4.1 On one scale, draw a box-and-whisker diagram for each league. (8)

4.2 Use this information to answer the following questions.

   4.2.1 How many teams, in total, taking both leagues into account, scored more than 127 runs? (2)

   4.2.2 On the strength of the information contained in these box-and-whisker diagrams, which league (measured in terms of home runs scored) performed more consistently? Support your answer. (2)

Total 50

Test C: Content and breakdown as for final exam

Note

In Science, our weight is referred to as mass. Weight is the term more generally used.

1. Obesity in teenagers is becoming a worldwide problem. The sport science at Hillside High decided to run a survey to investigate the situation at Hillside High.

   A sample of 48 learners was selected at random. Their weights in kilograms are reflected in the table below.

					Hillside High: Learners' weights						
			Key: Boys' weights are in bold print. Girls' weights are shaded grey.								
56	49	79	80	47	68	61	90	55	67	77	87
48	65	72	89	62	49	44	72	92	65	57	68
56	57	69	81	45	67	80	43	54	48	56	91
101	57	44	92	41	62	51	42	39	67	80	85

   1.1 Set up a neat, ordered back-to-back stem-and-leaf plot to rank this data. (7)

   1.2 Use the stem on your stem-and-leaf plot, or any other method, to set up a five-number summary for the boys' weights. (5)

   1.3 Calculate the mean weight for the boys at Hillside High, correct to two decimal places. (3)

   1.4 What is the difference between a mean and a median? (2)

1.5 Which measure do you think is more suitable when discussing the boys' weight distribution, the mean or the median? Support your answer. (3)

1.6 The weights of 10 additional male learners are added to this data set. The mean weight for these 10 learners is 76,3 kg. Calculate the new mean weight for the boys at Hillside High. (3)

2. The box-and-whisker diagram for the girls' weights in question 1 is given below. Use this to answer the questions that follow.

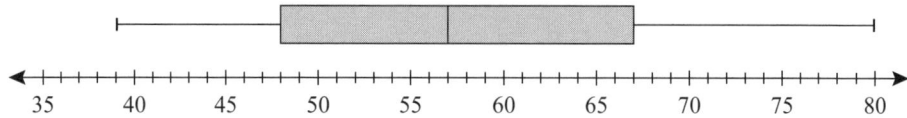

Complete the sentences by filling in the blanks with appropriate words and or figures.

2.1 The lower quartile tells us that (2.1.1)% of the girls weigh less than (2.1.2) kg. (2)

2.2 The box part of the plot tells us that _____. (2)

2.3 What percentage of the girls weigh more than 57 kg? (1)

2.4 If there are 300 girls at Hillside High, how many girls, in total, would you expect to weigh more than 57 kg? (1)

2.5 Calculate the IQR and the semi-IQR for the girls. (2)

2.6 What does the interquartile range actually measure? What does it actually tell us about the dispersion of this data set? (3)

3. The sports department now request that you organise all the raw data, i.e. for both boys and girls given in the table in question 1, into a frequency table using the intervals given in the table below.

Interval (Weight in kg)	Tally	Frequency	Midpoint of class interval
$30 \leq w \leq 39$			
$40 \leq w \leq 49$			
$50 \leq w \leq 59$			
$60 \leq w \leq 69$			
$70 \leq w \leq 79$			
$80 \leq w \leq 89$			
$90 \leq w \leq 99$			
$100 \leq w \leq 109$			

3.1 Use the raw data in question 1 to complete the tally and the frequency columns in the given table. (4)

3.2 Work out the midpoint for each interval and use this to help you to calculate the mean for this grouped data. You may find the additional column useful in calculating the mean. (5)

3.3 Now use your frequency table to set up an accurate frequency polygon for this grouped data, using the interval provided in the table. (5)

3.4 This data has two modal intervals, $40 \leq w \leq 49$ and $60 \leq w \leq 69$. Discuss a possible reason for this. (2)

Total 50

In this topic we will work with right-angled triangles in all four quadrants in the Cartesian plane. We will then apply this knowledge in practical problems.

Knowledge and skills for this topic

If you struggle with any of the work listed below, revise it before continuing with this Topic:

- the theorem of Pythagoras
- ratio and similarity.
- the Cartesian plane and point plotting
- addition, subtraction, multiplication and division of fractions
- solving basic trigonometric equations
- geometry of triangles including congruence.

Content of final exam

- Solve right-angled triangles in the Cartesian plane, using diagrams to determine the numerical values of trig ratios (lollipop questions).
- Solve problems in two dimensions using trigonometry.
- Determine areas of triangles and quadrilaterals.

Trigonometry in the Cartesian plane

In Topic 6 on trigonometry we worked in the first quadrant. These ideas will now be extended into all four quadrants in which $0° \leq \theta \leq 360°$.

First quadrant

A diagram in the 'standard position' in the first quadrant looks like the diagram below:

- A point, $P(x; y)$ is given.
- The distance from O, the origin, to P is equal to r, the radius.
- The rotation from the positive x-axis to OP (the radius, r) is θ, and θ is acute.
- A perpendicular from $P(x; y)$ is dropped onto the x-axis at M.
- OM $= x$ and PM $= y$ ∴ $x^2 + y^2 = r^2$ (Pythagoras).

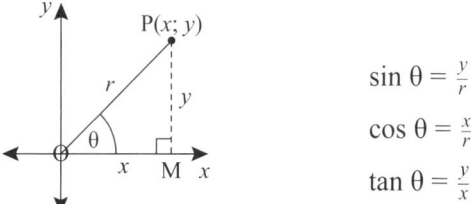

$$\sin \theta = \frac{y}{r}$$

$$\cos \theta = \frac{x}{r}$$

$$\tan \theta = \frac{y}{x}$$

We will refer to these questions as 'lollipop' questions, where OP $= r$ is the stick of the lollipop and the point P is the round sweet part. Whenever we work with a lollipop question, two pieces of information will be given. Using the information you are given, all other required information in the triangle can be found using Pythagoras and the trig ratios.

Second quadrant

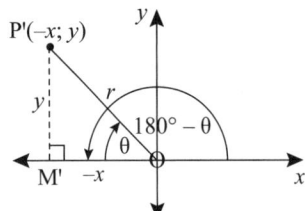

$$\sin(180° - \theta) = \frac{y}{r} = \sin\theta$$

$$\cos(180° - \theta) = \frac{-x}{r} = -\cos\theta$$

$$\tan(180° - \theta) = \frac{y}{-x} = -\tan\theta$$

If we draw a triangle in the second quadrant that is congruent to (≡) the triangle in the first quadrant, the angle of rotation (from the *x*-axis) is $(180° - \theta)$.

- Drop P'M' ⊥ OM'.
- △P'OM' (in second quadrant) ≡ △POM (in first quadrant).
- The coordinates of the new P are P'(–*x*; *y*), and this will result in any trig ratio containing an *x* being negative in the second quadrant.

Third quadrant

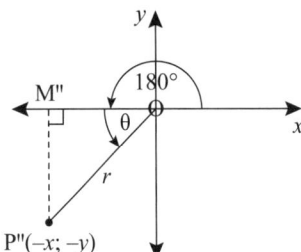

$$\sin(180° + \theta) = \frac{-y}{r} = -\sin\theta$$

$$\cos(180° + \theta) = \frac{-x}{r} = -\cos\theta$$

$$\tan(180° + \theta) = \frac{-y}{-x} = \frac{y}{x} = \tan\theta$$

If we draw a triangle in the third quadrant that is congruent to the first quadrant triangle, the angle of rotation (from the *x*-axis) is $(180° + \theta)$.

- Drop P"M" ⊥ OM".
- △P"OM" ≡ △POM.
- The coordinates of the new P are P"(–*x*; –*y*), and this will result in the trig ratios that contain either *x* and *r* or *y* and *r* being negative in the third quadrant. However, $\frac{-y}{-x}$ becomes positive, therefore the tan or cot ratios are positive.

Fourth quadrant

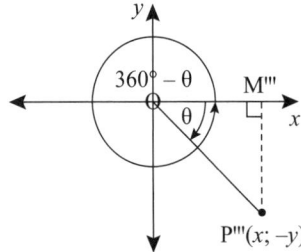

$$\sin(360° - \theta) = \frac{-y}{r} = -\sin\theta$$

$$\cos(360° - \theta) = \frac{x}{r} = \cos\theta$$

$$\tan(360° - \theta) = \frac{-y}{x} = -\tan\theta$$

If we draw a triangle in the fourth quadrant that is congruent to the triangle in the first quadrant, the angle of rotation is $(360° - \theta)$.

Drop P'''M''' ⊥ OM'''.

△P'''OM''' ≡ △POM.

The coordinates of the new P are P'''(*x*; –*y*), and this will result in any trig ratio containing a *y* being negative in the fourth quadrant.

CAST diagram

The definitions in terms of x, y and r can be used to solve simple problems and to determine trigonometric ratios in all four quadrants. We can summarise this information using the CAST diagram. This diagram indicates which trig ratio is positive in each quadrant. It is very important to know where the radius is situated in any given situation.

When solving problems, it is very important to draw the right-angled triangle in the correct quadrant, i.e. to determine where the radius will lie.

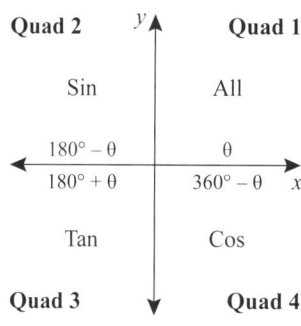

The CAST diagram

Worked examples

Give the quadrant in which the radius will lie for the following:

1. $\cos \theta = -\frac{5}{13}$ and $\theta \in [0°; 180°]$

2. $\tan \theta = \frac{4}{3}$ and $\theta \in [180°; 360°]$

3. $\sin \theta = -\frac{12}{13}$ and $\theta \in [0°; 270°]$

4. $\sin > 0$ and $\cos \theta < 0$

Solutions

1.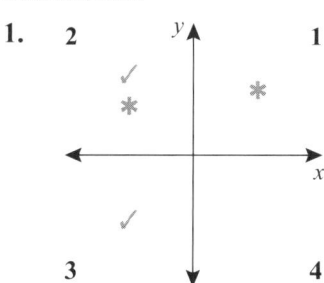

Where both conditions are valid in the same quadrant, that is the correct quadrant. These are indicated in each sketch.

$\cos \theta < 0$ ✓

$\theta \in [0°; 180°]$ ✳

Answer: 2nd quadrant

2.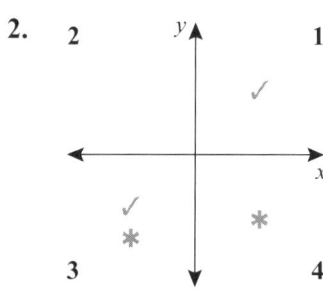

$\tan \theta > 0$ ✓

$\theta \in [180°; 360°]$ ✳

Answer: 3rd quadrant

3.

$\sin \theta < 0$ ✓

$\theta \in [0°; 270°]$ ✳

Answer: 3rd quadrant

4.

$\sin A > 0$ ✓

$\cos A < 0$ ✳

Answer: The radius lies in the quadrant in which both conditions occur, i.e. the 2nd quadrant.

Exercise 1

Determine the quadrant in which the radius will lie for each of the following:

1. $\sin \theta = \frac{5}{13}$ and $\theta \in [90°; 270°]$
2. $\tan \theta = -\frac{3}{4}$ and $\theta \in [0°; 180°]$
3. $\cos \theta = \frac{4}{5}$ and $\theta \in [180°; 360°]$
4. $\sec \theta = \frac{5}{-3}$ and $\theta \in [0°; 180°]$
5. $\cot \theta = -\frac{4}{3}$ and $\theta \in [0°; 180°]$

Type 1: Solving problems in which a specific point is given

Worked examples

Consider a point P(3: –4) that lies in the 4th quadrant, forming the angle $\hat{XOP} = \theta$.

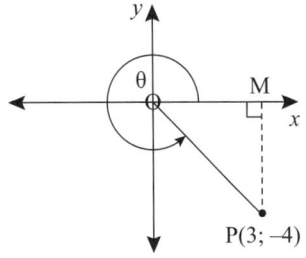

P(3; –4)

1. Use the information given in the diagram to find the value of $\sin \theta$, $\cos \theta$ and $\tan \theta$.

2. Use your results to show that $\tan \theta = \frac{\sin \theta}{\cos \theta}$.

Solutions

1. Given: point P has $x = 3$ and $y = -4$.

   Using Pythagoras: $r^2 = x^2 + y^2$

   $\therefore r = \sqrt{3^2 + (-4)^2} = 5$

   $\sin \theta = \frac{y}{r} = \frac{-4}{5}$, $\cos \theta = \frac{x}{r} = \frac{3}{5}$ and $\tan \theta = \frac{y}{x} = \frac{-4}{3}$

2. RHS $= \frac{\sin \theta}{\cos \theta} = \frac{-4}{5} \div \frac{3}{5} = \frac{-4}{5} \times \frac{5}{3} = \frac{-4}{3} = \tan \theta$

   $\therefore \tan \theta = \frac{\sin \theta}{\cos \theta}$

Type 2: Solving problems in which an equation is given

Steps for solving a lollipop question in which an equation is given

1. Determine the position of the radius, *r*.
2. Draw a right-angled triangle in this quadrant.
3. Find the value of all the sides of the triangle in order to answer the questions.

Worked examples

Given: $\sin \theta = -\frac{3}{5}$, $\theta \in [90°; 270°]$.

1. In which quadrant(s) can the radius can be found?

2. Determine without using a calculator, the value of :

    2.1 $\cos \theta$ **2.2** $\tan \theta$

    2.3 $\frac{\sin \theta}{\cos \theta}$ **2.4** Show that $\sin^2 \theta + \cos^2 \theta = 1$.

Solutions

1.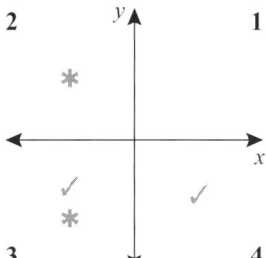

$\sin \theta < 0$ ✓

$\theta \in [90°; 270°]$ ✶

Answer: 3rd quadrant

2.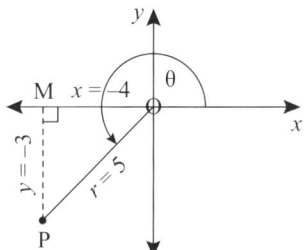

Given: $\sin \theta = -\frac{3}{5} = \frac{y}{r}$

$x = -\sqrt{5^2 + (-3)^2}$ Pythagoras

$\therefore x = -4$

    2.1 $\cos \theta = \frac{x}{r} = \frac{-4}{5}$ **2.2** $\tan \theta = \frac{y}{x} = \frac{-3}{-4} = \frac{3}{4}$ **2.3** $\frac{\sin \theta}{\cos \theta} = \frac{\frac{-3}{5}}{\frac{-4}{5}} = \frac{3}{4}$

    2.4 $\text{LHS} = \sin^2 \theta + \cos^2 \theta$

                $= \left(\frac{-3}{5}\right)^2 + \left(\frac{-4}{5}\right)^2$

                 $= \frac{9}{25} + \frac{16}{25}$

                 $= \frac{25}{25} = 1 = \text{RHS}$

Steps for solving a lollipop question in which you need to find values

1. Rewrite the equation in the form 'ratio of an angle' = 'fraction', e.g. $\cos \theta = \frac{3}{5}$.
2. Determine the position of the radius, r.
3. Draw a right-angled triangle in the relevant quadrant, and label the sides x, y and r.
4. Find the value of all sides of the triangle using Pythagoras, in order to answer the questions.

Worked examples

If $4 \cos \theta = -3$ and $\sin \theta > 0$, calculate the value of :

1. $\operatorname{cosec} \theta$ **2.** $\tan \theta$

3. $\sin^2 \theta + \cos^2 \theta$ **4.** $1 - \sec^2 \theta$

Solutions

1.

Step 1
Write $\cos \theta$ as a fraction (or ratio):
$\cos \theta = \frac{-3}{4}$ ✓

Step 2
Find the position of the radius:
$\sin \theta > 0$ ✶

Radius situated in second quad.

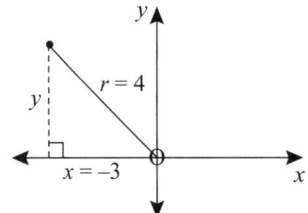

Step 3

Draw and label a right-angled △ in the correct quadrant:

$$y = \sqrt{(4)^2 + (-3)^2} \qquad \text{(Pythagoras)}$$

Step 4

Find the value of all sides of the △:

$$y = \sqrt{7}$$

$$\text{cosec}\,\theta = \frac{r}{y} = \frac{4}{\sqrt{7}}$$

2. $\tan \theta = \frac{y}{x} = \frac{\sqrt{7}}{-3}$

3. $\sin^2 \theta + \cos^2 \theta$

$\quad = \left(\frac{\sqrt{7}}{4}\right)^2 + \left(\frac{-3}{4}\right)^2$

$\quad = \frac{7}{16} + \frac{9}{16} = \frac{16}{16} = 1$

4. $1 - \sec^2 \theta$

$\quad = 1 - \left(\frac{4}{-3}\right)^2$

$\quad = 1 - \frac{16}{9} = -\frac{7}{9}$

Exercise 2

1. In which quadrant is $\sin \theta < 0$ and $\tan \theta > 0$?

2. In which quadrant is $\cos \theta > 0$ and $180° \leq \theta \leq 360°$?

3. If $\tan \theta = \frac{-3}{4}$, $90° \leq \theta \leq 270°$, determine the value of $\sin \theta - \cos \theta$.

4. In the diagram, P(–5; –12) and $X\hat{O}P = \theta$.
    Determine without using a calculator:

    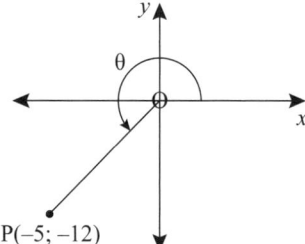

    4.1 $\tan \theta$

    4.2 $\sin \theta$

    4.3 $1 - \cos^2 \theta$

    4.4 $\frac{\sin \theta}{\cos \theta}$

    4.5 What conclusion can you draw from your answers to 4.1 and 4.4?

5. In the diagram, T(2; $-2\sqrt{3}$) and $X\hat{O}T = \theta$.
    Determine:

    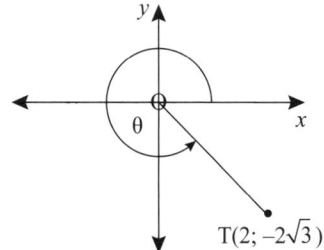

    5.1 $\sin \theta$

    5.2 $\cos^2 \theta + \sin^2 \theta$

    5.3 $\sqrt{3} \cdot \tan \theta$

    5.4 $\cos \theta$ and hence θ

6. P is the point (–5; $\sqrt{11}$). Determine:

    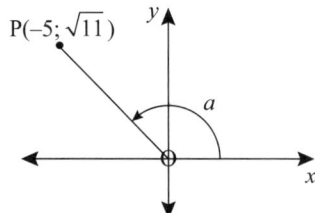

    6.1 OP

    6.2 $1 - \cos^2 \alpha$

    6.3 $\sqrt{11(\tan \alpha + 2 \sin \alpha)}$

7. If $\tan \theta = \frac{12}{5}$ and $180° < \theta < 360°$, calculate the
    value of:

    7.1 $\cos \theta$

    7.2 $\tan \theta \sin \theta$

    7.3 $\cos^2 \theta + \sin^2 \theta$

    7.4 $13 \sin \theta - 5 \tan \theta$

8. If $5 \cos \beta - 3 = 0$ and $0° < \beta < 90°$, calculate:

8.1 $\tan \beta$ **8.2** $\cos (90° - \beta)$

8.3 $\tan \beta \cos \beta$ **8.4** $\dfrac{\sin \beta}{\cos \beta}$

Solving word problems involving trigonometry in two dimensions

Quite often in two dimensional problems, areas need to be calculated. Here is a summary of some useful formulae.

Shape	Formula of area
triangle	$\frac{1}{2} \times$ base $\times$ height
rectangle	length $\times$ breadth
rhombus or square or kite	$\frac{d_1 \times d_2}{2}$, $d =$ diagonal
trapezium	$\frac{1}{2}$(sum of parallel sides) $\times$ perp. height between parallel lines

When triangles are **not in the standard position**, use the definitions as follows:

$\sin \theta = \dfrac{y}{r} = \dfrac{\text{opposite side}}{\text{hypotenuse}}$

$\cos \theta = \dfrac{x}{r} = \dfrac{\text{adjacent side}}{\text{hypotenuse}}$

$\tan \theta = \dfrac{y}{x} = \dfrac{\text{opposite side}}{\text{adjacent side}}$

Note that if you place your finger on the angle, it will always touch the adjacent side and the hypotenuse.

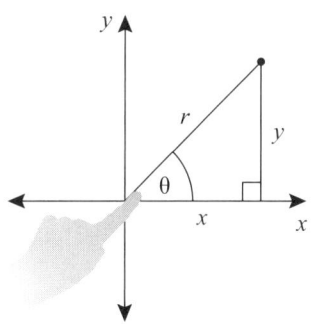

In exams the words 'angle of elevation' or 'angle of depression' may also be used for θ. 'Elevation' means going up, and 'depression' means going down, as shown in the diagram.

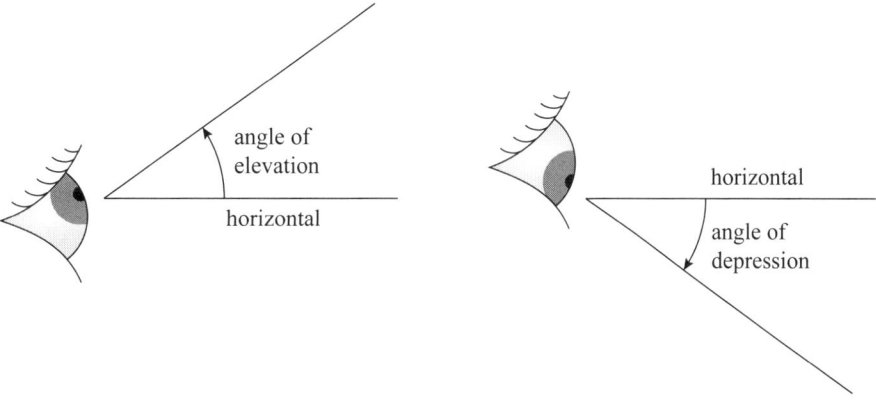

Worked examples

1. Write down three ratios for sin A.

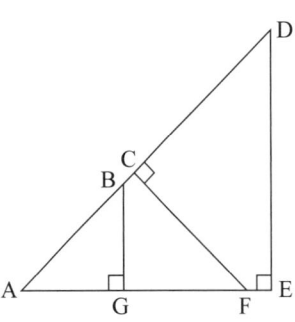

2. In $\triangle A\hat{B}C$, $\hat{A} = 90°$, $\hat{B} = 63,4°$, $AD \perp BC$ and $DC = 31$ mm. Calculate:

   2.1 $\hat{C}$ **2.2** AC   **2.3** AB

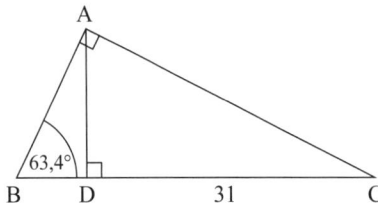

3. From the top of a perpendicular cliff, AB, that is 60 m above sea level, the angles of depression to two boats at C and D, in the same horizontal plane as B, are 20° and 25° respectively. Calculate the distance between the two boats at C and D.

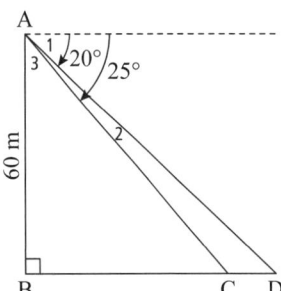

Solutions

1. In $\triangle ABG$: $\sin A = \frac{BG}{AB}$ In $\triangle ACF$: $\sin A = \frac{CF}{AF}$ In $\triangle ADE$: $\sin A = \frac{DE}{AD}$

2. **2.1** $\hat{C} = 180° - 90° - 63,4°$ sum ∠s △ABC

         $= 26,6°$

   2.2 AC is contained in $\triangle ADC$ and $\triangle ABC$, but more information is given in $\triangle ADC$, so work in this triangle. Use $\frac{\text{unknown side}}{\text{known side}} = $ trig ratio.

   $\frac{AC}{DC} = \frac{\text{hyp}}{\text{adj}} = \frac{1}{\cos C}$

   $AC = \frac{DC}{\cos C} = \frac{31°}{\cos 26,6°} = 34,7$ mm

   2.3 AB is determined in both $\triangle ABD$ and $\triangle ABC$. $\triangle ABC$ contains more information, so use this triangle. Use $\frac{\text{unknown side}}{\text{known side}} = $ trig ratio.

   $\frac{AB}{AC} = \frac{\text{adj}}{\text{opp}} = \frac{1}{\tan 63,4°}$

        $= \frac{34,7}{\tan 63,4°}$

        $= 17,38$ mm

3.

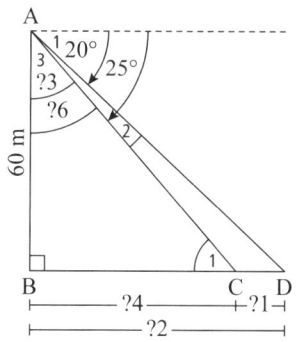

Method:

- Write in questions marks with a number below the sides and angles that you need to calculate in order to answer the question. Start with ?1 on CD, which is the distance required.
- To find CD, we need BD (?2) and BC (?3).
- To find BD we need $(\hat{A}_3 + \hat{A}_2)$ = ?4 and to find BC we need $\hat{A}_3$ = ?5.
- Now start finding all angles and sides containing questions marks. Start with the one with the highest number and work back towards ?1.

$?5 = \hat{A}_3 = 90° - 25° = 65°$ $B\hat{A}E = 90°$

?4 find length BC:

In $\triangle BAC$ use $\frac{\text{unknown}}{\text{known}}$ = trig ratio:

$\frac{BC}{60} = \tan 65°$

$BC = 60 \tan 65° = 128{,}67 \text{ m}$

?3 find $B\hat{A}C$:

$\hat{A}_2 = 25° - 20° = 5°$

$\therefore \hat{A}_3 + \hat{A}_2 = 70°$

?2 find BD:

In $\triangle BAD$

$\frac{BD}{AB} = \tan 70°$

$BD = 60 \tan 70° = 164{,}85 \text{ m}$

?1 find distance DC:

$164{,}85 - 128{,}67 = 36{,}18 \text{ m}$

Exercise 3

Round off all dimensions to one decimal place.

1. The length of the string of a kite, V, is 50 m and makes an angle of 68° with the ground. How high is the kite above the ground?

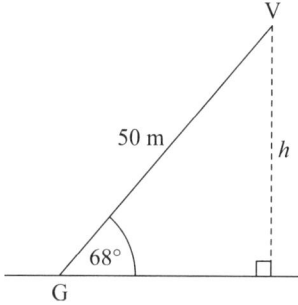

2. The angles of elevation from two points of observation on either side of a tower are 70° and 42°. The person nearest to the tower is 15 m from the foot of the tower. Calculate:

   2.1 the height of the tower

   2.2 the distance between the two points of observation.

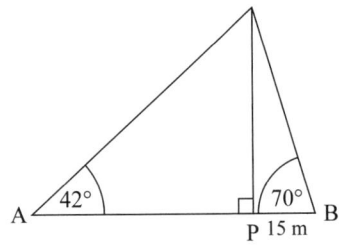

3. In the diagram, $\hat{B} = 65°$, BM = 80 mm, DC = 60 mm, MC = 160 mm and AE = 110 mm. Calculate:

   3.1 AM

   3.2 Area APDE

   3.3 $E\hat{D}P$ if ED = 150 mm.

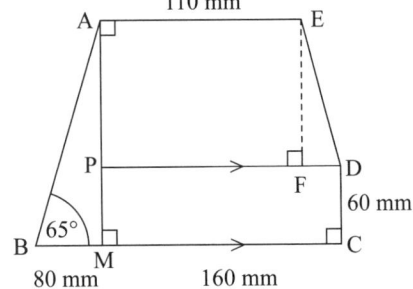

4. From point A, 40 m from a building, TP, the angle of elevation to the top of the building, is 35,4°. From point B, further away from the building, the angle of elevation to T is 22,2°. Determine:

   4.1 the height of the building TP

   4.2 the distance between A and B.

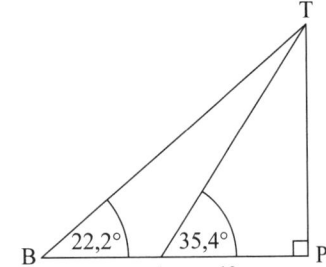

5. A river flows from north to south, and P and T are trees on either sides of the river, with P due west of T. A person at A, in line with P and T, walks 50 m due south to B.

   If $A\hat{B}P = 48°$ and $A\hat{B}T = 66°$, determine the width of the river, PT.

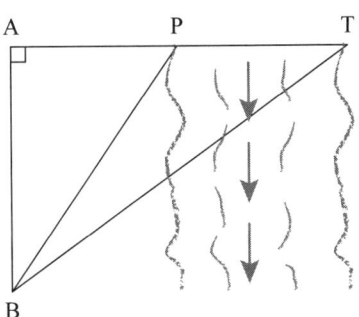

6. ABCD is a square, with MB = 80 mm, $A\hat{B}M = 33,4°$ and $F\hat{B}C = 27,2°$. Calculate:

   6.1 a side of the square

   6.2 DF

   6.3 the area of ABFD.

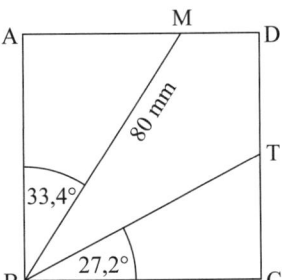

7. ABCD is a quadrilateral with $\hat{B} = \hat{D} = 90°$, AB = 80 mm, AD = 70 mm and $B\hat{A}C = 33,5°$. Calculate:

   7.1 AC

   7.2 $C\hat{A}D$

   7.3 CD.

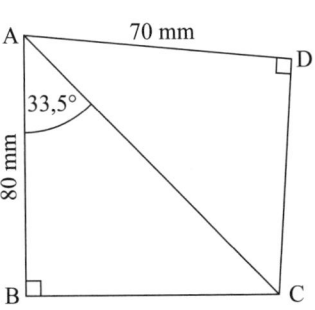

8. A parachutist, P, descends vertically at a constant speed. When she is directly above point A, her angle of elevation from point M, which is 750 m from A and in the same horizontal plane as A, is 64,9°.

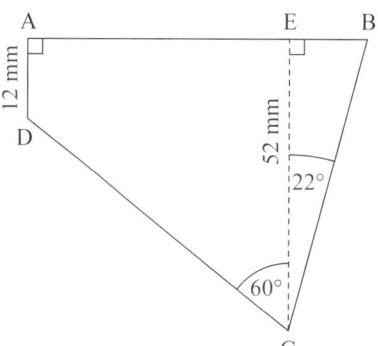

   8.1 Determine the height of P, i.e. P_1, at that exact moment.

   8.2 After an interval of 60 seconds, the angle of elevation of P, as seen from M, is 58°. How much longer will it take the parachutist to reach the ground? (Give the answer to the nearest minute.)

9. In the diagram, CE ⊥ AB, AD = 12 mm, CE = 52 mm, Â = 90°, DĈE = 60°, and EB̂C = 22°. Determine AB.
   Hint: Construct DT ⊥ EC, so that AETD is a rectangle.

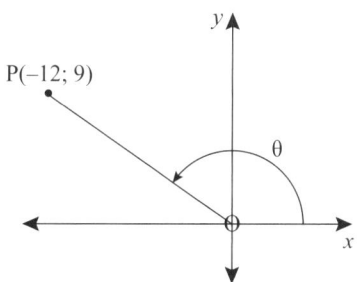

Test

1. In this question you may not use a calculator.

   1.1 P is the point $(-12; 9)$ and PÔX = θ. Show all your calculations and determine the value of:

   1.1.1 $\tan \theta$ (2)

   1.1.2 $1 - \cos^2 \theta$ (3)

   1.2 $\tan \theta = p$ and $180° \leq \theta \leq 270°$. Make a sketch and determine the value of:

   1.2.1 $\sin \theta$ in terms of p (3)

   1.2.2 $\cos \theta$ in terms of $\tan \theta$.   (3)

2. If $\cos A = \dfrac{2m}{m^2 + 1}$, determine the value of sin A in terms of m, for A < 90°.   (7)

3. $5 \cos \theta = 2$, $7 \tan \alpha = 3$ and θ and $\alpha \geq 180°$. Determine the value of $\dfrac{2}{21} \tan^2 \theta + 58 \sin^2 \alpha$.   (6)

4. A helicopter at H flies directly above a hotel at point A, at a height of 1 600 metres. The pilot is heading towards point T to pick up a group of tourists, who are staying at the hotel. The angle of depression from H to T is 28. D is a drop-off point between point A and point T.

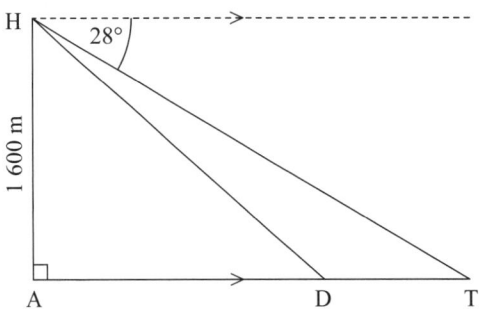

4.1 Determine the size of $\hat{T}$. (1)

4.2 Calculate the distance AT from the hotel to the landing position of the helicopter. (3)

4.3 Determine the distance the tourists will need to walk to get to the helicopter pick-up point, T, from the drop-off point D, if DT : AD = 1 : 5. (3)

4.4 Determine the angle of elevation of the helicopter from the drop off point, D. (4)

5. The sketch shows one side of the elevation of a house. Some dimensions are indicated on the sketch in metres.

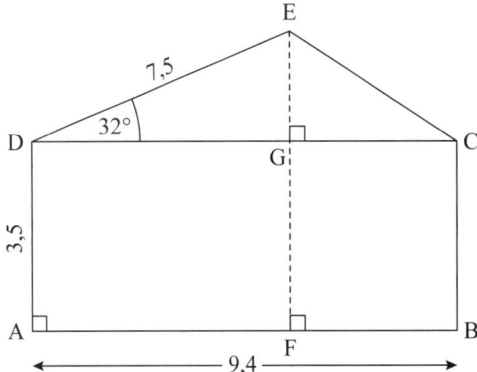

Determine, correct to two decimal places:

5.1 EC (8)

5.2 $D\hat{C}E$ (3)

5.3 area $\triangle DEC$ (2)

5.4 height of EF. (2)

Total 50

In Grades 8 and 9 you worked with and solved problems involving the volume and total surface area of rectangular prisms and cylinders. In Grade 10 you are introduced to cones, spheres and pyramids and are required to solve problems involving total surface area and volume for these objects. In this topic, we review volume and surface area formulae you worked with in Grades 8 and 9 before moving on to surface area and volume of pyramids, cones and spheres.

Knowledge and skills for this topic

If you struggle with any of the work listed below, revise it before continuing with this Topic:

- all formulae for the areas of quadrilaterals, triangles and circles
- all total surface area and volume formulae for right prisms
- basic understanding and working knowledge of conversions.

Content of final exam

Solve problems involving:

- volume and total surface area of right prisms and cylinders
- volume and surface area of spheres, right pyramids and right cones
- volume and surface area of composite shapes
- the effect on volume and surface of multiplying any dimension by a constant factor of k.

In Grade 10 you will only work with pyramids that have equilateral triangular bases and square bases. Rectangular bases and other shapes are extension work.

Working with cones is restricted to right cones.

Composite shapes are examinable.

Formulae:
All formulae must be memorised. Formulae are not given.

Revision of right prisms and cylinders

Area formulae

Note

From now on you must use the π key on your calculator for an accurate value for π, and not an approximation such as $\frac{22}{7}$.

You should remember how to calculate the area of each 2D shape in the table.

Shape		Formula
Rectangle		Area = length × breadth $A = lb$
Square		Area = side length squared $A = s^2$
Triangle		Area = half base × perpendicular height $A = \frac{1}{2}bh$
Circle		$A = \pi r^2$

Remember

2D stands for two dimensional.
3D stands for three dimensional.

Volume and surface area

Hints

- It helps to think of **volume** as the amount of liquid or air that a 3D object can hold.

- **Total surface area (TSA)** is the external surface of an object. You can find the TSA of any 3D object by adding together all the 2D shapes that make up the outside of the object.

Here is a reminder of what you learnt in earlier grades. Remember that this work is examinable in Grade 10 together with the new work involving cones, spheres and pyramids.

The general formula for the volume of a prism is:

Volume = base area × height

$\quad V = Ah$

The general formula for the surface area of a prism is:

Surface area = 2 × base area + perimeter × height

$\quad SA = 2A + Ph$

If a prism is hollow and open-ended or has one closed end you will need to modify the formula slightly:

- open-ended hollow prism: Surface area = perimeter of base × height
- one end closed: Surface area = perimeter of base × height + base area

These surface area formulae are for the **external** surface area. This means they are only for the outside of the prism.

Prism	Volume	Surface area
Right circular prism (cylinder)	$V = Ah$ $= \pi r^2 h$	Surface area $= 2A + Ph$ $= 2\pi r^2 + 2\pi r h$ $= 2\pi r(r + h)$
Rectangular prism	$V = Ah$ $= lbh$	Surface area $= 2A + Ph$ $= 2lb + 2(l + b)h$ $= 2lb + 2lh + 2bh$ $= 2(lb + lh + bh)$
Cube	$V = Ah$ $= s^3$	Surface area $= 2A + Ph$ $= 2s^2 + 4s \times s$ $= 6s^2$
Triangular prism	$V = Ah$ $= \frac{1}{2}bHh$	Surface area $= 2A + Ph$ $= 2\left(\frac{1}{2}bH\right) + (a + b + c)h$ $= bH + h(a + b + c)$

Multiplying by a scale factor

Look at what happens when we increase the dimensions of 2D shapes and 3D objects by a factor of k.

One-dimensional shape	$\underline{\qquad 1 \qquad}$ Length = 1 unit	$\underline{\qquad k \times 1 \qquad}$ Length = k units
Two-dimensional shape	Area = 1 unit^2	Area $= k \times 1 \times k \times 1$ $= k^2$ unit^2
Three-dimensional object	Volume = 1 unit^3	Volume $= k \times 1 \times k \times 1 \times k$ $= k^3$ unit^3

The diagrams show that multiplying the lengths or sides of a shape by a factor of k has the effects given in the table.

Length	$k \times$ original length
Area	$k^2 \times$ original area
Volume	$k^3 \times$ original volume

Note:
- If you multiply only one length or side of a shape by a factor of k, then the volume and area increase by a factor of k.
- If you multiply two lengths or sides of a shape by a factor of k, then the volume and area increase by a factor of k^2.

Worked examples

1. **1.1** Calculate the volume and total external area of the cylinder.

    1.2 What will be the surface area of the cylinder if all its dimensions are doubled?

    1.3 What will be the volume of the cylinder if its height is doubled?

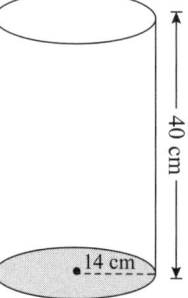

2. Calculate the following for the prism:

    2.1 the volume

    2.2 the total external area.

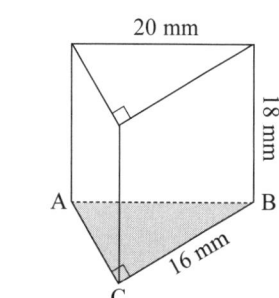

3. The dimensions on the prism are in centimetres. Calculate the following:

    3.1 the volume

    3.2 the external area

    3.3 the volume, if the dimensions of the shaded part are multiplied by a factor of 3.

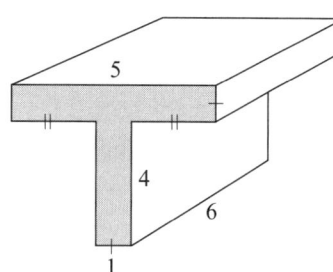

Solutions

1. **1.1** Volume = base area × height

    $V = \pi r^2 h$

    $= \pi \times 14^2 \times 40$

    $= 24\,630{,}09 \text{ cm}^3$

    Surface area = 2 × base area + perimeter × height

    $= 2\pi r^2 + 2\pi rh$

    $= 2\pi \times 14^2 + 2\pi \times 14 \times 40$

    $= 4\,750{,}09 \text{ cm}^2$

1.2 All dimensions are multiplied by a factor of 2. Doubling all dimensions will increase the area by a factor of $2^2 = 4$.

$\therefore$ Surface area $= 4\,750{,}09 \times 4$

$\qquad\qquad\qquad = 19\,000{,}35$ cm^2

1.3 Height is multiplied by a factor of 2. Increasing the length by a factor of 2 will increase the volume by a factor of 2.

$\therefore$ Volume $= 2 \times 24\,630{,}09$

$\qquad\qquad = 49\,260{,}17$ cm^3

2. In $\triangle ABC$, $\hat{C} = 90°$:

$AC^2 = 20^2 - 16^2$ $\qquad\qquad$ Pythagoras

$\qquad = 400 - 256$

$\qquad = 144$

$\therefore AC = 12$ mm

2.1 Volume $=$ base area $\times$ height

$\qquad\qquad = \frac{1}{2}b\text{H} \times h$

$\qquad\qquad = \frac{1}{2} \times 16 \times 12 \times 18$

$\qquad\qquad = 1\,728$ mm^3

2.2 Surface area $= 2 \times$ base area $+$ perimeter $\times$ height

$\qquad\qquad = 2\left(\frac{1}{2}b\text{H}\right) + (a + b + c)h$

$\qquad\qquad = b\text{H} + (a + b + c)h$

$\qquad\qquad = 16 \times 12 + (12 + 16 + 20)18$

$\qquad\qquad = 1\,056$ mm^2

3. **3.1** Volume $=$ base area $\times$ height

$\qquad V = \text{A}h$

$\qquad\quad = (5 \times 1 + 4 \times 1) \times 6$

$\qquad\quad = 54$ cm^3

3.2 Surface area $= 2 \times$ base area $+$ perimeter $\times$ height

$\qquad\qquad = 2(5 \times 1 + 4 \times 1) + (5 + 1 + 2 + 4 + 1 + 4 + 2 + 1) \times 6$

$\qquad\qquad = 2 \times 9 + 20 \times 6$

$\qquad\qquad = 138$ cm^2

3.3 Base dimensions have been multiplied by a factor of 3, so the area will increase by a factor of $3^2 = 9$. Since the height has not changed, the volume will increase by a factor of 9.

$\therefore$ Volume $= 9 \times 54$

$\qquad\qquad = 486$ cm^3

Exercise 1

1. Write down a formula for: (a) the volume and (b) the surface area of each prism in terms of the given variables.

1.1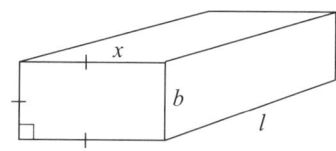

1.2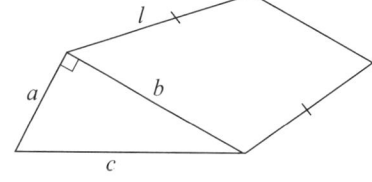

1.3

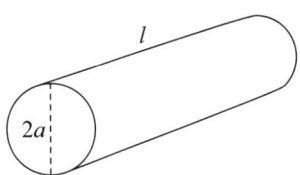

1.4

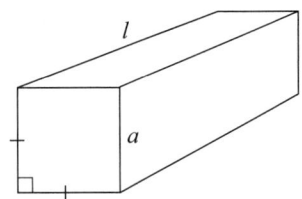

2. Calculate (a) the volume and (b) the total external area of each figure.

2.1

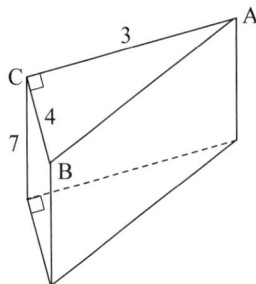

2.2

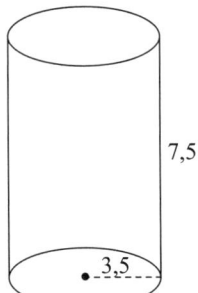

2.3

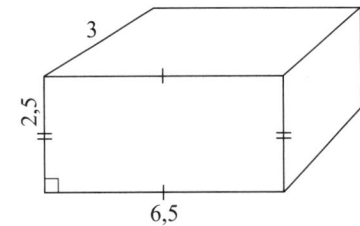

2.4

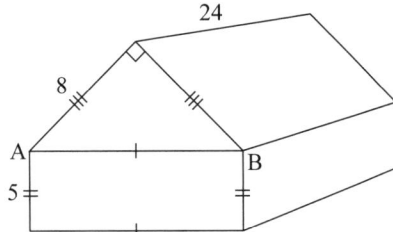

2.5

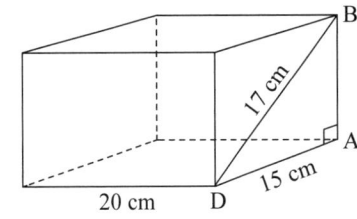

2.6

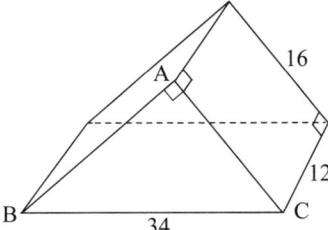

2.7

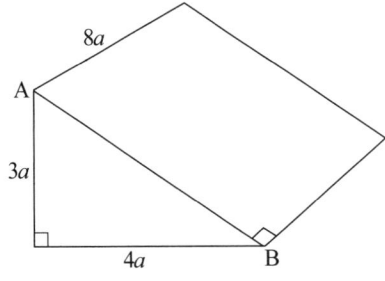

2.8

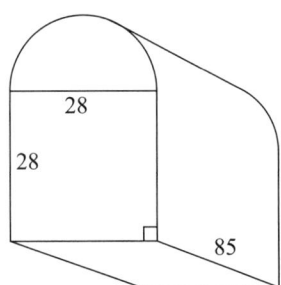

2.9

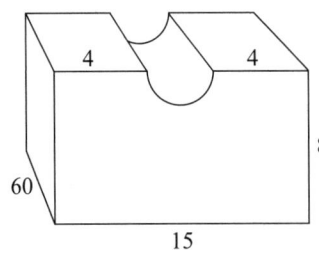

2.10

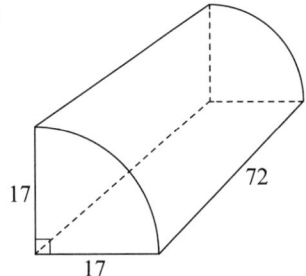

3. A wooden cylinder has a square hole bored through it as shown in the diagram. Calculate the volume of the remaining wood.

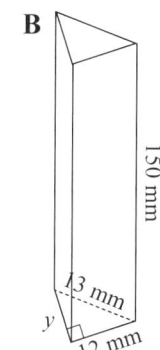

4. Study the two diagrams.

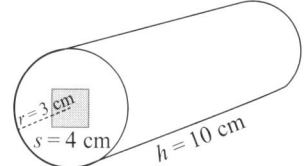

   4.1 Calculate x and y.

   4.2 Which container has the greater volume, A or B?

   4.3 If the height of A is doubled, what will its volume be?

   4.4 If the base dimensions of B are doubled, what will its volume be?

5. The diagram represents a tent. Determine the approximate area of canvas (material) needed to make the tent by calculating its total surface area (including the floor).

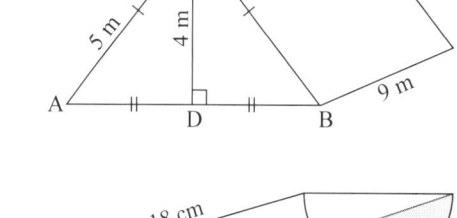

6. The figure illustrates a water trough for cattle. Calculate the capacity of the trough in cubic metres.

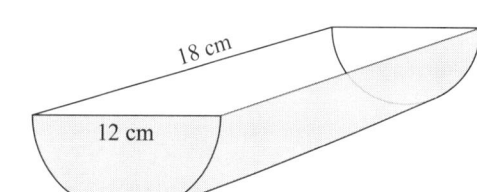

7. The sketch shows a metal coffee can with a plastic lid. Calculate:

   7.1 the volume of the can

   7.2 the area of the metal needed to manufacture the can

   7.3 the surface area of the plastic lid.

8. Find the total external area of the figure.

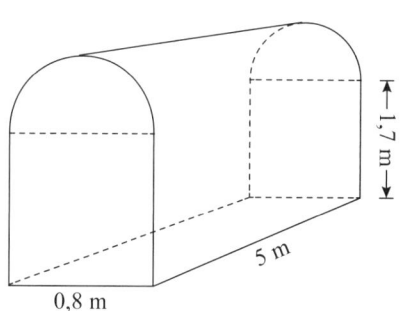

9. The diagram shows a set of stairs with a carpet covering the middle and front section of each stair.

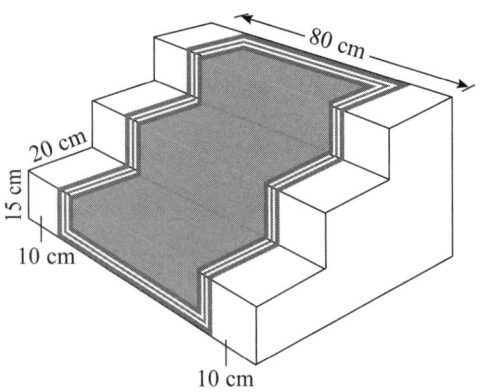

   9.1 Determine the area of the carpet.

   9.2 Determine the volume under the stairs.

10. Twelve cans of condensed milk are packed into a box as illustrated. Each can has a diameter and a height of 4 cm. Find the volume of the remaining space in the box.

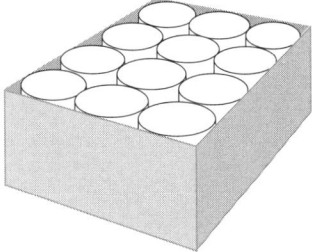

Cones

Definition: A **cone** is a **three-dimensional geometric object** that tapers smoothly from a **flat circular base** to a point called the apex or vertex.

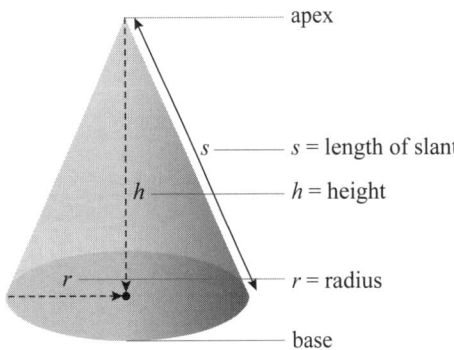

apex

s —— s = length of slant

h —— h = height

r —— r = radius

base

Note that a cone has **two different types of height**; a perpendicular height and a slant height.

When calculating **volume** we will use the **perpendicular height** (h) and when calculating **surface area** we use the **slant height** (s).

Volume of a cone

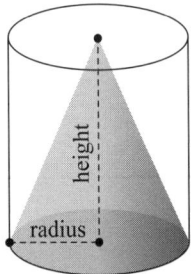

$= \frac{1}{3}$ of the volume of the cylinder on the same base and with the same height

$= \frac{1}{3} \times$ area of base $\times$ height

$\therefore V = \frac{1}{3}\pi r^2 h$

Worked examples

1. Find the volume of a cylinder with a radius of 3 cm and a height of 12 cm.

2. Find the volume of a cone with a radius of 3 cm and a perpendicular height of 12 cm.

Solutions

1. Volume of cylinder
   $$= \pi r^2 h = \pi (3)^2 (12) = 108\pi = 339{,}29 \text{ cm}^3$$

2. Volume of cone
   $$= \tfrac{1}{3}\pi r^2 h = \tfrac{1}{3}\pi (3)^2 (12) = 36\pi = 113{,}1 \text{ cm}^3$$

Surface area of a cone

To help you to understand the formulae for total surface area, consider how we would construct these shapes.

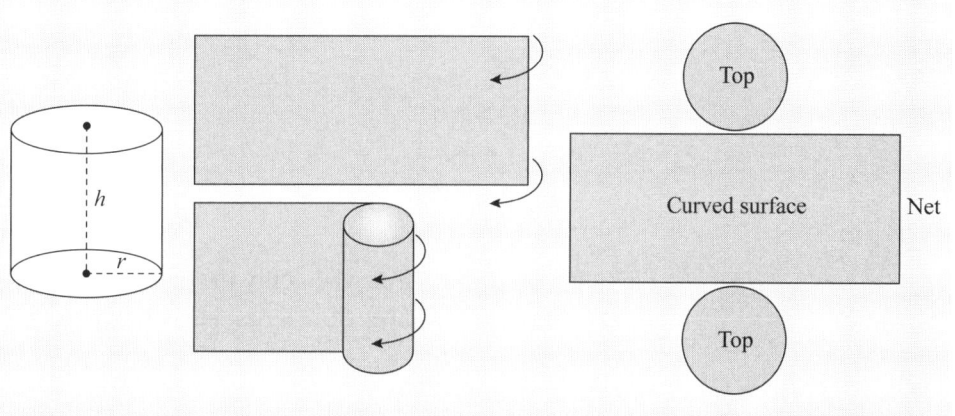

To construct a closed cylinder we need two circles and a rectangle which, when rolled up, forms a cylinder.

Recall that the formula for the total surface area of a cylinder is given by:

Total surface area for a cylinder $= 2\pi r^2 + 2\pi r$

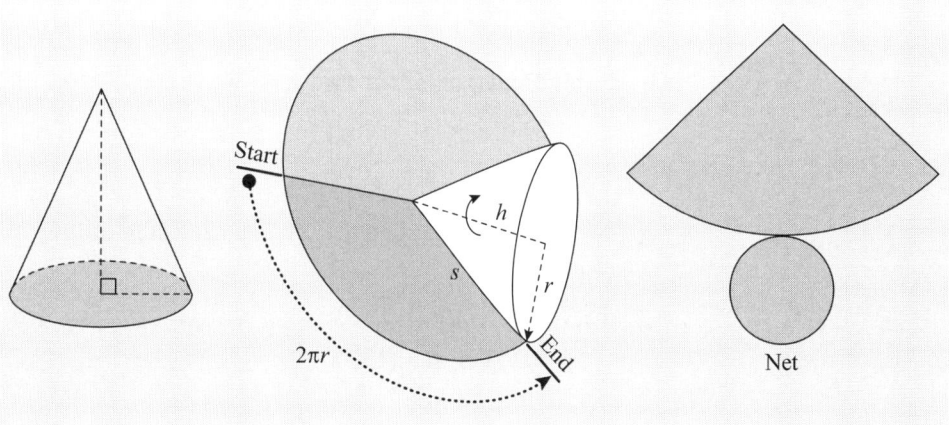

To construct a closed cone we need a circle and a curved area which wraps around to form the conical shape of the cone.

Calculating the surface area of a cone is more complicated:

Total surface area of a **closed cone** = area of circular base + curved area
$$= \pi r^2 + \pi rs \qquad \text{where } s \text{ represents the slant height}$$
$$= \pi r(r + s)$$

Total surface area of an **open cone** = curved area (i.e. a closed cone has no 'lid')
$$= \pi rs \qquad \text{where } s \text{ represents the slant height}$$

Worked examples

1. Find the total surface area of a **closed cone** with a radius of 3,5 m and a slant height of 5 m, correct to two decimal places.

2. Find the surface area of an **open cone** with a radius of 3,5 m and a slant height of 5 m, correct to two decimal places.

Solutions

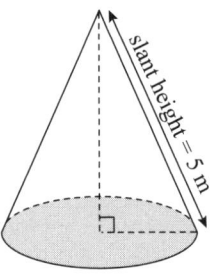

1. TSA $= \pi r^2 + \pi r s$

   $= \pi(3,5)^2 + \pi(3,5)(5)$

   $= 93,46 \ \text{m}^2$

2. TSA $= \pi r s$

   $= \pi(3,5)(5)$

   $= 54,9778 \ldots$

   $= 54,98 \ \text{m}^2$

Finding total surface area when only the perpendicular height is known takes some additional working, since we will first have to work out the slant height.

Worked examples

1. Find the total surface area of a **closed cone** with a radius of 5 m and a **perpendicular height** of 12 m, correct to two decimal places.

2. Find the total surface area of an **open cone** with a radius of 5 m and a **perpendicular height** of 12 m, correct to two decimal places.

Solutions

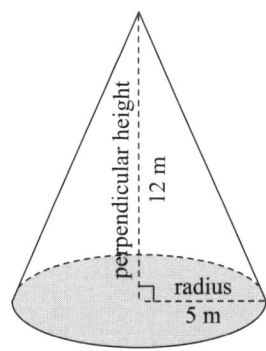

1. We first need to calculate the slant height:

   Using Pythagoras: $s = \sqrt{h^2 + r^2}$

   $= \sqrt{12^2 + 5^2}$

   $s = 13 \ \text{m}$

   Now we can calculate the total surface area of the closed cone:

   TSA $= \pi r^2 + \pi r s$

   $= \pi(5)^2 + \pi(5)(13)$

   $= 282,74 \ \text{m}^2$

2. TSA $= \pi r s$

   $= \pi(5)(13)$

   $= 65\pi$

   $= 204,20 \ \text{m}^2$

Pyramids

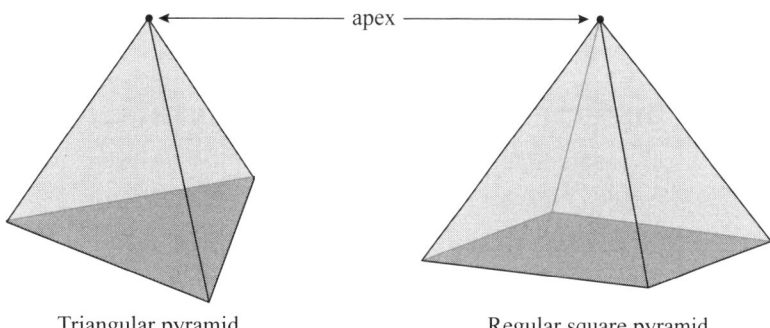

Triangular pyramid   Regular square pyramid

The base of a pyramid can take various forms. The most common pyramids have square, rectangular or triangular bases. In Grade 10, we will only study pyramids with equilateral triangular bases and square bases.

Square pyramids

Definition: In geometry, a **square pyramid** is a pyramid having a square base with the apex perpendicularly above the centre of the square.

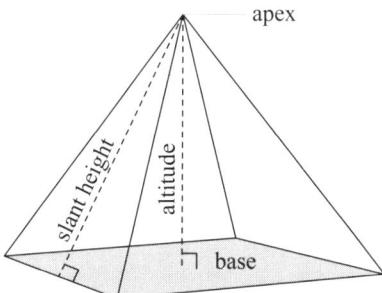

Note: A pyramid has two height lines: the perpendicular height and the height of the triangular face above its base.

Volume of a square pyramid

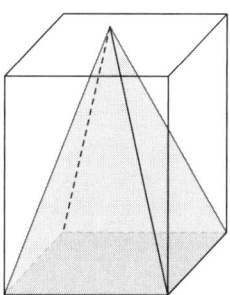

Volume = $\frac{1}{3}$ of the volume of the rectangular prism with the same base and perpendicular height

$= \frac{1}{3} \times$ area of base $\times$ perpendicular height

Volume = $\frac{1}{3}lbh$ for a rectangular prism

Volume = $\frac{1}{3}b^2h$ for a square prism

Worked examples

1. Find the total volume of a rectangular prism with a square base of 5 cm and a height of 16 cm, correct to two decimal places.

2. Find the total volume of a pyramid with a square base of 5 cm and a height of 16 cm, correct to two decimal places.

Solutions

1. Volume of rectangular prism
    $= lbh = (5)(5)(16) = 400 \text{ cm}^3$

2. Volume of square pyramid
    $= \frac{1}{3}b^2h = \frac{1}{3}(5)^2(16) = 133,33 \text{ cm}^3$

Surface area of a square pyramid

To help you to understand the formula for the total surface area, consider how we would construct a pyramid. i.e. draw the net.

To construct a square pyramid, we need a square base and four congruent triangles. The triangles all have the same height and the same base.

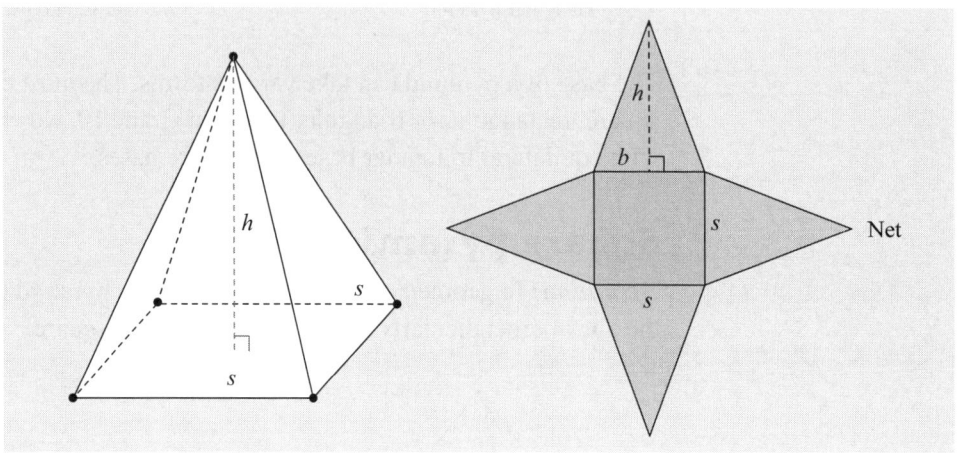

Net

Total surface area = area of the square base + area of four triangles

$\text{TSA} = s^2 + 4 \times \frac{1}{2}sh$

$\therefore \text{TSA} = s^2 + 2sh = s(s + 2h)$

Note

The **slant height** of each triangular face is used when calculating surface area, not the perpendicular height.

Worked examples

1. Find the TSA of a rectangular prism with a square base of 5 cm and a height of 16 cm, correct to two decimal places.

2. Find the TSA of a pyramid with a square base of 5 cm and a perpendicular height of 16 cm, correct to two decimal places.

Solutions

1. $\text{TSA} = (5)^2 + 2 \times (5)(16) = 185 \text{ cm}^2$

2. Begin by calculating the slant height:

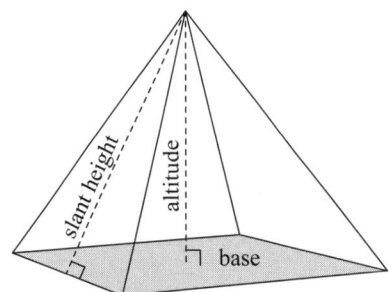

In $\triangle$ABC: AB = 16 cm, CB = $\frac{5}{2}$ = 2,5 cm

$\therefore \text{CA} = \sqrt{16^2 + 2,5^2} \approx 16,194 \text{ cm}$

$\text{TSA} = s^2 + 2sh = 5^2 + 2(5)\left(\sqrt{16^2 + 2,5^2}\right) = 186,94 \text{ cm}^2$

Triangular pyramids

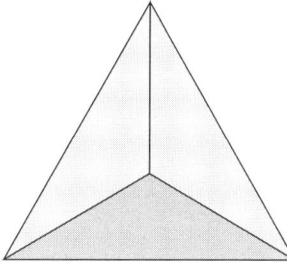 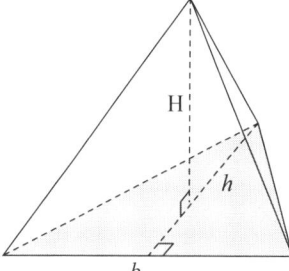

A triangular pyramid has a triangle as its base. A triangular pyramid with four equilateral triangular faces is known as a tetrahedron.

Note that a triangular pyramid has two different heights: one for the base traingle (h) and one for the perpendicular height (H) as shown in the diagram.

Volume of a triangular pyramid

The general formula for finding the volume of a triangular pyramid is the same as that for finding the area of a square pyramid.

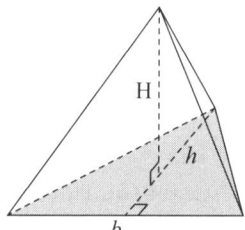

Volume $= \frac{1}{3} \times$ area of base $\times$ perpendicular height

$$= \frac{1}{3} \times \frac{bh}{2} \times H$$

$\therefore$ Volume $= \frac{bhH}{6}$ for a triangular pyramid

Worked examples

1. Find the volume of the tetrahedron with $h = \sqrt{300}$ cm, $b = 2$ m and H $= 1{,}63$ m. Leave your answer correct to the nearest cubic centimetre.

2. The triangular faces of a tetrahedron are 10 cm in length and the perpendicular height 8,16 cm. Find the volume of this tetrahedron.

3. The area of the triangular base of a pyramid is $10\sqrt{3}$ cm^2 and its perpendicular height is 16,33 cm. Find the volume of this pyramid.

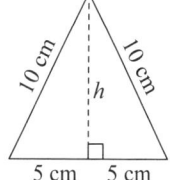

Solutions

Remember

Convert all measurements into cm.
$b = 200$ cm
H $= 163$ cm

1. Vol $= \frac{bhH}{6} = \frac{(200)(\sqrt{300})(163)}{6}$

     $= 94\ 108{,}09$ cm$^3 \approx 94\ 108$ cm^3

Note

Use Pythagoras to calculate the height (h) of the base triangle.

2. $h^2 = 10^2 - 5^2$ Pythagoras

    $h = \sqrt{10^2 - 5^2} = 5\sqrt{3}$

    $\therefore$ Vol $= \frac{bhH}{6} = \frac{10(5\sqrt{3})(8{,}16)}{6} = 117{,}78$ cm^3

Be careful!

The height for the base triangle is not the perpendicular height.

3. Vol $= \frac{bhH}{6} = \frac{1}{3} \times \frac{bh}{2} \times H$

    Vol $= \frac{1}{3} \times 10\sqrt{3} \times 16{,}33 = 94{,}28$ cm^3

Surface area of a triangular pyramid

Surface area = the sum the areas of all the triangular faces

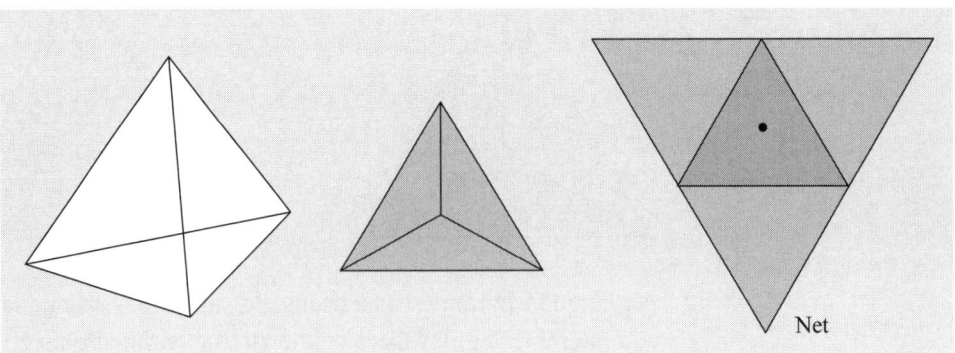
Net

To construct a triangular pyramid, we need a triangular base and three congruent triangular faces.

A tetrahedron has four congruent triangular faces, all with the same height and the same base.

Total surface area = 4 × area of base triangle

$$\text{TSA} = 4 \times \frac{bh}{2}$$

$$\text{TSA} = 2bh$$

Worked examples

1. Find the total surface area of the **tetrahedron** (triangular pyramid with four equilateral triangular faces) if the base and the height of each triangular face is 2 cm and $\sqrt{3}$ cm respectively, correct to two decimal places.

2. Find the total surface area of the tetrahedron made up of four equilateral triangles with sides measuring 25 m each, correct to the nearest metre squared.

3. Find the total surface area of the pyramid made up of an equilateral triangular base with a side length of 8 cm and three isosceles triangular faces each with equal sides of 18 cm. The net for this pyramid is given here.

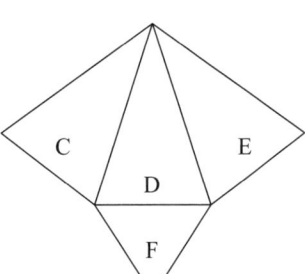

Solutions

1. TSA $= 4 \times \frac{bh}{2} = 2bh$

    $= 2(2)\left(\sqrt{3}\right) = 4\sqrt{3} = 6{,}93 \text{ cm}^2$

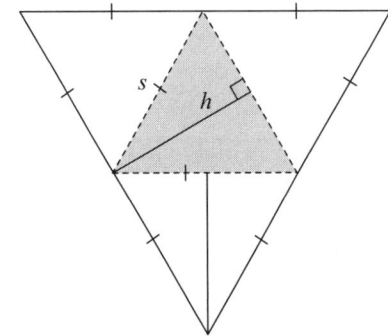

2. $h^2 = 25^2 - 12,5^2$ Pythagoras

   $h = \sqrt{25^2 - 12,5^2} = 21,65\ m = 22\ m$

   Base for each triangle = 25 m and height = 22 m

   $\therefore$ TSA $= 4 \times \frac{bh}{2} = 2bh = 2(25)(22) = 1\ 100\ m^2$

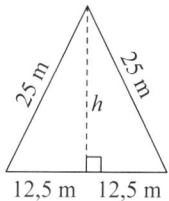

3. Total surface area for a triangular pyramid with an equilateral triangle as base and three isosceles triangular sides is given by:

   Total surface area = area of base triangle + 3 × area of triangular faces

$$\text{TSA} = \frac{b_1 h_1}{2} + 3 \times \frac{b_2 h_2}{2}$$

Height for the base triangle:

$h^2 = 8^2 - 4^2$ Pythagoras

$h = \sqrt{8^2 - 4^2} = 4\sqrt{3}\ cm$

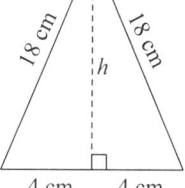

Height for each isosceles triangle:

$h^2 = 18^2 - 4^2$ Pythagoras

$h = \sqrt{18^2 - 4^2} = 2\sqrt{77}\ cm$

$\text{TSA} = \frac{b_1 h_1}{2} + 3\frac{b_2 h_2}{2}$

     $= \frac{8 \times 4\sqrt{3}}{2} + 3\frac{8 \times 2\sqrt{77}}{2}$

     $= 238,31\ cm^2$

Spheres

A sphere is perfectly symmetrical. It has no vertices or edges.

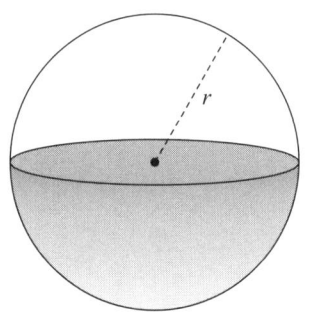

Volume of a sphere

Volume $= \frac{4}{3}\pi r^3$

Total surface area of a sphere

Total surface area $= 4\pi r^2$

Hemispheres

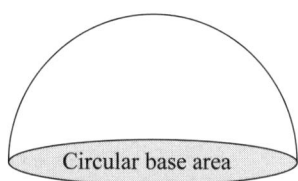

A hemisphere is half a sphere.

Volume of a hemisphere

Volume $= \frac{4}{3}\pi r^3 \div 2 = \frac{2}{3}\pi r^3$

Total surface area of a hemisphere

The total surface area of an **open hemisphere**:

Total surface area $= 4\pi r^2 \div 2 = 2\pi r^2$

The total surface area of a **closed hemisphere**:

(A closed hemisphere includes the circular base area.)

$\therefore$ Total surface area $= 2\pi r^2 + \pi r^2 = 3\pi r^2$

Worked examples

1. Find the volume and the surface area for a sphere with a radius of 5 cm, correct to one decimal place.

2. Find the volume and the surface area for an open hemisphere with a radius of 5 cm, correct to two decimal places.

3. Find the surface area for a closed hemisphere with a radius of 5 cm, correct to the nearest whole number.

Solutions

1. Volume $= \frac{4}{3}\pi r^3 = \frac{4}{3}\pi(5)^3 = \frac{500\pi}{3} = 523{,}6 \text{ cm}^3$

   Total surface area $= 4\pi r^2 = 4\pi(5)^2 = 100\pi = 314{,}2 \text{ cm}^2$

2. Volume $= \frac{4}{3}\pi r^3 \div 2 = \frac{2}{3}\pi r^3 = \frac{2}{3}\pi(5)^3 = \frac{250\pi}{3} = 261{,}80 \text{ cm}^3$

   Total surface area $= 4\pi r^2 \div 2 = 2\pi r^2 = 2\pi(5)^2 = 50\pi = 157{,}08 \text{ cm}^2$

3. Total surface area $= 2\pi r^2 + \pi r^2 = 3\pi r^2$

   $= 75\pi = 235{,}619 = 236 \text{ cm}^2$

Exercise 2

1. Find the volume and the total surface area for each shape.

   1.1 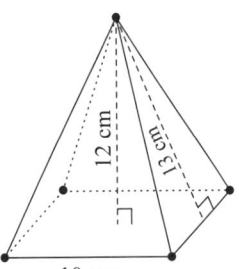 **1.2** 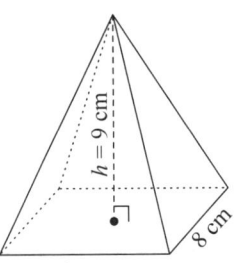 **1.3**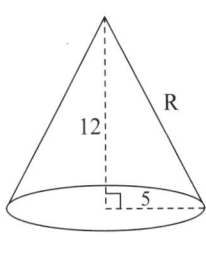

2. Calculate the total surface area and the volume for a spherical Christmas decoration with a diameter of 8 cm, to the nearest cm.

3. The centre of a sphere has a circumference of 50π m. If this sphere is split exactly in half, find the total surface area and volume for each hemisphere.

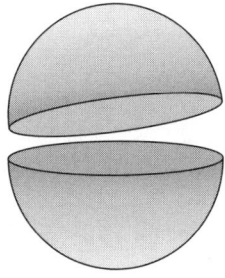

4. Calculate the volume and the total surface area of a steel ball that has a radius of 3,75 cm.

5. Calculate the volume and the total surface area of a cone with a radius of 12 mm and a perpendicular height of 28 mm.

6. The base of a pyramid 12 m high is an equilateral triangle of side $10\sqrt{3}$ m. Find the volume of the pyramid.

7. If the length of each side of the square base of a pyramid is 15 cm and each slant edge of each triangular face is 45 cm long, calculate the total surface area of the pyramid of the pyramid, including its base.

8. The tetrahedron with a volume of 398 cm³ is constructed from plastic straws. The perpendicular height of the tetrahedron is $5\sqrt{6}$ cm and the height of the base triangle 13 cm.

   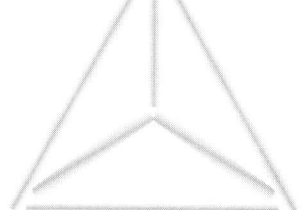

   8.1 Find the length of each of the straws, correct to the nearest cm.

   8.2 If you were to cover each triangular face with kite paper, how much kite paper would you need, correct to the nearest cm?

9. A birthday hat shaped like a cone is made out of cardboard. The hat has a perpendicular height of 25 cm and the diameter measures 18 cm. How much cardboard is needed to make this hat? Leave your answer correct to two decimal places.

10. The square pyramid as shown has an area of 500 m³ and a perpendicular height of 15 m. Use this information to calculate the surface area for one of the triangular faces, correct to one decimal place.

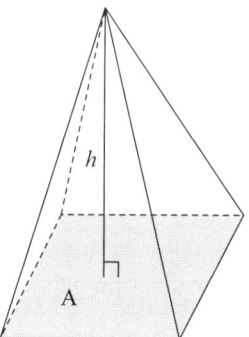

11. Find the surface area for the pyramid with the equilateral triangle as its base.

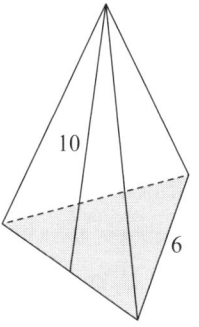

Working with composite figures

Composite figures are figures that are made up of more than one shape. To find the area of a composite figure, you divide the figure given into simple shapes such as triangles, rectangles etc. Find the areas of the individual shapes and then add up the areas or volumes as required by the problem.

Worked examples

A model is constructed by topping a cylindrical can with height of 20 m and base radius 10 m with a hemispherical solid as shown in the diagram.

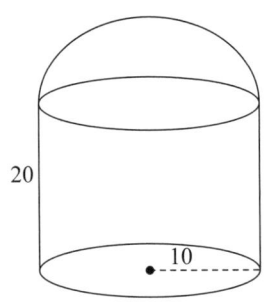

1. Find the approximate volume of this model.

2. Find the approximate surface area of this model.

Round your answers to the nearest whole numbers.

Solutions

1. Volume of the cylinder $= \pi r^2 h = \pi 10^2 \times 20 = 2\,000\pi$

   Volume for an open hemisphere $= \frac{4}{3}\pi r^3 \div 2 = \frac{2}{3}\pi r^3 = \frac{2}{3}\pi(10)^3 = \frac{2\,000}{3}\pi$

   Now add the two volumes to arrive at the total volume:

   Total volume = volume of cylinder + volume of hemisphere

   $= 2\,000\pi + \frac{2\,000}{3}\pi = \frac{8\,000\pi}{3} = 8\,377{,}58 = 8\,378 \text{ m}^3$

2. The surface area for a cylinder closed on only one end:

   Surface area $= 2\pi rh + \pi r^2 = 2\pi(10)(20) + \pi(10)^2 = 500\pi$

   The surface area for an open hemisphere is:

   Surface area $= 4\pi r^2 \div 2 = 2\pi r^2 = 2\pi(10)^2 = 200\pi$

   Now add the two surface areas together to arrive at the total surface area:

   Total surface area $= 500\pi + 200\pi = 700\pi = 2\,199{,}11 = 2\,199 \text{ m}^2$

Worked examples

The plan for a rain gauge is given in the diagram. Use the information given to calculate:

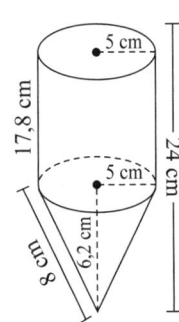

1. how much water (to the nearest cubic cm) this rain gauge will hold when totally full

2. the total external surface area for this rain gauge, correct to the nearest square cm.

Solutions

1. Volume for a cylinder: $V = \pi r^2 h = \pi(5)^2(17{,}8) = 445\pi$

   Volume for a cone: $V = \frac{1}{3}\pi r^2 h = \frac{1}{3}\pi(5)^2(6{,}2) = \frac{155\pi}{3}$

   Total volume: $V = 455\pi + \frac{155\pi}{3} = 1\,591{,}74 \approx 1\,592 \text{ cm}^3$

2. Surface area for the sides of the cylinder: $A = 2\pi rh = 2\pi(5)(17{,}8) = 178\pi$

   Surface area for the open cone: $A = \pi rh = \pi(5)(8) = 40\pi$

   Total surface area: $A = 178\pi + 40\pi = 218\pi = 684{,}87 = 685 \text{ cm}^2$

Problem solving

Worked example

A plastic bucket has an open circumference of 40π and a depth of 30 cm. The base of the bucket has a diameter of 20 cm. Calculate the maximum volume of the bucket, correct to the nearest litre.

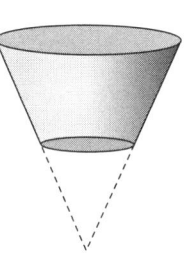

The frustum of a cone is the shape left when its top, or its bottom, has been cut off. To calculate the volume of a frustum, find the volume of the original cone, find the volume of the smaller, missing cone, and subtract the volume of the small missing cone from the original cone.

Solution

The bucket is the lower half of an upside-down cone.

The circumference of the top of the bucket $= 40\pi$.

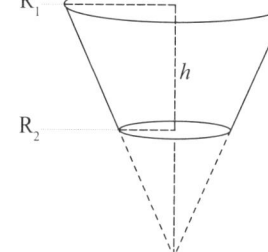

$\therefore 40\pi = 2\pi R_1$

$\therefore R_1 = 20$ cm

The radius of the bottom of the bucket, R_2, is 10 cm, and the depth of the bucket is 30 cm.

Using the midpoint theorem:

The depth of the original cone = 60 cm.

Volume of original cone $= \frac{1}{3}\pi R_1^2 h = \frac{1}{3}\pi(20)^2 \cdot 60 = 8\,000\pi$ cm^3

Volume of missing part of cone $= \frac{1}{3}\pi R_2^2 h = \frac{1}{3}\pi(10)^2 \cdot 30 = 1\,000\pi$ cm^3

$\therefore$ Volume of bucket $= 8\,000\pi - 1\,000\pi = 7\,000\pi = 21\,991{,}15$ cm^3

$= 21{,}99\ \ell = 22\ \ell$

Exercise 3

1. The diagram illustrates a capsule used by a pharmaceutical company. The capsule is made up of a cylinder and two perfect hemispheres. The capsule has an overall length of 16 mm and each hemisphere has a diameter of 6 mm. Calculate the volume and the surface area of one capsule.

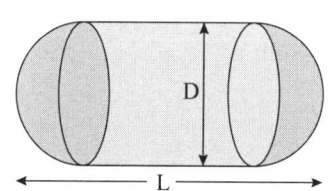

2. A tower is formed by attaching a cone of height 30 m onto a cylinder that has radius 40 m and height 50 m. Find the volume of the tower.

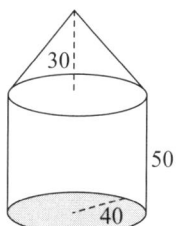

3. A party hat is made out of a cylindrical band 2 cm in height and a cone with a slant height of 13 cm and a radius of 5 cm. Calculate the total external surface area of this party hat.

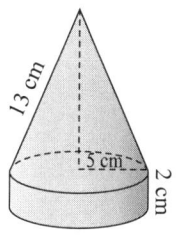

4. A conic shape is fits inside a cylinder with a diameter of 22 cm and a height of 53 cm. The area around the cone is filled with a liquid. Calculate the volume of liquid required.

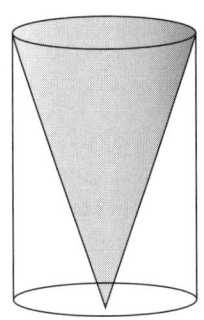

5. Jan is making lead sinkers out of lead. The sinker below is made up of two identical tetrahedrons each with a perpendicular height of $\frac{2\sqrt{6}}{3}$ cm and a slant edge of 2 cm.

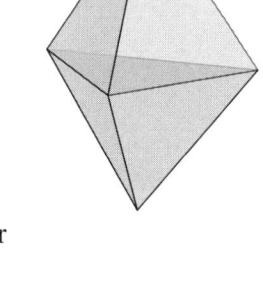

   5.1 Calculate the volume of the sinker.

   5.2 If Jan wants to pack each sinker into a small rectangular box, calculate the dimensions of the smallest box that Jan could use, leaving your answer in mm.

   5.3 Round these dimensions off to ensure that the sinker would fit into the box.

6. A glass ball is packed tightly into a cylinder, as shown, for transportation. If the radius of the glass ball is 2,1 m, calculate the total surface area of the cylinder and the volume of the packing material needed to hold the glass ball firmly in place.

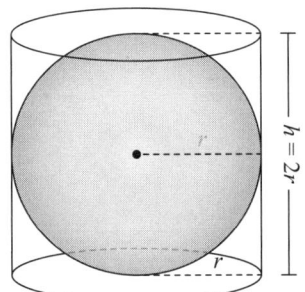

7. The roof of a canvas tent is in the shape of a right pyramid with a perpendicular height of 0,8 m on a square base. The length of one side of the base is 3 m.

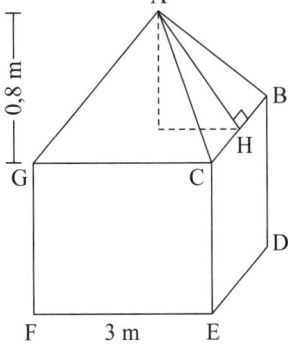

   7.1 Calculate the length of AH.

   7.2 Calculate the surface area of the roof.

   7.3 If the height of the walls of the tent is 2,1 m, calculate the total amount of canvas required to make the tent if the floor is excluded.

8. A metal ball has a radius of 8 mm.

   8.1 Calculate the volume of metal used to make this ball, correct to two decimal places.

   8.2 If the radius of the ball is doubled, write down the ratio of the new volume : the original volume.

   8.3 You want this ball to be silver-plated to a thickness of 1 mm. What volume of silver is required? Give your answer correct to two decimal places.

9. This festive cracker is made out of an open cylinder, with a diameter of 18 cm and a height of 38 cm, and an open cone with a radius of 11 cm and a slant height of 17 cm. Calculate how much cardboard in square metres is needed to make 10 of these festive crackers.

10. One-centimetre cubes are stacked as shown in the diagram.

    10.1 Calculate the total volume of the stack of cubes, correct to the nearest cubic cm.

    10.2 What is the total surface area of the stack of cubes, to the nearest cubic cm?

11. Brandon made the models shown for his design project. The smaller cones have a radius of 2 cm, and a depth of 4,5 cm. The bigger cones have a radius of 3 cm and a height of 9 cm. The silver balls have a radius of 1,8 cm.

    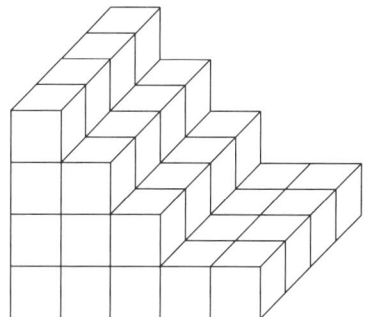

    11.1 If the shapes were totally hollow, calculate the total volume of each shape.

    11.2 The silver coating for the balls is very expensive at R55/cm². Calculate what it would cost to give both spherical shapes a double coating.

    11.3 The blue metallic paint needed for the cones is also expensive. The top cones, since their insides are visible, need to be coated both inside and out. Find the total area, correct to the nearest cm, that would need to be coated.

Test A: Knowledge and routine procedures

1.

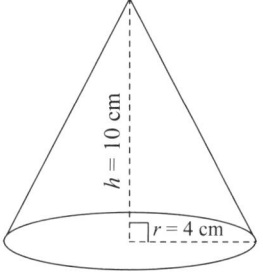

   1.1 Write down the formula for the volume of a cone. (1)

   1.2 Calculate the volume of the cone in the diagram, correct to one decimal place if necessary. (3)

   1.3 Write down the formula for the total surface area of a cone. (1)

   1.4 Calculate the total surface area of the cone in the diagram, correct to one decimal place if necessary. Show all your working. (4)

1.5 If you wanted to coat only the inside surface, without the circular base, with two coats of waterproof paint, how much paint would you need, in square centimetres? Round your answer off the nearest whole number.

(3)

2.

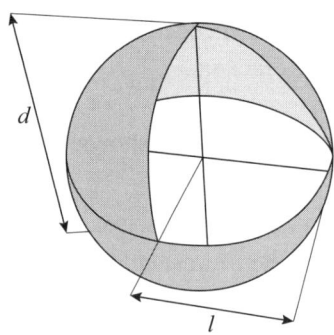

2.1 Write down the formula for the volume of a sphere. (1)

2.2 Calculate the volume of a sphere with a radius of 25 m, correct to two decimal places. (2)

2.3 Write down the formula for the total surface area of a sphere. (1)

2.4 Calculate the total surface area of a sphere with a radius of 25 m, correct to two decimal places. (2)

2.5 Write down the formula for the total surface area of a hemisphere. (1)

2.6 Calculate the total surface area of a closed hemisphere with a radius of 25 m, correct to two decimal places. (3)

3. Two different pyramids, A and B, are given.

Both pyramids have a perpendicular height of 12 m. Pyramid A has a square base with a length of 4 m. Pyramid B has a rectangular base, with a length of 8 m and a breadth of 4 m.

A 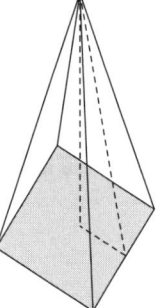   **B**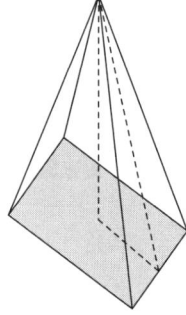

3.1 Write down a formula that you could use to find the volume of a pyramid. (1)

3.2 Calculate the volume of pyramid A, correct to one decimal place. (4)

3.3 Use Pythagoras to find the height of one of the triangular faces of pyramid A. (2)

3.4 Calculate the total surface area of pyramid A, correct to the nearest cm^2. (3)

3.5 Draw a neat net for the surface area of pyramid B. Fill in all the dimensions. (4)

3.6 Calculate any other heights you will need to find the total surface area of this pyramid. Show all of these measurements neatly on the net you drew. (5)

4. The diagram shows a wooden block that has had a hole drilled in it. The diameter of the hole is 2 cm.

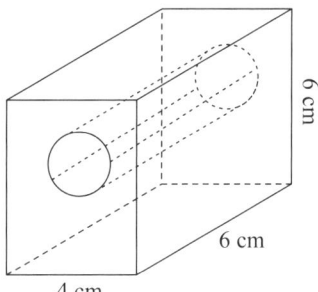

4.1 Calculate the volume of this solid, giving your answer correct to two decimal places. (5)

4.2 The inside surface of the hole has to be coated with special rubber paint that costs R115/m². Two coats of this special paint are required. Calculate the cost involved. (4)

Total 50

Test B: Complex procedures and problem solving

1. An owner of an ice cream parlour wants to install a steel model of an ice cream cone outside the entrance of the parlour. The shape of the model of the cone is constructed by using a hemisphere and a cone as shown in the diagram.

   The total height of the model is 1,4 m and the radius of the cone is 40 cm.

   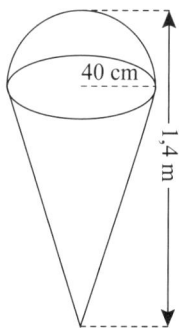

   Calculate:

   1.1 the volume of the model in cm³ (5)

   1.2 the total exterior surface area of the model in m² (5)

   1.3 the mass of the steel model if 1 m² has a mass of 2,5 kg (1)

2. The rough diagrams for three glasses are shown. All measurements are in centimetres.

   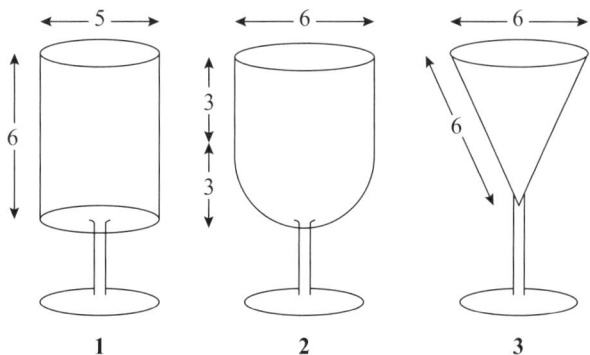

   Which glass holds the least liquid and which holds the most liquid?
   Show all your working to support your answer. (12)

3. Jenna wants to redecorate her bedroom. She decides to keep her existing bedside lamp (as shown in the diagram), which she plans to re-cover in keeping with the colour scheme of her new bedroom.

The cover for this bedside lamp is actually the lower half of a truncated cone. The fabric of the lampshade has a slant height of 13 cm as shown.

3.1 If the perpendicular height for the full cone is 24 cm, show that the radius R_2 = 5 cm and that R_1 = 10 cm. (4)

3.2 Use all of this information to calculate the exact amount of fabric required to recover the lampshade on the outside only. Do not leave any fabric for seams or overlays since it will be bonded onto the old lampshade with a special heat process, so no seams or overlaps are required. (3)

4. A tennis ball has a total surface area of 100π square cm. Tennis balls are commonly sold in cylindrical containers containing 3 tennis balls.

Three tennis balls are packed tightly into each container. Use the information given to work out the total surface area available for the labelling on this type of packaging, i.e. calculate the total surface area of the curved surface, without the lid or the base, correct to two decimal places. (5)

5. A cone with a radius of 5 cm and a depth of 15 cm is filled with fine beach sand. The sand from the cone is tipped into a cylindrical tin with a diameter of 12 cm and a height of 17 cm. Calculate the height of the sand in the cylinder and express it as a percentage of the total volume of the cylindrical tin. (8)

6. Cubes are stacked as shown in the diagram. Calculate the total surface area of the shape. (7)

= 1 cubic unit

Total 50

Test C: Content and breakdown as for exam

1. Eskom recently ran into problems supplying power to the entire country, which resulted in load shedding. The positive offshoot of this is that South Africans have become more energy efficient.

   Our geysers use large amounts of electricity and it is now advisable to use a geyser blanket to keep our geysers warm, thereby using less electricity.

   This geyser is **cylindrical** in shape. It has a radius of 30 cm and a length of 1,2 metres. Allow for an **extra 2 cm on the overall length and 1 cm extra on the radius** to ensure that the blanket will cover the geyser completely.

   1.1 Calculate the amount of fabric needed to make this blanket, in square metres, correct to the nearest square metre. (5)

   1.2 If 1 000 cm^3 = 1 ℓ, how many litres does this geyser hold when full? (2)

2. A concrete beam is to rest on two concrete pillars. The beam is a cuboid with sides of length 0,5 m, 3 m and 0,4 m. The pillars have diameter 0,4 m and height 2 m. Calculate the total volume of concrete needed to make the beam and the pillars. Round your answer to a sensible level of accuracy. (7)

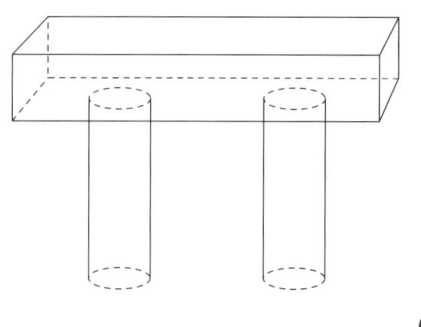

3. **3.1** The diagram shows a storage tank in which a farmer stores his grain. The tank is made up of a right cylinder with a hemisphere on top. The perpendicular height of the tank to the top is 75 m and the radius of the tank is 10 m.

   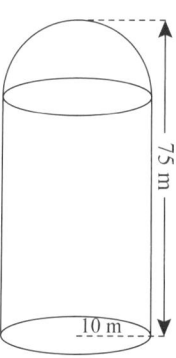

   Calculate the total exterior surface area of the tank. (6)

   3.2 Two identical cement pillars are placed at the entrance to a building. They are made up of a rectangular prism and a right pyramid placed upon it. The rectangular prism is 2 m high, 0,6 m long and 0,5 m wide. The perpendicular height of the pyramid is 0,8 m.

   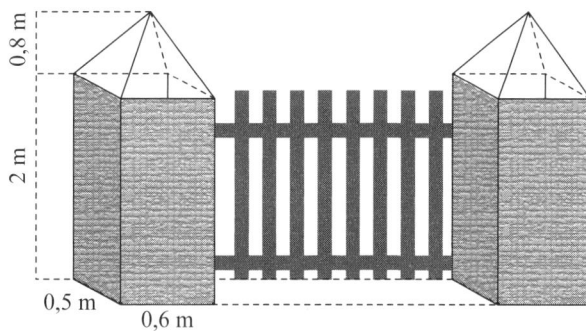

   Determine the volume of cement needed to make up the two pillars. (6)

4. The Grade 10 class are selling ice cream cones to raise funds for their SAVE THE RHINO campaign. They use sugar cones with a diameter of 4 cm and a slant height of 12 cm.

4.1 Use this information to work out and to fill in the missing measurements in the diagram given below, i.e. Radius$_1$ = ?? cm; Radius$_2$ = ?? cm, and the respective perpendicular heights. (4)

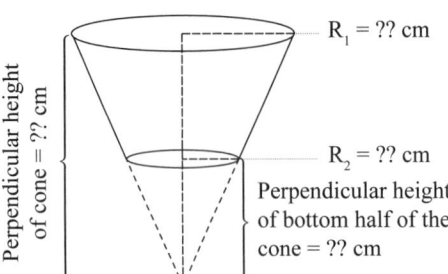

4.2 The bottom half of the sugar cone is filled with melted chocolate. Calculate the volume of melted chocolate in cm^3. (3)

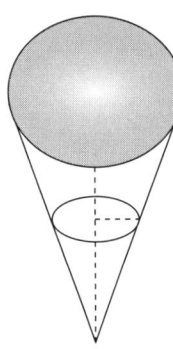

4.3 A perfectly spherical scoop of ice cream with a radius of 2 cm is placed on top of the cone. Calculate the volume of this scoop of ice cream, correct to two decimal places. (3)

4.4 Now calculate the space left inside the cone, i.e. the space between the ball of ice cream and the melted chocolate.
Hint: Start by calculating the volume of the hemisphere of ice cream that is actually inside the cone. (8)

5. A box contains two standard golf balls that fit snugly inside. The box is 85 mm long. What percentage of the space inside the box is air? Leave your final answer correct to one decimal place. (8)

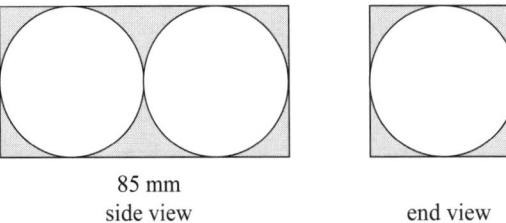

85 mm
side view   end view

Total 50

In Grade 10 you will formalise basic concepts and ideas in the study of probability. In earlier grades, most of the probability work you did was on finding relative frequencies through trials and experiments, which is known as empirical probability.

In Grade 10 you will use a probability model (a Venn diagram) to compare the relative frequency of events with the theoretical probability.

You will learn how to find probabilities and use specific rules in order to find probabilities using Venn diagrams.

Knowledge and skills for this topic

If you struggle with any of the work listed below, revise it before continuing with this Topic.

- simplifying fractions
- how to represent a ratio correctly
- being precise when using notation
- how to use the formula for calculating the probability of an event, which is:
  $$P(A) = \frac{\text{number of times even A occurred}}{\text{number of times the experiment was carried out}}$$

Content of final exam

- Write down the probability of any event occurring.
- Draw or use a given Venn diagram to solve probability questions.
- Define and work with mutually exclusive and complementary events.
- Use the rule for the identity of two events A and B, where
  $P(A \text{ or } B) = P(A) + P(B) - P(A \text{ and } B)$.

Properties of probability

1. Probabilities lie between 0 and 1 inclusive: $0 \leq P(E) \leq 1$.

   - 0 indicates that an event cannot occur; it is impossible.
   - 1 indicates the certainty of an event.
   - A score of 0,5 indicates equally likely outcomes.

2. The sample space (S) is the set of all possible outcomes. $P(S) = 1$.
   (This indicates the probability of a certain event or the sample space.)

   - $n(S)$ represents the number of elements in the sample space.

3. $P(\varnothing) = 0$ indicates the probability of an impossible event.

4. $P(A \text{ or } B) = P(A \cup B)$ – the symbol '$\cup$'is used to combine sets.

5. $P(A \text{ and } B) = P(A \cap B)$– the symbol '$\cap$' is used to indicate the common elements.

Using probability models to compare the relative frequency (empirical probability) of events with the theoretical probability

A **trial or experiment** is a performance of a random experiment. Each **trial** results in one or more **outcome**.

An **outcome or event** is a specific set of outcomes within the sample space. These events are usually denoted with capital letters. So, we say that an event is a subset of the sample space, and that $n(E)$ represents the number of outcomes in the event (E). The event will be specified in the question.

For example:

- In an **experiment**, three marbles are selected from a bag containing 100 marbles, 40 of which are blue and 60 yellow.
- We consider the selection of each marble to be an **event**.
- We say there are three **trials** in the experiment.

Worked examples

1. A coin is tossed in an experiment. There are two possible outcomes:
    - Event H, which is obtaining a head: $P(H) = \frac{1}{2}$
    - Event T, which is obtaining a tail: $P(T) = \frac{1}{2}$
2. A die is rolled in an experiment. There are six possible outcomes in the sample space (S) but various events can be stated in the question.
    - Event E, which is rolling an even number: $P(E) = \frac{3}{6} = \frac{1}{2}$
    - Event A, which is obtaining a number divisible by 3: $P(A) = \frac{2}{6} = \frac{1}{3}$

Relative (empirical or observed) probability

The actual observation of the probabilities of a random event during an experiment is called **empirical probability**. We call the actual physical tossing of a coin, rolling a die, selecting a card from a pack of 52 cards (excluding the jokers) or any other stated selection of elements from an event within a sample space, **experimental (empirical or relative) probability**.

Because these physical events would need to be performed many times in order to yield a credible result and are very time-consuming, we use **theoretical** probability to predict probabilities.

If E is the event that can occur when an experiment is performed, then the empirical probability of the event (E) is:

$$P(E) = \frac{\text{number of times event E occurred}}{\text{number of times the experiment occurred}}$$

If the experiment is performed a large number of times, the empirical probability will become closer and closer to the theoretical probability.

Worked example

A survey was conducted to determine learners' favourite ice cream. The results are recorded in the following table.

Ice cream	vanilla	chocolate	strawberry	rum and raisin	peppermint	other
Number of learners	15	35	10	8	5	12

What is the probability that a learner's favourite ice cream is vanilla?

Solution

$(E) = \frac{n(E)}{n(S)} = \frac{15}{85} = 0,18$

Theoretical probability

To define theoretical probability, if the outcomes of a particular sample space S are equally likely to occur, then:

$$P(Event) = \frac{\text{no. of favourable outcomes in the event}}{\text{no. of all possible outcomes in the sample space}}$$

Or using symbols:

$$P(E) = \frac{n(E)}{n(S)}$$

Exercise 1

1. Give your answer in (a) words, (b) decimals, (c) as a common fraction, and (d) as a percentage.

   1.1 What is the probability that if today is Monday, tomorrow is Friday?

   1.2 What is the probability that if today is Monday, tomorrow is Tuesday?

2. Give your answer as a (a) common fraction and (b) decimal, correct to two decimal places. If a die is rolled, what is the probability of rolling the following?

   2.1 a six   **2.2** no six

   2.3 an even number   **2.4** a composite number

3. Give your answer as a (a) common fraction and (b) percentage, correct to two decimal places. If one card is drawn from a standard pack of 52 cards, what is the probability of drawing the following cards?

   3.1 a black card

   3.2 a face card (a card with a picture of a face on it – a Jack, Queen or King)

   3.3 a card with the number 5 on it

Venn diagrams

Venn diagrams are used as a method to solve problems where the information collected often overlaps and needs to be sorted out. The probability of various events occurring can then be calculated fairly easily.

We also use tree diagrams and contingency tables to represent probabilities but these will only be used in Grade 11. For Grade 10, we will only focus on Venn diagrams.

Venn diagrams show:

- a visual summary of the probabilities for different events
- a visual summary of exclusive and complementary events.

Deriving and applying certain methods for any two events A and B in a sample space S

Mutually exclusive events

If we have two events A and B, and event A happens, then event B cannot happen and vice versa. We write this as: P(A and B) = P(A ∩ B) = 0

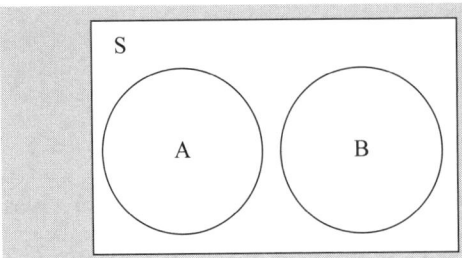 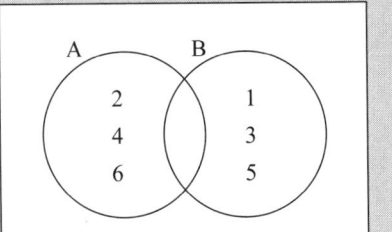

For example, when selecting (drawing) one card from a pack of 52 cards, the probability of drawing a King or an Ace is mutually exclusive, because the events cannot occur at the same time.

Complementary events

If an event has two possible outcomes A and B, and there are only two possible outcomes of the event, the outcomes are said to be complementary.

We write this as P(A ∩ B) = 0 and P(A) = 1 – P(B).

> **Note**
> - When you toss a coin (event), the two possible outcomes A (heads) or B(tails) are complementary.
> - All complementary events are mutually exclusive, but not all mutually exclusive events are complementary.

Rules for probability

> **Note**
> P(A ∪ B) = 1

1. **General addition rule:** If A and B are two events in a sample space S, then:
    - P(A or B) = P(A) + P(B) –P(A and B)
      i.e. P(A ∪ B) = P(A) + P(B) –P(A ∩ B)
    - A and B are mutually exclusive if P(A and B) = 0.

> **Note**
> The symbol used for the complement of A is Ā.

2. **Complementary rule:** If A and B are mutually exclusive, then they are complementary and P(A) + P(B) = 1. Then P(B) = P(not A) = P($\bar{A}$) = 1 – P(A).

How to deal with probabilities and mutually exclusive events

Worked example

What is the probability of rolling a die such that we obtain a 3 and 6 in any order?

Solution

It does not matter whether the first roll is a 3 or a 6.
However, if the first roll is a 3, clearly this rules out the possibility of a 6. This means that the two events rolling a 3 or a 6 are exclusive, as one result directly affects the other.

So obtaining a 3 or 6 with the first roll is the sum of the probabilities: $\frac{1}{6} + \frac{1}{6} = \frac{1}{3}$.

The probability of the second roll being successful (favourable) is $\frac{1}{6}$. The second roll can only be one specific number 6 if the first roll is a 3, and vice versa.

So the probability of throwing a 3 or a 6 in any order with two rolls of the die is:

$\frac{1}{3} \times \frac{1}{6} = \frac{1}{18}$.

Worked examples

Mikaela calculated that there is an 80% chance of receiving an SMS from her best friend Juliana after school on any given day (event A). She also calculated that the probability of receiving an SMS from her mother is 0,48 (event B) and the probability of receiving an SMS from both her mother and Juliana on the same day is 0,36.

1. Draw a Venn diagram to illustrate the outcomes clearly.

2. Are events A and B mutually exclusive? Validate your answer.

3. Determine the probability that neither her mother nor Juliana will SMS her on a Friday.

Solutions

1.

- Start by drawing a Venn diagram.
- Indicate the sample space with 'S' and $n(S) = 1$.
- Let one event be A and the other event be B. Draw these as overlapping circles. Label the different categories.
- Fill in the number of SMSs she receives from her mother, from Juliana and from both her mother and Juliana.
- Fill in the remaining numbers:
  - the number of SMSs from Juliana only = (0,8 – 0,36 = 0,44)
  - the number of SMSs from her mother only = (0,48 – 0,36 = 0,12).
- Let the probability that neither will send an SMS be x.

2. Not mutually exclusive because P(A and B) = $P(A \cap B) \neq 0$

3. P(neither will send SMS) = 1 – P(A or B)

$$= 1 - P(A \cup B)$$
$$= 1 - (P(A) + P(B) - P(A \cap B))$$
$$= 1 - (0,8 + 0,48 - 0,36)$$
$$= 0,08$$

Worked examples

A study was done to test how effective three different drugs – A, B and C – were in relieving headaches. Over the period covered by the study, 80 patients were given the opportunity to use all three drugs. The following results were obtained:

- 40 reported relief from drug A.
- 35 reported relief from drug B.
- 40 reported relief from drug C.
- 21 reported relief from both drugs A and C.
- 18 reported relief from drugs B and C.
- 68 reported relief from at least one of the drugs.
- 7 reported relief from all three drugs.

1. Record this information in a Venn diagram.
2. How many subjects got no relief from any of the drugs?
3. How many subjects got relief from drugs A and B but not C?
4. What is the probability that a randomly chosen subject got relief from at least one of the drugs?

Solutions

Hint

Read, read and read again before starting.

Here we have a study of the effectiveness of three different drugs. There are three events: A, B and C.

1. Draw a Venn diagram with three circles overlapping as indicated.

   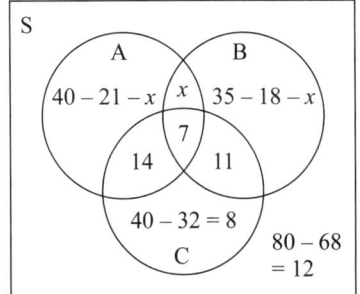

   - Fill in the overlap of all three events first, i.e. 7.
   - Fill in the overlapping areas for every combination of two events. There is no value for the effectiveness of A and B, so let A and B but not C be x.
   - Fill in the details for the effectiveness of A only and then B only in terms of x.
   - Calculate how many people experienced no effectiveness from all three drugs.

2. 12 subjects

3. A and B but not C is x.

   So $68 = 19 - x + x + 7 + 14 + 17 - x + 11 + 8$

   $\therefore x = 8$ subjects

4. P(at least one of the drugs is effective) $= \frac{68}{80} = 0,85$

Exercise 2

1. The probability that event A will occur is 0,4, and the probability that event B will occur is 0,48. The probability that event A or event B will occur is 0,73.

   1.1 Draw a Venn diagram to illustrate the probabilities.

   1.2 Calculate the probability that both events A and B will occur.

   1.3 Are events A and B mutually exclusive?

2. From a class survey conducted by 120 Grade 10 learners, it was found that 90 learners owned cell phones, 54 owned iPods, and 42 owned both cell phones and iPods.

   2.1 Draw a Venn diagram and show all possible probabilities.

2.2 How many learners owned either a cell phone or an iPod?

2.3 How many learners did not own either cell phone or an iPod?

3. At a particular school in a class of learners:

   - 9 learners play cricket and water polo
   - 15 learners play cricket
   - 21 learners play water polo

   3.1 Illustrate this information using a Venn diagram. Let the event 'play cricket' be C and the event 'play water polo' be W.

   3.2 How many learners are in this class?

   3.3 What is the probability that a learner plays cricket only?

   3.4 What is the probability that a learner plays cricket and water polo?

   3.5 What is the probability that a learner plays cricket or water polo?

   3.6 Are events C and W mutually exclusive? Validate your answer.

   3.7 Is event C the complement of event W? Validate your answer.

Test A: Knowledge and routine procedures

1. Estimate the probability for each event. State if the probability is *below average*, *a 50-50 chance*, *above average*, *certain*, *impossible* or if you cannot tell with the given information.

   1.1 Event A: The sun will rise tomorrow. (2)

   1.2 Event B: You will captain the national U23 soccer team this year. (2)

   1.3 Event C: The next baby to be born will be a girl. (2)

   1.4 Event D: Bafana-Bafana will win the next FIFA Soccer World Cup tournament. (2)

   1.5 Event E: You will get a tail when you toss a coin. (2)

   1.6 Event F: You will select a number card from a pack of 52 cards. (2)

   1.7 Event G: On rolling a die, the number on the top of the die will be divisible by 3. (2)

2. Determine the following theoretical probability for the following events, on rolling a die. Give you answer as a (a) fraction, (b) decimal and (c) percentage for each event.

   2.1 Event A: Obtaining an odd number (3)

   2.2 Event B: Obtaining a number that is a factor of 12 (3)

   2.3 Event C: Obtaining a 6 (3)

   2.4 Event D: Not obtaining a 6 (3)

   2.5 Event E: Obtaining a 9 (2)

3. In a sports-loving family of seven, which includes both parents, everybody plays either hockey or tennis or both sports. Three family members play only hockey and two family members play both sports.

   3.1 How many family members play only tennis? (1)

3.2 Illustrate the above information using a Venn diagram. Let T represent the tennis players and H represent the hockey players. (3)

3.3 What is the probability that a family member plays tennis or hockey? (1)

3.4 Are the events T and H complementary events? Validate your answer. (2)

4. In a class of 36 learners:

   - 24 learners like Mathematics
   - 18 learners like Biology
   - 8 learners don't like Mathematics or Biology.

4.1 Illustrate the above information using a Venn diagram. Let the number of learners who like both Mathematics (event M) and Biology (event B) be x. (4)

4.2 Determine how many learners like both Biology and Mathematics. (2)

4.3 What is the probability that a learner likes only Mathematics? (2)

4.4 What is the probability that a learner likes either Mathematics or Biology? (3)

4.5 Are the events M and B:

   4.5.1 mutually exclusive? (1)

   4.5.2 complementary? (Validate your answers.) (2)

Total 50

Test B: Complex procedures and problem solving

1. What is the probability of drawing the following card/s from a pack of 52 cards (jokers excluded)?

   Example set of 52 playing cards; 13 of each suit, namely clubs, diamonds, hearts, and spades:

	Ace	2	3	4	5	6	7	8	9	10	Jack	Queen	King
Clubs													
Diamonds													
Hearts													
Spades													

1.1 Drawing a black card (2)

1.2 Drawing a face card (Jack, Queen or King) (2)

1.3 Drawing a King (2)

1.4 Only one King does not have a moustache, the King of hearts. What is the probability of drawing this card from the set of face cards? (2)

2. In a particular school, there are 160 Grade 10 learners. Eighty-four learners take Mathematics and sixty take Business Studies. Thirty-six learners take both Mathematics and Business Studies.

   2.1 Draw a Venn diagram to illustrate the given data. (4)

   2.2 Hence, calculate the probability that a learner in Grade 10, chosen at random:

       2.2.1 takes Business Studies (2)

       2.2.2 takes Mathematics but not Business Studies (2)

       2.2.3 takes Mathematics or Business Studies. (2)

   2.3 If event A is the Mathematics learners and event B the Business Studies learners, are the events A and B mutually exclusive? (2)

   2.4 Using your answers above, in your opinion, does this school encourage learners to take Mathematics or Business Studies?
   Validate your answer. (3)

3. Pamela has a pencil case containing 32 items. She states that the probability of selecting certain items from the pencil case is as follows:

   - P(blue pen) = 0,25
   - P(green pen) = 0,75
   - P(ruler) = 0
   - P(pen) = 1

   3.1 Make a list of all the items in her pencil case and their quantities – in other words, the sample space. (2)

   3.2 Is P(blue pen) the complement of P(green pen)? Validate your answer. (2)

   3.3 Is P(blue pen) mutually exclusive from the P(green pen)? Validate your answer. (2)

4. The following Venn diagrams represent the winter sports played by a Grade 10 class. They can play soccer (S), hockey (H) or rugby (R).

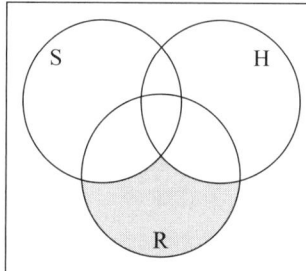 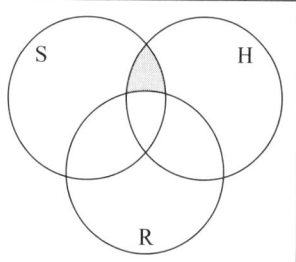 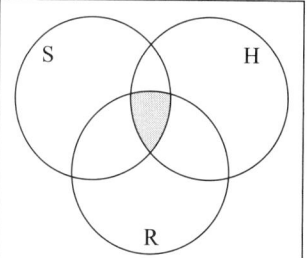

Diagram A   Diagram B   Diagram C

Examine the diagrams and decide which diagram represents the shading for the following:

   4.1 Those who play soccer and hockey but not rugby (2)

   4.2 Those who play all three sports (2)

   4.3 Those who play only rugby (2)

5. In each of the following cases, draw the Venn diagram and illustrate the probabilities for the information given and answer the question asked.

    5.1 In a class of 50 learners, 28 take Dance (event A), 16 take Drama (event B) and 6 learners take both Drama and Dance. How many learners in the class do not take either Dance or Drama? (5)

    5.2 In a class of 34 learners, Kim does a survey to find out how many learners have either a cat or a dog as a pet. She discovers that 22 learners have dogs, 16 have cats and 7 have both dogs and cats. How many learners have no pets? (5)

    5.3 A survey was conducted on a group of 24 learners to find out which sport they liked. The results of the survey were as follows: 6 learners said they liked soccer, 12 said they liked cricket, and 15 liked rugby. Only one of the learners said she liked soccer, cricket and rugby. Two liked soccer and cricket by not rugby, and two liked soccer and rugby but not cricket. If all the learners liked at least one of the sports mentioned, how many learners liked cricket and rugby but not soccer? (5)

Total 50

Test C: Content and breakdown as for exam

1. Shooting goals is how points are scored in netball game. Rebecca is a keen netball goal shooter who is diligent at practising her shooting. During a practice she shoots 32 balls from a total of 40 balls into the net.

    1.1 Calculate the approximate probability that she will not get a ball into the net. (2)

    1.2 At a follow-up practice, Rebecca decides to shoot a total of 100 balls. How many times does she expect to get the ball into the net? (2)

    1.3 Explain how the approximate answer to question 1.1 can be reduced. (2)

2. A bag contains 12 green, 6 red, 4 yellow and 2 blue marbles. A marble is selected from the bag at random. What is the probability that it is:

    2.1 green (2)

    2.2 red or yellow (2)

    2.3 not red (2)

    2.4 not yellow or blue? (2)

3. Mrs James has a CD player that she wants to award to one of the learners in her class who took part in a competition. The names of the learners who participated and their ages are as follows:

Boys	Ages	Girls	Ages
Nathan	15	Delia	17
Manie	16	Francina	17
Jonathan	15	Nomsa	16
Mutumi	15	Gracie	15
John	15	Petro	16
David	17	Violet	17
Rafiek	16	Morwesi	16
		Charlise	15

3.1 Give $n(S)$. (1)

3.2 Events are defined as follows:

E$_1$: A boy is selected

E$_2$: A 16-year-old must win

E$_3$: The winner's name starts with an M

E$_4$: A 17-year-old must win

   3.2.1 Determine $P(E_1)$. (1)

   3.2.2 Determine $P(E_4)$. (1)

   3.2.3 Determine $P(\bar{E}_2)$. (1)

   3.2.4 Determine $P(\bar{E}_4)$. (1)

   3.2.5 Determine $P(E_1 \text{ or } E_4)$. (3)

   3.2.6 Determine $P(E_2 \text{ and } E_3)$. (3)

4. Given that events A and B are mutually exclusive such that $P(A \cup B) = 0{,}75$ and $P(A) = 4P(B)$. Determine $P(B)$. (4)

5. A survey was conducted on 100 people to find out what category of movie they preferred: comedy, action or science fiction. The following results were obtained:

- 50 liked comedy
- 66 liked action
- 40 liked science fiction
- 27 liked comedy and action but did not like science fiction
- 13 liked action and science fiction but not comedy
- 4 liked all three
- 94 liked at least one.

5.1 How many people did not like any of the three categories of movies? (2)

5.2 Illustrate the above information using a Venn diagram. (8)

5.3 How many people liked comedy and science fiction, but not action? (3)

5.4 Determine the following probabilities:

   5.4.1 P(liked action movies) (2)

   5.4.2 P(liked only action movies) (2)

   5.4.3 P(liked only action and science fiction movies but not comedy) (2)

   5.4.4 P(liked all three categories of movie) (2)

Total 50

TOPIC 1   Real numbers and surds   (page 1)

Exercise 1   (page 3)

1. 12   rational

2. $\sqrt{12} = 3{,}464$ irrational

3. $\sqrt[3]{12} = 2{,}289$ irrational

4. $-0{,}36$ rational

5. $-\sqrt{0{,}36} = -0{,}6$ rational

6. $-\sqrt[4]{0{,}36} \approx -0{,}7745\ldots \approx -0{,}775$ irrational

7. $\frac{9}{25}$ rational

8. $\sqrt{\frac{9}{25}} = \frac{3}{5}$ rational

9. $\sqrt[6]{\frac{9}{25}} \approx 0{,}843432\ldots \approx 0{,}843$ irrational

10. $0{,}45$ rational

11. $0{,}4\dot{5} = \frac{41}{90}$ rational

12. $3\sqrt{81} = 3 \times 9 = 27$ rational

13. $-3\sqrt{81} = -3 \times 9 = -27$ rational

14. $-3\sqrt{-81}$ neither

15. $-3\sqrt[3]{81} \approx -3 \times -4{,}3267\ldots \approx 12{,}980$ irrational

16. $25 + 144 = 169$ rational

17. $\sqrt{25 + 144} = \sqrt{169} = 13$ rational

18. $\sqrt{25} + \sqrt{144} = 5 + 12 = 17$ rational

19. $36 - 169 = -133$ rational

20. $\sqrt{36 - 169} = \sqrt{-133}$ neither

21. $\sqrt{36} - \sqrt{169} = 6 - 13 = -7$ rational

22. $9\pi \approx 28{,}2743338\ldots \approx 28{,}274$ irrational

23. $\sqrt[3]{9\pi} \approx 3{,}046463\ldots \approx 3{,}046$ irrational

24. $4\pi + 5\pi = 9\pi \approx 28{,}274$ irrational

25. C

Exercise 2   (page 5)

1. $\sqrt{50} = 5\sqrt{2}$ 2. $\sqrt{125} = 5\sqrt{5}$

3. $\sqrt{56} = 2\sqrt{14}$

4. $\sqrt{128} = 8\sqrt{2}$

5. $3\sqrt{48} = 12\sqrt{3}$

6. $-2\sqrt{32} = -8\sqrt{2}$

7. $2\sqrt{18} \times 3\sqrt{12} = 36\sqrt{6}$

8. $2\sqrt{18} + 3\sqrt{12} = 6\sqrt{2} + 6\sqrt{3}$

9. $2\sqrt{18} \div 3\sqrt{12} = \frac{2\,.\,3\sqrt{2}}{3\,.\,2\sqrt{3}} = \frac{\sqrt{2}}{\sqrt{3}} \times \frac{\sqrt{3}}{\sqrt{3}} = \frac{\sqrt{6}}{3}$

10. $\sqrt{\frac{24}{6}} = 2$

11. $3\sqrt{\frac{24}{6}} = 6$

12. $\sqrt{\frac{24}{6}} + \frac{3}{4}\sqrt{\frac{16}{9}} = 2 + \frac{3}{4} \times \frac{4}{3} = 2 + 1 = 3$

13. $\sqrt{8} + \sqrt{162} - \sqrt{18} + \sqrt{50}$
    $= 2\sqrt{2} + 9\sqrt{2} - 3\sqrt{2} + 5\sqrt{2} = 13\sqrt{2}$

14. $8\sqrt{50} - 3\sqrt{24} - \sqrt{96} = 40\sqrt{2} - 6\sqrt{6} - 4\sqrt{6}$
    $\qquad\qquad\qquad\qquad = 40\sqrt{2} - 10\sqrt{6}$

15. $\sqrt[3]{8} = 2$

16. $\sqrt[3]{125} = 5$

17. $\sqrt[4]{16} = 2$

18. $\sqrt[5]{128} = \sqrt[5]{2^7} = \sqrt[5]{2^5 \, . \, 2^2} = 2\sqrt[5]{4}$

19. $\sqrt[3]{8} - 3\sqrt[3]{27} + 5\sqrt[5]{32} = 2 - 9 + 10 = 3$

Test A   (page 5)

1.1 irrational   1.2 rational
1.3 rational   1.4 irrational
1.5 rational

2.1 D   2.2 C

3.1 $5\sqrt{16} = 5 \times 4 = 20$

3.2 $-8\sqrt{8} = -8 \times 2\sqrt{2} = -16\sqrt{2}$

3.3 $3\sqrt{6} \times 5\sqrt{3} = 15\sqrt{18}$
     $\qquad\qquad\qquad = 15 \times 3\sqrt{2} = 45\sqrt{2}$

3.4 $3\sqrt{6} \div 5\sqrt{3} = \frac{3\sqrt{2}}{5}$

4.1 $3\sqrt{12} + 5\sqrt{3} = 3 \times 2\sqrt{3} + 5\sqrt{3} = 11\sqrt{3}$

4.2 $3\sqrt{12} - 5\sqrt{27} = 6\sqrt{3} - 15\sqrt{3} = -9\sqrt{3}$

4.3 $\dfrac{-3\sqrt{40}}{\sqrt{5}} = \dfrac{-3\sqrt{5.4.2}}{\sqrt{5}} = \dfrac{-3\sqrt{5}.\sqrt{4}.\sqrt{2}}{\sqrt{5}}$
$$= -3 \times 2 \times \sqrt{2} = -6\sqrt{2}$$

4.4 $\sqrt{72} - \sqrt{162} - 3\sqrt{75} = 6\sqrt{2} - 9\sqrt{2} - 15\sqrt{3}$
$$= -3\sqrt{2} - 15\sqrt{3}$$

5.1 $\sqrt{2} - \sqrt[3]{3} \approx -0,0280... \approx 0,028$

5.2 $\dfrac{\sqrt{41}}{-3} \approx -2,1343... \approx -2,134$

5.3 $\dfrac{4}{3} \times \sqrt{21^2} = 28$

6.1 $\sqrt{78}$ lies between $\sqrt{64}$ and $\sqrt{81}$
$\therefore \sqrt{78}$ lies between 8 and 9

6.2 $\sqrt{78} \approx 8,8$

7.1 $\sqrt[3]{4}$ and -4π **7.2** $\dfrac{4}{0}$

7.3 $\dfrac{4}{0}$ which equals 0, and $\sqrt{4}$ which equals 2

Test B   (page 6)

1.   E: $\sqrt[5]{64} = \sqrt[5]{2^6} = 2\sqrt[5]{2}$

2.1 $5\sqrt[3]{16} = 5\sqrt[3]{8 \times 2} = 10\sqrt[3]{2}$

2.2 $-8\sqrt[3]{8} = -8 \times 2 = -16$

3.1 $21^2 = 441$ and $22^2 = 484$
$\therefore \sqrt{470}$ lies between 21 and 22

3.2 $\sqrt{470}$ lies closer to $\sqrt{484} \therefore \sqrt{470} \approx 21,7$

4.1 $6^3 = 216$ and $7^3 = 343$
$\therefore \sqrt[3]{225}$ lies between 6 and 7

4.2 $\sqrt[3]{225}$ lies far closer to $\sqrt[3]{216} \therefore \sqrt[3]{225} \approx 6,08$

5.1 $3\sqrt[3]{27} = 3 \times 3 = 9$

5.2 $-3\sqrt[3]{-27} = -3 \times -3 = 9$

5.3 $-3\sqrt[4]{-81}$ non-real, neither rational nor irrational

6.1 $\sqrt{225 - 289}$
$= \sqrt{-64}$ non-real, neither rational nor irrational

6.2 $\sqrt{225} - \sqrt{289} = 15 - 17 = -2$ rational

7.1 $4\sqrt[3]{16} - 7\sqrt[4]{32}$
$= 4\sqrt[3]{2^3 . 2} - 7\sqrt[4]{2^4 . 2}$
$= 4 \times 2\sqrt[3]{2} - 7 . 2\sqrt[4]{2}$
$= 8\sqrt[3]{2} - 14\sqrt[4]{2}$

7.2 $3\sqrt[3]{54} + 5\sqrt{8} - \sqrt{32} + \sqrt[3]{8}$
$= 3\sqrt[3]{27 . 2} + 5\sqrt{2^2 . 2} - \sqrt{16} . 2 + 2$
$= 3 . \sqrt[3]{3^3 . 2} + 5 . 2\sqrt{2} - 4\sqrt{2} + 2$
$= 3 . 3 . \sqrt[3]{2} + 10\sqrt{2} - 4\sqrt{2} + 2$
$= 9\sqrt[3]{2} + 6\sqrt{2} + 2$

8.1 $\sqrt[5]{10}; \dfrac{\pi^2}{2}$ **8.2** $\sqrt[8]{-8}$

8.3 $\sqrt[3]{-27} = -3$ and $12 \times 0,\dot{3} = 12 \times \dfrac{1}{3} = 4$

9.1 $-\sqrt{\pi^2} = -\pi \approx -3,142$ irrational

9.2 $\sqrt{(1 + 2\pi)^2} = 1 + 2\pi \approx 7,283$ irrational

Test C   (page 7)

1.1   C   **1.2**   B

2.1 $3\sqrt{27} = 3 \times 3\sqrt{3} = 9\sqrt{3}$ irrational

2.2 $-\dfrac{\sqrt{27}}{3} = -\dfrac{3\sqrt{3}}{3} = -\sqrt{32}$ irrational

2.3 $3,78888 = 3\dfrac{71}{90}$ rational

3.1 $\sqrt{56} = 2\sqrt{14}$

3.2 $-3\sqrt{128} = -3 \times 8\sqrt{2} = -24\sqrt{2}$

3.3 $2\sqrt{18} \times 3\sqrt{12} = 2 \times 3\sqrt{2} \times 3 \times 2\sqrt{3} = 26\sqrt{6}$

3.4 $2\sqrt{18} - 3\sqrt{12} = 2 \times 3\sqrt{2} - 3 \times 2\sqrt{3}$
$$= 6\sqrt{2} - 6\sqrt{3}$$

4.1 $\sqrt[3]{27} = 3$

4.2 $\sqrt[4]{2\,592} = \sqrt[4]{2^5 \times 3^4} = \sqrt[4]{2^4 \times 2 \times 3^4}$
$$= 2 \times \sqrt[4]{2} \times 3 = 6\sqrt[4]{2}$$

5.1 $\sqrt[3]{33}$ lies between $\sqrt[3]{27}$ and $\sqrt[3]{64}$
$\therefore \sqrt[3]{33}$ lies between 3 and 4

5.2 $\sqrt[3]{33} \approx 3,2$

6.1 $36 + 169 = 205$ rational

6.2 $\sqrt{36 + 169} = \sqrt{205} \approx 14,32$ irrational

6.3 $\sqrt{36} + \sqrt{169} = 6 + 13 = 19$ rational

7.1 $\pi^2 \approx 9,87$ irrational

7.2 $\sqrt{\pi^2} = \pi$ irrational

7.3 $\sqrt[3]{\pi^2} \approx 2,145$ irrational

8.1 False: all rational numbers are real numbers.

8.2 True.

8.3 False: an undefined number cannot be written as zero, it is simply undefined and cannot be simplified.

TOPIC 2 Algebraic expressions (page 9)

Exercise 1 (page 12)

1. $f^2(f-2g)(f+g)$
$= f^2(f^2 - fg - 2g^2)$
$= f^4 - f^3g - 2f^2g^2$

2. $-(x-3y)(x+5y)$
$= -(x^2 + 2xy - 15y^2)$
$= -x^2 - 2xy + 15y^2$

3. $-5a^3(a^2-1)(a^2+1)$
$= -5a^3(a^4 - 1)$
$= -5a^7 + 5a^3$

4. $\frac{1}{x^2}(\frac{1}{x^2} - 2x)^2$
$= \frac{1}{x^2}\left(\frac{1}{x^2} - 4 + 4x^2\right)$
$= \frac{1}{x^4} - 2 + 2x^2$

5. $(9-a)(a+3) - (2a-5)^2$
$= (6a - a^2 + 27) - (4a^2 - 20a + 25)$
$= 6a - a^2 + 27 - 4a^2 + 20a - 25$
$= -5a^2 + 26a + 2$

6. $y(xy - z^2) + xy(x-y)^2 - (x^2 - z^2)y$
$= xy^2 - yz^2 + xy(x^2 - 2xy + y^2) - x^2y + yz^2$
$= xy^2 - yz^2 + x^3y - 2x^2y^2 + xy^3 - x^2y + yz^2$
$= xy^2 + x^3y - 2x^2y^2 + xy^3 - x^2y$

7. $2\left(3a - \frac{1}{a}\right)\left(2a + \frac{3}{a}\right) - \left(-\frac{2}{a}\right)^2$
$= 2\left(6a^2 + 7 - \frac{3}{a^2}\right) - \frac{4}{a^2}$
$= 12a^2 + 14 - \frac{6}{a^2} - \frac{4}{a^2}$
$= 12a^2 + 14 - \frac{10}{a^2}$

8. $(p-2)^2 - p^2(2p+1) + (3p-2)(-2p^2)$
$\quad - (4p+3) - p^2$
$= p^2 - 4p + 4 - 2p^3 - p^2 - 6p^3 + 4p^2 - 4p$
$\quad - 3 - p^2$
$= -8p^3 + 3p^2 - 8p + 1$

9. $(a^2-4)^2 + (a^2+1)(a^2-1) - (2a^2)^2$
$= a^4 - 8a^2 + 16 + a^4 - 1 - 4a^4$
$= -2a^4 - 8a^2 + 15$

Exercise 2 (page 13)

1. $A = (2x^2 - 3x + 4)$, $B = (3x - 7)$
and $C = (3 - 2x)$

1.1 $B - A = (3x - 7) - (2x^2 - 3x + 4)$
$= 3x - 7 - 2x^2 + 3x - 4$
$= -2x^2 + 6x - 11$

1.2 $B^2 = (3x - 7)^2$
$= 9x^2 - 42x + 49$

1.3 $BC = (3x - 7)(3 - 2x)$
$= -6x^2 + 23x - 21$

1.4 $AB = (2x^2 - 3x + 4)(3x - 7)$
$= 6x^3 - 9x^2 + 12x - 14x^2 + 21x - 28$
$= 6x^3 - 23x^2 + 33x - 28$

1.5 $AB - C = (2x^2 - 3x + 4)(3x - 7) - (3 - 2x)$
$= 6x^3 - 23x^2 + 33x - 28 - 3 + 2x$
$= 6x^3 - 23x^2 + 35x - 31$

1.6 $A + 2B - BC$
$= (2x^2 - 3x + 4) + 2(3x - 7) - (3x - 7)(3 - 2x)$
$= 2x^2 - 3x + 4 + 6x - 14 - (-6x^2 + 23x - 21)$
$= 2x^2 - 3x + 4 + 6x - 14 + 6x^2 - 23x + 21$
$= 8x^2 - 20x + 11$

1.7 $AC - C^2 - 3C$
$= (2x^2 - 3x + 4)(3 - 2x) - (3 - 2x)^2 - 3(3 - 2x)$
$= 6x^2 - 9x + 12 - 4x^3 + 6x^2 - 8x - (9 - 12x$
$\quad + 4x^2) - 9 + 6x$
$= 6x^2 - 9x + 12 - 4x^3 + 6x^2 - 8x - 9 + 12x$
$\quad - 4x^2 - 9 + 6x$
$= -4x^3 + 8x^2 + x - 6$

2.1 $(3x^2 - 3x)(x^2 - 3x - 4)$
$= 3x^4 - 9x^3 - 12x^2 - 3x^3 + 9x^2 + 12x$
$= 3x^4 - 12x^3 - 3x^2 + 12x$

2.2 $-2x(5x - 2x^3)(4x^2 + 2x - 7)$
$= -2x(20x^3 + 10x^2 - 35x - 8x^5 - 4x^4 + 14x^3)$
$= -2x(-8x^5 - 4x^4 + 34x^3 + 10x^2 - 35x)$
$= 16x^6 + 8x^5 - 68x^4 - 20x^3 + 70x^2$

2.3 $(2x - 3y)^3$
$= (2x - 3y)(4x^2 - 12xy + 9y^2)$
$= 8x^3 - 24x^2y + 18xy^2 - 12x^2y + 36xy^2 - 27y^3$
$= 8x^3 - 36x^2y + 54xy^2 - 27y^3$

3. $(3x - 1)(5x^2 - 2x - 1) - (2x^3 - x + 4)$

$= 15x^3 - 6x^2 - 3x - 5x^2 + 2x + 1 - 2x^3 + x - 4$

$= 13x^3 - 11x^2 - 3$

4.1 $\left(\frac{1}{x^2}x - 3\right)\left(\frac{3}{2}x^2 + \frac{3}{2}x + 9\right)$

$= \frac{1}{8}x^3 + \frac{3}{4}x^2 + \frac{9}{2}x - \frac{3}{4}x^2 - \frac{9}{2}x - 27$

$= \frac{1}{8}x^3 - 27$

4.2 $-\left(\frac{1}{x^2}x - 3\right)^3$

$= -\left(\frac{1}{x^2}x - 3\right)\left(\frac{x^2}{4} - 3x + 9\right)$

$= -\left(\frac{x^3}{8} - \frac{3x^2}{2} + \frac{9x}{2} - \frac{3x^2}{4} + 9x - 27\right)$

$= -\frac{x^3}{8} + \frac{3x^2}{2} - \frac{9x}{2} + \frac{3x^2}{4} - 9x + 27$

$= -\frac{x^3}{8} + \frac{9x^2}{4} - \frac{27x}{2} + 27$

Exercise 3 (page 15)

1. $4p^2y - 16py^2 = 4py(p - 4y)$

2. $p^2(x + y) - 16(y + x) = (x + y)(p^2 - 16)$

$\qquad\qquad\qquad\qquad\quad = (x + y)(p - 4)(p + 4)$

3. $25x^2 - 9y^2 = (5x - 3y)(5x + 3y)$

4. $3a(a - b) + 9b(b - a) = 3a(a - b) - 9b(a - b)$

$\qquad\qquad\qquad\qquad\quad = (a - b)(3a - 9b)$

$\qquad\qquad\qquad\qquad\quad = 3(a - b)(a - 3b)$

5. $x^2 - 2xy - 3y^2$

$= (x - 3y)(x + y)$

6. $\frac{9}{16}x^2 - \frac{3}{2}y^2$

$= \left(\frac{3}{4}x - \frac{1}{x^2}y\right)\left(\frac{3}{4}x + \frac{1}{x^2}y\right)$

7. $4x^2(2x - 1) + y^2(1 - 2x)$

$= 4x^2(2x - 1) - y^2(2x - 1)$

$= (2x - 1)(4x^2 - y^2)$

$= (2x - 1)(2x - y)(2x + y)$

8. $2x^2 + 10x + 12$

$= 2(x^2 + 5x + 6)$

$= 2(x + 3)(x + 2)$

9. $12y^2 - 7xy + x^2$

$= (3y - x)(4y - x)$

10. $\frac{x^2}{2} - \frac{1}{x^2}$

$= \frac{1}{x^2}(x^2 - 1)$

$= \frac{1}{x^2}(x - 1)(x + 1)$

Exercise 4 (page 16)

1. $3y + 9z + by + 3bz$

$= 3(y + 3z) + b(y + 3z)$

$= (y + 3z)(3 + b)$

2. $40p + 60r - 20pq - 30qr$

$= 10(4p + 6r - 2pq - 3qr)$

$= 10[2(2p + 3r) - q(2p + 3r)]$

$= 10(2p + 3r)(2 - q)$

3. $ac - bc - ad + bd$

$= c(a - b) - d(a - b)$

$= (a - b)(c - d)$

4. $b^3 + 6 + 3b^2 + 2b$

$= b^3 + 3b^2 + 2b + 6$

$= b^2(b + 3) + 2(b + 3)$

$= (b + 3)(b^2 + 2)$

5. $6ac - 15ax + 10xb - 4bc$

$= 3a(2c - 5x) + 2b(5x - 2c)$

$= 3a(2c - 5x) - 2b(2c - 5x)$

$= (2c - 5x)(3a - 2b)$

6. $5(a + b) - (a + b)^2$

$= (a + b)(5 - a - b)$

7. $3ax^2 + 2ab^2 - 2x^3 - 4xb^2$

$= a(3x^2 + 2b^2) - 2x(3x^2 + 2b^2)$

$= (3x^2 + 2b^2)(a - 2x)$

8. $1 - p - p^2 + p^3$

$= (1 - p) - p^2(1 - p)$

$= (1 - p)(1 - p^2)$

$= (1 - p)(1 - p)(1 + p)$

9. $x^2(p - q) - 2x(p - q) - p + q$

$= x^2(p - q) - (p - q)$

$= (p - q)(x^2 - 2x - 1)$

Exercise 5 (page 17)

1. $x^3 + 2x^2y - 15xy^2$

$= x(x^2 + 2xy - 15y^2)$

$= x(x + 5y)(x - 3y)$

2. $3x^2 + 24x + 45$

$= 3(x^2 + 8x + 15)$

$= 3(x + 5)(x + 3)$

3. $2x + 15 - x^2$

$= -x^2 + 2x + 15$

$= -(x^2 - 2x - 15)$

$= -(x - 5)(x + 3)$

4. $30 + 2x^2 - 16x$
$= 2x^2 - 16x + 30$
$= 2(x^2 - 8x + 15)$
$= 2(x - 5)(x - 3)$

5. $3a^2 + 2ab - 8b^2$
$= (3a - 4b)(a + 2b)$

6. $12x^2 - 14x + 4$
$= 2(6x^2 - 7x + 2)$
$= 2(3x - 2)(2x - 1)$

7. $11x + 6 + 3x^2$
$= 3x^2 + 11x + 6$
$= (3x + 2)(x + 3)$

8. $27x + 6 - 15x^2$
$= -15x^2 + 27x + 6$
$= -3(5x^2 - 9x - 2)$
$= -3(5x + 1)(x - 2)$

9. $22a^2 - 37a + 6$
$= (11a - 2)(2a - 3)$

10. $78x^2 - 111xy - 9y^2$
$= 3(26x^2 - 37xy - 3y^2)$
$= 3(2x - 3y)(13x + y)$

Exercise 6 (page 18)

1. $8b^3 - 343$
$= (2b - 7)(4b^2 + 14b + 49)$

2. $64b^3 + 27c^3$
$= (4b + 3c)(16b^2 - 12bc + 9c^2)$

3. $81ab^4 + 24a^7b$
$= 3ab(27b^3 + 8a^6)$
$= 3ab(3b + 2a^2)(9b^2 - 6a^2b + 4a^4)$

4. $-1\,000x^6 + 125y^3$
$= -125(8x^6 - y^3)$
$= -125(2x^2 - y)(4x^4 + 2x^2y + y^2)$

5. $64b^3 - 125c^3$
$= (4b - 5c)(16b^2 + 20bc + 25c^2)$

6. $(x - 1)^3 - 27$
$= [(x - 1) - 3][(x - 1)^2 + 3(x - 1) + 9]$
$= (x - 4)(x^2 - 2x + 1 + 3x - 3 + 9)$
$= (x - 4)(x^2 + x + 7)$

7. $\dfrac{a^3}{8} - \dfrac{b^3}{27}$
$= \left(\dfrac{a}{2} - \dfrac{b}{3}\right)\left(\dfrac{a^2}{4} + \dfrac{ab}{6} + \dfrac{b^2}{9}\right)$

8. $\dfrac{1\,000}{x^6} + \dfrac{y^3}{27}$
$= \left(\dfrac{10}{x^2} + \dfrac{y}{3}\right)\left(\dfrac{100}{x^4} - \dfrac{10y}{3x^2} + \dfrac{y^2}{9}\right)$

Exercise 7 (page 19)

1. $3x^2 - 7x - 6$
$= (x - 3)(3x + 2)$

2. $6x^2 - 9ax + 8cx - 12ac$
$= (6x^2 - 9ax) + (8cx - 12ac)$
$= 3x(2x - 3a) + 4c(2x - 3a)$
$= (2x - 3a)(3x + 4c)$

3. $4x^4 - 36$
$= 4(x^4 - 9)$
$= 4(x^2 - 3)(x^2 + 3)$

4. $x^4 - 13x^2 + 36$
$= (x^2 - 9)(x^2 - 4)$
$= (x - 3)(x + 3)(x - 2)(x - 2)$

5. $\dfrac{2}{5}x^2y^4 - \dfrac{4}{5}py^3$
$= \dfrac{2}{5}y^3(x^2y - \dfrac{2}{5}p)$

6. $8 - f^3$
$= (2 - f)(4 + 2f + f^2)$

7. $6x^2 + 4x - 16$
$= 2(3x^2 + 2x - 8)$
$= 2(3x - 4)(x + 2)$

8. $4x^2 - 12xy + 9y^2$
$= (2x - 3y)(2x - 3y)$

9. $18a^2 - 32b^2$
$= 2(9a^2 - 16b^2)$
$= 2(3a + 4b)(3a - 4b)$

10. $(2z - 1)^2 - 7(2z - 1)$
$= (2z - 1)(2z - 1 - 7)$
$= (2z - 1)(2z - 8)$

11. $(5c + 1)(a - b) - 2c(a - b)$
$= (a - b)(5c + 1 - 2c)$

12. $3ay - 5y + y^2 - 15a$

$= (3ay + y^2) + (-5y - 15a)$

$= y(3a + y) - 5(y + 3a)$

$= (3a + y)(y - 5)$

13. $\frac{1}{x^2}f^3 - 4g^3$

$= \frac{1}{x^2}(f^3 - 8g^3)$

$= \frac{1}{x^2}(f - 2g)(f^2 + 2fg + 4g^2)$

14. $2x^2 - 3 - \frac{2}{x^2}$

$= \frac{2x^4 - 3x^2 - 2}{x^2}$

$= \frac{(2x^2 + 1)(x^2 - 2)}{x^2}$

15. $3x^6 - 3y^6$

$= 3(x^6 - y^6)$

$= 3(x^3 - y^3)(x^3 + y^3)$

$= 3(x - y)(x^2 + xy + y^2)(x + y)(x^2 - xy + y^2)$

OR

$3x^6 - 3y^6$

$= 3(x^2 - y^2)(x^4 + x^2y^2 + y^4)$

$= 3(x - y)(x + y)(x^4 + x^2y^2 + y^4)$

Exercise 8 (page 22)

1.1 $x = 0$ **1.2** $x = -8$

1.3 $x = -\frac{3}{2}$ **1.4** $x = -3$ or $x = 5$

2. $\frac{x^2 - 4}{x^2 - 5x + 6}$

$= \frac{(x - 2)(x + 2)}{(x - 2)(x - 3)}$

$= \frac{(x + 2)}{(x - 3)}$

3. $\frac{(x - y)}{(x + y)} \times \frac{xy + y^2}{x^2 - xy}$

$= \frac{(x - y)}{(x + y)} \times \frac{y(x + y)}{x(x - y)}$

$= \frac{y}{x}$

4. $\frac{x}{x - y} \div \frac{y}{y - x}$

$= \frac{x}{(x - y)} \times \frac{(y - x)}{y}$

$= \frac{x}{(x - y)} \times \frac{-(y - x)}{y}$

$= -\frac{x}{y}$

5. $\frac{(3z - 2)}{z^2 + 4z} \times \frac{3z}{3z^2 + 10z - 8}$

$= \frac{(3z - 2)}{z(z + 4)} \times \frac{3z}{(3z - 2)(z + 4)}$

$= \frac{3}{(z + 4)^2}$

$= \frac{3}{z^2 + 8z + 16}$

6. $\frac{x^3 - 4x}{21x} \times \frac{7x^2}{x^4 - 8x}$

$= \frac{x(x^2 - 4)}{21x} \times \frac{7x^2}{x(x^3 - 8)}$

$= \frac{x(x - 2)(x + 2)}{21x} \times \frac{7x^2}{x(x - 2)(x^2 + 2x + 4)}$

$= \frac{(x + 2)}{3} \times \frac{x}{(x^2 + 2x + 4)}$

$= \frac{x(x + 2)}{3(x^2 + 2x + 4)}$

Exercise 9 (page 24)

1. $\frac{1}{a} + \frac{1}{3a} + \frac{1}{2a}$

$= \frac{1(3)(2) + 1(2) + 1(3)}{6a}$

$= \frac{6 + 2 + 3}{6a}$

$= \frac{11}{6a}$

2. $\frac{2y - 1}{3y} - \frac{1}{6}$

$= \frac{2(2y - 1) - 1(y)}{6y}$

$= \frac{4y - 2 - y}{6y}$

$= \frac{3y - 2}{6y}$

3. $\frac{1}{b + 1} + \frac{1}{b - 1}$

$= \frac{1(b + 1) + 1(b + 1)}{(b + 1)(b - 1)}$

$= \frac{b - 1 + b + 1}{(b + 1)(b - 1)}$

$= \frac{2b}{(b + 1)(b - 1)}$

4. $\frac{6}{2w - 3} - \frac{3}{3 - 2w}$

$= \frac{6}{2w - 3} + \frac{3}{2w - 3}$

$= \frac{6 + 3}{2w - 3}$

$= \frac{9}{2w - 3}$

5. $\frac{1}{x - 2} + \frac{2x}{x} - \frac{x - 1}{x + 2}$

Common denominator: $x(x - 2)(x + 2)$

$= \frac{1(x)(x + 2) + 2x(x - 2)(x + 2) - (x - 1)(x)(x - 2)}{x(x - 2)(x + 2)}$

$= \frac{x(x + 2) + 2x(x^2 - 4) - x(x^2 - 3x + 2)}{x(x - 2)(x + 2)}$

$= \frac{x^2 + 2x + 2x^3 - 8x - x^3 + 3x^2 - 2x}{x(x - 2)(x + 2)}$

$= \frac{x^3 + 4x^2 - 8x}{x(x - 2)(x + 2)}$

$= \frac{x(x^2 + 4x - 8)}{x(x - 2)(x + 2)}$

$= \frac{x^2 + 4x - 8}{(x - 2)(x + 2)}$

Be careful, the numerator cannot be factorised.

6.

$$\frac{3}{1-c} + \frac{4}{1-2c+c^2}$$

$$= \frac{3}{(1-c)} + \frac{4}{(1-c)^2}$$

$$= \frac{3(1-c)+4}{(1-c)^2}$$

$$= \frac{3-3c+4}{(1-c)^2}$$

$$= \frac{7-3c}{(1-c)^2}$$

7.

$$\frac{4}{r^2+2r} - \frac{3r}{r+2} + \frac{3r-2}{r}$$

$$= \frac{4}{r(r+2)} - \frac{3r}{(r+2)} + \frac{3r-2}{r}$$

$$= \frac{4-3r(r)+(3r-2)(r+2)}{r(r+2)}$$

$$= \frac{4-3r^2+3r^2+4r-4}{r(r+2)}$$

$$= \frac{4r}{r(r+2)}$$

$$= \frac{4}{(r+2)}$$

8.

$$\frac{1}{2z-4} + \frac{1}{z^2-5z+6} + \frac{1}{(z-2)(z^2-5z+6)}$$

$$= \frac{1}{2(z-2)} + \frac{1}{z^2-5z+6} + \frac{1}{(z-2)(z^2-5z+6)}$$

$$= \frac{1(z^2-5z+6)+2(z-2)+1(2)}{2(z-2)(z^2-5z+6)}$$

$$= \frac{z^2-5z+6+2z-4+2}{2(z-2)(z^2-5z+6)}$$

$$= \frac{z^2-3z+4}{2(z-2)(z^2-5z+6)}$$

Exercise 10   (page 24)

1. $x+y$

$$= \frac{t-2}{t+2} + \frac{t+2}{t-2}$$

$$= \frac{(t-2)(t-2)+(t+2)(t+2)}{(t+2)(t-2)}$$

$$= \frac{t^2-4t+4+t^2+4t+4}{(t+2)(t-2)}$$

$$= \frac{2t^2+8}{(t+2)(t-2)}$$

2. $x-y$

$$= \frac{t-2}{t+2} - \frac{t+2}{t-2}$$

$$= \frac{(t-2(t-2)-(t+2)(t+2)}{(t+2)(t-2)}$$

$$= \frac{t^2-4t+4-t^2-4t-4}{(t+2)(t-2)}$$

$$= \frac{-8t}{(t+2)(t-2)}$$

3. xy

$$= \frac{(t-2)}{(t+2)} \times \frac{(t+2)}{(t-2)}$$

$$= 1$$

4. $x \div y$

$$= \frac{(t-2)}{(t+2)} \div \frac{(t+2)}{(t-2)}$$

$$= \frac{(t-2)}{(t+2)} \times \frac{(t-2)}{(t+2)}$$

$$= \frac{(t-2)^2}{(t+2)^2}$$

$$= \frac{t^2-4t+4}{t^2+4t+4}$$

5. $(x+y) \div (x-y)$

$$= \frac{2t^2+8}{(t+2)(t-2)} \div \frac{-8t}{(t+2)(t-2)}$$

$$= \frac{2t^2+8}{(t+2)(t-2)} \times \frac{(t+2)(t-2)}{-8t}$$

$$= \frac{2(t^2+4)}{-8t} - \frac{t^2+4}{4t}$$

6. $x+y \div x-y$

$$= \frac{t-2}{t+2} + \frac{t+2}{t-2} \div \frac{t-2}{t+2} - \frac{t+2}{2-2}$$

Remember to multiply and divide before adding and subtracting.

$$= \frac{t-2}{t+2} + \left(\frac{t+2}{t-2} \div \frac{t-2}{t+2}\right) - \frac{t+2}{t-2}$$

$$= \frac{t-2}{t+2} + \left(\frac{t+2}{t-2} \times \frac{t+2}{t-2}\right) - \frac{t+2}{t-2}$$

$$= \frac{t-2}{t+2} + \frac{(t+2)^2}{(t-2)^2} - \frac{t+2}{t-2}$$

$$= \frac{(t-2)(t-2)^2+(t+2)^2(t+2)-(t+2)(t+2)(t-2)}{(t+2)(t-2)^2}$$

LCD $= (t+2)(t-2)^2$

$$= \frac{(t-2)(t^2-4t+4)+(t^2+4t+4)(t+2)-(t+2)(t^2-4)}{(t+2)(t-2)^2}$$

$$= \frac{t^3-6t^2+12t-8+t^3+6t^2+12t+8-(t^3+2t^2-4t-8)}{(t+2)(t-2)^2}$$

$$= \frac{t^3-6t^2+12t-8+t^3+6t^2+12t+8-t^3-2t^2+4t+8}{(t+2)(t-2)^2}$$

$$= \frac{t^3-2t^2+28t+8}{(t+2)(t-2)^2}$$

Exercise 11   (page 27)

1.1 $(a-b-3-x)(a-b+3+x)$

$$= [a-b-(3+x)][a-b+(3+x)]$$

$$= (a-b)^2-(3+x)^2$$

$$= (a^2-2ab+b^2)-(9+6x+x^2)$$

$$= a^2-2ab+b^2-9-6x-x^2$$

1.2 $(8x^3+1)(2x-1)(4x^2+2x+1)$

$$= (8x^3+1)(8x^3-1)$$

$$= 64x^6-1$$

1.3 $(a-b)(a+b)(a^2+b^2)(a^4+b^4)$

$$= (a^2-b^2)(a^2+b^2)(a^4+b^4)$$

$$= (a^4-b^4)(a^4+b^4)$$

$$= a^8-b^8$$

2.1 $x^2 - 2x - y^2 - 2y$
$= x^2 - y^2 - 2x - 2y$
$= (x - y)(x + y) - 2(x + y)$
$= (x + y)(x - y - 2)$

2.2 $a^2(1 + b) - b^2(1 - a)$
$= a^2 + a^2b - b^2 + ab^2$
$= a^2 - b^2 + a^2b + ab^2$
$= (a - b)(a + b) + ab(a + b)$
$= (a + b)(a - b + ab)$

2.3 $p^2 + 4m - 4m^2 - 1$
$= p^2 + (-4m^2 + 4m - 1)$
$= p^2 - (4m^2 - 4m + 1)$
$= p^2 - (2m - 1)^2$
$= [p - (2m - 1)][p + (2m - 1)]$
$= (p - 2m + 1)(p + 2m - 1)$

2.4 $a^2 - ab + 2a + b - 3$
$= a^2 + 2a - 3 - ab + b$
$= (a^2 + 2a - 3) + (-ab + b)$
$= (a + 3)(a - 1) - b(a - 1)$
$= (a - 1)(a + 3 - b)$

2.5 $16 - 2(p - 3)^3$
$= 2[8 - (p - 3)^3]$
$= 2[2 - (p - 3)][4 + 2(p - 3) + (p - 3)^2]$
$= 2(2 - p + 3)(4 + 2p - 6 + p^2 - 6p + 9)$
$= 2(5 - p)(7 - 4p + p^2)$

2.6 $a^2 - 16 - 2ab + b^2$
$= a^2 - 2ab + b^2 - 16$
$= (a - b)^2 - 16$
$= (a - b - 4)(a - b + 4)$

2.7 $6(2x - 1)^2 + (2x - 1) - 2$
Let K $= (2x - 1)$:
$\therefore 6(2x - 1)^2 + (2x - 1) - 2 = 6K^2 + K - 2$
$= (2K - 1)(3K + 2)$

But K $= (2x - 1)$
$\therefore (2K - 1)(3K + 2)$
$= [2(2x - 1) - 1][3(2x - 1) + 2]$
$= (4x - 2 - 1)(6x - 3 + 2)$
$= (4x - 3)(6x - 1)$

2.8 $(2x - 1)^2 - 4(a - 1)^2$
Let K $= (2x - 1)$ and P $= (a - 1)$,
then $K^2 - 4P^2 = (K - 2P)(K + 2P)$

But K $= (2x - 1)$ and P $= (a - 1)$
$\therefore (K - 2P)(K + 2P)$
$= [2x - 1 - 2(a - 1)][2x - 1 + 2(a - 1)]$
$= (2x - 1 - 2a + 2)(2x - 1 + 2a - 2)$
$= (2x - 2a + 1)(2x + 2a - 3)$

2.9 $2(a - b)^2 - 2(a - b) - 12$
Let K $= (a - b)$,
then $2K^2 - 2K - 12 = 2(K^2 - K - 6)$
$= 2(K - 3)(K + 2)$

But K $= (a - b)$
$\therefore 2(K - 3)(K + 2)$
$= 2[(a - b) - 3][(a - b) + 2]$
$= 2(a - b - 3)(a - b + 2)$

2.10 $2(2x - y)^2 - 8(a + b)^2$
Let K $= (2x - y)$ and P $= (a + b)$,
then $2K^2 - 8P^2 = 2(K^2 - 4P^2)$
$= 2(K - 2P)(K + 2P)$

But K $= (2x - y)$ and P $= (a + b)$
$\therefore 2(K - 2P)(K + 2P)$
$= 2[(2x - y) - 2(a + b)][(2x - y) + 2(a + b)]$
$= 2(2x - y - 2a - 2b)(2x - y + 2a + 2b)$

3. $\left(\dfrac{2(3 + x)}{2 - x} + 1 \right) \div \left(\dfrac{3 + x}{2 - x} - 1 \right)$
$= \left(\dfrac{2(3 + x) + 1(2 - x)}{2 - x} \right) \div \left(\dfrac{3 + x - 1(2 - x)}{2 - x} \right)$
$= \left(\dfrac{6 + 2x + 2 - x}{2 - x} \right) \div \left(\dfrac{3 + x - 2 + x}{2 - x} \right)$
$= \left(\dfrac{8 + x}{2 - x} \right) \div \left(\dfrac{1 + 2x}{2 - x} \right)$
$= \left(\dfrac{8 + x}{2 - x} \right) \times \left(\dfrac{2 - x}{2 + 2x} \right)$
$= \dfrac{8 + x}{1 + 2x}$

4. $\dfrac{\frac{1}{y - 1} + \frac{1}{y}}{\frac{2}{y} - \frac{3}{y} - 1}$
$= \dfrac{\frac{y + (y - 1)}{y(y - 1)}}{\frac{2 - 3 - y}{y}}$
$= \dfrac{\frac{2y - 1}{y(y - 1)}}{\frac{(-1 - y)}{y}}$
$= \dfrac{2y - 1}{y(y - 1)} \times \dfrac{y}{-1 - y}$
$= \dfrac{2y - 1}{-(y - 1)(y + 1)}$
$= \dfrac{2y^2 - 1}{-y^2 + 1}$

Test A (page 27)

1.1 $(3 - x)(3 + 2x) = 9 + 3x - 2x^2$

1.2 $(2a + 3b)(a^2 - 4ab + 3b^2)$
$= 2a^3 - 5a^2b - 6ab^2 + 9b^3$

1.3 $\left(\dfrac{2}{x} + \dfrac{4}{x^2} \right)^2 = \dfrac{4}{x^2} + 2 + \dfrac{x^2}{4}$

2. $(z + 2t)^2 = z^2 + 4zt + 4t^2$
$\therefore -4zt$ must be added.

3.1 $A - B$

$= \dfrac{1}{m+2} - \dfrac{1}{m+1}$

$= \dfrac{1(m+1) - 1(m+2)}{(m+2)(m+1)}$

$= \dfrac{m+1-m-2}{(m+2)(m+1)}$

$= \dfrac{-1}{(m+2)(m+1)}$

3.2 $A \div B$

$= \dfrac{1}{m+2} \div \dfrac{1}{m+1}$

$= \dfrac{1}{m+2} \times \dfrac{m+1}{1}$

$= \dfrac{m+1}{m+2}$

3.3 $m = -2$

4.1 $\dfrac{(m-3)^2}{m^2-9}$

$= \dfrac{(m-3)^2}{(m-3)(m+3)}$

$= \dfrac{m-3}{m+3}$

4.2 $\dfrac{2x-3y}{xy-y^2} \div \dfrac{9xy-6x^2}{2x-2y}$

$= \dfrac{2x-3y}{y(x-y)} \times \dfrac{2(x-y)}{3x(3y-2x)}$

$= \dfrac{-(3y-2x)}{y(x-y)} \times \dfrac{2(x-y)}{3x(3y-2x)}$

$= \dfrac{-1}{y} \times \dfrac{2}{3x}$

$= -\dfrac{2}{3xy}$

4.3 $\dfrac{6x}{(x+2)(x-2)} - \dfrac{2}{x+2} + \dfrac{3}{2-x}$

$= \dfrac{6x}{(x+2)(x-2)} - \dfrac{2}{(x+2)} - \dfrac{3}{(x-2)}$

$= \dfrac{6x - 2(x-2) - 3(x+2)}{(x+2)(x-2)}$

$= \dfrac{6x - 2x + 4 - 3x - 6}{(x+2)(x-2)}$

$= \dfrac{(x-2)}{(x+2)(x-2)}$

$= \dfrac{1}{(x+2)}$

4.4 $\left(1 + \dfrac{1}{r}\right) \div (1+r)$

$= \dfrac{r+1}{r} \times \dfrac{1}{1+r}$

$= \dfrac{1}{r}$

5.1 $5a^3 - 15a^2 - 20a$

$= 5a(a^2 - 3a - 4)$

$= 5a(a-4)(a+1)$

5.2 $F^3 - 125 = (F-5)(F^2 + 5F + 25)$

5.3 $5 - \dfrac{45}{z^2}$

$= 5\left(1 - \dfrac{9}{z^2}\right)$

$= 5\left(1 - \dfrac{3}{z}\right)\left(1 + \dfrac{3}{z}\right)$

5.4 $x^2(a-3) + 7(3-a) - 10(3-a)$

$= x^2(a-3) - 3(3-a)$

$= x^2(a-3) + 3(a-3)$

$= (a-3)(x^2+3)$

5.5 $6ab - 15ac + 4b^2 - 25c^2$

$= 3a(2b-5c) + (2b-5c)(2b+5c)$

$= (2b-5c)(3a+2b+5c)$

Test B (page 28)

1.1 $(3a+b)(3a-b)(81a^4 + 9a^2b^2 + b^4)$

$= (9a^2 - b^2)(81a^4 + 9a^2b^2 + b^4)$

$= 729a^6 - b^6$

1.2 $(x-2)(x+2)(x^2+4)(x^4+16)$

$= (x^2-4)(x^2+4)(x^4+16)$

$= (x^4-16)(x^4+16)$

$= x^8 - 256$

2.1 $5a + 5b - a^2 - 2ab - b^2$

$= 5(a+b) - (a^2 + 2ab + b^2)$

$= 5(a+b) - (a+b)^2$

$= (a+b)(5-a-b)$

2.2 $4x^2(3y+1) - 9y^2(2x+1)$

$= 12x^2y + 4x^2 - 18xy^2 - 9y^2$

$= 12x^2y - 18xy^2 + 4x^2 - 9y^2$

$= 6xy(2x-3y) + (2x-3y)(2x+3y)$

$= (2x-3y)(6xy+2x+3y)$

2.3 Let $K = (7x-3)$,

then $6(7x-3)^2 - 7(7x-3) - 3$

$\quad = 6K^2 - 7K - 3$

$\quad = (2K-3)(3K+1)$

But $K = (7x-3)$

$\therefore (2K-3)(3K+1)$

$\quad = [2(7x-3)-3][3(7x-3)+1]$

$\quad = (14x-6-3)(21x-9+1)$

$\quad = (14x-9)(21x-8)$

2.4 $p^7 - 27p(x+y)^3$

$= p[p^6 - 27(x+y)^3]$

$= p[p^2 - 3(x+y)][p^4 + 3p^2(x+y) + 9(x+y)^2]$

$= p(p^2 - 3x - 3y)[p^4 + 3p^2x + 3p^2y$
$\qquad + 9(x^2 + 2xy + y^2)]$

$= p(p^2 - 3x - 3y)(p^4 + 3p^2x + 3p^2y + 9x^2$
$\qquad + 18xy + 9y^2)$

2.5 $a^2 - 2ab + b^2 - m^2 + 2mn - n^2$

$= (a^2 - 2ab + b^2) + (-m^2 + 2mn - n^2)$

$= (a^2 - 2ab + b^2) - (m^2 - 2mn + n^2)$

$= (a-b)^2 - (m-n)^2$

$= [(a-b)-(m-n)][(a-b)+(m-n)]$

$= (a-b-m+n)(a-b+m-n)$

3.1

$$\frac{3x-4}{3x^2-x-4} - \frac{4x}{x^2-2x-3}$$

$$= \frac{3x-4}{(3x-4)(x+1)} - \frac{4x}{(x-3)(x+1)}$$

$$= \frac{1}{(x+1)} - \frac{4x}{(x-3)(x+1)}$$

$$= \frac{1(x-3)-4x}{(x-3)(x+1)}$$

$$= \frac{-3-3x}{(x-3)(x+1)}$$

$$= \frac{-3(1+x)}{(x-3)(x+1)}$$

$$= \frac{-3}{x-3}$$

3.2

$$\left(2x-3+\frac{7}{x+3}\right) \div \left(x+1-\frac{3}{2x+1}\right)$$

$$= \left(\frac{(2x-3)(x+3)+7}{x+3}\right) \div \left(\frac{(x+1)(2x+1)-3}{2x+1}\right)$$

$$= \left(\frac{(2x^2+3x-2)}{x+3}\right) \times \left(\frac{2x+1}{(2x^2+3x-2)}\right)$$

$$= \frac{2x+1}{x+3}$$

4. $(3x-5)(3x-5) = 9x^2 - 30x + 25 \therefore k = 25$

5. $(4-3x)(2-3x) = 8 - 18x + 9x^2 \therefore f = -18$

6. $(2x-5)^2 - (3-x)^2$

$$= [(2x-5)-(3-x)][(2x-5)+(3-x)]$$

$$= (2x-5-3+x)(2x-5+3-x)$$

$$= (3x-8)(x-2)$$

Test C (page 28)

1.1 True

1.2 False: $(x-2)^2 = x^2 - 4x + 4$

1.3 True

1.4 False: $(x-3)(x+2) = x^2 - x - 6$

2.1 $(3 - \frac{4}{x^2})^2 = 9 - 3x + \frac{x^2}{4}$

2.2 $(x-y)(2x^2-x+2)$

$$= 2x^3 - 2x^2y - x^2 + xy + 2x - 2y$$

2.3 LHS: $2(2p-1)^2 + 5(2p-1) - 3$

Let $K = (2p-1)$,

then $2(2p-1)^2 + 5(2p-1) - 3$

$$= 2K^2 + 5K - 3$$

$$= (2K-1)(K+3)$$

But $K = (2p-1)$

$\therefore (2K-1)(K+3)$

$$= [2(2p-1)-1][(2p-1)+3]$$

$$= (4p-2-1)(2p-1+3)$$

$$= (4p-3)(2p+2)$$

$$= 2(4p-3)(p+1)$$

$\therefore 2(2p-1)^2 + 5(2p-1) - 3 = 2(4p-3)(p+1)$

3.1

$$A + B = \frac{n}{1-n^2} + \frac{1}{1-2n+n^2}$$

$$= \frac{n}{(1-n)(1+n)} + \frac{1}{(1-n)^2}$$

$$= \frac{n(1-n)+1(1+n)}{(1-n)^2(1+n)}$$

$$= \frac{n-n^2+1+n}{(1-n)^2(1+n)}$$

$$= \frac{1+2n-n^2}{(1-n)^2(1+n)}$$

3.2

$$B \div A = \frac{1}{1-2n+n^2} \div \frac{n}{1-n^2}$$

$$= \frac{1}{(1-n)^2} \div \frac{n}{(1-n)(1+n)}$$

$$= \frac{1}{(1-n)^2} \times \frac{(1-n)(1+n)}{n}$$

$$= \frac{1}{(1-n)} \times \frac{(1+n)}{n}$$

$$= \frac{(1+n)}{n(1-n)}$$

3.3 $1 - 2n + n^2 = (1-n)(1-n)$

$\therefore$ undefined for $n = 1$

4. $\left(x + \frac{1}{x}\right) = 7$

$\left(x + \frac{1}{x}\right)^2 = 7^2$

$x^2 + 2 + \frac{1}{x^2} = 49$

$\therefore x^2 + \frac{1}{x^2} = 49 - 2 = 47$

5.1 $(x-y)^3 - 125$

$$= [(x-y)-5][(x-y)^2+5(x-y)+25]$$

$$= (x-y-5)(x^2-2xy+y^2+5x-5y+25)$$

5.2 $c^2 - b^2 - 2ab - 2ac$

$$= (c^2-b^2) + (-2ab-2ac)$$

$$= (c-b)(c+b) - 2a(b+c)$$

$$= (c+b)[(c-b)-2a]$$

$$= (c+b)(c-b-2a)$$

5.3 $x^2(p+q) - 2x(p+q) + p + q$

$$= x^2(p+q) - 2x(p+q) + (p+q)$$

$$= (p+q)(x^2-2x+1)$$

$$= (p+q)(x-1)^2$$

6.1 $\left(1 - \frac{1}{u}\right) \div (1-u)$

$$= \frac{(u-1)}{u} \times \frac{1}{(1-u)}$$

$$= \frac{(u-1)}{u} \times \frac{1}{-(u-1)}$$

$$= -\frac{1}{u}$$

6.2 $u = 0$ or $u = 1$

Topic 3   Exponents (page 30)

Exercise 1   (page 31)

1. $2x^2 \times 3x^3 \times 4x = 24x^6$

2. $2^2 \times 3^3 \times 2 \times 3^{-2} = 2^3 \times 3^1 = 8 \times 3 = 24$

3. $\dfrac{3x^5 y^2}{6xy^4} = \dfrac{x^4}{2y^2}$

4. $-5f^7 \div 10ef^9 = \dfrac{-5f^7}{10ef^9} = -\dfrac{1}{2ef^2}$

5. $(-2x^3 y)^3 = -2^3 x^9 y^3 = -8x^9 y^3$

6. $\left(\dfrac{21a^3}{7a^2}\right)^2 = \left(-\dfrac{3a}{1}\right)^2 = 9a^2$

7. $\dfrac{1}{(3)^4} = \dfrac{1}{81}$

8. $2^{-3} \times 5 \times 2^4 \times 5^{-2} = 2 \cdot 5^{-1} = \dfrac{2}{5}$

9. $(-xy^3)^0 = 1$

10. $\left(\dfrac{3a}{v}\right)^0 + 2b^0 = 1 + 2(1) = 3$

11. $2^3 \times 5^3 = 8 \times 125 = 1\,000$

12. $\dfrac{4^2 \cdot 3^2}{2^3 \cdot 5^3} = \dfrac{16 \cdot 9}{8 \cdot 125} = \dfrac{2 \cdot 9}{125} = \dfrac{18}{125}$

13. $\dfrac{2ab^{-3}}{c^{-3}} = \dfrac{2ac^3}{b^3}$

14. $\left(\dfrac{(3x)^{-2}}{3x^{-2}}\right)^{-2} = \left(\dfrac{3^{-2}x^{-2}}{3x^{-2}}\right)^{-2} = \left(\dfrac{1}{3^3}\right)^{-2} = \dfrac{1}{3^{-6}} = 3^6 = 729$

15. $\left(\dfrac{a^2 b^{-3}}{a^4 b^2}\right)^{-2} = \left(\dfrac{1}{a^2 b^5}\right)^{-2} = \dfrac{1}{a^{-4}b^{-10}} = a^4 b^{10}$

16. $\left(\dfrac{p}{2}\right)^2 \times \left(\dfrac{p^{-2}}{t^{-3}}\right)^2 = \dfrac{p^2}{2^2} \times \dfrac{p^{-4}}{t^{-6}} = \dfrac{p^{-2}}{4t^{-6}} = \dfrac{t^6}{4p^2}$

17. $\left(\dfrac{1}{x^{-1}} + \dfrac{2}{x^{-1}} - \dfrac{10}{x^{-1}}\right)^2 = [x + 2x - 10x]^2 = [-7x]^2$
$$= 49x^2$$

18. $\dfrac{q}{p^2} \times \left(\dfrac{q^3}{p}\right)^2 \times p^{-4} = \dfrac{q}{p^2} \times \dfrac{q^6}{p^2} \times \dfrac{1}{p^4} = \dfrac{q^7}{p^8}$

Exercise 2   (page 33)

1.1 $32^{\frac{4}{5}} = (2^5)^{\frac{4}{5}} = 2^4 = 16$

1.2 $(-16)^{\frac{3}{4}} = (-2^4)^{\frac{3}{4}} = (-2)^3 = -8$

1.3 $64^{-\frac{2}{3}} = (2^6)^{-\frac{2}{3}} = 2^{-4} = \dfrac{1}{16}$

1.4 $\left(\dfrac{16}{9}\right)^{\frac{1}{2}} = \left(\dfrac{2^4}{3^2}\right)^{\frac{1}{2}} = \dfrac{2^2}{3} = \dfrac{4}{3}$

1.5 $(16^{\frac{1}{2}})(16^{-\frac{1}{2}}) = (2^4)^{\frac{1}{2}} \cdot (2^4)^{-\frac{1}{2}} = 2^2 \cdot 2^{-2} = 2^0 = 1$

1.6 $(-32)^{\frac{3}{5}} \div (-32)^{\frac{2}{5}} = \dfrac{(-2^5)^{\frac{3}{5}}}{(-2^5)^{\frac{2}{5}}} = \dfrac{(-2)^3}{(-2)^2} = -2$

1.7 $81^{\frac{3}{4}} \times 27^{-\frac{2}{3}} = (3^4)^{\frac{3}{4}} \times (3^3)^{-\frac{2}{3}} = 3^3 \cdot 3^{-2} = 3$

2.1 $2^n \cdot 3^n \cdot 4^{n+1} \cdot 3^{n-2}$
$= 2^n \cdot 3^n \cdot (2^2)^{n+1} \cdot 3^{n-2}$
$= 2^n \cdot 3^n \cdot 2^{2n+2} \cdot 3^{n-2}$
$= 2^{n+2n+2} \cdot 3^{n+n-2}$
$= 2^{3n+2} \cdot 3^{2n-2}$

2.2 $6^{2n-1} \cdot 9^{1-n} \cdot 3^{2n+1}$
$= (2 \cdot 3)^{2n-1} \cdot (3^2)^{1-n} \cdot 3^{2n+1}$
$= 2^{2n-1} \cdot 3^{2n-1} \cdot 3^{2-2n} \cdot 3^{2n+1}$
$= 2^{2n-1} \cdot 3^{2n-1+2-2n+2n+1}$
$= 2^{2n-1} \cdot 3^{2n+2}$

2.3 $25^n \cdot 5^{1-n} \cdot 15^n \cdot 9^{n-2}$
$= (5^2)^n \cdot 5^{1-n} \cdot (3 \cdot 5)^n \cdot (3^2)^{n-2}$
$= 5^{2n} \cdot 5^{1-n} \cdot 3^n \cdot 5^n \cdot 3^{2n-4}$
$= 5^{2n+1-n+n} \cdot 3^{n+2n-4}$
$= 5^{2n+1} \cdot 3^{3n-4}$

2.4 $27^{n+1} \cdot 18^{2n} \cdot 4^{n-1}$
$= (3^3)^{n+1} \cdot (2 \cdot 3^2)^{2n} \cdot (2^2)^{n-1}$
$= 3^{3n+3} \cdot 2^{2n} \cdot 3^{4n} \cdot 2^{2n-2}$
$= 3^{3n+3+4n} \cdot 2^{2n-2+2n}$
$= 3^{7n+3} \cdot 2^{4n-2}$

Exercise 3   (page 34)

1. $\dfrac{9^n \times 12^{n+1}}{4 \times 6^n}$

$= \dfrac{3^{2n} \cdot 2^{2n+2} \cdot 3^{n+1}}{2^2 \cdot 2^n \cdot 3^n}$

$= 3^{2n+n+1-n} \cdot 2^{2n+2-2-n}$

$= 3^{2n+1} \cdot 2^n$

2. $\dfrac{2^{n-2} \cdot 4^{n+3}}{8^{n+2}}$

$= \dfrac{2^{n-2} \cdot 2^{2n+6}}{2^{3n+6}}$

$= 2^{n-2+2n+6-3n-6}$

$= 2^{-2}$

$= \dfrac{1}{4}$

3.

$$\frac{2^n \cdot 8^{n+2} \cdot 4^{-3n}}{2^{-2n}}$$

$$= \frac{2^n \cdot 2^{3n+6} \cdot 2^{-6n}}{2^{-2n}}$$

$$= 2^{n+3n+6-6n+2n}$$

$$= 2^6$$

$$= 64$$

4.

$$\frac{2^{x+1} \cdot 2^{2x+1}}{(2^2)^{2x+1}}$$

$$= \frac{2^{x+1} \cdot 2^{2x+1}}{2^{4x+2}}$$

$$= 2^{x+1+2x+1-4x-2}$$

$$= 2^{-x}$$

$$= \frac{1}{2^x}$$

5.

$$\frac{5^{2x-1} \cdot (3^2)^{x-2}}{(5 \cdot 3)^{2x-3}}$$

$$= \frac{5^{2x-1} \cdot 3^{2x-4}}{5^{2x-3} \cdot 3^{2x-3}}$$

$$= 5^{2x-1-2x+3} \cdot 3^{2x-4-2x+3}$$

$$= 5^2 \cdot 3^{-1}$$

$$= \frac{25}{3}$$

6.

$$\frac{4^{x+1} \cdot 8^{x-1}}{16^{x-2}}$$

$$= \frac{2^{2x+2} \cdot 2^{3x-3}}{2^{4x-8}}$$

$$= 2^{2x+2+3x-3-4x+8}$$

$$= 2^{x+7}$$

7.

$$\frac{45^{x-3} \cdot 3 \cdot 75^{4-x}}{25^{-x} \cdot 15^{x+2}}$$

$$= \frac{5^{x-3} \cdot 3^{2x-6} \cdot 3 \cdot 3^{4-x} \cdot 5^{8-2x}}{5^{-2x} \cdot 5^{x+2} \cdot 3^{x+2}}$$

$$= 5^{x-3+8-2x+2x-x-2} \cdot 3^{2x-6+1+4-x-x-2}$$

$$= 5^3 \cdot 3^{-3}$$

$$= \frac{125}{27}$$

8.

$$\frac{(98^{x+1})^2 \cdot (2^x)^{-2}}{49^x}$$

$$= \frac{98^{2x+2} \cdot 2^{-2x}}{49^x}$$

$$= \frac{(2 \cdot 7^2)^{2x+2} \cdot 2^{-2x}}{7^{2x}}$$

$$= \frac{7^{4x+4} \cdot 2^{2x+2} \cdot 2^{-2x}}{7^{2x}}$$

$$= 7^{4x+4-2x} \cdot 2^{2x+2-2x}$$

$$= 7^{2x+4} \cdot 2^2$$

$$= 4 \cdot 7^{2x+4}$$

Exercise 4 (page 36)

1.1

$$x^{\frac{3}{2}} = 8$$

$$(x^{\frac{3}{2}})^{\frac{2}{3}} = (2^3)^{\frac{2}{3}}$$

$$x = 2^2$$

$$\therefore x = 4$$

1.2

$$-3x^{\frac{1}{3}} = 15$$

$$x^{\frac{1}{3}} = \frac{15}{-3} = -5$$

$$(x^{\frac{1}{3}})^3 = (-5)^3$$

$$\therefore x = -125$$

1.3

$$10x^{\frac{3}{4}} = 270$$

$$x^{\frac{3}{4}} = 27$$

$$(x^{\frac{3}{4}})^{\frac{4}{3}} = (3^3)^{\frac{4}{3}}$$

$$x = 3^4$$

$$\therefore x = 81$$

1.4

$$x^{\frac{1}{4}} - 2 = 0$$

$$x^{\frac{1}{4}} = 2$$

$$(x^{\frac{1}{4}})^4 = 2^4$$

$$\therefore x = 16$$

1.5

$$x^{-1} = \frac{4}{7}$$

$$(x^{-1})^{-1} = \left(\frac{4}{7}\right)^{-1}$$

$$\therefore x = \frac{7}{4}$$

2.1

$$16^x = 32$$

$$2^{4x} = 2^5$$

$$4x = 5$$

$$\therefore x = \frac{5}{4}$$

2.2

$$16^{x-3} = 1$$

$$(2^4)^{x-3} = 2^0$$

$$2^{4x-12} = 2^0$$

$$4x - 12 = 0$$

$$4x = 12$$

$$\therefore x = 3$$

2.3

$$3 \cdot 5^x = 75$$

$$5^x = \frac{75}{5}$$

$$5^x = 25$$

$$5^x = 5^2$$

$$\therefore x = 2$$

2.4 $2 \cdot 3^{x+2} = 486$

$3^{x+2} = \frac{486}{2}$

$3^{x+2} = 243$

$3^{x+2} = 3^5$

$x + 2 = 5$

$\therefore x = 3$

2.5 $4 \times 2^{x-1} = \frac{1}{4}$

$2^{x-1} = \frac{1}{4} \times \frac{1}{4}$

$2^{x-1} = \frac{1}{16}$

$2^{x-1} = \frac{1}{2^4}$

$2^{x-1} = 2^{-4}$

$x - 1 = -4$

$\therefore x = -3$

2.6 $4 \times 3^{2x+1} = 108$

$3^{2x+1} = \frac{108}{4}$

$3^{2x+1} = 27$

$3^{2x+1} = 3^3$

$2x + 1 = 3$

$2x = 2$

$\therefore x = 1$

2.7 $5^{2x+1} = 125$

$5^{2x+1} = 5^3$

$2x + 1 = 3$

$2x = 2$

$\therefore x = 1$

2.8 $2^{x-1} = \frac{1}{16}$

$2^{x-1} = \frac{1}{2^4}$

$2^{x-1} = 2^{-4}$

$x - 1 = -4$

$\therefore x = -3$

2.9 $2^{x+12} = 16^x$

$2^{x+12} = 2^{4x}$

$x + 12 = 4x$

$-3x = -12$

$\therefore x = 4$

2.10 $3 \cdot 9^{2x-3} - 81 = 0$

$3 \cdot 9^{2x-3} = 81$

$9^{2x-3} = 27$

$(3^2)^{2x-3} = 3^3$

$4x - 6 = 3$

$4x = 9$

$\therefore x = \frac{9}{4}$

Exercise 5   (page 37)

1. $2^{2n+2} - 2^{2n}$

$= 2^{2n}(2^2 - 1) = 2^{2n}(3)$

2. $4^n \cdot 2^{2n+2} - 8^n$

$= 2^{2n} \cdot 2^{2n+2} - 2^{3n}$

$= 2^{4n} \cdot 2^2 - 2^{3n}$

$= 2^{3n}(2^n \cdot 2^2 - 1)$

$= 2^{3n}(2^{n+1} - 1)$

3. $2 \cdot 2^x + 6 \cdot 2^{x-1}$

$= 2^x(2 + 6 \cdot 2^{-1})$

$= 2^x(2 + 3)$

$= 2^x(5)$

4. $3^{x+1} + 3^{x-1}$

$= 3^x \cdot 3^1 + 3^x \cdot 3^{-1}$

$= 3^x(3^1 + 3^{-1})$

$= 3^x\left(3^1 + \frac{1}{3}\right)$

$= 3^x\left(\frac{10}{3}\right)$

5. $3^{3n} - 3^{n+1}$

$= 3^n \cdot 3^n \cdot 3^n \cdot -3^n \cdot 3^1$

$= 3^n(3^n \cdot 3^n - 3^1)$

$= 3^n(3^{2n} - 3)$

6. $2^{2n} \cdot 3^n - 2^{n+1} \cdot 3^n + 2^n \cdot 3^{2n}$

$= 2^n \cdot 2^n \cdot 3^n - 2^n \cdot 2^1 \cdot 3^n + 2^n \cdot 3^n \cdot 3^n$

$= 2^n \cdot 3^n(2^n - 2 + 3^n)$

Exercise 6   (page 38)

1. $\frac{2^{n+2} - 2^{2n}}{2^{2n}}$

$= \frac{2^{2n}(2^2 - 1)}{2^{2n}}$

$= 2^2 - 1$

$= 3$

2. $\dfrac{3^n \cdot 5 - 3^{n+1}}{3^n \cdot 4}$

$= \dfrac{3^n(5 - 3^1)}{3^n \cdot 4}$

$= \dfrac{(5 - 3)}{4}$

$= \dfrac{1}{2}$

3. $\dfrac{3^{3n} - 3^{n+1}}{3^{n+1}} = \dfrac{3^n(3^{2n} - 3^1)}{3^n \cdot 3^1}$

$= \dfrac{(3^{2n} - 3)}{3}$

4. $\dfrac{2^n \cdot 2^{2n+2} - 8^n}{8^{n+1}}$

$= \dfrac{2^{3n+2} - 2^{3n}}{2^{3n+3}}$

$= \dfrac{2^{3n}(2^2 - 1)}{2^{3n} \cdot 2^3}$

$= \dfrac{3}{8}$

5. $\dfrac{2 \cdot 2^x + 6 \cdot 2^{x-1}}{10^x}$

$= \dfrac{2^x(2 + 6 \cdot 2^{-1})}{2^x \cdot 5^x}$

$= \dfrac{(2 + 3)}{5^x}$

$= \dfrac{5}{5^x}$

6. $\dfrac{3^n \cdot 2 + 2^2 \cdot 3^n}{3^n \cdot 5 - 3^n \cdot 2}$

$= \dfrac{3^n(2 + 2^2)}{3^n(5 - 2)}$

$= \dfrac{(2 + 2^2)}{(5 - 2)}$

$= \dfrac{6}{3}$

$= 2$

7. $\dfrac{5^{n+1} + 5^{n-1}}{5^n \cdot 10 - 5^{n+1}}$

$= \dfrac{5^n(5^1 + 5^{-1})}{5^n(10 - 5^1)}$

$= \dfrac{(5^1 + 5^{-1})}{(10 - 5^1)}$

$= \dfrac{\left(\frac{26}{5}\right)}{5}$

$= \dfrac{26}{25}$

Exercise 7 (page 38)

1. $2^x \cdot 2^{x+2} = 64$

$2^{2x+2} = 2^6$

$2x + 2 = 6$

$2x = 4$

$\therefore x = 2$

2. $2^{x+2} + 2^x = 40$

$2^x(2^2 + 1) = 40$

$2^x(5) = 40$

$2^x = 8$

$2^x = 2^3$

$\therefore x = 3$

3. $2^x \cdot 2^{x+2} = \dfrac{25}{100}$

$2^{2x+2} = \dfrac{1}{4}$

$2^{2x+2} = 2^{-2}$

$2x + 2 = -2$

$2x = -4$

$\therefore x = -2$

4. $2^{x+2} - 2^x = \dfrac{3}{4}$

$2^x(2^2 - 1) = \dfrac{3}{4}$

$2^x(3) = \dfrac{3}{4}$

$2^x = \dfrac{1}{4}$

$2^x = 2^{-2}$

$\therefore x = -2$

5. $8^{x+1} = 2^x \cdot 16^x$

$2^{3x+3} = 2^x \cdot 2^{4x}$

$3x + 3 = 5x$

$-2x = -3$

$\therefore x = \dfrac{3}{2}$

6. $8^{x+1} - 2^x \cdot 4^x = 112$

$2^{3x+3} - 2^x \cdot 2^{2x} = 112$

$2^{3x}(2^3 - 1) = 112$

$2^{3x}(7) = 112$

$2^{3x} = 16$

$2^{3x} = 2^4$

$\therefore x = \dfrac{4}{3}$

7. $(2^x)^{x+1} = 4$

$2^{x^2+x} = 2^2$

$x^2 + x - 2 = 0$

$(x + 2)(x - 1) = 0$

$\therefore x = -2 \text{ or } x = 1$

8. $(2^x)^x + 2^{x^2} = 32$

$2^{x^2} + 2^{x^2} = 32$

$2^{x^2}(2) = 32$

$2^{x^2} = 16$

$2^{x^2} = 2^4$

$x^2 = 4$

$\therefore x = \pm 2$

9. $3^x - 10 = -3^{x+2}$

$3^x + 3^{x+2} = 10$

$3^x(1 + 3^2) = 10$

$3^x(10) = 10$

$3^x = 1$

$3^x = 3^0$

$\therefore x = 0$

10. $3 \cdot 5^{x-1} + 4 \cdot 5^x = \frac{23}{25}$

$5^x(3 \cdot 5^{-1} + 4) = \frac{23}{25}$

$5^x\left(\frac{3}{5} + 4\right) = \frac{23}{25}$

$5^x\left(\frac{23}{5}\right) = \frac{23}{25}$

$5^x = \frac{23}{25} \times \frac{5}{23}$

$5^x = 5^{-1}$

$\therefore x = -1$

11. $3^{x+1} - \left(\frac{1}{3}\right)^{-x-3} = -11(3^{x+2}) + 25$

$3^{x+1} - 3^{x+3} + 11(3^{x+2}) = 25$

$3^x(3 - 3^3 + 11 \cdot 3^2) = 25$

$3^x(75) = 25$

$3^x = 3^{-1}$

$\therefore x = -1$

Test A (page 39)

1.1 $C\left(2^{-12} = \frac{1}{2^{12}}\right)$

1.2 $B\left(\frac{1}{8^3} = 8^{-3}\right)$

1.3 $D\left(\frac{1}{3^8} = 3^{-8}\right)$

2.1 $r^4 \times r^{-3} = r$

2.2 $2r^{-7} \times 3r^9 = 6r^2$

2.3 $\frac{12t^{-7}}{3t} = 4t^{-8} = \frac{4}{t^8}$

2.4 $(x^{-3})^3 = x^{-9} = \frac{1}{x^9}$

2.5 $\frac{9^{a-2} \cdot 10^{a-2}}{6^{a-4} \cdot 15^a}$

$= \frac{(3^2)^{a-2} \cdot (2 \cdot 5)^{a-2}}{(2 \cdot 3)^{a-4} \cdot (3 \cdot 5)^a}$

$= \frac{3^{2a-4} \cdot 2^{a-2} \cdot 5^{a-2}}{2^{a-4} \cdot 3^{a-4} \cdot 3^a \cdot 5^a}$

$= 3^{2a-4-a+4-a} \cdot 2^{a-2-a+4} \cdot 5^{a-2-a}$

$= 3^0 \cdot 2^2 \cdot 5^{-2}$

$= \frac{4}{25}$

3.

3^6	$\times$	3^0	$=$	3^6
$\times$		$\times$		$\div$
3^2	$\div$	3^2	$=$	3^0
$=$		$=$		$=$
3^8	$\div$	3^2	$=$	3^6

4.1 $\left(\frac{16}{81}\right)^{\frac{3}{4}} = \left(\frac{2^4}{3^4}\right)^{\frac{3}{4}} = \left(\frac{2^3}{3^3}\right) = \frac{8}{27}$

4.2 $3^{\frac{5}{4}} \div 48^{\frac{1}{4}} = \frac{3^{\frac{5}{4}}}{(2^4 \cdot 3)^{\frac{1}{4}}} = \frac{3^{\frac{5}{4}}}{2 \cdot 3^{\frac{1}{4}}} = \frac{3}{2}$

4.3 $(2^0 + 4^0 + 6^0)^{-2} = (1 + 1 + 1)^{-2} = 3^{-2} = \frac{1}{3^2} = \frac{1}{9}$

5.1 $2^x = 32$

$2^x = 2^5$

$\therefore x = 5$

5.2 $x^{-2} = \frac{1}{49}$

$x^{-2} = 7^{-2}$

$\therefore x = 7$

5.3 $x^{\frac{3}{4}} = 27$

$x = (3^3)^{\frac{4}{3}}$

$x = 3^4$

$\therefore x = 81$

5.4 $3^{3x-3} = 3^{x-4}$

$3x - 3 = x - 4$

$2x = -1$

$\therefore x = -\frac{1}{2}$

5.5 $81^{x-2} = 27^4$

$(3^4)^{x-2} = (3^3)^4$

$3^{4x-8} = 3^{12}$

$4x - 8 = 12$

$4x = 20$

$\therefore x = 5$

6.1 $a^0 \times a^0 \times a^0 \times a^0 > 2$ is never true, because $a^0 = 1$ for all values of a.

$\therefore a^0 \times a^0 \times a^0 \times a^0 = 1 \times 1 \times 1 \times 1 = 1$

6.2 $(2a)^{-3} = \frac{8}{a^3}$ is never true, because

$(2a)^{-3} = \frac{1}{(2a)^3} = \frac{1}{8a^3} \neq \frac{8}{a^3}$.

6.3 $2a^2 = a$ is sometimes true, but only for $a = \frac{1}{2}$, because $2\left(\frac{1}{2}\right)^2 = 2\left(\frac{1}{4}\right) = \frac{1}{2}$.

Note: The statement is also true for $a = 0$, but a was given as any integer except 0.

Test B (page 40)

1. $\frac{3^2}{3^{-2} - 6^{-1}} = \frac{9}{\frac{1}{9} - \frac{1}{6}} = \frac{9}{\frac{-3}{54}} = \frac{9}{1} \times -\frac{54}{3} = -162$

2. $\dfrac{(3x^3)^{-2}(-2x^{-1})^3}{\left(\frac{3x^2}{-2}\right)^{-2}}$

$= \dfrac{3^{-2} \cdot x^{-6} \cdot (-2)^3 \cdot x^{-3}}{\left(\frac{3^{-2}x^{-4}}{(-2)^{-2}}\right)}$

$= \dfrac{3^{-2} \cdot x^{-6} \cdot (-2)^3 \cdot x^{-3}}{1} \times \dfrac{(-2)^{-2}}{3^{-2}x^{-4}}$

$= \dfrac{x^{-5} \cdot (-2)^1}{1} = \dfrac{-2}{x^5}$

3.1 $(0,5)^{x-1} = 4^{1,5}$

$\left(\frac{1}{2}\right)^{x-1} = (2^2)^{\frac{3}{2}}$

$2^{-x+1} = 2^3$

$-x + 1 = 3$

$-x = 3 - 1$

$\therefore x = -2$

3.2 $2^{2x} - 4^{x-1} = 6$

$2^{2x} - 2^{2x-2} = 6$

$2^{2x}(1 - 2^{-2}) = 6$

$2^{2x}\left(\frac{3}{4}\right) = 6$

$2^{2x} = 8$

$2^{2x} = 2^3$

$\therefore x = \frac{3}{2}$

4.1 $\dfrac{3^{a+1} - 3^{a-1}}{3^a - 3^{a-2}}$

$= \dfrac{3^a(3 - 3^{-1})}{3^a(1 - 3^{-2})}$

$= \dfrac{(3 - 3^{-1})}{(1 - 3^{-2})}$

$= \dfrac{\left(3 - \frac{1}{3}\right)}{\left(1 - \frac{1}{9}\right)} = 3$

4.2.1 $9^x - 1 = 3^{2x} - 1 = (3^x - 1)(3^x + 1)$

4.2.2 $\dfrac{9^x - 1}{3^x + 1} = \dfrac{(3^x - 1)(3^x + 1)}{(3x + 1)} = 3^x - 1$

5.1 False, for example if $x = \frac{1}{2}$, then $\left(\frac{1}{2}\right)^{-4} = 2^4 = 16$, and $\left(\frac{1}{2}\right)^{-3} = 2^3 = 8$. The statement is therefore false because $16 > 8$.

5.2 False: for example if $x = -\frac{1}{2}$, then $\left(-\frac{1}{2}\right)^4 = \frac{1}{16}$, and $\left(-\frac{1}{2}\right)^3 = -\frac{1}{8}$. The statement is therefore false because $\frac{1}{16} > -\frac{1}{8}$.

6. 2^x can never equal zero because no real value exists which will make $2^x = 0$.

7. There are 5^3 cats and 5^4 kittens.

Per day: cats catch $5^3 \times 5 = 5^4$ mice and kittens catch $5^4 \times 1 = 5^4$ mice.

Per week: cats catch $5^4 \times 7$ mice and kittens catch $5^4 \times 7$ mice.

In seven weeks: cats catch $5^4 \times 7^2$ mice and kittens catch $5^4 \times 7^2$ mice.

$\therefore$ Together they catch: $5^4 \times 7^2 \times 2$

$= 61\ 250$ mice.

8.1

n	1	2	3	4	5	6
n^5	$1^5 = 1$	$2^5 = 32$	$3^5 = 243$	$4^5 = 1\ 024$	$5^5 = 3\ 125$	$6^5 = 7\ 776$
Last digit of n^5	1	2	3	4	5	6

The last digit of 412^5 will be 2.

n	1	2	3	4	5
9^n	$9^1 = 9$	$9^2 = 81$	$9^3 = 729$	$9^4 = 6\,561$	$9^5 = 59\,049$
Last digit of n^5	9	1	9	1	9

Every odd power of 9 ends in a 9 and every even power of 9 ends in a 1.

$\therefore$ the last digit of 9^{999} will be a 9.

Test C (page 40)

1. $A = I \left(3^{-2} = \frac{1}{9}\right)$ $\qquad B = H \left(2^{-3} = \frac{1}{8}\right)$

$C = G \left(4^{-2} = \frac{1}{16}\right)$ $\qquad D = F \left(6^{-1} = \frac{1}{6}\right)$

2.1 $x(-x)^{-1} = \frac{x}{-x} = -1$

2.2 $\dfrac{(2+x)^0}{3(-3)^{-1}} = \dfrac{1}{3\left(-\frac{1}{3}\right)} = \dfrac{1}{-1} = -1$

3.

x^5	$\times$	x^{-2}	$=$	x^3
$\times$		$\times$		$\div$
x^{-5}	$\times$	x^{-3}	$=$	x^{-2}
$=$		$=$		$=$
1	$\div$	x^{-5}	$=$	x^5

4.1 $10^x = 1$

$10^x = 10^0$

$\therefore x = 0$

4.2 $12^{-1} = \frac{1}{x}$

$\frac{1}{12} = \frac{1}{x}$

$\therefore x = 12$

4.3 $10^{x-1} = 100\,000$

$10^{x-1} = 10^5$

$x - 1 = 5$

$\therefore x = 6$

4.4 $4^x = 0{,}0625$

$4^x = \dfrac{625}{10\,000}$

$4^x = \dfrac{1}{16}$

$4^x = \dfrac{1}{4^2}$

$4^x = 4^{-2}$

$\therefore x = -2$

4.5 $3 \cdot 5^{x-1} - 75 = 0$

$5^{x-1} = \dfrac{75}{3}$

$5^{x-1} = 25$

$5^{x-1} = 5^2$

$x - 1 = 2$

$\therefore x = 3$

4.6 $3^{x-2} + 3^{x+1} = 28$

$3^x(3^{-2} + 3^1) = 28$

$3^x\left(\frac{1}{9} + 3\right) = 28$

$3^x\left(\frac{28}{9}\right) = 28$

$3^x = 9$

$3^x = 3^2$

$\therefore x = 2$

5.1 $\dfrac{-3a^{-1}}{2a^{-2} - (a^3)^{-1}}$

$= \dfrac{\frac{-3}{a}}{\frac{2}{a^2} - \frac{1}{a^3}}$

$= \dfrac{\frac{-3}{a}}{\frac{2a-1}{a^3}}$

$= \dfrac{-3}{a} \times \dfrac{a^3}{2a-1}$

$= \dfrac{-3a^2}{2a-1}$

5.2 $\left(\dfrac{x}{y}\right)^{a+b} \times \left(\dfrac{y}{x}\right)^{a-b}$

$= \dfrac{x^{a+b} \cdot y^{a-b}}{y^{a+b} \cdot x^{a-b}}$

$= x^{a+b-a+b} \cdot y^{a-b-a-b}$

$= \dfrac{x^{2b}}{y^{2b}}$

5.3 $\dfrac{3^{x+2} - 3^{x-2}}{4 \cdot 3^{x-3}}$

$= \dfrac{3^x \cdot 3^2 - 3^x \cdot 3^{-2}}{4 \cdot 3^x \cdot 3^{-3}}$

$= \dfrac{3^x(3^2 - 3^{-2})}{4 \cdot 3^x \cdot 3^{-3}}$

$= \dfrac{\frac{8}{9}}{4 \cdot \frac{1}{27}}$

$= 6$

6.1 Always true, because $(0{,}5)^n = \left(\frac{1}{2}\right)^n = 2^{-n}$ for all integer values of n.

6.2 Never true, because n^3 has two other factors besides 1 and n^3, which are n^2 and n.

7.1

n	1	2	3	4	5	6	7	8	9	10
2^n	2	4	8	16	32	64	128	256	512	1 024
Last digit of 2^n	2	4	8	6	2	4	8	6	2	4

7.2 $100 \div 4 = 25$

∴ the last digit of 2^{100} will be a 6.

Topic 4   Number patterns (page 42)

Exercise 1 (page 43)

1.   3;   6;   9;   12; ...
           +3   +3   +3

$T_1 = 1 \times 3$
$T_2 = 2 \times 3$
$T_3 = 3 \times 3$
$T_4 = 4 \times 3$
$\therefore T_n = n \times 3$
$\therefore T_n = 3n$

2.   1;   4;   7;   10; ...
           +3   +3   +3

$T_1 = 1 \times 3 - 2$
$T_2 = 2 \times 3 - 2$
$T_3 = 3 \times 3 - 2$
$T_4 = 4 \times 3 - 2$
$\therefore T_n = n \times 3 - 2$
$\therefore T_n = 3n - 2$

3.   −5;   −10;   −15;   −20; ...
            −5   −5   −5

$T_1 = 1 \times -5$
$T_2 = 2 \times -5$
$T_3 = 3 \times -5$
$T_4 = 4 \times -5$
$\therefore T_n = n \times -5$
$\therefore T_n = -5n$

4.   −2;   −7;   −13;   −18; ...
            −5   −5   −5

$T_1 = 1 \times -5 + 3$
$T_2 = 2 \times -5 + 3$
$T_3 = 3 \times -5 + 3$
$T_4 = 4 \times -5 + 3$
$\therefore T_n = n \times -5 + 3$
$\therefore T_n = -5 + 3$

5.   1;   3;   5;   7; ...
           +2   +2   +2

$T_1 = 1 \times 2 - 1$
$T_2 = 2 \times 2 - 1$
$T_3 = 3 \times 2 - 1$
$T_4 = 4 \times 2 - 1$
$\therefore T_n = n \times 2 - 1$
$\therefore T_n = 2n - 1$

6.   −7;   −5;   −13;   −1; ...
            +2   +2   +2

$T_1 = 1 \times 2 - 9$
$T_2 = 2 \times 2 - 9$
$T_3 = 3 \times 2 - 9$
$T_4 = 4 \times 2 - 9$
$\therefore T_n = n \times 2 - 9$
$\therefore T_n = 2n - 9$

7.   1;   2;   3;   4; ...
           +1   +1   +1

$T_1 = 1 \times 1$
$T_2 = 2 \times 1$
$T_3 = 3 \times 1$
$T_4 = 4 \times 1$
$\therefore T_n = n \times 1$
$\therefore T_n = n$

8.   −5;   −4;   −3;   −2; ...
            +1   +1   +1

$T_1 = 1 \times 1 - 6$
$T_2 = 2 \times 1 - 6$
$T_3 = 3 \times 1 - 6$
$T_4 = 4 \times 1 - 6$
$\therefore T_n = n \times 1 - 6$
$\therefore T_n = n - 6$

9. $\frac{1}{2}, \frac{1}{3}, \frac{1}{4}, \frac{1}{5}, \ldots$

Consider the denominators as a sequence of their own:

   2;   3;   4;   5; ...
       +1   +1   +1

$T_1 = 1 \times 1 + 1$
$T_2 = 2 \times 1 + 1$
$T_3 = 3 \times 1 + 1$
$T_4 = 4 \times 1 + 1$
$\therefore T_n = n \times 1 + 1$
$\therefore T_n = n + 1$

So the general term for the actual sequence:
$\frac{1}{2}, \frac{1}{3}, \frac{1}{4}, \frac{1}{5} \ldots$ is $\therefore T_n = \frac{1}{n+1}$.

10. $1; \frac{1}{2}, \frac{1}{3}, \frac{1}{4}, \ldots = \frac{1}{1}, \frac{1}{2}, \frac{1}{3}, \frac{1}{4}, \ldots$

In this case the denominators are the natural numbers starting at 1, $\therefore T_n = \frac{1}{n}$.

11.   4;   6;   8; ...
        +2   +2

$T_1 = 1 \times 2 + 2$
$T_2 = 2 \times 2 + 2$
$T_3 = 3 \times 2 + 2$
$\therefore T_n = n \times 2 + 2$
$\therefore T_n = 2n + 2$

12.   3;   5;   7; ...
        +2   +2

$T_1 = 1 \times 2 + 1$
$T_2 = 2 \times 2 + 1$
$T_3 = 3 \times 2 + 1$
$\therefore T_n = n \times 2 + 1$
$\therefore T_n = 2n + 1$

Exercise 2   (page 45)

1.1 $T_n = 8n + 5$
$T_1 = 8(1) + 5 = 13$, so the first term is 13.
$T_2 = 8(2) + 5 = 21$
The common difference is: $T_2 - T_1 = 21 - 13$
$= 8.$

1.2 $T_n = 7n - 1$
$T_1 = 7(1) - 1 = 6$, so the first term is 6.
$T_2 = 7(2) - 1 = 13$
The common difference is: $T_2 - T_1 = 13 - 6$
$= 7.$

1.3 $T_n = 6 - 2n$
$T_1 = 6 - 2(1) = 4$. The first term is 4.
$T_2 = 6 - 2(2) = 2$
The common difference is : $T_2 - T_1 = 2 - 4$
$= -2.$

1.4 $T_n = -4n$
$T_1 = -4(1) = -4$. The first term is -4.
$T_2 = -4(2) = -8$
The common difference is: $T_2 - T_1 = -8 - (-4)$
$= -4.$

2.1 $T_1 = 4; T_2 = 4 + (-7) = -3,$
$T_3 = -3 + (-7) = -10$
Sequence: $4; -3; -10; ...$
$T_n = -7n + 11$

2.2 $T_1 = -1; T_2 = -1 + 12 = 11, T_3 = 11 + 12 = 23$
Sequence: $-1; 11; 23; ...$
$T_n = 12n - 13$

2.3 $T_1 = 0; T_2 = 0 - 23 = -23, T_3 = -23 - 23 = -46.$
Sequence: $0; -23; -46; ...$
$T_n = -23n + 23$

2.4 $T_1 = 1; T_2 = 1 + 0 = 1, T_3 = 1 + 0 = 1$
Sequence: $1; -1; 1; ...$
$T_n = 1$ (this is simply a sequence of 1s)

2.5 $T_1 = -13; T_2 = -13 - 5 = -18,$
$T_3 = -18 - 5 = -23$
Sequence: $-13; -18; -23; ...$
$T_n = -5n - 8$

2.6 $T_1 = \frac{1}{10}; T_2 = \frac{1}{10} - \frac{1}{2} = -\frac{2}{5}, T_3 = -\frac{2}{5} - \frac{1}{2} = -\frac{9}{10}$
Sequence: $\frac{1}{10}; -\frac{2}{5}; -\frac{9}{10}; ...$
$T_n = -\frac{1}{2}n + \frac{3}{5}$

3.1 $T_n = 3n + 2, T_1 = 3(1) + 2 = 5$
The first term is 5 and the common difference is 3.

3.2 $T_n = \frac{1}{3}n - 6, T_1 = \frac{1}{3}(1) - 6 = -\frac{17}{3}$
The first term is $-\frac{17}{3}$ and the common difference is $\frac{1}{3}$.

3.3 $T_n = 5 - n, T_1 = 5 - 1 = 4$
The first term is 4 and the common difference is -1.

4. $T_n = 8 - 3n, T_1 = 8 - 3(1) = 5$
The first term is 5 and the common difference is -3.

5. $T_n = -2n - 1, T_1 = -2(1) - 1 = -3$
The first term is -3 and the common difference is -2.

6. $T_n = 7 - \frac{n}{3}, T_1 = 7 - \frac{1}{3} = \frac{20}{3}$
The first term is $\frac{20}{3}$ and the common difference is $-\frac{1}{3}$.

7. $Tn = \frac{3}{4}n + \frac{5}{4}, T_1 = \frac{3}{4}(1) + \frac{5}{4} = 2$
The first term is 2 and the common difference is $\frac{3}{4}$.

8. $T_n = (n + 1)(n - 2) - n^2,$
$T_1 = (1 + 1)(1 - 2) - 1^2 = -3,$
$T_2 = (2 + 1)(2 - 2) - 2^2 = -4$
The first term is -3 and the common difference is -1.

9.1 $T_n = 7n$

$294 = 7n$

$n = 42$

∴ The 42nd term is 294.

9.2 $T_n = -5n$

$-195 = -5n$

$n = 39$

∴ The 39th term is −195.

9.3 $T_n = 4 - 3n$

$-149 = 4 - 3n$

$-153 = -3n$

∴ $n = 51$

9.4 $T_n = \dfrac{n+5}{n-3}$

$2 = \dfrac{n+5}{n-3}$

$2(n-3) = \dfrac{n+5}{n-3} \times n - 3$

$2n - 6 = n + 5$

$n = 11$

∴ The 11th term is 2.

Exercise 3   (page 47)

1.1 $T_n = 7n$

$T_2 = 7(2) = 14$

$T_4 = 7(4) = 28$

∴ −7; 14; −21; 28; −35

1.2 $T_n = n^3 - n^2$

$T_4 = 4^3 - 4^2 = 64 - 16 \qquad = 48$

∴ 0; 4; 18; 48; 100

1.3 $T_n = n^2 + n - 4$

$T_1 = 1^2 + 1 - 4 = -2$

$T_4 = 4^2 + 4 - 4 = 16$

∴ −2; 2; 8; 16; 26

1.4 $T_n = n(n + 1)$

$T_2 = 2(2 + 1) = 6$

$T_4 = 4(4 + 1) = 20$

$T_5 = 5(5 + 1) = 30$

∴ 2; 6; 12; 20; 30

1.5 $T_n = \dfrac{n+1}{n}$

$T_1 = \dfrac{1+1}{1} = 2$

$T_3 = \dfrac{3+1}{3} = \dfrac{4}{3}$

∴ $2; \dfrac{3}{2}; \dfrac{4}{3}; \dfrac{5}{4}; \dfrac{6}{5}$

1.6 $T_n = -n + \dfrac{1}{n}$

$T_4 = -4 + \dfrac{1}{4} = -\dfrac{15}{4}$

$T_5 = -5 + \dfrac{1}{5} = -\dfrac{24}{5}$

∴ $0; -\dfrac{3}{2}; -\dfrac{8}{3}; -\dfrac{15}{4}; -\dfrac{24}{5}$

2.1   **(a)**   9; 6; 3

(b) $T_{12} = 12 - 3(12) = -24$

2.2   **(a)**   6; 5; 4

(b) $T_{12} = -12 + 7 = -5$

2.3   **(a)** $\dfrac{1}{2}; 2; \dfrac{9}{2}$

(b) $T_{12} = \dfrac{1}{2}(12)^2 = 72$

2.4   **(a)**   0; 6; 24

(b) $T_{12} = 12(13)(11) = 1\,1726$

2.5   **(a)** $1; \dfrac{4}{3}; \dfrac{3}{2}$

(b) $T_{12} = \dfrac{2(12)}{1+12} = \dfrac{24}{13}$

2.6   **(a)** $0; \dfrac{3}{8}; \dfrac{8}{15}$

(b) $T_{12} = \dfrac{(13)(11)}{12(14)} = \dfrac{143}{168}$

3.1 $T_n = -n$

3.2 $T_n = 5n$

3.3 $T_n = \dfrac{n}{1+n}$

3.4 $T_n = 2n^2$

3.5 $T_n = 17 + n$

3.6 $T_n = n(n - 1)$ or $n^2 - n$

4.1 $1\,056 = n + 7$

∴ $n = 1\,049$

4.2 $138 = 6n$

∴ $n = 23$

4.3 $255 = (n - 1)(n + 1)$

$255 = n^2 - 1$

∴ $256 = n^2$

∴ $n = \pm 16$ (Remember, $(16)^2$ and $(-16)^2$ will give 256.)

∴ $n = 16$ (n cannot be < 0)

4.4 $\dfrac{1}{2} = \dfrac{3}{1+n}$

∴ $1 + n = 6$ ($\times 2(1 + n), n \neq -1$)

∴ $n = 5$

5. $96 = 15n - 2$

$\therefore 98 = 15n$

$\therefore n = 6,533\ldots$

Because the answer is not a whole number, this means that 96 is not a term in the given sequence.

Test A (page 48)

1.

Shape number	1	2	3	4	15	20	n
Number of matchsticks	6	11	16	21	76	101	$T_n = 5n + 1$

2.1 $T_{10} = -5(10) + 7 = -43$

2.2 $T_1 = -5(1) + 7 = 2$

2.3 $-5(-103) + 7 = 515 + 7 = 522$

$\therefore$ Term 522 will be -103

3. $p - 3 = -6 - p$

$2p = -3$

$\therefore p = -1,5$

4.1 $-6; \quad -4; \quad -2; \quad 0; \ldots$

$\qquad +2 \qquad +2 \qquad +2$

$T_n = 2n - 8$

4.2 $\dfrac{11}{3}, \qquad \dfrac{9}{3}, \qquad \dfrac{7}{3}, \qquad \dfrac{5}{3}, \ldots$

$\qquad -\dfrac{2}{3} \qquad -\dfrac{2}{3} \qquad -\dfrac{2}{3}$

$T_n = -\dfrac{2}{3}n + \dfrac{13}{3}$

4.3 $T_n = 3n^2$

5.1 $0 + 2 + 4 = -1 + 2 + 5$

$5 + 7 + 9 = 4 + 7 + 10$

5.2 $-1 + 1 + 3 = -2 + 1 + 4$

5.3 $n + n + 2 + n + 4 = n - 1 + n + 2 + n + 5$

5.4 LHS: $n + n + 2 + n + 4 = 3n + 6$

RHS: $n - 1 + n + 2 + n + 5 = 3n + 6$

$\therefore$ LHS = RHS

We can conclude that the pattern will be true for all values of n.

6. $T_n = 18n - 3$

$T_1 = 18(1) - 3 = 15$

$T_2 = 18(2) - 3 = 33$

$T_2 - T_1 = 33 - 15 = 18$

$\therefore a = 15$ and $d = 18$

7.1 **(a)** 28   **(b)** 248   **(c)** $-2 + 10n$

7.2 $1\,870 = 10n$

$\therefore n = 187,2$

$\therefore$ 187 layers are completely full.

$187 \div 7 = 26,71\ldots$ boxes

$\therefore$ 26 boxes are completely filled.

Test B (page 49)

1.1 Pattern for the dark tiles: 1; 3; 5; 7; …
Pattern no. 8 will have 15 tiles.

1.2 Pattern for the light tiles: 0; 1; 4; 9; …
Pattern no. 8 will have 49 tiles.

1.3 $T_n = 2n - 1$

1.4 $T_n = (n - 1)^2$

1.5 $\qquad (n - 1)^2 = 841$

$\qquad n^2 - 2n + 1 = 841$

$\qquad n^2 - 2n - 840 = 0$

$(n - 30)(n + 28) = 0$

$\therefore n = 30$ ($n = -28$ is not applicable)

The 30th pattern will have 841 light tiles.

1.6 $T_n = 2(30) - 1 = 59$

There will be 59 dark tiles in the 30th pattern.

2.1 $6^3 - 5^3 = 6^2 + 5^2 + 6 \times 5$

$7^3 - 6^3 = 7^2 + 6^2 + 7 \times 6$

2.2 $(n + 1)^3 - n^3 = (n + 1)^2 + n^2 + (n + 1) \times n$

2.3 LHS: $(n + 1)^3 - n^3 = (n + 1)(n^2 + 2n + 1) - n^3$

$\qquad\qquad = (n^3 + 3n^2 + 3n + 1) - n^3$

$\qquad\qquad = 3n^2 + 3n + 1$

RHS: $(n + 1)^2 + n^2 + (n + 1) \times n$

$\qquad = (n^2 + 2n + 1) + n^2 + n^2 + n$

$\qquad = 3n^2 + 3n + 1$

2.4 LHS = RHS. This proves that this pattern will be true for all values of n where $n \in \mathbb{N}$.

3.1 $0,11; \qquad 0,22; \qquad 0,33; \qquad 0,44; \ldots$

$\qquad 0,11 \qquad 0,11 \qquad 0,11$

$T_n = 0,11n$

3.2 Denominators: 8; 27; 64; 125; …

Cubic sequence: $T_n = (n + 1)^3$

$\therefore$ general term for the sequence is $T_n = \dfrac{1}{(n + 1)^3}$

3.3 The sequence 1; 3; 5; 7… has $T_n = 2n - 1$.

$\therefore 1^2; 3^2; 5^2; 7^2; \ldots$ has $T_n = (2n - 1)^2$

4.1 $T_7 = 7^2 + 4(7) - 6$

$\qquad = 49 + 28 - 6$

$\qquad = 71$

4.2
$$15 = n^2 + 4n - 6$$
$$0 = n^2 + 4n - 21$$
$$= (n - 3)(n + 7)$$
∴ $n = 3$ (or −7, but n must be positive)
∴ the third term will be 15.

5.1 This is an exponential pattern, so the bacteria is growing exponentially: $T_n = 2^n$.

5.2 $T_{18} = 2_{18} = 262\ 144$

6. Four consecutive odd numbers are $2n - 1$; $2n + 1$; $2n + 3$ and $2n + 5$
$$(2n + 1)(2n + 3) - (2n - 1)(2n + 5)$$
$$= 4n^2 + 2n + 6n + 3 - (4n^2 - 2n + 10n - 5)$$
$$= 4n^2 + 8n + 3 - 4n^2 - 8n + 5$$
$$= 8$$

Test C (page 50)

1. $T_4 = -24$ and $T_n = -7n + 4$

2.1 $T_8 = 7$ and $T_9 = 10$
2.2 $T_7 = 16 + 10 = 26$ and $T_8 = 26 + 16 = 42$

3.1 $T_n = \frac{n}{2} + 3$
$$T_1 = \frac{1}{2} + 3 = 3\frac{1}{2}$$
3.2 $T_2 = \frac{2}{2} + 3 = 4$, common difference is $\frac{1}{2}$
3.3 $T_{11} + T_{13} = \left(\frac{11}{2} + 3\right) + \left(\frac{13}{2} + 3\right) = 18$

4.1 The answer will always be 2 for any set of 4 consecutive numbers.
For example: 20; 21; 22; 23:
The product of the middle two numbers is: $21 \times 22 = 462$.
The product of the first and last numbers is: $20 \times 23 = 460$.
The difference between the two numbers is: $462 - 460 = 2$.

4.2 Multiplying the middle numbers of four consecutive numbers and subtracting the answer from the product of the first and last numbers always gives an answer of 2.

4.3 Let the four consecutive numbers be n; $n + 1$; $n + 2$; $n + 3$:
∴ $(n + 1)(n + 2) = n^2 + 3n + 2$
$n(n + 3) = n^2 + 3n$

Difference: $n^2 + 3n + 2 - (n^2 + 3n)$
$$= n^2 + 3n + 2 - n^2 - 3n$$
$$= 2$$

5. $3x - 7$; $2x$; $3x + 1$ are the first three terms of a linear pattern.

5.1 $T_2 - T_1 = 2x - (3x - 7) = -x + 7$
$T_3 - T_2 = 3x + 1 - 2x = x + 1$
∴ $-x + 7 = x + 1$
$-2x = -6$
∴ $x = 3$

5.2 $T_1 = 3(3) - 7 = 2$
$T_2 = 2(3) = 6$
∴ $d = 4$, $T_n = 4n - 2$
$31 = 4n - 2$
$33 = 4n$
∴ $n = 8,25$
T_9 will be the first term greater than 31.

6.1 **(a)** 1   **(b)** 9   **(c)** 7   **(d)** 9

6.2 Pattern 1 has a common difference of 1, and is therefore linear.
Pattern 2 has a common difference of 4, and is therefore linear.
Pattern 3 has a common difference of 2, and is therefore linear.

6.3 Pattern 1: $T_n = n$
Pattern 2: $T_n = 4n + 1$
Pattern 3: $T_n = 2n + 1$

6.4 $T_{86} = 4(86) + 1 = 345$

6.5 To complete 59 rows Franco will need:
$T_{59} = 4n + 1(59) + 1 = 237$ stars.

6.6 $T_n = 2n + 1$
$105 = 2n + 1$
$104 = 2n$
∴ $n = 51$
∴ Row number 51 will have 105 rows.
Step number 52 requires
$T_{51} = 4(51) + 1 - 205$ stars.
He will have enough stars to complete 105 rows.

Topic 5   Equations and inequalities (page 52)

Exercise 1   (page 55)

1.
$5 - x - 2(x - 5) = 3(x - 3) - 3x$
$5 - x - 2x + 10 = 3x - 9 - 3x$
$15 - 3x = -9$
$-3x = -24$
$\therefore x = 8$

2.
$\frac{2x + 1}{5} = -\frac{1}{7}(x - 1)$
$\frac{7(2x + 1)}{35} = -\frac{5(x - 1)}{35}$
$14x + 7 = -5x + 5$
$19x = -2$
$\therefore x = -\frac{2}{19}$

3.
$6 - 3[2x - 4(x - 2)] = 0$
$6 - 3(2x - 4x + 8) = 0$
$6 - 3(-2x + 8) = 0$
$6 + 6x - 24 = 0$
$6x = 18$
$\therefore x = \frac{18}{6} = 3$

4.
$5\frac{1}{2} - 4(x + 1) + \frac{1}{3}(x - 2) = \frac{1}{3}\left(3x + 1\frac{1}{2}\right)$
$\frac{11}{2} - 4(x + 1) + \frac{1}{3}(x - 2) = \frac{1}{3}\left(3x + \frac{3}{2}\right)$
$\frac{11(3) - 4(6)(x + 1) + 2(x - 2)}{6} = \frac{2\left(3x + \frac{3}{2}\right)}{6}$
$33 - 24x - 24 + 2x - 4 = 6x + 3$
$5 - 22x = 6x + 3$
$-28x = -2$
$\therefore x = -\frac{2}{-28} = \frac{1}{14}$

5.
$(x + 3)^2 - 5x = (x + 1)^2 + 6$
$x^2 + 6x + 9 - 5x = x^2 + 2x + 1 + 6$
$x + 9 = 2x + 7$
$-x = -2$
$\therefore x = 2$

6.
$2 - \frac{4x - 3}{6} = \frac{3}{2}(x + 3) - \frac{2(x - 1)}{3}$
$\frac{2(6) - (4x - 3)}{6} = \frac{3(3)(x + 3) - 2(2)(x - 1)}{6}$
$12 - 4x + 3 = 9x + 27 - 4x + 4$
$-4x + 15 = 5x + 31$
$-9x = 16$
$\therefore x = \frac{16}{9}$

7.
$\frac{1 - 5x}{6} = \frac{2x - 1}{3} - \frac{3x - 1}{2}$
$\frac{1 - 5x}{6} = \frac{2(2x - 1) - 3(3x - 1)}{6}$
$1 - 5x = 4x - 2 - 9x + 3$
$1 - 5x = -5x + 1$
$0x = 0$
$\therefore x \in \mathbb{R}$, an unlimited number of solutions exist.

8.
$x - \frac{2}{3}(x - 2) = \frac{3x + 1}{4} - \frac{x + 1}{6}$
$\frac{12x - 2(4)(x - 2)}{12} = \frac{3(3x + 1) - 2(x + 1)}{12}$
$12x - 8x + 16 = 9x + 3 - 2x - 2$
$4x + 16 = 7x + 1$
$-3x = -15$
$\therefore x = \frac{15}{3} = 5$

Exercise 2   (page 57)

1.
$(x - 3)(6x + 5) = 0$
$\therefore x = 3$ or $x = -\frac{6}{5}$

2.
$x(3x - 4) = 0$
$\therefore x = 0$ or $x = \frac{4}{3}$

3.
$(x^2 + 4) = 49$
$x^2 + 8x + 16 = 49$
$x^2 + 8x - 33 = 0$
$(x + 11)(x - 3) = 0$
$\therefore x = -11$ or $x = 3$

4.
$36 - 4x^2 = 0$
$4(9 - x^2) = 0$
$4(3 - x)(3 + x) = 0$
$\therefore x = -3$ or $x = 3$

5.
$5x^2 + 6 = 17x$
$5x^2 - 17x + 6 = 0$
$(5x - 2)(x - 3) = 0$
$5x = 2$ or $x = 3$
$\therefore x = \frac{2}{5}$ or $x = 3$

6.
$$2x^2 + 6 = -6 - 11x$$
$$2x^2 + 11x + 12 = 0$$
$$(2x + 3)(x + 4) = 0$$
$$2x = -3 \text{ or } x = -4$$
$$\therefore x = -\tfrac{3}{2} \text{ or } x = -4$$

7.
$$(3x - 2)(x - 2) = 20$$
$$3x^2 - 8x + 4 = 20$$
$$3x^2 - 8x - 16 = 0$$
$$(3x + 4)(x - 4) = 0$$
$$3x = -4 \text{ or } x = 4$$
$$\therefore x = -\tfrac{4}{3} \text{ or } x = 4$$

Exercise 3 (page 58)

1. $x - \tfrac{5}{2} = \tfrac{6}{x}$: restriction $x \neq 0$
$$\frac{2x(x) - 5(x)}{2x} = \frac{2(6)}{2x}$$
$$2x^2 - 5x = 12$$
$$2x^2 - 5x - 12 = 0$$
$$(2x + 3)(x - 4) = 0$$
$$2x + 3 = 0 \text{ or } x - 4 = 0$$
$$2x = -3 \text{ or } x = 4$$
$$\therefore x = -\tfrac{3}{2} \text{ or } x = 4$$

2. $\frac{4}{3x - 2} - \frac{3}{2x - 3} = \frac{1}{2x - 1}$

Restrictions: $x \neq \tfrac{2}{3}, x \neq \tfrac{3}{2}, x \neq \tfrac{1}{2}$
$$\frac{4(2x - 3)(2x - 1) - 3(3x - 2)(2x - 1)}{(3x - 2)(2x - 3)(2x - 1)}$$
$$= \frac{1(3x - 2)(2x - 3)}{(3x - 2)(2x - 3)(2x - 1)}$$
$$4(4x^2 - 8x + 3) - 3(6x^2 - 7x + 2)$$
$$= 1(6x^2 - 13x + 6)$$
$$16x^2 - 32x + 12 - 18x^2 + 21x - 6$$
$$= 6x^2 - 13x + 6$$
$$-2x^2 - 11x + 6 = 6x^2 - 13x + 6$$
$$0 = 8x^2 - 2x$$
$$0 = 2x(4x - 1)$$
$$2x = 0 \text{ or } 4x - 1 = 0$$
$$\therefore x = 0 \text{ or } x = \tfrac{1}{4}$$

3. $\frac{2}{x - 1} - \frac{1}{x} = \frac{1}{x + 2}$

Restrictions: $x \neq 1, x \neq 0, x \neq -2$
$$\frac{2x(x + 2) - 1(x - 1)(x + 2)}{x(x - 1)(x + 2)} = \frac{1(x - 1)}{x(x - 1)(x + 2)}$$
$$2x^2 + 4x - 1(x^2 + x - 2) = x^2 - x$$

$$2x^2 + 4x - x^2 - x + 2 = x^2 - x$$
$$4x + 2 = 0$$
$$4x = -2$$
$$\therefore x = -\tfrac{1}{2}$$

4. $\frac{x - 2}{x - 1} = \frac{2x - 1}{x + 7}$

Restrictions: $x \neq 1, x \neq -7$
$$\frac{(x - 2)(x + 7)}{(x - 1)(x + 7)} = \frac{(2x - 1)(x - 2)}{(x - 1)(x + 7)}$$
$$x^2 + 5x - 14 = 2x^2 - 5x + 2$$
$$-x^2 + 10x - 16 = 0$$
$$x^2 - 10x + 16 = 0$$
$$(x - 8)(x - 2) = 0$$
$$\therefore x = 8 \text{ or } x = 2$$

5. $\frac{5}{x - 2} - \frac{4}{x} = \frac{3}{x + 6}$

Restrictions: $x \neq 2, x \neq 0, x \neq -6$
$$\frac{5x(x + 6) - 4(x - 2)(x + 6)}{x(x - 2)(x + 6)} = \frac{3x(x - 2)}{x(x - 2)(x + 6)}$$
$$5x(x + 6) - 4(x - 2)(x + 6) = 3x(x - 2)$$
$$5x^2 + 30x - 4(x^2 + 4x - 12) = 3x^2 - 6x$$
$$5x^2 + 30x - 4x^2 - 16x + 48 = 3x^2 - 6x$$
$$-2x^2 + 20x + 48 = 0$$
$$x^2 - 10x - 24 = 0$$
$$(x - 12)(x + 2) = 0$$
$$\therefore x = 12 \text{ or } x = -2$$

Exercise 4 (page 61)

1.1
$$4x - 3y = 10 \qquad \ldots \text{①}$$
$$4x + y = 2$$
$$\therefore y = 2 - 4x \qquad \ldots \text{②}$$

Substitute $2 - 4x$ for y in ①:
$$4x - 3(2 - 4x) = 10$$
$$4x - 6 + 12x = 10$$
$$16x = 16$$
$$x = 1$$

Substitute 1 for x in ②:
$$y = 2 - 4(1)$$
$$y = -2$$
$$\therefore x = 1 \text{ and } y = -2$$

1.2
$$x + 2y = 4 \qquad \ldots \text{①}$$
$$3x + y = 7$$
$$\therefore y = 7 - 3x \qquad \ldots \text{②}$$

Substitute $7 - 3x$ for y in ①:

$x + 2(7 - 3x) = 4$

$x + 14 - 6x = 4$

$-5x = -10$

$x = 2$

Substitute 2 for x in ②:

$y = 7 - 3(2)$

$y = 1$

$\therefore x = 2$ and $y = 1$

1.3 $2x + y + 3 = 0$ ... ①

$y = x + 1$ ... ②

Substitute $x + 1$ for y in ①:

$2x + (x + 1) + 3 = 0$

$2x + x + 1 + 3 = 0$

$3x + 4 = 0$

$3x = -4$

$x = -\frac{4}{3}$

Substitute $-\frac{4}{3}$ for x in ②:

$y = -\frac{4}{3} + 1 = -\frac{1}{3}$

$\therefore x = -\frac{4}{3}$ and $y = -\frac{1}{3}$

2.1 $2x + 2y = 14$ ... ①

$5x - 2y = 21$ ... ②

Add ① and ②:

$$\begin{array}{r} 2x + 2y = 14 \\ + \quad 5x - 2y = 21 \\ \hline 7x - 0y = 35 \end{array}$$

$\therefore x = 5$

Substitute 5 for x in ①:

$2(5) + 2y = 14$

$10 + 2y = 14$

$2y = 4$

$y = 2$

$\therefore x = 5$ and $y = 2$

2.2 $3x - 4y = -8$ ... ①

$3x - y = 10$ ... ②

Subtract ② from ①:

$$\begin{array}{r} 3x - 4y = -8 \\ -3x - y = 10 \\ \hline 0x - 3y = -18 \end{array}$$

$\therefore y = 6$

Substitute 6 for y in ②:

$3x - 6 = 10$

$3x = 16$

$x = \frac{16}{3}$

$\therefore x = \frac{16}{3}$ and $y = 6$

2.3 $2x - 5y = -12$ ... ①

$4x + 5y = 6$ ... ②

Add ① and ②:

$$\begin{array}{r} 2x - 5y = -12 \\ + \quad 4x + 5y = \quad 6 \\ \hline 6x - 0y = -6 \end{array}$$

$\therefore x = -1$

Substitute -1 for x in ①:

$2(-1) - 5y = -12$

$-5y = -10$

$y = 2$

$\therefore x = -1$ and $y = 2$

3.1 $3x + 5y = -1$ ... ①

$4x + 7y = 4$ ... ②

Add ① × 4 and ② × –3:

$$\begin{array}{lr} ① \times 4: & 12x + 20y = -4 \\ + \quad ② \times -3: & -12x - 21y = -12 \\ \hline & 0x - \quad y = -16 \end{array}$$

$\therefore y = 16$

Substitute 16 for y in ①:

$3x + 5(16) = -1$

$3x + 80 = -1$

$3x = -81$

$x = -27$

$\therefore x = -27$ and $y = 16$

3.2 $y = 2x - 1$ ... ①

$y = -5x + 2$ ... ②

Substitute $2x - 1$ for y in ②:

$2x - 1 = -5x + 2$

$7x = 3$

$x = \frac{3}{7}$

Substitute $\frac{3}{7}$ for x in ①:

$y = 2(\frac{3}{7}) - 1$

$y = -\frac{1}{7}$

$\therefore x = \frac{3}{7}$ and $y = -\frac{1}{7}$

3.3 $\frac{2x+y}{3} = 11$

$2x + y = 33$

$y = 33 - 2x$ … ①

$\frac{1}{6}x - 2y - 34 = 0$

$x - 12y - 204 = 0$ … ②

Substitute $33 - 2x$ for y in ②:

$x - 12(33 - 2x) - 204 = 0$

$x - 396 + 24x - 204 = 0$

$25x = 600$

$x = 24$

Substitute 22 for x in ①:

$y = 33 - 2(24)$

$y = -15$

$\therefore x = 22$ and $y = -15$

3.4 $x + 2y + 3 = 7x - y$ and $2x - 3y + 16 = 7x - y$

$2x - 3y + 16 = 7x - y$

$-5x - 2y = -16$ … ①

$x + 2y + 3 = 7x - y$

$-6x + 3y = -3$

$3y = -3 + 6x$

$y = 2x - 1$ … ②

Substitute $2x - 1$ for y in ①:

$-5x - 2(2x - 1) = -16$

$-5x - 4x + 2 = -16$

$-9x = -18$

$x = 2$

Substitute 2 for x in ②:

$y = 2(2) - 1$

$y = 3$

$\therefore x = 2$ and $y = 3$

Exercise 5   (page 63)

1.1 $ax + bx = c$

$x(a + b) = c$

$\therefore x = \frac{c}{(a+b)}$

1.2 $k - \frac{m}{x} = t$

$\frac{kx - m}{x} = \frac{tx}{x}$

$kx - m = tx$

$kx - tx = m$

$x(k - t) = m$

$\therefore x = \frac{m}{(k-t)}$

1.3 $\frac{x-a}{x+b} = \frac{x}{x+a}$

$\frac{(x-a)(x+a)}{(x+b)(x+a)} = \frac{x(x+b)}{(x+b)(x+a)}$

$(x - a)(x + a) = x(x + b)$

$x^2 - a^2 = x^2 + xb$

$-a^2 = xb$

$\therefore x = -\frac{a^2}{b}$

1.4 $(x - a)^2 = (x - b)^2$

$x^2 - 2ax + a^2 = x^2 - 2bx + b^2$

$-2ax + a^2 = -2bx + b^2$

$2bx - 2ax = b^2 - a^2$

$2x(b - a) = (b - a)(b + a)$

$2x = (b + a)$

$\therefore x = \frac{(b+a)}{2}$

2.1 $\frac{x-m}{2a} = 3$

$x - m = 6a$

$\therefore x = 6a + m$

2.2 $\frac{x-m}{2a} = 3$

Restriction: $a \neq 0$

$x - m = 6a$

$-6a = m - x$

$a = \frac{m-x}{6}$

$\therefore a = \frac{-m+x}{6}$

3.1 $xy = p + xt$

$xy - xt = p$

$x(y - t) = p$

$\therefore x = \frac{p}{(y-t)}$

3.2 $xy = p + xt$

$xt = xy - p$

$\therefore t = \frac{xy-p}{x}$

4.1
$$p = \frac{1}{x} + v + \frac{1}{y}$$
$$\frac{pxy}{xy} = \frac{y + vxy + x}{xy}$$
$$pxy = y + vxy + x$$
$$-y = vxy + x - pxy$$
$$x(vy + 1 - py) = -y$$
$$\therefore x = -\frac{y}{(vy + 1 - py)}$$

4.2
$$p = \frac{1}{x} + v + \frac{1}{y}$$
$$\frac{pxy}{xy} = \frac{y + vxy + x}{xy}$$
$$pxy = y + vxy + x$$
$$pxy - y - vxy = x$$
$$y(px - 1 - vx) = x$$
$$\therefore y = \frac{x}{(px - 1 - vx)}$$

Exercise 6   (page 65)

1.
$$5x - 3(x + 1) < 2 + 3x$$
$$5x - 3x - 3 < 2 + 3x$$
$$-x < 5$$
$$\therefore x > -5$$
$$(-5; \infty)$$

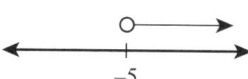

2.
$$x - \frac{x-3}{2} \geq 2, 5 + \frac{5x}{6}$$
$$\frac{6x - 3(x-3)}{6} \geq \frac{2,5(6) + 5x}{6}$$
$$6x - 3x + 9 \geq 15 + 5x$$
$$-2x \geq 6$$
$$\therefore x \leq -3$$
$$(-\infty; -3]$$

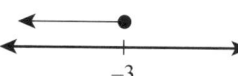

3.
$$\frac{1}{3}(x - 2) \geq 1\frac{1}{2} - \frac{1 - 2x}{2}$$
$$\frac{(x-2)}{3} \geq \frac{3}{2} - \frac{1-2x}{2}$$
$$\frac{2(x-2)}{6} \geq \frac{3(3) - 3(1-2x)}{6}$$
$$2x - 4 \geq 9 - 3 + 6x$$
$$-4x \geq 10$$
$$\therefore x \leq -\frac{5}{2}$$
$$(-\infty; -\frac{5}{2}]$$

4.
$$\frac{x-2}{4} - \frac{x-4}{6} \geq 1\frac{2}{3}$$
$$\frac{x-2}{4} - \frac{x-4}{6} \geq \frac{5}{3}$$
$$\frac{3(x-2) - 2(x-4)}{12} \geq \frac{5(4)}{12}$$
$$3x - 6 - 2x + 8 \geq 20$$
$$\therefore x \geq 18$$
$$[18; \infty)$$

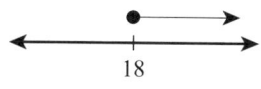

5.
$$0 < \frac{x}{3} + 1 \leq 3$$
$$-1 < \frac{x}{3} \leq 2$$
$$-3 < x \leq 6$$
$$(-3; 6]$$

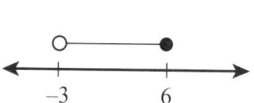

Exercise 7   (page 67)

1.
$$\frac{x}{x - 3} - \frac{6}{3x - x^2} = 1$$

Restrictions: $x \neq 3, x \neq 0$

$$\frac{x}{x - 3} - \frac{6}{x(3 - x)} = 1$$
$$\frac{x}{(x - 3)} + \frac{6}{x(x - 3)} = 1$$
$$\frac{x(x) + 6}{x(x - 3)} = \frac{x(x - 3)}{x(x - 3)}$$
$$x^2 + 6 = x^2 - 3x$$
$$3x = -6$$
$$\therefore x = -2$$

2.
$$\frac{3}{x} + \frac{3}{x^2 - x} = \frac{1}{x^2 - 1}$$

Restrictions: $x \neq 0, x \neq 1, x \neq -1$

$$\frac{3}{x} + \frac{3}{x(x - 1)} = \frac{1}{(x - 1)(x + 1)}$$
$$\frac{3(x - 1)(x + 1) + 3(x + 1)}{x(x - 1)(x + 1)} = \frac{1x}{x(x - 1)(x + 1)}$$
$$3(x - 1)(x + 1) + 3(x + 1) = x$$
$$3(x^2 - 1) + 3(x + 1) = x$$
$$3x^2 - 3 + 3x + 3 = x$$
$$3x^2 + 2x = 0$$
$$x(3x + 2) = 0$$
$$x = 0 \text{ or } x = -\frac{2}{3}$$

But $x \neq 0, x \neq 1, x \neq -1$
$$\therefore x = -\frac{2}{3}$$

3.
$$\frac{1}{x + 3} + \frac{x}{x - 1} = \frac{x^2 + 7}{x^2 + 2x - 3}$$
$$\frac{1}{(x + 3)} + \frac{x}{(x - 1)} = \frac{x^2 + 7}{(x + 3)(x - 1)}$$

Restrictions: $x \neq -3, x \neq 1$

$$\frac{1(x - 1) + x(x + 3)}{(x + 3)(x - 1)} = \frac{x^2 + 7}{(x + 3)(x - 1)}$$
$$x - 1 + x^2 + 3x = x^2 + 7$$
$$4x = 8$$
$$\therefore x = 2$$

4.
$$\frac{x+2}{x^2-3x-4}=\frac{3}{x-4}-\frac{1}{2x+2}$$

$$\frac{x+2}{(x-4)(x+1)}=\frac{3}{(x-4)}-\frac{1}{2(x+1)}$$

Restrictions: $x \neq 4$, $x \neq -1$

$$\frac{(x+2)(2)}{2(x-4)(x+1)}=\frac{3(2)(x+1)-1(x-4)}{2(x-4)(x+1)}$$

$$2x+4=6x+6-1x+4$$

$$2x+4=5x+10$$

$$-3x=6$$

$$\therefore x=-2$$

5.
$$\frac{3+8}{x^2+x-2}-\frac{2x-3}{x^2+6x+8}=\frac{x-4}{x^2+3x-4}$$

$$\frac{3x+8}{(x+2)(x-1)}-\frac{2x-3}{(x+4)(x+2)}=\frac{x-4}{(x+4)(x-1)}$$

Restrictions: $x \neq -2$, $x \neq 1$, $x \neq -4$

$$\frac{(3x+8)(x+4)-(2x-3)(x-1)}{(x+2)(x-1)(x+4)}=\frac{(x-4)(x+2)}{(x+2)(x-1)(x+4)}$$

$$(3x+8)(x+4)-(2x-3)(x-1)$$
$$=(x-4)(x+2)$$

$$3x^2+20x+32-(2x^2-5x+3)=x^2-2x-8$$

$$3x^2+20x+32-2x^2+5x-3=x^2-2x-8$$

$$x^2+25x+29=x^2-2x-8$$

$$27x=-37$$

$$x=-\frac{37}{27}$$

Exercise 8 (page 68)

1.
$$\frac{x+y}{3-y}=\frac{1}{2} \qquad \qquad \dots ①$$

$$\frac{2(x+y)}{2(3-y)}=\frac{1(3-y)}{2(3-y)}$$

$$2x+2y=3-y$$

$$2x+3y=3 \qquad \qquad \dots ③$$

$$\frac{x-y}{3+2y}=4 \qquad \qquad \dots ②$$

$$\frac{x-y}{3+2y}=\frac{4(3+2y)}{3+2y}$$

$$x-y=12+8y$$

$$x-9y=12 \qquad \qquad \dots ④$$

Add ③ × 3 and ④:

$$③ \times 3: \quad 6x + 9y = 9$$
$$+ \quad ④: \quad \underline{x - 9y = 12}$$
$$7x - 0y = 21$$

$$x=3$$

Substitute 3 for x in ④:

$$3-9y=12$$

$$-9y=9$$

$$y=-1$$

$$\therefore x=3 \text{ and } y=-1$$

2.
$$\frac{2(x-y)}{3}+\frac{3(x+y)}{10}=\frac{10}{3} \qquad \dots ①$$

$$\frac{2(10)x-y)}{30}+\frac{3(3)(x+y)}{30}=\frac{10(10)}{30}$$

$$20x-20y+9x+9y=100$$

$$29x-11y=100 \qquad \qquad \dots ③$$

$$\frac{x-y}{4}+\frac{x+y}{5}=3 \qquad \qquad \dots ②$$

$$\frac{5(x-y)}{20}+\frac{4(x+y)}{20}=\frac{3(20)}{20}$$

$$5x-5y+4x+4y=60$$

$$9x-y=60$$

$$y=9x-60 \qquad \qquad \dots ④$$

Substitute $9x-60$ for y in ③:

$$29x-11(9x-60)=100$$

$$29x-99x+660=100$$

$$-70x=-560$$

$$\therefore x=8$$

Substitute 8 for x in ④:

$$y=9(8)-60$$

$$\therefore y=12$$

$$\therefore x=8 \text{ and } y=12$$

3.
$$3x-\frac{5-y}{2}=\frac{5x-2}{3} \qquad \qquad \dots ①$$

$$\frac{3x(6)-3(5-y)}{6}=\frac{2(5x-2)}{6}$$

$$18x-15+3y=10x-4$$

$$8x+3y=11 \qquad \qquad \dots ③$$

$$\frac{2x-3}{5}=y-\frac{6}{5} \qquad \qquad \dots ②$$

$$\frac{2x-3}{5}=\frac{5y-6}{5}$$

$$2x-3=5y-6$$

$$2x-5y=-3 \qquad \qquad \dots ④$$

$$③: \qquad \quad 8x + 3y = 11$$
$$+ \quad ④ \times -4: \quad \underline{-8x + 20y = 12}$$
$$23y = 23$$

$$y=1$$

Substitute 1 for y in ④:

$$2x+5(1)=-3$$

$$2x=2$$

$$\therefore x=1$$

$$\therefore x=1 \text{ and } y=1$$

4.
$$\frac{6}{x}-\frac{1}{y}=4 \qquad \qquad \dots ①$$

$$\frac{9}{x}+1=-\frac{2}{y} \qquad \qquad \dots ②$$

Let $\frac{1}{x} = t$ and $\frac{1}{y} = p$:

$6t - p = 4$ … ③

$9t + 1 = -2p$

$9t + 2p = -1$ … ④

$$
\begin{array}{rrrrr}
③ \times 2: & 12t & - & 2p & = & 8 \\
+ \quad ④: & 9t & + & 2p & = & -1 \\
\hline
& 21t & + & 0p & = & 7
\end{array}
$$

$t = \frac{7}{21} = \frac{1}{3}$

$\therefore x = 3$

Substitute $\frac{1}{3}$ for t in ③:

$6\left(\frac{1}{3}\right) - p = 4$

$\quad 2 - p = 4$

$\quad\quad\quad p = -2$

$\therefore y = -\frac{1}{2}$

$\therefore x = 3$ and $y = -\frac{1}{2}$

Exercise 9 (page 68)

1.1

$\frac{x-3}{b} = 5 - \frac{2x-1}{2c}$

$\frac{2c(x-3)}{2bc} = \frac{5(2bc) - b(2x-1)}{2bc}$

$2c(x-3) = 5(2bc) - b(2x-1)$

$2cx - 6c = 10bc - 2bx + b$

$2cx + 2bx = 10bc + b + 6c$

$x(2c + 2b) = 10bc + b + 6c$

$\therefore x = \frac{10bc + b + 6c}{2c + 2b}$

1.2

$\frac{9ax}{b} - \frac{4bx}{a} = 3a - 2b$

$\frac{9a^2 - 4b^2 x}{ab} = \frac{3a^2 b - 2ab^2}{ab}$

$9a^2 - 4b^2 x = 3a^2 b - 2ab^2$

$x(9a^2 - 4b^2) = 3a^2 b - 2ab^2$

$x(3a - 2b)(3a + 2b) = ab(3a - 2b)$

$x(3a + 2b) = ab$

$\therefore x = \frac{ab}{(3a + 2b)}$

1.3

$(x - a)(x - b) = (x + a)(x - 3b)$

$x^2 - ax - bx + ab = x^2 + ax - 3ab - 3bx$

$-2ax + 2bx = -4ab$

$-2x(a - b) = -4ab$

$x = \frac{-4ab}{-2(a - b)}$

$\therefore x = \frac{2ab}{(a - b)}$

1.4

$(x - a)^2 - (x - b)^2 = a - b$

$x^2 - 2ax + a^2 - (x^2 - 2bx + b^2) = a - b$

$x^2 - 2ax + a^2 - x^2 + 2bx - b^2 = a - b$

$-2ax + 2bx = a - b - a^2 + b^2$

$-2x(a - b) = a - b - (a^2 - b^2)$

$-2x(a - b) = (a - b) - (a - b)(a + b)$

$-2x(a - b) = (a - b)(1 - a - b)$

$-2x = (1 - a - b)$

$\therefore x = \frac{(1 - a - b)}{2} = \frac{a + b - 1}{2}$

1.5

$\frac{1 - ax}{1 + bx} = \frac{b}{a}$

$a(1 - ax) = b(1 + bx)$

$a - a^2 x = b + b^2 x$

$-a^2 x - b^2 x = b - a$

$-x(a^2 + b^2) = (b - a)$

$-x = \frac{(b - a)}{(a^2 + b^2)}$

$\therefore x = \frac{(a - b)}{(a^2 + b^2)}$

2.

$t = \frac{m}{2}\sqrt{\frac{f}{g}}$

$\frac{m}{2}\sqrt{\frac{f}{g}} = t$

$\sqrt{\frac{f}{g}} = \frac{2t}{m}$

$\frac{f}{g} = \left(\frac{2t}{m}\right)^2$

$f = \left(\frac{2t}{m}\right)^2 \times g$

$\therefore f = \frac{4gt^2}{m^2}$

Exercise 10 (page 71)

1. Let the number be x:

$\frac{1}{2}x + \frac{1}{3}x + \frac{1}{4}x - 10 = x$

$\times 12:\ 6x + 4x + 3x - 120 = 12x$

$\therefore x = 120$

$\therefore$ The number is 120.

2. Let the number be x:

$2x - \frac{1}{2}x = 30$

$\times 2:\ 4x - x = 60$

$3x = 60$

$\therefore x = 20$

$\therefore$ The number is 20.

3. Let the smaller number be x:

$\therefore$ larger number $= 50 - x$

$\therefore 50 - x - x = 10$

$\quad 50 - 2x = 10$

$\quad\quad -2x = -40$

$\therefore x = 20$

$\therefore$ The numbers are 20 and 30.

4. Let the smaller number be x:

$\therefore$ numbers are: x; $x + 2$; $x + 4$; $x + 6$

$\therefore x + x + 2 + x + 4 + x + 6 = 68$

$\quad\quad\quad\quad\quad\quad\quad 4x = 56$

$\therefore x = 14$

$\therefore$ The numbers are 14, 16, 18 and 20.

5. Let the daughter be x years old:

$\therefore$ the mother is $x + 30$ years old

$\therefore 2(x - 10) = x + 30 - 10$

$\quad 2x - 20 = x + 20$

$\therefore x = 40$

$\therefore$ The mother is 70 years old.

6. Let the daughter be x years old:

$\therefore$ the mother is $3x$ years old

$\therefore 3x + 12 = 2(x + 12)$

$\quad 3x + 12 = 2x + 24$

$\therefore x = 12$

$\therefore$ The mother is 36 years old.

7. Say it happens in x years' time:

$\therefore 35 + x = 2(7 + x)$

$\quad 35 + x = 14 + 2x$

$\therefore 21 = x$

$\therefore$ In 21 years' time the man will be twice the age of his son.

8. Let x be the breadth:

$\therefore$ length $= x + 5$

$x + x + 5 = 15$

$\quad\quad 2x = 10$

$\therefore x = 5$

$\therefore$ Length is 10 and breadth is 5.

9. Let x be the number of R4 tickets:

$\therefore$ number of R5 tickets $= 2x$

and number of R3 tickets $= 22 - 3x$

$4x + 5(2x) + 3(22 - 3x) = 96$

$\quad 4x + 10x + 66 - 9x = 96$

$\quad\quad\quad\quad\quad\quad 5x = 30$

$\therefore x = 6$

$\therefore$ She sold 12 R5 tickets.

10. Let x be the number of 5c coins:

$\therefore$ number of 10c coins $= 2x$

and number of 20c coins $= 4x$

$5x + 10(2x) + 4x(2) = 315$

$\quad\quad 5x + 20x + 80x = 315$

$\therefore x = 3$

$\therefore$ three 5c coins, six 10c coins, twelve 20c coins.

11. $6{,}5x + 8{,}2y = 261{,}2$ $\quad\quad\quad$ ... ①

$7{,}8(x + y) = 280{,}8$ $\quad$ <small>$261{,}20 + 19{,}60 = 280{,}80$</small>

$\quad\quad\therefore 7{,}8x + 7{,}8y = 280{,}8$ $\quad$ ... ②

② $\div 7{,}8$: $\quad\quad x + y = 36$ $\quad\quad$ ... ③

③ $\times 6{,}5$: $\quad 6{,}5x + 6{,}5y = 234$ $\quad\quad$ ... ④

① $-$ ④: $\quad\quad\quad 1{,}7y = 27{,}2$

$\therefore y = 16$

Substitute 16 for y in ③:

$x + 16 = 36$

$\therefore x = 20$

$\therefore$ 20 kg of the first brand and 16 kg of the second brand.

12. Let the digits be x and y:

$\therefore$ number: $\quad 10x + y + 2x = 33$

$\quad\quad\quad\quad\quad 12x + y = 33$ $\quad\quad$ ... ①

$\quad\quad (10y + x) - 63 = (10x + y)$

$\quad\quad\quad\quad\quad -9x + 9y = 63$

$\div 9$: $\quad\quad\quad\quad -x + y = 7$ $\quad\quad$ ... ②

① $-$ ②: $\quad\quad\quad\quad 13x = 26$

$\therefore x = 2$

Substitute 2 for x in ②:

$-2 + y = 7$

$\therefore y = 9$

$\therefore$ The number is 29.

13. $\quad\quad\quad (l + 20)(b - 10) = lb$

$\quad\quad \therefore lb + 20b - 10l - 200 = lb$

$\div 10$: $\quad\quad\quad\quad 2b - l = 20$ $\quad\quad$ ... ①

$\quad\quad\quad (l + 10)(b + 10) = lb + 800$

$\quad\quad\quad\quad\quad\quad$ <small>$8\ cm^2 = 800\ mm^2$</small>

$\quad lb + 10b + 10l + 100 = lb + 800$

$\div 10$: $\quad\quad\quad\quad b + l = 70$ $\quad\quad$ ... ②

① $+$ ②: $\quad\quad\quad\quad 3b = 90$

$\therefore b = 30$

Substitute 30 for b in ①:

$60 - l = 20$

$\therefore l = 40$

$\therefore$ Rectangle: length $= 40$ mm and breadth $= 30$ mm.

14. Let the smaller number be x:

∴ numbers are: x, $x + 2$, $x + 4$ and $x + 6$

$x + x + 2 + x + 4 + x + 6 = 68$

$\qquad\qquad\qquad 4x = 56$

∴ $x = 14$

∴ The numbers are 14, 16, 18 and 20.

15. Let B be x years old:

∴ A is $2x$ years old

$3(x - 8) = 2x - 8$

$3x - 24 = 2x - 8$

∴ $x = 16$

∴ A is 32 years old and B is 16 years old.

16. Let Thabo be x years old:

∴ Sipho is $72 - x$ years old

∴ $6(x - 21) = 72 - x + 12$

$\qquad 6x - 126 = 72 - x + 12$

$\qquad\qquad\qquad 7x = 210$

∴ $x = 30$

16.1 ∴ Thabo is 30 years old and Sipho is 42 years old.

16.2 12 years

17. Let John be x years old:

∴ Bill is $x + 10$ years old

∴ $2(x + 18) = 3(x + 8)$

$\qquad 2x + 36 = 3x + 24$

∴ $x = 12$

∴ John is 12 years old and Bill is 22 years old.

18. Let x be the number of sheep:

∴ number of cattle $= 100 - x$

∴ $250x + 730(100 - x) = 53\,800$

$250x + 73\,000 - 730x = 53\,800$

$\qquad\qquad\qquad -480x = -19\,200$

∴ $x = 40$

∴ He bought 40 sheep and 60 cattle.

19. Let x be the number of bull's-eyes:

∴ number of misses $= 25 - x$

∴ $5x - 2\frac{1}{2} \times (25 - x) = 5$

$\times 2$: $10x - 125 + 5x = 10$

$\qquad\qquad 15x = 135$

∴ $x = 9$

∴ He hit nine bull's-eyes.

20.

	Upstream	Downstream
D	x	x
S	3 km/h	7 km/h
T	$\frac{x}{3}$	$\frac{x}{7}$

In the table, to work out the speed, when he rows upstream against the flow, his speed will be slower by 2 km/h: $5 - 2 = 3$ km/h. Downstream he rows with the flow, so his speed will be faster by 2 km/h: $5 + 2 = 7$ km/h.

Let the distance to the bridge be x km:

$\frac{x}{3} + \frac{x}{7} = 4$

$\times 21$: $7x + 3x = 84$

$10x = 84$

∴ $x = 8{,}4$ km

∴ Distance to bridge $= 8{,}4$ km.

21.

	Cyclist 1	Cyclist 2
D	150	150
S	x km/h	$(x + 5)$ km/h
T	$\frac{150}{x}$	$\frac{150}{x + 5}$

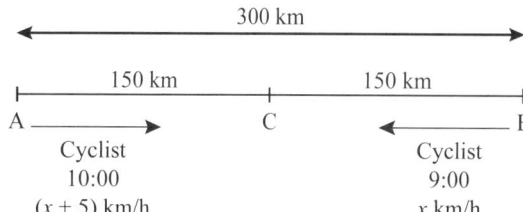

21.1 Cyclist 1 takes $\frac{150}{x}$ hours

Cyclist 2 takes $\frac{150}{x + 5}$ hours

21.2 $\frac{150}{x} = \frac{150}{x + 5} + 1$

$\times x(x + 5)$: $\qquad 150\,(x + 5) = 150x + x(x + 5)$

$\qquad\qquad\qquad 150x + 750 = 150x + x^2 + 5x$

$\qquad\qquad x^2 + 5x - 750 = 0$

$\qquad\qquad (x + 30)(x - 25) = 0$

∴ $x = -30$ or $x = 25$

$x = -30$ must be discarded

∴ $x = 25$

∴ Cyclist 1 cycles at 21 km/h.

Time of cyclist $1 = \frac{150}{25} = 6$ hours.

∴ They arrive at C at 15:00.

Test A (page 73)

1.1
$$\frac{2x-3}{3} - 3x = \frac{2x}{6}$$
$$\frac{2(2x-3) - 6(3x)}{6} = \frac{2x}{6}$$
$$4x - 6 - 18x = 2x$$
$$4x - 18x - 2x = 6$$
$$16x = 6$$
$$\therefore x = \frac{6}{16} = \frac{3}{8}$$

1.2
$$4(3x+1) - 15 \geq 5(3x-1)$$
$$12x + 4 - 15 \geq 15x - 5$$
$$-6 \geq 3x$$
$$-2 \geq x$$
$$\therefore x \leq -2$$

1.3
$$tx - b = d - qx$$
$$tx + qx = d + b$$
$$x(t+q) = d + b$$
$$\therefore x = \frac{d+b}{t+q}$$

1.4
$$2x^2 - x = 1$$
$$2x^2 - x - 1 = 0$$
$$(2x+1)(x-1) = 0$$
$$2x + 1 = 0 \quad \text{or} \quad x - 1 = 0$$
$$\therefore x = -\tfrac{1}{2} \quad \text{or} \quad x = 1$$

1.5
$$\frac{x-2}{x-1} = \frac{2x-1}{x+7}$$
$$\frac{(x-2)(x+7)}{(x-1)(x+7)} = \frac{(2x-1)(x-1)}{(x-1)(x+7)}$$
Restrictions: $x \neq 1$, $x \neq -7$
$$(x-2)(x+7) = (2x-1)(x-1)$$
$$x^2 + 5x - 14 = 2x^2 - 3x + 1$$
$$-x^2 + 8x - 15 = 0$$
$$x^2 - 8x + 15 = 0$$
$$(x-5)(x-3) = 0$$
$$\therefore x = 5 \text{ or } x = 3$$

2.
$$x + y = -3 \qquad \ldots ①$$
$$4x + 3y = -8 \qquad \ldots ②$$
Using elimination:
$$① \times -4: \quad -4x - 4y = 12 \qquad \ldots ③$$
$$② + ③: \quad -4x - 4y = 12$$
$$+ \qquad \quad 4x + 3y = -8$$
$$\overline{\qquad\qquad -y = 4}$$
$$\therefore y = -4$$

Substitute -4 for y in ①:
$$x - 4 = -3$$
$$\therefore x = 1$$
$$\therefore x = 1 \text{ and } y = -4$$

3.
$$V = \pi r^2 h$$
$$\pi r^2 h = V$$
$$r^2 = \frac{V}{\pi h}$$
$$r = \sqrt{\frac{V}{\pi h}}$$
Technically, $r = \pm\sqrt{\frac{V}{\pi h}}$, but because r is the radius it can never have a negative value, so $r =$ **only** the positive root.

4. $(-2,5; 5]$

5.1
$$2x + 7 \geq 4x - 10$$
$$-2x \geq -17$$
$$x \leq \frac{17}{2}$$

5.2
$$3 < -2x + 1 \leq 6$$
$$2 < -2x \leq 5$$
$$-1 > x \geq -\tfrac{5}{2}$$
$$-\tfrac{5}{2} \leq x < -1$$

6.1
$$(3x-2)(x-2) = 0$$
$$\therefore x = \tfrac{2}{3} \text{ or } x = 2$$

6.2
$$(3x-2)(x-2) = 20$$
$$3x^2 - 8x + 4 = 20$$
$$3x^2 - 8x - 16 = 0$$
$$(3x+4)(x-4) = 0$$
$$\therefore x = -\tfrac{4}{3} \text{ or } x = 4$$

Test B (page 73)

1.1
$$1\tfrac{1}{2} - \tfrac{5x}{3} > \tfrac{1}{4}(5 - 3x) + \tfrac{1}{3}(3 - 5x)$$
$$\frac{3}{2} - \frac{5x}{3} > \frac{(5-3x)}{4} + \frac{(3-5x)}{3}$$
$$\frac{18 - 20x}{12} > \frac{3(5-3x) + 4(3-5x)}{12}$$
$$18 - 20x > 15 - 9x + 12 - 20x$$
$$9x > 9$$
$$\therefore x > 1$$

1.2
$$\frac{1}{6x^2 + x - 15} - \frac{1}{9 - 4x^2} = \frac{1}{6x^2 + 19x + 15}$$

$$\frac{1}{(2x - 3)(3x + 5)} - \frac{1}{(3 - 2x)(3 + 2x)} = \frac{1}{(2x + 3)(3x + 5)}$$

$$\frac{1}{(2x - 3)(3x + 5)} + \frac{1}{(2x - 3)(3 + 2x)} = \frac{1}{(2x + 3)(3x + 5)}$$

$$\frac{1(3 + 2x) + 1(3x + 5)}{(2x - 3)(3x + 5)(3 + 2x)} = \frac{1(2x - 3)}{(2x + 3)(3x + 5)(2x - 3)}$$

$$3 + 2x + 3x + 5 = 2x - 3$$

$$5x + 8 = 2x - 3$$

$$3x = -11$$

$$\therefore x = -\frac{11}{3}$$

2.
$$4x^2 + 22xy = 12y^2$$

$$4x^2 + 22xy - 12y^2 = 0$$

$$2(2x^2 + 11xy - 6y^2) = 0$$

$$2x^2 + 11xy - 6y^2 = 0$$

$$(2x - y)(x + 6y) = 0$$

$$x = \frac{y}{2} \quad \text{or} \quad x = -6y$$

$$\therefore \frac{x}{y} = \frac{1}{2} \quad \text{or} \quad \frac{x}{y} = -6$$

3.
$$\frac{3(x - y)}{4} = \frac{x + y}{2} - 5$$

$$\frac{3(x - y)}{4} = \frac{2(x + y) - 5(4)}{2}$$

$$3x - 3y = 2x + 2y - 20$$

$$x - 5y = -20 \qquad \text{... ①}$$

$$\frac{x + y}{2} = 2(x - y)$$

$$\frac{x + y}{2} = \frac{2(2)(x - y)}{2}$$

$$x + y = 4x - 4y$$

$$-3x + 5y = 0 \qquad \text{... ②}$$

$$\begin{array}{rl} ① - ②: & x - 5y = -20 \\ & \underline{-\ 3x + 5y = 0} \\ & -2x = -20 \end{array}$$

$$\therefore x = 10$$

Substitute 10 for x in ①:

$$10 - 5y = -20$$

$$-5y = -30$$

$$\therefore y = 6$$

$$\therefore x = 10 \text{ and } y = 6$$

4.
$$\frac{mx + n}{n} - \frac{nx + m}{m} = m + n$$

$$\frac{m(mx + n)}{nm} - \frac{n(nx + m)}{mn} = \frac{mn(m + n)}{mn}$$

$$m(mx + n) - n(nx + m) = mn(m + n)$$

$$m^2x + mn - n^2x - nm = m^2n + mn^2$$

$$m^2x - n^2x = m^2n + mn^2$$

$$x(m^2 - n^2) = mn(m + n)$$

$$x(m - n)(m + n) = mn(m + n)$$

$$x(m - n) = mn$$

$$\therefore x = \frac{mn}{m - n}$$

5.
$$-4x + 12 = 4x^2 - 8x - 3$$

$$4x^2 - 4x - 15 = 0$$

$$(2x + 3)(2x - 5) = 0$$

$$2x = -3 \quad \text{or} \quad 2x = 5$$

$$\therefore x = -\frac{3}{2} \quad \text{or} \quad x = \frac{5}{2}$$

Substitute for each value of x in $y = -4x + 12$:

$$y = -4\left(-\frac{3}{2}\right) + 12 \quad \text{or} \quad y = -4\left(\frac{5}{2}\right) + 12$$

$$y = 18 \qquad\qquad\qquad y = 2$$

$$\therefore x = -\frac{3}{2} \text{ and } y = 18 \text{ or } x = \frac{5}{2} \text{ and } y = 2$$

6. Let the numbers be x and $x + 1$:

$$x^2 + (x + 1)^2 = 85$$

$$x^2 + x^2 + 2x + 1 = 85$$

$$2x^2 + 2x - 84 = 0$$

$$\div 2: \quad x^2 + x - 42 = 0$$

$$(x + 7)(x - 6) = 0$$

$$\therefore x = -7 \text{ or } x = 6$$

Only natural numbers, ∴ numbers are 6 and 7.

7.

	Motor cycle	Motor car
D	³ 320	³ 320
S	¹ x	² $2x$
T	⁴ $\frac{320}{x}$	⁵ $\frac{320}{2x}$

Let the speed of the motor cycle be x km/h:

∴ Speed of the motor car = $2x$ km/h

Car's time + $2\frac{1}{2}$ h = motor cycle's time

$$\therefore \frac{320}{2x} + \frac{5}{2} = \frac{320}{x}$$

$$\times 2x: \quad 320 + 5x = 640$$

$$5x = 320$$

$$\therefore x = 64$$

Motor cycle's time = $\frac{320}{64}$ = 5 hours

∴ Speed of motor cycle = 64 km/h and speed of motor car = 128 km/h

∴ They meet at 13:00.

Test C (page 74)

1.1
$$3(2 - x) - 2x = 6 - 5x$$

$$6 - 3x - 2x = 6 - 5x$$

$$6 - 5x = 6 - 5x$$

$$0x = 0$$

$$\therefore x \in \mathbb{R}, \text{ because } x \text{ can be any real number.}$$

1.2
$$(2x - 7)(y + 2) = 0$$

$$x = \frac{7}{2}$$

1.3

$$k - \frac{m}{x} = t$$

$$\frac{kx - m}{x} = \frac{tx}{x}$$

$$kx - m = tx$$

$$kx - tx = m$$

$$x(k - t) = m$$

$$\therefore x = \frac{m}{k - t}$$

2.1

$$-1 \leq 2 - 3x < 8$$

$-1 \leq 2 - 3x$	and	$2 - 3x < 8$
$3x \leq 2 + 1$		$2 - 8 < 3x$
$3x \leq 3$		$-6 < 3x$
$x \leq 1$		$-2 < x$

$$\therefore -2 < x \leq 1$$

2.2 Interval notation: $(-2; 1]$

2.3

3.1 Let the number of multiple choice questions be x, and the number of short questions be y, then:

$$x + y = 22 \qquad \ldots ①$$
$$2x + 3y = 50 \qquad \ldots ②$$

3.2

$$x + y = 22 \qquad \ldots ①$$
$$\therefore y = 22 - x$$

Substitute $22 - x$ for y in ②:

$$2x + 3(22 - x) = 50$$
$$2x + 66 - 3x = 50$$
$$-x = 50 - 66$$
$$\therefore x = 16$$

$\therefore$ There were 16 multiple choice questions in this test.

4.1

$$\frac{3 - x}{1 - x^2} + \frac{2x + 4}{(x + 1)} = \frac{2x}{x - 1}$$

$$\frac{3 - x}{(1 - x)(1 + x)} + \frac{2x + 4}{(x + 1)} = \frac{2x}{(x - 1)}$$

Restrictions: $x \neq 1, \ x \neq -1$

$$\frac{-3 + x}{(x - 1)(1 + x)} + \frac{2x + 4}{(x + 1)} = \frac{2x}{(x - 1)}$$

$$\frac{-3 + x + (2x + 4)(x - 1)}{(x - 1)(1 + x)} = \frac{2x(1 + x)}{(x - 1)(1 + x)}$$

$$-3 + x + (2x + 4)(x - 1) = 2x(1 + x)$$

$$-3 + x + 2x^2 + 2x - 4 = 2x + 2x^2$$

$$\therefore x = 7$$

4.2

$$6x^2 = 13x + 5$$
$$6x^2 - 13x - 5 = 0$$
$$(2x - 5)(3x + 1) = 0$$
$$\therefore x = -\frac{5}{2} \text{ or } x = -\frac{1}{3}$$

4.3

$$px = p^2 - t^2 + tx$$
$$px - tx = p^2 - t^2$$
$$x(p - t) = (p - t)(p + t)$$
$$\therefore x = p + t$$

5.

$$x + 2y = 1$$
$$x = 1 - 2y \qquad \ldots ①$$
$$\frac{x}{3} + \frac{y}{2} = 1 \qquad \ldots ②$$

Substitute $1 - 2y$ for x in ②:

$$\frac{1 - 2y}{3} + \frac{y}{2} = 1$$

$$\frac{2(1 - 2y + 3(y))}{6} = \frac{1(6)}{6}$$

$$2 - 4y + 3y = 6$$

$$-y = 4$$

$$\therefore y = -4$$

Substitute -4 for y in ②:

$$x = 1 - 2(-4)$$
$$\therefore x = 9$$
$$\therefore x = 9 \text{ and } y = -4$$

6. Let the distance to the station be x:

	Distance	Speed	Time
Walk	x	4 km/h	$\frac{x}{4}$
Jog	x	8 km/h	$\frac{x}{8}$

$\therefore$ time walked $-7\frac{1}{2}$ minutes = time jogged

$$\therefore \frac{x}{4} - \frac{7\frac{1}{2}}{60} = \frac{x}{8} \qquad \text{LCD} = 120$$

$$30x - 15 = 15x$$

$$15x = 15$$

$$\therefore x = 1$$

$\therefore$ Distance to station = 1 km.

Topic 6   Trigonometric ratios and right-angled triangles (page 75)

Exercise 1   (page 78)

1.1 $\sin P = \frac{ST}{PT} = \frac{8}{17}$

$\sec P = \frac{PT}{PS} = \frac{17}{15}$

$\tan P = \frac{ST}{PS} = \frac{8}{15}$

1.2 $\cos T = \frac{ST}{PT} = \frac{8}{17}$

$\cot T = \frac{ST}{PS} = \frac{8}{15}$

$\operatorname{cosec} T = \frac{PT}{PS} = \frac{17}{15}$

2.1   In $\triangle BAD$:

$\sin B = \frac{DA}{DB}$, $\cot B = \frac{AB}{DA}$ and $\sec B = \frac{AB}{DB}$

In $\triangle BAC$:

$\sin B = \frac{CA}{CB}$, $\cot B = \frac{AB}{AC}$ and $\sec B = \frac{DB}{AB}$

2.2 $\tan B = \frac{b}{d}$, $\cos D = \frac{b}{a}$ and $\operatorname{cosec} C = \frac{a}{b}$

Exercise 2   (page 81)

1. $x = 5$, $y = 12$
Using Pythagoras: $r = \sqrt{5^2 + 12^2} = 13$

1.1 $\sin \beta = \frac{12}{13}$

1.2 $\cot \beta = \frac{5}{12}$

1.3 $(\sin \beta + \cos \beta)(\tan \beta + \cot \beta)$

$= \left(\frac{12}{13} + \frac{5}{13}\right)\left(\frac{12}{5} + \frac{5}{12}\right)$

$= \left(\frac{17}{13}\right)\left(\frac{169}{60}\right) = \frac{221}{60}$

2.   If $\sin \theta = \frac{3}{5}$ and $\theta \in [0°; 90°]$

$r = 5$, $y = 3$
Using Pythagoras: $x = \sqrt{5^2 + 3^2} = 4$

2.1 $\cos \theta = \frac{4}{5}$

2.2 $\tan \theta = \frac{3}{4}$

2.3 $\sec^2 \theta - 1$

$= \left(\frac{5}{4}\right)^2 - 1 = \frac{9}{16}$

3. $3 \tan \theta - 4 = 0$ and $0° \le \theta \le 90°$

$\therefore \tan \theta = \frac{4}{3}$

$x = 3$, $y = 4$

Using Pythagoras: $r = \sqrt{3^2 + 4^2} = 5$

3.1 $\sin \theta = \frac{4}{5}$

3.2 $\cos \theta = \frac{3}{5}$

3.3 $\sin^2 \theta + \cos^2 \theta$

$= \left(\frac{4}{5}\right)^2 + \left(\frac{3}{5}\right)^2 = 1$

4.1 $x = 7$, $y = 9$

Using Pythagoras: $r = \sqrt{7^2 + 9^2} = \sqrt{130}$

$\therefore OD = \sqrt{130}$

4.2 $\sin \theta \operatorname{cosec} \theta$

$= \left(\frac{9}{\sqrt{130}}\right) \cdot \left(\frac{\sqrt{130}}{9}\right) = 1$

4.3 $\cos \theta \sec \theta$

$= \left(\frac{7}{\sqrt{130}}\right) \cdot \left(\frac{\sqrt{130}}{7}\right) = 1$

4.4 $\sec^2 \theta - \tan^2 \theta$

$= \left(\frac{\sqrt{130}}{7}\right)^2 - \left(\frac{9}{7}\right)^2 = 1$

5. $\sec \theta = \frac{5}{3}$ and $0° \le \theta \le 90°$

$x = 3$, $r = 5$

Using Pythagoras: $y = \sqrt{5^2 - 3^2} = 4$

$\tan^2 \theta + 1 = \left(\frac{4}{3}\right)^2 + 1 = \frac{25}{9}$

6.   If $8 \cos \theta - 6 = 0$ and $\theta \in [0°; 90°]$

$\cos \theta = \frac{6}{8} = \frac{3}{4}$

$\therefore x = 3$ and $r = 4$

Using Pythagoras: $y = \sqrt{4^2 - 3^2} = \sqrt{7}$

6.1 $\operatorname{cosec} \theta = \frac{4}{\sqrt{7}}$

6.2 $1 + \cot^2 \theta = 1 + \left(\frac{3}{\sqrt{7}}\right)^2 = \frac{16}{7}$

7. $\therefore x = 24$ and $y = 8$

Using Pythagoras:
$r = \sqrt{24^2 + 8^2} = \sqrt{640} = 8\sqrt{10}$

7.1 $\sin \theta \cot \theta$

$\left(\frac{8}{8\sqrt{10}}\right) \cdot \left(\frac{24}{8}\right) = \frac{3\sqrt{10}}{10}$

7.2 $\cos \theta \tan \theta$

$= \left(\frac{24}{8\sqrt{10}}\right) \cdot \left(\frac{8}{24}\right) = \frac{\sqrt{10}}{10}$

7.3 $\operatorname{cosec}^2 \theta - \cot^2 \theta$

$= \left(\frac{8\sqrt{10}}{8}\right)^2 - \left(\frac{24}{8}\right)^2$

$= 10 - 9 = 1$

7.4 $\cos^2 \theta = \left(\frac{24}{8\sqrt{10}}\right)^2 = \frac{9}{10}$

$1 - \sin^2 \theta = 1 - \left(\frac{8}{8\sqrt{10}}\right)^2 = 1 - \frac{1}{10} = \frac{9}{10}$

$\therefore \cos^2 \theta = 1 - \sin^2 \theta$

8. $2 \operatorname{cosec} \theta = 10$

$\operatorname{cosec} \theta = 5$

$\therefore r = 5$ and $y = 1$

Using Pythagoras: $x = \sqrt{5^2 - 1^2} = \sqrt{24} = 2\sqrt{6}$

8.1 $\sin \theta = \frac{1}{5}$

8.2 $\cos \theta = \frac{2\sqrt{6}}{5}$

8.3 $2 \sin \theta \cos \theta = 2 \left(\frac{1}{5}\right)\left(\frac{2\sqrt{6}}{5}\right) = \frac{4\sqrt{6}}{25}$

9. $\frac{\cot \theta}{4} = \frac{3}{8}$

$\cot \theta = 4 \times \frac{3}{8} = \frac{12}{8} = \frac{3}{2}$

$\therefore x = 3$ and $y = 2$

Using Pythagoras: $r = \sqrt{3^2 + 2^2} = \sqrt{13}$

9.1 $\tan \theta = \frac{2}{3}$

9.2 $\sec \theta = \frac{\sqrt{13}}{3}$

9.3 $\sec^2 \theta - 1 = \left(\frac{\sqrt{13}}{3}\right)^2 - 1 = \frac{13}{9} - 1 = \frac{4}{9}$

$\tan^2 \theta = \left(\frac{2}{3}\right)^2 = \frac{4}{9}$

$\therefore \sec^2 \theta - 1 = \tan^2 \theta$

Exercise 3   (page 83)

1.1 $\sin^2 30° - \cos^2 30° = \left(\frac{1}{2}\right)^2 - \left(\frac{\sqrt{3}}{2}\right)^2 = -\frac{1}{2}$

1.2 $\sin 90° - (\tan 30° + \operatorname{cosec} 60°)^2$

$= 1 - \left(\frac{1}{\sqrt{3}} + \frac{2}{\sqrt{3}}\right)^2$

$= 1 - \left(\frac{3}{\sqrt{3}}\right)^2$

$= 1 - \frac{9}{3} = 1 - 3 = -2$

1.3 $\tan 0° + \tan 30° + \cot 60° - \operatorname{cosec} 60°$

$= 0 + \frac{1}{\sqrt{3}} + \frac{1}{\sqrt{3}} - \frac{2}{\sqrt{3}}$

$= \frac{2}{\sqrt{3}} - \frac{2}{\sqrt{3}} = 0$

1.4 $\frac{\sin 45°}{\cos 45°} - 5 \operatorname{cosec} 90° + 3 \tan^2 30°$

$= \frac{\frac{1}{\sqrt{2}}}{\frac{1}{\sqrt{2}}} - 5(1) + 3\left(\frac{1}{\sqrt{3}}\right)^2$

$= 1 - 5 + 3\left(\frac{1}{3}\right)$

$= 1 - 5 + 1 = -3$

1.5 $\sin^2 30° + \cos^2 30°$

$= \left(\frac{1}{2}\right)^2 + \left(\frac{\sqrt{3}}{2}\right)^2$

$= \frac{1}{4} + \frac{3}{4} = 1$

1.6 $\frac{\sin 30°}{\cos 60°} - 2 \sin 90° (\tan 45° + \sin 30°)$

$= \frac{\frac{1}{2}}{\frac{1}{2}} - 2(1)\left(1 + \frac{1}{2}\right)$

$= 1 - 2\left(\frac{3}{2}\right) = -2$

1.7 $\frac{2 \sin 45° \cos 30° \tan 60° \operatorname{cosec} 90°}{\sin 30° \cos 0° \cos 45°}$

$= \frac{2\left(\frac{1}{\sqrt{2}}\right)\left(\frac{\sqrt{3}}{2}\right)\left(\frac{\sqrt{3}}{1}\right)(1)}{\left(\frac{1}{2}\right)(1)\left(\frac{1}{\sqrt{2}}\right)}$

$= \frac{3}{\sqrt{2}} \div \frac{1}{2\sqrt{2}}$

$= \frac{3}{\sqrt{2}} \times \frac{2\sqrt{2}}{1} = 6$

2.1 $\sin \theta = \frac{1}{2}$ and $\theta = \in [0°; 90°]$

$\theta = 30°$

2.2 $2 \cos \theta = 1$ and $0° \leq \theta$ $90°$

$\cos \theta = \frac{1}{2}$

$\theta = 60°$

2.3 $2 \tan 3\theta - 2 = 0$ and $\theta = \in [0°; 90°]$

$\tan 3\theta = \frac{2}{2} = 1$

$3\theta = 45°$

$\theta = 15°$

2.4. $2 \cos (\theta + 12°) - \sqrt{3} = 0$ and $0° \leq \theta \leq 90°$

$\cos (\theta + 12°) = \frac{\sqrt{3}}{2}$

$\theta + 12° = 30°$

$\theta = 18°$

2.5 $10 \operatorname{cosec} (\theta - 21°) - 20 = 0$ and $0° \leq \theta \leq 90°$

$10 \operatorname{cosec} (\theta - 21°) = 20$

$\operatorname{cosec} (\theta - 21°) = 2$

$\theta - 21° = 30°$

$\theta = 51°$

Exercise 4   (page 84)

1. $\sin S = \frac{TR}{TS}$, $\sin S = \frac{PT}{PS}$, $\sin S = \frac{QP}{QS}$

2.1 $\cos 60° \cot 45° - \cos^2 45°$

$= \cos 60° \cdot \frac{1}{\tan 45°} - \cos^2 45°$

$= \left(\frac{1}{2}\right)(1) - \left(\frac{1}{\sqrt{2}}\right)^2$

$= \left(\frac{1}{2}\right) - \left(\frac{1}{2}\right)$

$= 0$

2.2 $\sin (60° + 30°)$

$= \sin 90°$

$= 1$

2.3 $\frac{\cos 45° \operatorname{cosec} 45° \tan^2 60°}{\sin 30°}$

$= \frac{\left(\frac{1}{\sqrt{2}}\right)\left(\frac{\sqrt{2}}{1}\right)\left(\frac{\sqrt{3}}{1}\right)^2}{\frac{1}{2}}$

$= \frac{3}{\frac{1}{2}} = 6$

3.1 $x \cos 30° = \sin 30°$

$$x\left(\tfrac{\sqrt{3}}{2}\right) = \tfrac{1}{2}$$

$$x = \tfrac{1}{2}\left(\tfrac{2}{\sqrt{3}}\right) = \tfrac{1}{\sqrt{3}}$$

3.2 $x + \operatorname{cosec}^2 45° = \cos^2 45° - \cos 90°$

$$x + \left(\tfrac{\sqrt{2}}{1}\right)^2 = \left(\tfrac{1}{\sqrt{2}}\right)^2 - 0$$

$$x + 2 = \tfrac{1}{2}$$

$$x = \tfrac{1}{2} - 2 = -\tfrac{3}{2}$$

4.1 $4 \tan \theta - 4 = 0$

$$4 \tan \theta = 4$$

$$\tan \theta = 1$$

$$\theta = 45°$$

4.2 $2 \sin \theta = \sqrt{3}$

$$\sin \theta = \tfrac{\sqrt{3}}{2}$$

$$\theta = 60°$$

4.3 $\tan 3\theta - \sqrt{3} = 0$

$$\tan 3\theta = \sqrt{3}$$

$$3\theta = 60°$$

$$\theta = 20°$$

4.4 $\sqrt{3} \cot 3\theta - 1 = 0$

$$\sqrt{3} \cot 3\theta = 1$$

$$\cot 3\theta = \tfrac{1}{\sqrt{3}}$$

$$3\theta = 60°$$

$$\theta = 20°$$

4.5 $2 \cos (2\theta - 20°) = 1$

$$\cos (2\theta - 20°) = \tfrac{1}{2}$$

$$2\theta - 20° = 60°$$

$$2\theta = 80°$$

$$\theta = 40°$$

4.6 $\dfrac{\sec (\theta - 10°)}{2} = 1$

$$\sec (\theta - 10°) = 2$$

$$\theta - 10° = 60$$

$$\theta = 70°$$

5.

$x = 12,\ y = 5,\ r = 13$

$$\cos^2 \theta - 1 = \left(\tfrac{12}{13}\right)^2 - 1 = -\tfrac{25}{169}$$

$$-\sin^2 \theta = -\left(\tfrac{5}{13}\right)^2 = -\tfrac{25}{169}$$

$$\therefore \cos^2 \theta - 1 = -\sin^2 \theta$$

6.1 $OQ = \sqrt{6^2 + 3^2} = 3\sqrt{5}$

6.2 $\sin \theta = \tfrac{3}{3\sqrt{5}} = \tfrac{1}{\sqrt{5}}$

6.3 $\sqrt{5} \cos \theta = \sqrt{5} \times \tfrac{6}{3\sqrt{5}} = 2$

6.4 $\sec^2 \theta - \tan \theta = \left(\tfrac{3\sqrt{5}}{6}\right)^2 - \tfrac{3}{6}$

$$= \tfrac{5}{4} - \tfrac{3}{6} = \tfrac{3}{4}$$

7. $\sin \theta = \tfrac{5}{8}$ $\therefore y = 5$ and $r = 8$

Using Pythagoras: $x = \sqrt{8^2 - 5^2} = \sqrt{39}$

7.1 $\cos \theta = \tfrac{\sqrt{39}}{8}$

7.2 $\tan \theta = \tfrac{5}{\sqrt{39}}$

7.3 $\cot \theta = \tfrac{\sqrt{39}}{5}$

$$\tfrac{\cos \theta}{\sin \theta} = \tfrac{\sqrt{39}}{8} \div \tfrac{5}{8}$$

$$= \tfrac{\sqrt{39}}{8} \times \tfrac{8}{5} = \tfrac{\sqrt{39}}{5}$$

$$\therefore \cot \theta = \tfrac{\cos \theta}{\sin \theta}$$

8. $\sin \beta = \tfrac{12}{13}$ $\therefore y = 12$ and $r = 13$

Using Pythagoras: $x = \sqrt{13^2 - 12^2} = 5$

$$\cos \beta + \cot \beta = \tfrac{5}{13} + \tfrac{5}{12} = \tfrac{125}{156}$$

Exercise 5   (page 87)

1.1 $\dfrac{\sin 2(31°)}{2} = 0{,}441$

1.2 $\dfrac{2 \sin (4 \times 31°)}{3} = 0{,}553$

1.3 $\cos^2 (31° + 78°) = 0{,}106$

1.4 $\sin^2 31° + \cos^2 78° = 0{,}308$

1.5 $\sec 2 (31°) = \dfrac{1}{\cos 2(31°)} = 2{,}130$

1.6 $\operatorname{cosec} (78° - 31°) = \dfrac{1}{\sin (78° - 31°)} = 1{,}058$

1.7 $\sqrt{\cot B - \tan \beta} = \sqrt{\dfrac{1}{\tan \beta}} - \tan \beta$

$$= \sqrt{\dfrac{1}{\tan 78°}} - \tan 78° = -4{,}244$$

1.8 $3\sqrt{\tan A - \operatorname{cosec}^2 B} = 3\sqrt{\tan 31° - \dfrac{1}{\sin^2 78°}}$

$$= 1{,}280$$

2.1 $\sin \beta = 0{,}45$ and $\theta \in [0°; 90°]$

$$\beta = 26{,}7°$$

2.2 $3 \cos \beta = 1{,}3$ and $0° \le \beta \le 90°$

$$\cos \beta = \tfrac{1{,}3}{3}$$

$$\beta = 64{,}3°$$

2.3 $2 \tan 3\beta = 10$ and $0° \leq \beta \leq 90°$

$\tan 3\beta = 5$

$3\beta = 78,7°$

$\beta = 26,2°$

2.4 $4 \sin \beta - 1 = 3$ and $\theta \in [0°; 90°]$

$\sin \beta = 1$

$\beta = 90°$

2.5 $\text{cosec } 2\beta = \frac{7}{4}$ and $\theta \in [0°; 90°]$

$\sin 2\beta = \frac{4}{7}$

$2\beta = 34,849$

$\beta = 17,4°$

2.6 $21 \cot (\beta - 5°) - 23 = 0$ and $\theta \in [0°; 90°]$

$\cot (\beta - 5°) = \frac{23}{21}$

$\tan (\beta - 5°) = \frac{21}{23}$

$\beta - 5° = 42,4$

$\beta = 47,4°$

Exercise 6 (page 90)

1.1 $\sin 44° = \frac{b}{c}$, $\cot 44° = \frac{a}{b}$, $\text{cosec } 44° = \frac{c}{b}$,

$\tan 46° = \frac{a}{b}$, $\sec 46° = \frac{c}{b}$, $\cos 46° = \frac{b}{c}$

1.2 $\cos \theta = \frac{p}{q}$, $\cot \theta = \frac{p}{r}$, $\text{cosec } \theta = \frac{q}{r}$,

$\sin (90° - \theta) = \frac{p}{q}$, $\sec (90° - \theta) = \frac{q}{r}$,

$\tan (90° - \theta) = \frac{p}{r}$

2.1 $\frac{a}{12} = \sin 38°$

$a = 12 \sin 38° = 7,39$ m

2.2 $\frac{13}{85} = \cos b$

$b = 81,2°$

2.3 $\frac{c}{45} = \sec 32°$

$\frac{c}{45} = \frac{1}{\cos 32°}$

$c = \frac{45}{\cos 32°} = 53,06$ mm

2.4 $\frac{29}{21} = \tan d$

$d = 54,1°$

2.5 $\frac{55}{61} = \sin e$

$e = 64,4°$

2.6 $\frac{f}{17} = \cot 62°$

$\frac{f}{17} = \frac{1}{\tan 62°}$

$f = \frac{17}{\tan 62°} = 9,04$ m

Exercise 7 (page 91)

1.1 $\frac{x}{20} = \tan 48°$

$x = 20 \tan 48°$

$x = 22,21$ cm

$\tan y = \frac{22,21}{48}$

$y = 24,83...° = 24,8°$

1.2 $\frac{x}{16} = \tan 25°$

$x = 16 \tan 25°$

$x = 7,46$ cm

$y = \sqrt{16^2 - 7,46^2} = 14,15$ cm

1.3 $\frac{x}{10} = \text{cosec } 28,5°$

$x = \frac{10}{\sin 28,5°}$

$x = 20,957 = 21$ km

$\frac{10}{13} = \cos y$

$y = 39,7°$

1.4 $x^2 + x^2 = (25\sqrt{2})^2$

$2x^2 = (25\sqrt{2})^2$

$x^2 = \frac{(25\sqrt{2})^2}{2}$

$x = \sqrt{\frac{(25\sqrt{2})^2}{2}}$

$x = 25$ cm

$\frac{y}{25\sqrt{2}} = \sec 45°$

$y = \frac{25\sqrt{2}}{\cos 45°} = 50$ cm

2.1 $\hat{S} = 37°$ sum $\angle$s of $\triangle = 180°$

$\frac{PR}{65} = \tan 37°$

$PR = 65 \tan 37°$

$PR = 48,98$ m

$\frac{RQ}{PR} = \cot 53°$

$RQ = \frac{48,98}{\tan 53°}$

$RQ = 36,91$ m

2.2 Area $\triangle PSQ = \frac{1}{2}(PR)(SQ)$

$= \frac{1}{2}PR(SR + RQ)$

$= \frac{1}{2}(48,98)(65 + 36,91)$

$= 2\,495,77$ m^2

$= 2\,496$ m^2

Exercise 8 (page 91)

1. $\frac{21}{17} = \tan x$

$\therefore x = 51°, y = 39°$

2.1 $\frac{DA}{AB} = \tan 20°$

$DA = 50 \tan 20° = 18,198 = 18,2$

2.2 $\frac{CA}{AB} = \tan 40°$

$CA = 50 \tan 40° = 41,954 = 42$

$CD = CA - DA = 40 - 18,2 = 21,8$

3. $\frac{x}{100} = \sec 42°$

$x = \frac{100}{\cos 42°} = 134,56$

$y = \sqrt{134,56^2 + 100^2} = 90,04$

4.1 $\tan (56,7° + 67,5°) = -1,47$

4.2 $2 \sin (67,5°) = 1,85$

4.3 $\cos (3(56,7°)) = -0,99$

4.4 $\frac{\sec (67,5°)}{3} = \frac{1}{3 \cos (67,5°)} = 0,87$

4.5 $\cot (56,7° + 23,6°) = \frac{1}{\tan (56,7° + 23,6°)} = 0,17$

4.6 $\sin^2 67,5° - \cos^2 56,7° = 0,55$

4.7 $3 \cos (3 \times 67,5° - 30°) - \operatorname{cosec}^2 56,7°$

$3 \cos (3 \times 67,5° - 30°) - \frac{1}{\sin^2 56,7°} = -4,41$

4.8 $\frac{1 - \tan (2 \times 67,5°)}{3 \cos 56,7°} = 1,21$

4.9 $\frac{2}{5} \tan 56,7° - \cot 67,5° + 2 \sin^2 (56,7°$ $+ 67,5°)$

$= \frac{2}{5} \tan 56,7° - \frac{1}{\tan 67,5°} + 2 \sin^2 (56,7°$ $+ 67,5°)$

$= 1,56$

4.10 $\sqrt{\tan^3 67,5° - \sin 56,7°} = 3,64$

5.1 $\tan \theta = 4,357$

$\therefore \theta = 77,1°$

5.2 $3 \cos \theta = 1,657$

$\cos \theta = \frac{1,657}{3}$

$\therefore \theta = 56,5°$

5.3 $\sin 3\theta = 0,677$

$3\theta = 42,61°$

$\therefore \theta = 14,2°$

5.4 $\sec \theta = 3,37$

$\therefore \cos \theta = \frac{1}{3,37}$

$\therefore \theta = 72,7°$

5.5 $5 \operatorname{cosec} \theta = 2,578$

$\operatorname{cosec} \theta = \frac{5}{2,578}$

$\sin \theta = \frac{2,578}{5}$

$\therefore \theta = 31,0°$

5.6 $\cot 3\theta = 0,677$

$\tan 3\theta = \frac{1}{0,677}$

$3\theta = 55,901°$

$\therefore \theta = 18,6°$

5.7 $5 \tan \theta - 4,357 = 0$

$\tan \theta = \frac{4,357}{5}$

$\therefore \theta = 41,1°$

5.8 $\cos (3\theta - 10°) = 0,6887$

$(3\theta - 10°) = 46,47°$

$3\theta = 56,47°$

$\therefore \theta = 18,8°$

5.9 $4 \sin^2 \theta = 1,234$

$\sin^2 \theta = \frac{1,234}{4}$

$\sin \theta = \sqrt{\frac{1,234}{4}}$

$\therefore \theta = 33,7°$

5.10 $\frac{2}{3} \sec \theta = 2,3887$

$\sec \theta = \frac{2,3887 \times 3}{2}$

$\cos \theta = \frac{2}{2,3887 \times 3}$

$\therefore \theta = 73,8°$

5.11 $\frac{\operatorname{cosec} \theta}{5} = \sin 34,5°$

$\operatorname{cosec} \theta = 5 \sin 34,5°$

$\sin \theta = \frac{1}{5 \sin 34,5°}$

$\therefore \theta = 20,7°$

5.12 $5 + \cot 2\theta = 6,677$

$\cot 2\theta = 6,677 - 5$

$\tan 2\theta = \frac{1}{6,677 - 5}$

$2\theta = 30,8°$

$\therefore \theta = 15,4°$

Test A   (page 92)

1.1 $\sin E = \frac{FD}{ED}, \sin E = \frac{GD}{EG}$

1.2 $\cos D = \frac{e}{f}, \cot E = \frac{d}{e}$

2.1 $\frac{\sin^2 35,6°}{2} = 0,169$

2.2 $5 \sec 75° = \frac{5}{\cos 75°} = 19,319$

3. $\frac{a}{35} = \cos 51°$

$a = 35 \cos 51° = 22,0 \text{ m}$

$\frac{b}{56} = \operatorname{cosec} 48°$

$b = \frac{56}{\sin 48°} = 75,4 \text{ cm}$

$\frac{24}{13} = \tan c$

$c = 61{,}6°$

$d = 45°$

$e^2 + e^2 = (12\sqrt{2})^2$

$2e^2 = 288$

$e^2 = 144$

$e = 12 \text{ cm}$

4.1 $OE = \sqrt{5^2 + 12^2} = 13$

$\sin B = \frac{12}{13}$

4.2 $\tan B + \sec \beta = \frac{12}{5} + \frac{13}{5} = \frac{25}{5} = 5$

5.1 $7 \tan \beta = 9$

$\tan \beta = \frac{9}{7}$

$\therefore \beta = 52{,}1°$

5.2.1 $\tan \beta = \frac{9}{7} \therefore O = 9, A = 7, H = \sqrt{130}$

5.2.2 $\sin \beta \csc \beta = \frac{9}{\sqrt{130}} \times \frac{\sqrt{130}}{9} = 1$

5.2.3 $\sec^2 \beta - \tan^2 \beta$

$= \left(\frac{\sqrt{130}}{7}\right)^2 - \left(\frac{9}{7}\right)^2$

$= \frac{130}{49} - \frac{81}{49} = \frac{49}{49} = 1$

6.1 $\sin 30° \tan 45° \cos 60°$

$= \frac{1}{2} \times 1 \times \frac{1}{2} = \frac{1}{4}$

6.2 $\sin 0° \cos 45° - \tan 60° \sin 45°$

$= 0 . \frac{1}{\sqrt{2}} - \sqrt{3} . \frac{1}{\sqrt{2}} = -\frac{\sqrt{3}}{\sqrt{2}}$

7. $\frac{x}{17} = \tan 63$

$x = 17 \tan 63°$

$x = 33{,}4 \text{ cm}$

$\frac{38}{33{,}4} = \tan y$

$y = 48{,}7°$

Test B (page 93)

1. $\sin \theta = \frac{AB}{AC}$

$\cos A = \cos (90° - \theta) = \frac{AB}{AC}$

$\therefore \sin \theta = \cos (90° - \theta)$

2.1 $\frac{\sec 20° + \csc 20°}{\tan^2 20°} = \frac{\frac{1}{\cos 20°} + \frac{1}{\sin 20°}}{(\tan 20°)^2}$

2.2 $\frac{3}{5} \csc (3x - 100°)$

$= 10 \text{ and } (3x - 100°) < 90°$

$\csc (3x - 100°) = \frac{50}{3}$

$\frac{1}{\sin (3x - 100°)} = \frac{50}{3}$

$\sin (3x - 100°) = \frac{3}{50}$

$(3x - 100°) = 3{,}4398...°$

$3x = 103{,}4398...°$

$x = 34{,}5°$

2.3 $\sec \left(\frac{\beta}{2} - 15°\right) = 5 \sin 75{,}9° \text{ and } \frac{\beta}{2} - 15° < 90°$

$\frac{1}{\cos \left(\frac{\beta}{2} - 15°\right)} = 5 \sin 75{,}9°$

$\cos \left(\frac{\beta}{2} - 15°\right) = \frac{1}{5 \sin 75{,}9°} = 0{,}206212...$

$\frac{\beta}{2} - 15° = 78{,}09949...°$

$\frac{\beta}{2} = 93{,}09949...°$

$\beta = 186{,}1988...° = 186°$

$\therefore \cot^2 (2\beta + 18°)$

$= \cot^2 (2 \times 186 + 18°)$

$= \cot^2 390°$

$= \frac{1}{\tan^2 390°} = 3$

3.1 $\left(\frac{2 \sin 45° \csc 90°}{\cos 60°}\right)^2$

$= \left(\frac{2\left(\frac{1}{\sqrt{2}}\right)(1)}{\frac{1}{2}}\right)^2 = \left(\frac{\frac{2}{\sqrt{2}}}{\frac{1}{2}}\right)^2 = (2\sqrt{2})^2 = 8$

3.2 $2 \sin 2\theta - \sec 0° = \sin 0° \text{ and } \theta \in [0°; 90°]$

$2 \sin 2\theta = \sec 0° + \sin 0°$

$\sin 2\theta = \frac{1}{2}$

$2\theta = 30°$

$\theta = 15°$

4. Draw a diagram to help you answer this question.

$\csc \beta = \frac{p}{2} \therefore r = p \text{ and } y = 2$

Using Pythagoras: $x = \sqrt{p^2 - 2^2} = \sqrt{p^2 - 4}$

$\cos^2 \beta = \left(\frac{\sqrt{p^2 - 4}}{p}\right)^2 = \frac{p^2 - 4}{p^2}$

5.1 In $\triangle ABC$: $\frac{AC}{100} = \sin 70°$

$AC = 100 \sin 70°$

$AC = 93{,}969 ... = 94 \text{ mm}$

In $\triangle ACD$: $\frac{AC}{CD} = \tan D$

$\frac{94}{85} = \tan (3x + 12°)$

$(3x + 12°) = 47{,}8783...°$

$3x = 35{,}87°$

$x = 11{,}959...° = 12°$

5.2 In $\triangle ABD$: $\hat{B} = 70°$ *given*

$\hat{D} = 47,9°$ calculated in 5.1

$B\hat{A}D = 180° - 70° - 47,9° = 62,1°$

$\therefore \triangle ABD$ is not a right-angled triangle

6.1 $\frac{PT}{UT} = \tan 60°$

$\frac{PT}{10} = \tan 60°$

$PT = 10 \tan 60° = 10\sqrt{3}$

$WQ = 10\sqrt{3} - 10$

6.2 $\frac{WQ}{WR} = \tan R$

$\frac{10\sqrt{3} - 10}{10} = \tan R$

$Q\hat{R}W = 36,2°$

Test C   (page 95)

1.1 $OP = \sqrt{4^2 + 3^2} = 5$

1.2 $\cos \theta = \frac{4}{5}$, $\cot \theta = \frac{4}{3}$

1.3 $\therefore \sec^2 \theta - 1 = \left(\frac{5}{4}\right)^2 - 1 = \frac{9}{16}$

$\tan^2 \theta = \left(\frac{3}{4}\right)^2 = \frac{9}{16}$

$\therefore \sec^2 \theta - 1 = \tan^2 \theta$

2.1 $\beta + M\hat{A}R = 90°$ *sum ∠s △*

$F\hat{A}M + M\hat{A}R = 90°$ *adj. compl. ∠s*

$\therefore F\hat{A}M = \beta$

2.2 In $\triangle MAR$: $\sin \beta = \frac{AM}{AR}$

In $\triangle FMA$: $\sin \beta = \frac{FM}{FA}$

In $\triangle FAR$: $\sin \beta = \frac{FA}{FR}$

2.3 $\frac{2,5}{10} = \sin R$

$\hat{R} = 14,48°$

3.1 $x = \cos^2 30° + \tan 45° - \sin 90°$

$x = \left(\frac{\sqrt{3}}{2}\right)^2 + 1 - 1 = \frac{3}{4}$

3.2 $2 \cos 2\theta - \sqrt{3} = 0$ and $\theta \in [0°; 90°]$

$\cos 2\theta = \frac{\sqrt{3}}{2}$

$2\theta = 30°$

$\theta = 15°$

4.1 $x = \sin^2 34,5° = \sqrt{\cos 65,43°}$

$x = 0,9656... = 0,966$

4.2 $x = \frac{\sec 34,5°}{3}$

$x = \frac{1}{3 \cos 34,5°} = 0,404$

5.1 $\operatorname{cosec} \theta = 3,456$ and $\theta \in [0°; 90°]$

$\sin \theta = \frac{1}{3,456}$

$\theta = 16,8°$

5.2 $10 \tan (\theta + 15°) = 234$ and $\theta \in [0°; 90°]$

$\tan (\theta + 15°) = \frac{234}{10}$

$(\theta + 15°) = 87,6°$

$\theta = 72,6°$

6. $\frac{DT}{AD} = \cos 35°$

$DT = 45 \cos 35° = 36,861 \text{ km}$

$\frac{AT}{AD} = \sin 35°$

$AT = 45 \sin 35° = 25,8109 \text{ km}$

$\frac{TR}{AT} = \tan 48°$

$TR = (45 \sin 35°) \tan 48° = 28,6659$

$DR = DT + TR = 36,861 + 28,665$

$\qquad = 65,53 \text{ km}$

7. $\frac{QR}{PQ} = \cot 40°$

$\frac{QR}{80} = \frac{1}{\tan 40°}$

$QR = \frac{80}{\tan 40°} = 95,340$

$\therefore QS = QR + RS = 95,34 + 20 = 115,340$

In $\triangle PQS$: $\frac{QS}{PQ} = \tan Q\hat{P}S$

$\frac{115,34}{80} = \tan Q\hat{P}S$

$Q\hat{P}S = 55,3°$

$Q\hat{P}R = 50°$

$R\hat{P}S = Q\hat{P}S - Q\hat{P}R = 55,3° - 50°$

$\qquad = 5,3°$

Topic 7   Functions (page 96)

Exercise 1 (page 97)

1. $g(x) = -2x + 1$
1.1 $g(3) = -2(3) + 1 = -5$
1.2 $g(-2) = -2(-2) + 1 = 5$
1.3 $g(x) = 0$
   $\therefore 0 = -2x + 1$
   $2x = 1$
   $\therefore x = \frac{1}{2}$
1.4 $g(x) = 5$
   $\therefore 5 = -2x + 1$
   $2x = 1 - 5$
   $2x = -4$
   $\therefore x = -2$

2. $h(x) = x^2 - 1$
2.1 $h(2) = 2^2 - 1 = 3$
2.2 $h(-3) = (-3)^2 - 1 = 8$
2.3 $h(x) = 0$
   $\therefore x^2 - 1 = 0$
   $x^2 = 1$
   $\sqrt{x^2} = \pm\sqrt{1}$
   $\therefore x = 1 \text{ or } x = -1$
2.4 $h(x) = 3$
   $\therefore x^2 - 1 = 3$
   $x^2 = 4$
   $\sqrt{x^2} = \pm\sqrt{4}$
   $\therefore x = -2 \text{ or } x = 2$

3. $f(x) = 2^x$
3.1 $f(0) = 2^0 = 1$
3.2 $f(1) = 2^1 = 2$
3.3 $f(x) = 0$
   $\therefore 2^x \neq 0$
   $\therefore$ no solution
3.4 $f(x) = 8$
   $\therefore 2^x = 8$
   $2^x = 2^3$
   $\therefore x = 3$

4. $p(x) = \frac{4}{x}$
4.1 $p(1) = \frac{4}{1} = 4$
4.2 $p(0) \neq \frac{4}{0}$
   $\therefore$ no solution

4.3 $p(x) = 16$
   $\therefore \frac{4}{x} = 16$
   $4 = 16x$
   $\therefore x = \frac{1}{4}$
4.4 $p(x) = 1$
   $\therefore 1 = \frac{4}{x}$
   $x = 4$

Exercise 2 (page 99)

1.   IN: Domain: $x \in \mathbb{R}$, Range: $y \geq -4$; $y \in \mathbb{R}$
   SBN: Domain: $y \in (-\infty; \infty)$, Range: $x \in [-4; \infty)$

2.   IN: Domain: $x \in \mathbb{R}$, Range: $y \leq 4$; $y \in \mathbb{R}$
   SBN: Domain: $x \in (-\infty; \infty)$, Range: $y \in (-\infty; 4]$

3.   IN: Domain: $x \in \mathbb{R}$, Range: $y \geq -2$; $y \in \mathbb{R}$
   SBN: Domain: $x \in (-\infty; \infty)$, Range: $y \in [-2; \infty)$

4.   IN: Domain: $x \in \mathbb{R}, x \neq 0$, Range: $y \in \mathbb{R}, y \neq -1$
   SBN: Domain: $x \in (-\infty; 0) \cup (0; \infty)$,
   Range: $y \in (-\infty; -1) \cup (-1; \infty)$

5.   IN: Domain: $x \in \mathbb{R}, x \neq 1$, Range: $y \in \mathbb{R}, y \neq -1$
   SBN: Domain: $x \in (-\infty; 1) \cup (1; \infty)$,
   Range: $y \in (-\infty; -1) \cup (-1; \infty)$

6.   IN: Domain: $x \in \mathbb{R}, x \neq -1$,
   Range: $y \in \mathbb{R}, y \neq -1$
   SBN: Domain: $x \in (-\infty; -1) \cup (-1; \infty)$,
   Range $y \in (-\infty; -1) \cup (-1; \infty)$

7.   IN: Domain: $x \in \mathbb{R}$, Range: $y > 1$; $y \in \mathbb{R}$
   SBN: Domain: $x \in (-\infty; \infty)$, Range: $y \in (1; \infty)$

8.   IN: Domain: $x \in \mathbb{R}$, Range: $y > 1$; $y \in \mathbb{R}$
   SBN: Domain: $x \in (-\infty; \infty)$, Range: $y \in (1; \infty)$

9.   IN: Domain: $x \in \mathbb{R}$, Range: $y < 1$; $y \in \mathbb{R}$
   SBN: Domain: $x \in (-\infty; \infty)$, Range: $y \in (-\infty; 1)$

10. IN: Domain: $x \in \mathbb{R}, x \neq 3$, Range: $y \in \mathbb{R}, y \neq -3$
   SBN: Domain: $x \in (-\infty; 3) \cup (3; \infty)$,
   Range $y \in (-\infty; -3) \cup (-3; \infty)$

11. IN: Domain: $x \in \mathbb{R}$, Range: $y > -3$; $y \in \mathbb{R}$
   SBN: Domain: $x \in (-\infty; \infty)$, Range: $y \in (-3; \infty)$

12. IN: Domain: $x \in \mathbb{R}$, Range: $y \leq 12$; $y \in \mathbb{R}$
   SBN: Domain: $x \in (-\infty; \infty)$,
   Range: $y \in (-\infty; 12]$

Exercise 3 (page 99)

1.1 $1 + 2x = 0$

$2x = -1$

$\therefore x = -\frac{1}{2}$

1.2 $x^2 - 9 = 0$

$(x - 3)(x + 3) = 0$

$\therefore x = 3$ or $x = -3$

1.3 $2^x - 8 = 0$

$2^x = 8$

$2^x = 2^3$

$\therefore x = 3$

1.4 $-x^2 + 16 = 0$

$-(x^2 - 16) = 0$

$(x - 4)(x + 4) = 0$

$\therefore x = 4$ or $x = -4$

1.5 $2 \cdot 3^x - 18 = 0$

$2 \cdot 3^x = 18$

$3^x = 9$

$3^x = 3^2$

$\therefore x = 2$

2. $f(x) = 2x - 3$

2.1 $f(-2) = 2(-2) - 3 = -7$

2.2 $f(5) = 2(5) - 3 \quad = 7$

2.3 $f(0) = 2(0) - 3 \quad = -3$

2.4 $3 \cdot f(-2) = 3[2(-2) - 3] = 3(-7) = -21$

2.5 $2x - 3 = 0$

$2x = 3$

$\therefore x = \frac{3}{2}$

2.6 Domain: $x \in (-\infty; \infty)$, Range: $y \in (-\infty; \infty)$

3. $f(x) = x^2 - 1$

3.1 $f(2) = (2)^2 - 1 = 3$

3.2 $f(-2) = (-2)^2 - 1 = 3$

3.3 $f(0) = (0)^2 - 1 = -1$

3.4 $f(2): 3(2^2 - 1) = 3(3) = 9$

3.5 $f(x) = 0$

$\therefore x^2 - 1 = 0$

$(x - 1)(x + 1) = 0$

$\therefore x = 1$ or $x = -1$

3.6 Domain: $x \in \mathbb{R}$, Range: $x \geq -1; x \in \mathbb{R}$

4. $f(x) = 5^x - 1$

4.1 $f(1) = 5^1 - 1 = 4$

4.2 $f(-1) = 5^{-1} - 1 = \frac{1}{5} - 1 = -\frac{4}{5}$

4.3 $f(0) = 5^0 - 1 = 1 - 1 = 0$

4.4 $3 \cdot f(1) = 3[5^1 - 1] = 3(4) = 12$

4.5 $f(x) = 0$

$\therefore 5^x - 1 = 0$

$5^x = 1$

$5^x = 5^0$

$\therefore x = 0$

4.6 Domain: $x \in \mathbb{R}$, Range: $y > -1; y \in \mathbb{R}$

Exercise 4 (page 107)

1.1 $y = -2x + 4$

$a = -2$: $\therefore$ negative gradient $\therefore$ slopes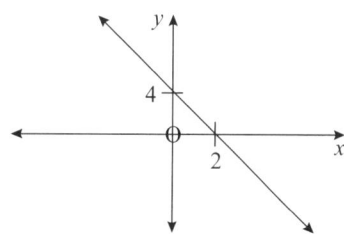

$q = 4$: shifts straight line vertically upwards by 4 units

x-intercept: let $y = 0$:

$0 = -2x + 4$

$2x = 4$

$\therefore x = 2$

y-intercept, let $y = 0$:

$y = -2(0) + 4$

$\therefore y = 4$

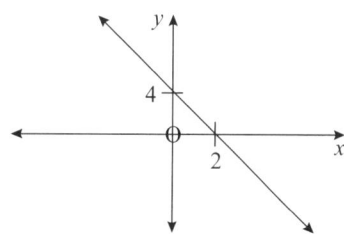

1.2 $y = 3x + 2$

$a = 3$ $\therefore$ positive gradient $\therefore$ slopes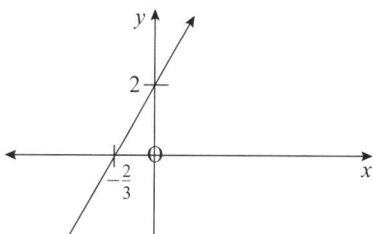

$q = 2$: vertical shift upwards by 2 units

x-intercept, let $y = 0$:

$0 = 3x + 2$

$-2 = 3x$

$\therefore x = -\frac{2}{3}$

y-intercept, let $x = 0$: $\therefore y = 2$

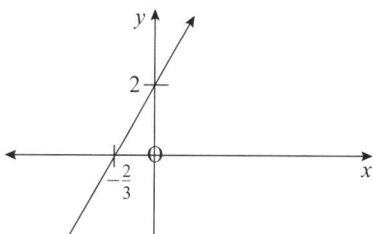

1.3 $y = -3x - 6$

$a = -3$: negative gradient ∴ slopes ↘

$q = -6$: vertical shift downwards by 6 units

x-intercept, let $y = 0$:

∴ $0 = -3x - 6$

$3x = 6$

∴ $x = -2$

y-intercept, let $x = 0$:

∴ $y = -3(0) - 6$

∴ $y = -6$

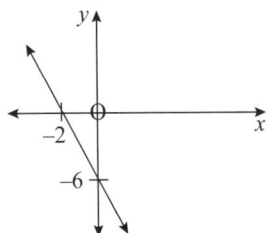

1.4 $y = x + 3$

$a = 1$: positive gradient ∴ slopes ↗

$q = 3$: vertical shift upwards by 3 units

x-intercept, let $y = 0$:

∴ $0 = x + 3$

∴ $-3 = x$

y-intercept, let $x = 0$: ∴ $y = 3$

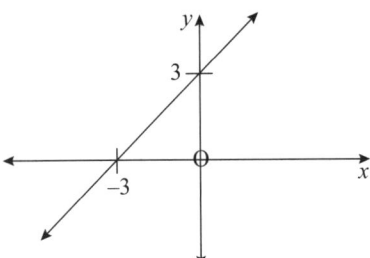

1.5 $y = -\frac{1}{2}x + 2$

$a = -\frac{1}{2}$: negative gradient ∴ slopes ↘

$q = 2$: vertical shift upwards by 2 units

x-intercept, let $y = 0$:

∴ $0 = -\frac{1}{2}x + 2$

$\frac{1}{2}x = 2$

∴ $x = 4$

y-intercept, let $x = 0$: ∴ $y = 2$

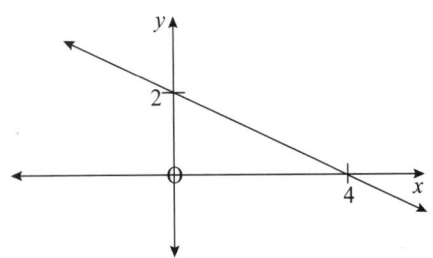

1.6 $2y = 4x - 2$

∴ $y = 2x - 1$

$a = 2$: positive gradient ∴ slopes ↗

$q = -1$ vertical shift downwards by 1 unit

x-intercept, let $y = 0$:

∴ $0 = 2x - 1$

$1 = 2x$

∴ $\frac{1}{2} = x$

y-intercept, let $x = 0$: ∴ $y = 2(0) - 1 = -1$

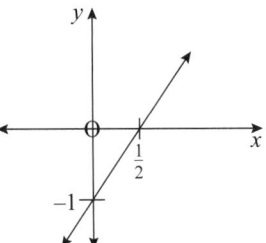

2.1 $y = 3x^2 - 3$

$a = 3$: smiley parabola and vertical stretch (becomes thinner) by a factor of 3

$q = -3$: vertical shift downwards by 3 units

x-intercept, let $y = 0$:

$$0 = 3x^2 - 3$$

$$3(x^2 - 1) = 0$$

$$(x - 1)(x + 1) = 0$$

∴ $x = 1$ or $x = -1$

y-intercept, let $x = 0$: ∴ $y = 3(0)^2 - 3 = -3$

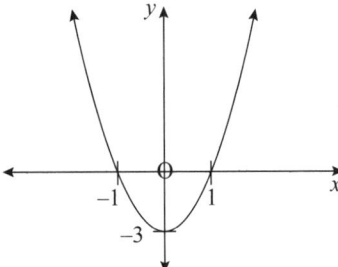

2.2 $y = \frac{1}{2}x^2 - 8$

$a = \frac{1}{2}$: smiley parabola and vertical stretch (becomes wider) by a factor of $\frac{1}{2}$

$q = -8$: vertical shift downwards by 8 units

x-intercept, let $y = 0$:

$$0 = \frac{1}{2}x^2 - 8$$

$$\frac{1}{2}(x^2 - 16) = 0$$

$$(x - 4)(x + 4) = 0$$

∴ $x = 4$ or $x = -4$

y-intercept, let $x = 0$: ∴ $y = \frac{1}{2}(0)^2 - 8 = -8$

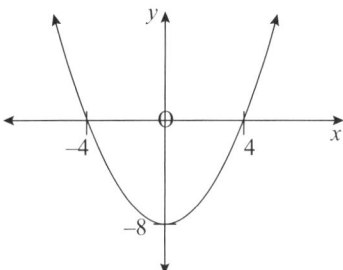

2.3 $y = -3x^2 + 1$

$a = -3$: frowny parabola and vertical stretch (becomes thinner) by a factor of -3

$q = 1$: vertical shift upwards by 1 unit

x-intercept, let $y = 0$:

$0 = -3x^2 + 1$

$3x^2 = 1$

$x^2 = \frac{1}{3}$

$\therefore x = \pm\sqrt{\frac{1}{3}}$

y-intercept, let $x = 0$: $\therefore y = -3(0)^2 + 1 = 1$

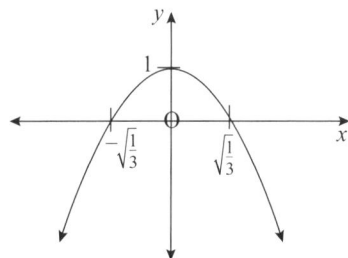

2.4 $y = x^2 - 3$

$a = 1$: smiley parabola and stays the same as the standard parabola $y = x^2$

$q = -3$: vertical shift downwards by 3 units

x-intercept, let $y = 0$:

$0 = x^2 - 3$

$3 = x^2$

$\therefore x = \pm\sqrt{3}$

y-intercept, let $x = 0$: $\therefore y = -3$

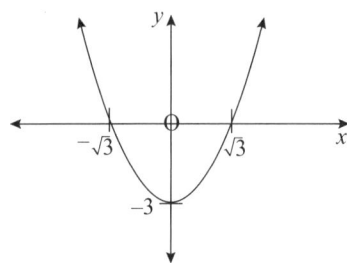

2.5 $y = 3 - 4x^2$

$a = -4$: frowny parabola and vertical stretch (becomes thinner) by a factor of -4

$q = 3$: vertical shift upwards by 3 units

x-intercept, let $y = 0$:

$0 = 3 - 4x^2$

$4x^2 = 3$

$x^2 = \frac{3}{4}$

$\therefore x = \pm\frac{\sqrt{3}}{2}$

y-intercept, let $x = 0$: $\therefore y = 3 - 4(0)^2 = 3$

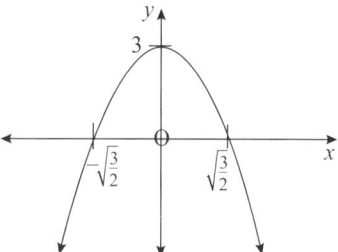

2.6 $y = -2x^2 - 1$

$a = -2$: frowny parabola and vertical stretch (becomes thinner) by a factor of -2

$q = -1$: vertical shift downwards by 1 unit

x-intercept, let $y = 0$:

$0 = -2x^2 - 1$

$2x^2 = -1$

$x^2 \neq -\frac{1}{2}$

$\therefore$ no solution (no x-intercepts)

y-intercepts: let $x = 0$ $\therefore y = -2(0)^2 - 1 = -1$

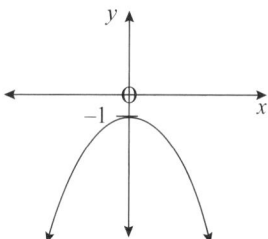

Exercise 5 (page 111)

1.1 $y = \frac{3}{x} - 1$

$a = 3$: hyperbola in quadrants 1 and 3, and vertical stretch by a factor of 3

$q = -1$: vertical shift downwards by 1 unit

Asymptotes: $x = 0$ (y-axis) and $y = -1$

x-intercept, let $y = 0$:

$0 = \frac{3}{x} - 1$

$\frac{3}{x} = 1$

$\therefore x = 3$

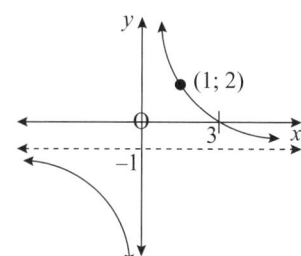

1.2 $y = -\frac{3}{x} - 1$

$a = -3$: hyperbola in quadrants 2 and 4, and vertical stretch by a factor of -3

$q = -1$: vertical shift downwards by 1 unit

Asymptotes: $x = 0$ (y–axis) and $y = -1$

x-intercept, let $y = 0$:

$$0 = -\frac{3}{x} - 1$$
$$\frac{3}{x} = -1$$
$$3 = -x$$
$$\therefore x = -3$$

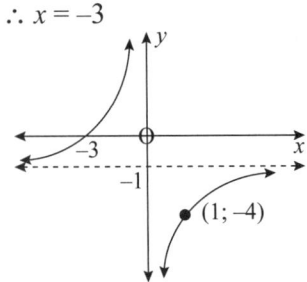

1.3 $y = \frac{3}{x} + 1$

$a = 3$: hyperbola in quadrants 1 and 3, and vertical stretch by a factor of 3

$q = 1$ vertical shift upwards by 1 unit

Asymptotes: $x = 0$ (y-axis) $\therefore y = 1$

x-intercept, let $y = 0$:

$$0 = \frac{3}{x} + 1$$
$$-1 = \frac{3}{x}$$
$$-x = 3$$
$$\therefore x = -3$$

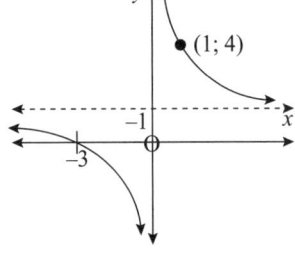

1.4 $y = -\frac{3}{x} + 1$

$a = -3$: hyperbola in quadrants 2 and 4, and vertical stretch by a factor of -3

$q = 1$: vertical shift upwards by 1 unit

Asymptotes: $x = 0$ (y-axis) $\therefore y = 1$

x-intercept, let $y = 0$:

$$0 = -\frac{3}{x} + 1$$
$$\frac{3}{x} = 1$$
$$\therefore x = 3$$

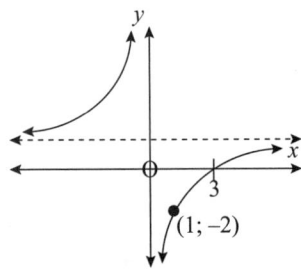

1.5 $y = \frac{9}{x} - 3$

$a = 9$: hyperbola in quadrants 1 and 3, and vertical stretch by a factor of 9

$q = -3$: vertical shift downwards by 3 units

Asymptotes: $x = 0$ (y-axis) $\therefore y = -3$

x-intercepts: let $y = 0$:

$$0 = \frac{9}{x} - 3$$
$$3 = \frac{9}{x}$$
$$3x = 9$$
$$\therefore x = 3$$

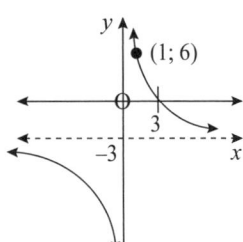

1.6 $y = \frac{9}{x} - 1$

$a = 9$: hyperbola in quadrants 1 and 3, and vertical stretch by a factor of 9

$q = -1$: vertical shift downwards by 1 unit

Asymptotes: $x = 0$ (y-axis) $\therefore y = -1$

x-intercepts,

let $y = 0$:

$$0 = \frac{9}{x} - 1$$
$$1 = \frac{9}{x}$$
$$\therefore x = 9$$

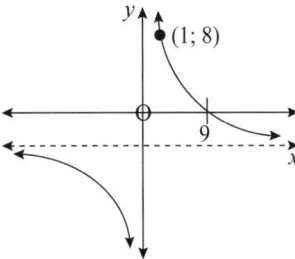

1.7 $y = -\frac{9}{x} + 3$

$a = -9$: hyperbola in quadrants 2 and 4, and vertical stretch by a factor of -9

$q = 3$: vertical shift upwards by 3 units

Asymptotes: $x = 0$ (y-axis) $\therefore y = 3$

x-intercept, let $y = 0$:

$$0 = -\frac{9}{x} + 3$$
$$\frac{9}{x} = 3$$
$$9 = 3x$$
$$\therefore x = 3$$

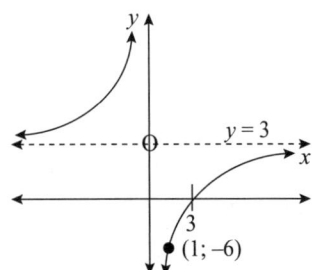

1.8 $y = \frac{5}{x} - 2$

$a = 5$: hyperbola in quadrants 1 and 3, and vertical stretch by a factor of 5

$q = -2$: vertical shift downwards by 2 units

Asymptotes: $x = 0$ (y-axis) $\therefore$ $y = -2$

x-intercept, let $y = 0$:

$$0 = \frac{5}{x} - 2$$
$$2 = \frac{5}{x}$$
$$2x = 5$$
$$\therefore x = \frac{5}{2}$$

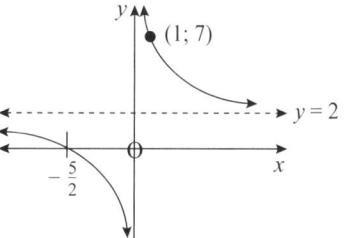

1.9 $y = \frac{5}{x} + 2$

$a = 5$: hyperbola in quadrants 1 and 3, and vertical stretch by a factor of 5

$q = 2$: vertical shift upwards by 2 units

Asymptotes: $x = 0$ (y-axis) $\therefore$ $y = 2$

x-intercept, let $y = 0$:

$$0 = \frac{5}{x} + 2$$
$$-2 = \frac{5}{x}$$
$$-2x = 5$$
$$\therefore x = -\frac{5}{2}$$

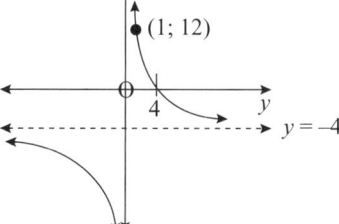

1.10 $y = \frac{16}{x} - 4$

$a = 16$: hyperbola in quadrants 1 and 3, and vertical stretch by a factor of 16

$q = -4$: vertical shift downwards by 4 units

Asymptotes: $x = 0$ (y-axis) $\therefore$ $y = -4$

x-intercept, let $y = 0$:

$$0 = \frac{16}{x} - 4$$
$$4 = \frac{16}{x}$$
$$4x = 16$$
$$\therefore x = 4$$

2. $f(x) = -\frac{4}{x} + 8$

2.1 $f(1) = -\frac{4}{1} + 8 = 4$

2.2 $f(x) = 0$

$$\therefore 0 = -\frac{4}{x} + 8$$
$$\frac{4}{x} = 8$$
$$4 = 8x$$
$$\therefore x = \frac{1}{2}$$

2.3 $y = 8$

Exercise 6 (page 114)

1.1 $y = 3^x - 1$

$a = 1$: no effect

$b = 3$: increasing exponential function

$q = -1$: vertical shift downwards by 1 unit

Asymptotes: $y = -1$

x-intercept: let $x = 0$:

$y = 3^0 - 1 = 0$

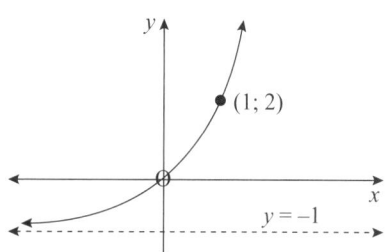

1.2 $y = -3^x - 1$

$a = -1$: reflection of $y = 3^x$ about the line $y = -1$, and decreasing function

$q = -1$: vertical shift downwards by 1 unit

Asymptotes: $y = -1$

x-intercept, let $y = 0$:

$$0 = -3^x - 1$$
$$3^x \neq -1$$

$\therefore$ no solution (no x-intercept)

y-intercept, let $x = 0$:

$y = -3^0 - 1 = -1 - 1 = -2$

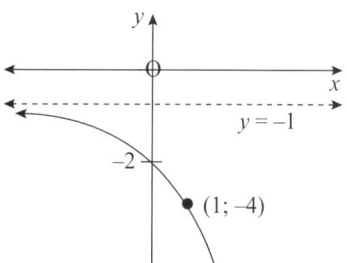

1.3 $y = 3^x + 1$

$b = 3$: increasing exponential function

$q = 1$: vertical shift upwards by 1 unit

Asymptotes: $y = 1$

No x-intercept

y-intercept, let $x = 0$:

$y = 3^0 + 1 = 2$

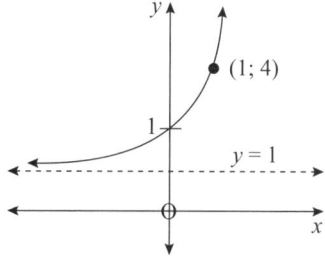

1.4 $y = -3^x + 1$

$a = -1$: decreasing function and reflection of $y = 3^x$ about the line $y = 1$

$q = 1$: vertical shift upwards by 1 unit

Asymptotes: $y = 1$

x-intercept, let $y = 0$:

$0 = -3^x + 1$

$3^x = 3^\circ$

$\therefore x = 0$

y-intercept, let $x = 0$:

$y = -3^\circ + 1 = -1 + 1 = 0$

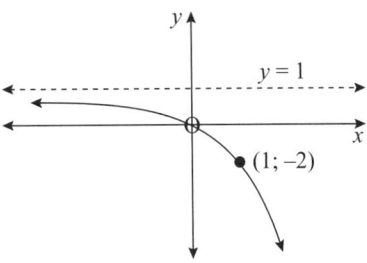

1.5 $y = 3 \cdot 2^x - 3$

$a = 3$: vertical stretch by a factor of 3

$b = 2$: increasing exponential function

$q = -3$: vertical shift downwards by 3 units

Asymptotes: $y = -3$

x-intercept, let $y = 0$:

$0 = 3 \cdot 2^x - 3$

$3 = 3 \cdot 2^x$

$1 = 2^x$

$\therefore 2^\circ = 2^x$

$\therefore x = 0$

y-intercept, let $x = 0$:

$y = 3 \cdot 2^\circ - 3 = 3 - 3 = 0$

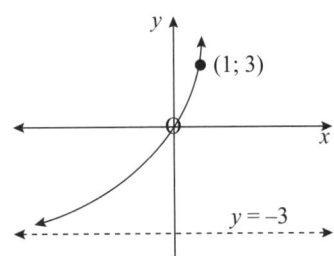

1.6 $y = \frac{1}{2} \cdot 5^x - 1$

$a = \frac{1}{2}$: vertical stretch by a factor of $\frac{1}{2}$

$b = 5$: increasing exponential function

$q = -1$: vertical shift downwards by 1 unit

Asymptotes: $y = -1$

y-intercept, let $x = 0$:

$y = \frac{1}{2} \cdot 5^\circ - 1 = -\frac{1}{2}$

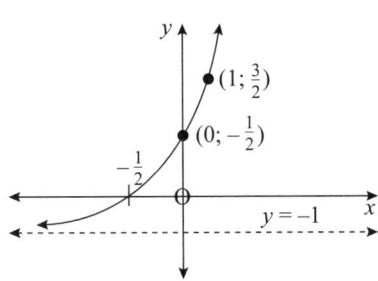

Exercise 7 (page 117)

1.1 $m = \frac{3-0}{0-2} = -\frac{3}{2}$

$\therefore y = -\frac{3}{2}x + 3$

1.2 $m = \frac{0-(-2)}{4-0} = \frac{1}{2}$

$\therefore y = \frac{1}{2}x - 2$

1.3 $x = 4$

1.4 $m = \frac{1-0}{-2-0} = -\frac{1}{2}$

$\therefore y = -\frac{1}{2}x$

1.5 $m = \frac{3-(-1)}{-3-0} = -\frac{4}{3}$

$\therefore y = -\frac{4}{3}x - 1$

1.6 $m = \tan \theta$

$\therefore m = \tan 45^\circ = 1$

$\therefore y = x - 4$

1.7 $y = 1$

1.8 $m = \frac{4-0}{1-(-1)} = 2$

$\therefore y = 2x + c$ ... ①

Substitute $(-1; 0)$ into ①:

$0 = 2(-1) + c$

$\therefore c = 2$

$\therefore y = 2x + 2$

1.9 $m = \frac{5-(-3)}{-1-3} = \frac{8}{-4} = -2$

$\therefore y = -2x + c$ ... ①

Substitute $(-1; 5)$ into ①:

$5 = -2(-1) + c$

$3 = c$

$\therefore y = -2x + 3$

2.1 $m = \frac{4-0}{0-2} = -2$

$\therefore y = -2x + 4$

2.2 $m = \frac{3-(-1)}{1-(-5)} = \frac{4}{6} = \frac{2}{3}$

$\therefore y = \frac{2}{3}x + c$ ... ①

Substitute (1; 3) into ①:

$3 = \frac{2}{3}(1) + c$

$\therefore c = \frac{7}{3}$

$\therefore y = \frac{2}{3}x + \frac{7}{3}$

2.3 $m_g = 3$

$\therefore y = 3x - 2$

2.4 $4y = 3x + 6$

$y = \frac{3}{4}x + \frac{3}{2}$

$\therefore m_g = -\frac{4}{3}$

$\therefore y = -\frac{4}{3}x + \frac{3}{2}$

3.1 $-3x + 6 = 2x + 1$

$-3x - 2x = 1 - 6$

$-5x = -5$

$\therefore x = 1$

$\therefore y = 2(1) + 1 = 3$

$\therefore$ point of intersection: (1; 3)

3.2 For $y = -3x + 6$:

x-intercept, let $y = 0$:

$0 = -3x + 6$

$3x = 6$

$\therefore x = 2$

y-intercept: $y = 6$

For $y = 2x + 1$:

x-intercept, let $y = 0$:

$0 = 2x + 1$

$-1 = 2x$

$\therefore x = -\frac{1}{2}$

y-intercept: $y = 1$

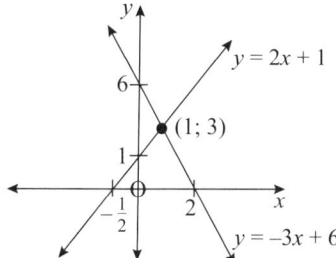

4.1 $m_{AC} = \frac{3-0}{0-(-2)} = \frac{3}{2}$

$\therefore y = \frac{3}{2}x + 3$

4.2 At $x = 1\frac{1}{2}$:

$y = \frac{3}{2}\left(\frac{3}{2}\right) + 3 = \frac{21}{4}$

$\therefore AB = \frac{21}{4} = 5{,}25$ units

At $y = -3$:

$-3 = \frac{3}{2}x + 3$

$-6 = \frac{3}{2}x$

$-12 = 3x$

$\therefore x = -4$

$\therefore CD = 4$ units

5.1.1 $y = -2x + 5$ ... ①

$4y - 5x + 6 = 0$ ... ②

Substitute ① into ②:

$4(-2x + 5) - 5x + 6 = 0$

$-8x + 20 - 5x + 6 = 0$

$-13x + 26 = 0$

$26 = 13x$

$\therefore x = 2$

$\therefore y = -2(2) + 5 = 1$

$\therefore$ T(2; 0) and P(2; 1)

$\therefore PT = 1$ unit

5.1.2 x-intercept of f: let $y = 0$

$4(0) - 5x + 6 = 0$

$-5x = -6$

$\therefore x = \frac{6}{5}$

At $x = \frac{6}{5}$: $y = -2\left(\frac{6}{5}\right) + 5 = \frac{13}{5}$

$\therefore$ A$\left(\frac{6}{5}; 0\right)$ and B$\left(\frac{6}{5}; \frac{13}{5}\right)$

$\therefore AB = \frac{13}{5} = 2{,}6$ units

5.1.3 y-intercept of f: let $x = 0$:

$4y - 5(0) + 6 = 0$

$4y = -6$

$y = -\frac{3}{2}$

At $y = -\frac{3}{2}$:

$-\frac{3}{2} = -2x + 5$

$2x = \frac{13}{2}$

$\therefore x = \frac{13}{4}$

$\therefore CD = \frac{13}{4} = 3{,}25$ units

5.1.4 At $x = \frac{13}{4}$:

$$4y - 5\left(\frac{13}{4}\right) + 6 = 0$$

$$4y = \frac{41}{4}$$

$$y = \frac{41}{16}$$

$\therefore$ E$\left(\frac{13}{4}; \frac{41}{16}\right)$ and D$\left(\frac{13}{4}; -\frac{3}{2}\right)$

$\therefore$ DE $= \frac{41}{16} + \frac{3}{2} = \frac{65}{16}$ units

5.2.1 $g: y = -2x + 5$

or $4y = -8x + 20$

$f: 4y - 5x + 6 = 0$

$4y = 5x - 6$

$\therefore 5x - 6 > 20 - 8x$

$\therefore f > g$

$\therefore x > 2; x \in \mathbb{R}$

5.2.2 $f(x) . g(x) < 0$

$\therefore x < \frac{6}{5}$ or $x > \frac{5}{2}; x \in \mathbb{R}$

Exercise 8 (page 122)

1.1 $y = ax^2 + 1$ ... ①

Substitute (1; 0) into ①

$0 = a(1)^2 + 1$

$-1 = a$

$\therefore y = -x^2 + 1$

1.2 $y = ax^2 + q$

$\therefore y = ax^2 + 2$ ... ①

Substitute (4; 4) into ①:

$4 = a(4)^2 + 2$

$4 - 2 = 16a$

$2 = 16a$

$\therefore a = \frac{1}{8}$

$\therefore y = \frac{1}{8}x^2 + 2$

1.3 $y = ax^2 + q$

$\therefore y = ax^2 + 0$ ... ①

Substitute (2; 4) into ①:

$4 = a(2)^2$

$4 = 4a$

$1 = a$

$\therefore y = x^2$

1.4 $y = ax^2 + q$ ... Ⓐ

Substitute (−1; 3) into Ⓐ:

$3 = a(-1)^2 + q$

$\therefore 3 = a + q$

$a = 3 - q$ ... ①

Substitute (3; −5) into Ⓐ:

$-5 = a(3)^2 + q$

$-5 = 9a + q$ ... ②

Substitute ① into ②:

$9(3 - q) + q = -5$

$27 - 8q = -5$

$32 = 8q$

$\therefore q = 4$ ... ③

Substitute ③ into ①:

$a = 3 - 4$

$\therefore a = -1$

$\therefore y = -x^2 + 4$

2.1 $y = ax^2 + q$

$\therefore y = ax^2 - 3$ ... ①

Substitute (2; 0) into ①:

$0 = a(2)^2 - 3$

$3 = 4a$

$\frac{3}{4} = a$

$\therefore y = \frac{3}{4}x^2 - 3$

2.2 $y = ax^2 + 0$ ... ①

Substitute (2; 1) into ①:

$1 = a(2)^2$

$1 = 4a$

$\therefore a = \frac{1}{4}$

$\therefore y = \frac{1}{4}x^2$

2.3 $y = ax^2 + 2$ ... ①

Substitute (−2; −4) into ①:

$-4 = a(-2)^2 + 2$

$-6 = 4a$

$-\frac{3}{2} = a$

$\therefore y = -\frac{3}{2}x^2 + 2$

2.4 $y = ax^2 - 8$ ... ①

Substitute (1; −5) into ①:

$-5 = a(1)^2 - 8$

$3 = a$

$\therefore y = 3x^2 - 8$

2.5 $y = ax^2 + 1$ ... ①

Substitute (2; 4) into ①:

$4 = a(2)^2 + 1$

$3 = 4a$

$\frac{3}{4} = a$

$\therefore y = \frac{3}{4}x^2 + 1$

2.6 $y = ax^2 + q$ ... Ⓐ

Substitute $(1; -1)$ into Ⓐ:

$-1 = a(1)^2 + q$

$\therefore a = -1 - q$ ... ①

Substitute $(-2; 5)$ into Ⓐ

$5 = a(-2)^2 + q$

$5 = 4a + q$ ... ②

Substitute ① into ②:

$5 = 4(-1 - q) + q$

$5 = -4 - 3q$

$9 = -3q$

$\therefore q = -3$

$\therefore a = 2$

$\therefore y = 2x^2 - 3$

3. $f(x) = -3x^2 + 12$: frowny graph

3.1 x-intercept, let $y = 0$:

$0 = -3x^2 + 12$

$3x^2 = 12$

$x^2 = 4$

$x = \pm\sqrt{4}$

$= \pm 2$

$\therefore x = -2$ or $x = 2$

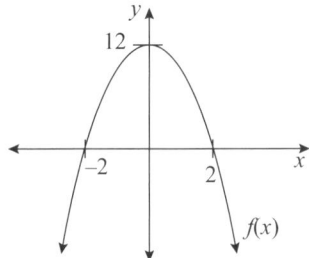

3.2 Maximum value at the turning point: $(0; 12)$

3.3.1 $f(3) = -3(3)^2 + 12 = k$

$-27 + 12 = k$

$\therefore k = -15$

3.3.2 $f(k) = -3k^2 + 12 = 3$

$-3k^2 = 3 - 12$

$-3k^2 = -9$

$k^2 = 3$

$k = \pm\sqrt{3}$

3.4 Domain: $x \in \mathbb{R}$, Range: $y \in (-\infty; 12]$

4.1 $f: y = ax^2 + 2$ ... ①

Substitute $(-3; -4)$ into ①:

$-4 = a(-3)^2 + 2$

$-6 = 9a$

$\therefore a = -\frac{2}{3}$

$\therefore f(x) = -\frac{2}{3}x^2 + 2$

4.2 x-intercept: let $f(x) = 0$:

$0 = -\frac{2}{3}x^2 + 2$

$\frac{2}{3}x^2 = 2$

$2x^2 = 6$

$x^2 = 3$

$x = \pm\sqrt{3}$

$\therefore C(-\sqrt{3}; 0)$

4.3 $CD = 2\sqrt{3}$ units

4.4 Range: $y \in (-\infty; 2]$

4.5 $m = \frac{2 + 4}{0 + 3} = 2$

$\therefore y = 2x + 2$

4.6 $y = 2$

4.7 $y = -\frac{1}{2}x$

5. $f(x) = -x^2 + 4$ and $g(x) = 2x + 4$

f: x-intercept: let $f(x) = 0$:

$0 = -x^2 + 4$

$x^2 = 4$

$x = \pm 2$

g: x-intercept, let $y = 0$:

$0 = 2x + 4$

$-4 = 2x$

$\therefore x = -2$

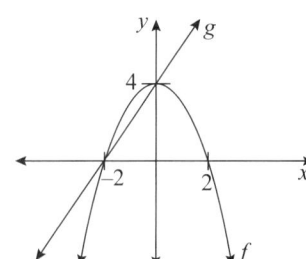

5.1 Range: $y \in (-\infty; 4]$

5.2 $f(x) = g(x)$

$\therefore -x^2 + 4 = 2x + 4$

$-x^2 - 2x = 0$

$-x(x + 2) = 0$

$\therefore x = 0$ or $x = -2$

5.3 $x \in (-2; 0)$

5.4 $x \in [-2; 2]$

5.5 Substitute $(2a; -5)$ into $y = -x^2 + 4$

$-5 = -(2a)^2 + 4$

$-9 = -4a^2$

$a^2 = \frac{9}{4}$

$\therefore a = \pm\frac{3}{2}$

5.6 $f(1) - g(0)$

$\quad = -(1)^2 + 4 - [2(0) + 4]$

$\quad = 3 - 4$

$\quad = -1$

5.7.1 $f(-3) = -(-3)^2 + 4 = a$

$\quad -9 + 4 = a$

$\quad \therefore a = -5$

5.7.2 $f(a) = -a^2 + 4 = 2$

$\quad a^2 = 2$

$\quad a = \pm\sqrt{2}$

5.8 $x < 0; x \in \mathbb{R}$

6. $f(x) = x^2 - 1$

6.1.1 x-intercept, let $y = 0$:

$\quad 0 = x^2 - 1$

$\quad\quad = (x - 1)(x + 1)$

$\quad \therefore x = 1 \text{ or } x = -1$

$\quad \therefore A(-1; 0) \text{ and } B(1; 0)$

$\quad \therefore OA = 1 \text{ unit}$

6.1.2 $OB = 1$ unit

6.1.3 $OC = 1$ unit

6.2 $m = \dfrac{0 - (-1)}{1 - 0} = 1$

6.3 $y = x - 1$

6.4 At $x = -1$:

$\quad y = -1 - 1 = -2$

$\quad \therefore E(-1; -2)$

$\quad \therefore AE = 2 \text{ units}$

7. $f(x) = x^2 - 9$

7.1 $Q(0; -9)$

7.2 x-intercept: let $f(x) = 0$:

$\quad\quad x^2 - 9 = 0$

$\quad (x - 3)(x + 3) = 0$

$\quad \therefore x = 3 \text{ or } x = -3$

$\quad \therefore A(3; 0)$

$\quad \therefore OA = 3 \text{ units}$

7.3 $m_g = \dfrac{3 - 0}{0 - 3} = -1$

$\quad \therefore g(x) = -x + 3$

7.4.1 $NQ = 3 + 9 = 12$ units

7.4.2 $g(-3) = -(-3) + 3 = 6$

$\quad \therefore BP = 6 \text{ units}$

7.4.3 $AP^2 = 6^2 + 6^2$ $\quad\quad$ (Pythagoras)

$\quad\quad = 72$

$\quad \therefore AP = \sqrt{72}$

$\quad\quad = 6\sqrt{2} \text{ units}$

7.5 Range: $y \in [-9; \infty)$

7.6 $x < 0; x \in \mathbb{R}$

7.7 $x = -4$

Exercise 9 (page 127)

1.1 Exponential function

$\quad \therefore y = ab^x + q$

$\quad a = 1$

$\quad q = 0$ $\quad\quad$ (*x*-axis is asymptote)

$\quad \therefore y = b^x$ $\quad\quad\quad\quad$ ... ①

$\quad$ Substitute B(1; 3) into ①:

$\quad 3 = b^1$

$\quad \therefore b = 3$

$\quad \therefore f(x) = 3^x$

1.2 Hyperbolic function

$\quad q = 0$ $\quad\quad$ (*x*-axis is asymptote)

$\quad \therefore y = \dfrac{a}{x}$

$\quad 3 = \dfrac{a}{-1}$

$\quad \therefore a = -3$

$\quad \therefore g(x) = -\dfrac{3}{x}$

1.3 Exponential function

$\quad q = -2$

$\quad \therefore y = ab^x - 2$ $\quad\quad\quad$ ... ①

$\quad$ Substitute (0; 0) into ①:

$\quad 0 = ab^0 - 2$

$\quad \therefore a = 2$

$\quad \therefore y = 2b^x - 2$ $\quad\quad\quad$ ... ②

$\quad$ Substitute P(2; 16) into ②

$\quad\quad 16 = 2b^2 - 2$

$\quad\quad 18 = 2b^2$

$\quad\quad b^2 = 9$

$\quad \therefore b = 3 \text{ only } (b > 0)$

$\quad \therefore f(x) = 2 \cdot 3^x - 2$

1.4 Hyperbolic function

$\quad q = -2$

$\quad \therefore y = \dfrac{a}{x} - 2$ $\quad\quad\quad$ ... ①

$\quad$ Substitute A(1; 4) into ①:

$\quad 4 = \dfrac{a}{1} - 2$

$\quad 6 = a$

$\quad \therefore g(x) = \dfrac{6}{x} - 2$

1.5 Exponential function

$q = 1$

$\therefore y = a \cdot b^x + 1$ ... ①

Substitute A(0; –2) into ①:

$-2 = a \cdot b^0 + 1$

$-3 = a$

$\therefore y = -3 \cdot b^x + 1$ ... ②

Substitute B(1; –5) into ②:

$-5 = -3b^1 + 1$

$-6 = -3b$

$2 = b$

$\therefore f(x) = -3.2^x + 1$

1.6 Exponential function

$q = -1$

$\therefore y = a \cdot b^x - 1$ ... ①

Substitute A(1; 1) into ①:

$1 = a \cdot b^1 - 1$

$2 = ab$

$\therefore a = \frac{2}{b}$ ... ②

Substitute B(2; 7) into ①

$7 = ab^2 - 1$

$\therefore ab^2 = 8$ ... ③

Substitute ② into ③:

$\frac{2}{b}(b^2) = 8$

$2b = 8$

$b = 4$ ... ④

Substitute ④ into ②:

$a = \frac{2}{4} = \frac{1}{2}$

$\therefore f(x) = \frac{4^x}{2} - 1$

2.1 $f(x) = 3^x$

2.1.1 $m = \frac{3-1}{1-0} = 2$

2.1.2 $f(-2) = 3^{-2} = \frac{1}{9}$

2.2 $g(x) = -\frac{3}{x}$

2.2.1 $6 = -\frac{3}{x}$

$6x = -3$

$x = -\frac{1}{2}$

$\therefore B\left(-\frac{1}{2}; 6\right)$

2.2.2 A(–1; 3) and $B\left(\frac{-1}{2}; 6\right)$

$\therefore m = \frac{6-3}{-\frac{1}{2} - (-1)} = 6$

2.2.3 Increasing

2.3 $f(x) = 2 \cdot 3^x - 2$

2.3.1 O(0; 0) and P(2; 16)

$\therefore m = \frac{16 - 0}{2 - 0} = 8$

2.3.2 $4 = 2 \cdot 3^x - 2$

$6 = 2 \cdot 3^x$

$\therefore 3^x = 3$

$\therefore x = 1$

2.3.3 Not symmetrical $\therefore$ no axis of symmetry

2.4 $g(x) = \frac{6}{x} - 2$

2.4.1 x-intercept, let $y = 0$:

$0 = \frac{6}{x} - 2$

$\frac{6}{x} = 2$

$2x = 6$

$\therefore x = 3$

$\therefore$ B(3; 0)

2.4.2 A(1; 4) and B(3; 0)

$\therefore m = \frac{4 - 0}{1 - 3} = -2$

2.4.3 Domain: $x \in \mathbb{R}, x \neq 0$, Range: $y \in \mathbb{R}, y \neq -2$

2.4.4 $x < 0$ or $x > 3; x \in \mathbb{R}$

2.4.5 $y = x - 2$ or $y = -x - 2$

2.5 $f(x) = -3 \cdot 2^x + 1$

2.5.1 A(0; –2) and B(1; –5)

$\therefore m = \frac{-2 - (-5)}{0 - 1} = -3$

2.5.2 x-intercept, let $y = 0$:

$0 = -3 \cdot 2^x + 1$

$-1 = -3 \cdot 2^x$

$\therefore 2^x = \frac{1}{3}$

$\therefore x \approx -1,58$ (trial and error)

2.5.3 $f(3) = -3 \cdot 2^3 + 1$

$= -3(8) + 1$

$= -23$

2.5.4 $-\frac{1}{2} = -3 \cdot 2^x + 1$

$-\frac{3}{2} = -3 \cdot 2^x$

$\frac{1}{2} = 2^x$

$\therefore 2^x = 2^{-1}$

$\therefore x = -1$

2.6 $f(x) = \frac{4^x}{2} - 1$

2.6.1 x-intercept: let $f(x) = 0$:

$$0 = \frac{4^x}{2} - 1$$

$$1 = \frac{4^x}{2}$$

$$4^x = 2$$

$$2^{2x} = 2$$

$$2x = 1$$

$$\therefore x = \frac{1}{2}$$

$\therefore$ x-intercept: $\left(\frac{1}{2}; 0\right)$

y-intercept, let $x = 0$:

$$y = \frac{4^0}{2} - 1 = -\frac{1}{2}$$

$\therefore$ y-intercept: $\left(0; -\frac{1}{2}\right)$

2.6.2 A(1; 1) and B(2; 7)

$$m = \frac{7-1}{2-1} = 6$$

2.6.3 $-\frac{7}{8} = \frac{4^x}{2} - 1$

$$\frac{4^x}{2} = \frac{1}{8}$$

$$4^x = \frac{1}{4}$$

$$4^x = 4^{-1}$$

$$\therefore x = -1$$

3. $h(x) = a \cdot 2^x + q$

$\qquad q = 1$

$\therefore h(x) = a \cdot 2^x + 1$ ... ①

Substitute P(0; −2) into ①:

$$-2 = a \cdot 2^0 + 1$$

$$\therefore a = -3$$

4. f: x-intercept, let $y = 0$:

$$\therefore 2x^2 - 8 = 0$$

$$2(x-2)(x+2) = 0$$

$$\therefore x = 2 \text{ or } x = -2$$

$\therefore$ A(−2; 0); B(2; 0) and C(0; −8)

g: x-intercept: let $g(x) = 0$:

$$\therefore 2^x - 8 = 0$$

$$2^x = 2^3$$

$$\therefore x = 3 \therefore \text{D}(3; 0)$$

4.2 $y = -8$

4.3 $(x; y) \rightarrow (x; -y)$

$$\therefore -y = 2^x - 8$$

$$y = -2^x + 8$$

$$\therefore h(x) = -2^x + 8$$

Exercise 10 (page 131)

1. **Set 1:**

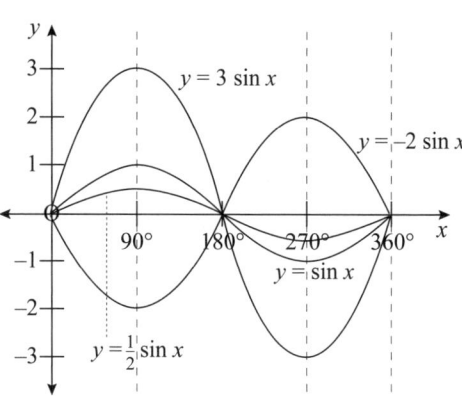

Set 2:

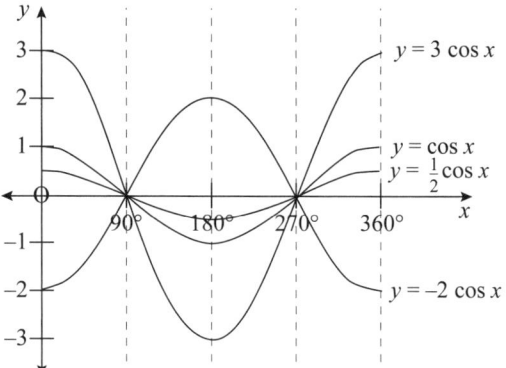

Set 3:

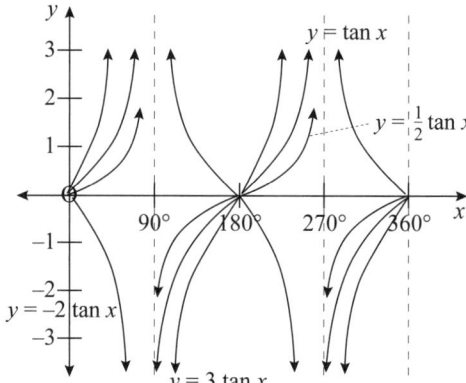

2. **Set 1:**

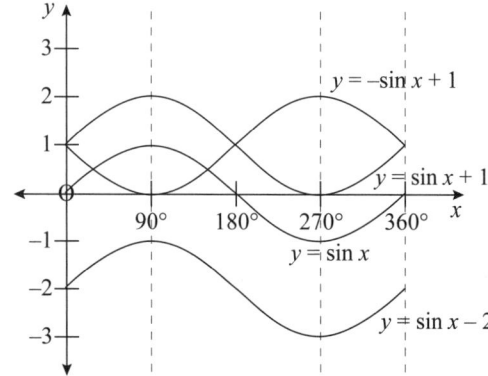

Set 2:

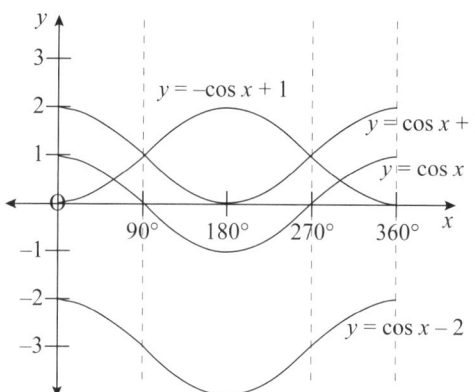

Set 3:

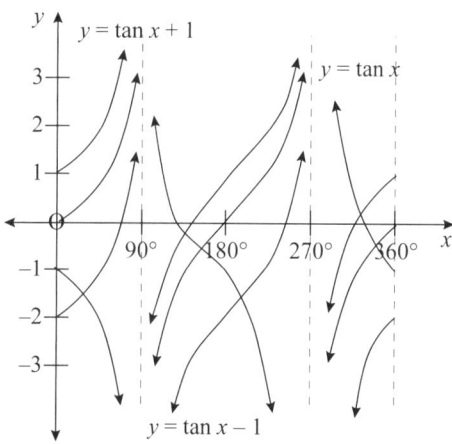

Exercise 11 (page 132)

1.1 A(90°; 3), B(180°; 0), C(270°; –3), D(360°; 0)

1.2 Period = 360°

1.3 Amplitude = 3

1.4 Range: $y \in [-3; 3]$

1.5 A(90°; 3) and C(270°; –3)

1.6 $x = 0°$ or $x = 180°$ or $x = 360°$

1.7 $x \in (0°; 180°)$

2.1 $a = -1$

2.2 A(90°; 0), B(90°; –1), C(270°; 1), D(270°; 0), E(360°; 1)

2.3 Amplitude = 1

2.4 $x = 0°$ or $x = 180°$ or $x = 360°$

2.5 $x \in (90°; 270°)$

2.6 Two solutions

3.

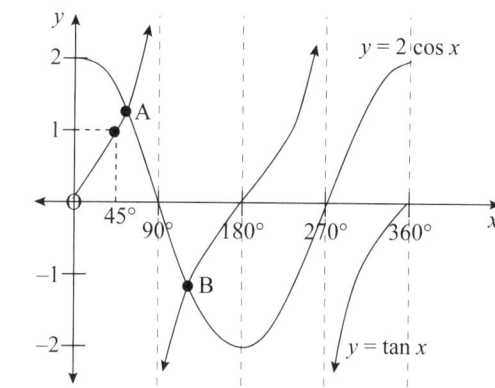

3.1 $x \in (180°; 360°)$

3.2 Min. value: $y = -2$

3.3 $x = 90°$ and $x = 270°$

3.4 Period = 180°

3.5 Points A and B on graph

3.6 $x = 0°$ or $x = 360°$

4.1 $a = 3$

4.2.1 Amplitude = 3

4.2.2 Period = 180°

4.2.3 Max. value: $y = 3$

4.2.4 Two solutions

4.3 $x = 0°$ or $x = 360°$

4.4.1 $x = 180°$

4.4.2 $x = 90°$ or $x = 270°$

4.4.3 $x \in (0°; 90°)$ and (90°; 180°)

4.4.4 $x = 0°$ or $x = 180°$ or $x = 360°$

Test A (page 134)

1.1 $3x - 2y = 5$

$3x - 5 = 2y$

$\therefore y = \frac{3}{2}x - \frac{5}{2}$

1.2 x-intercept, let $y = 0$:

$3x - 2(0) = 5$

$3x = 5$

$\therefore x = \frac{5}{3}$

y-intercept, let $x = 0$:

$3(0) - 2y = 5$

$-2y = 5$

$y = -\frac{5}{2}$

2. $4x - 3y - 12 = 0$

2.1 y-intercept, let $x = 0$:

$$4(0) - 3y - 12 = 0$$
$$3y = -12$$
$$\therefore y = -4$$

2.2 x-intercept, let $y = 0$:

$$4x - 3(0) - 12 = 0$$
$$4x = 12$$
$$x = 3$$

2.3 $4x - 12 = 3y$

$$\therefore y = \frac{4}{3}x - 4$$

3.1 $y = -3x + 1$

x-intercept, let $y = 0$:

$$0 = -3x + 1$$
$$3x = 1$$
$$\therefore x = \frac{1}{3}$$

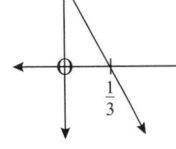

3.2 $y = \frac{2}{x} + 1$

x-intercept, let $y = 0$:

$$0 = \frac{2}{x} + 1$$
$$-1 = \frac{2}{x}$$
$$\therefore x = -2$$
$$\therefore y = 1$$

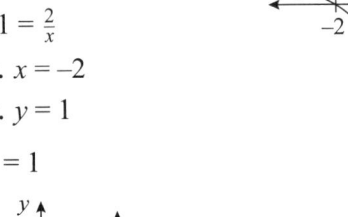

3.3 $x = 1$

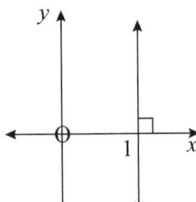

3.4 $y = 2x^2 - 2$

x-intercept, let $y = 0$:

$$0 = 2x^2 - 2$$
$$0 = 2(x - 1)(x + 1)$$
$$\therefore x = 1 \text{ or } x = -1$$

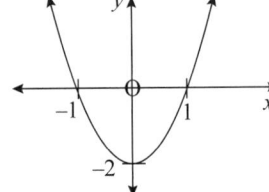

4.1 $m = \frac{1 - 0}{0 - 2} = \frac{1}{2}$

$$\therefore y = -\frac{1}{2}x + 1$$

4.2 $y = \frac{9}{x}$ ... ①

Substitute C(3; 3) into ①:

$$3 = \frac{9}{3}$$
$$\therefore a = 9$$
$$\therefore y = \frac{9}{x}$$

4.3 $y = ax^2 + q$

$$\therefore y = ax^2 - 9 \qquad \text{... ①}$$

Substitute (3; 0) into ①:

$$0 = a(3)^2 - 9$$
$$9a = 9$$
$$\therefore a = 1$$
$$\therefore y = x^2 - 9$$

4.4 $y = a \cdot b^x + q$

$$q = -2$$
$$\therefore y = a \cdot b^x - 2 \qquad \text{... ①}$$

Substitute G(1; 0) into ①:

$$0 = a \cdot b^1 - 2$$
$$2 = ab$$
$$\therefore a = \frac{2}{b} \qquad \text{... ②}$$

Substitute H(0; −1) into ①:

$$-1 = a \cdot b^0 - 2$$
$$1 = a \qquad \text{... ③}$$

Substitute ③ into ②:

$$1 = \frac{2}{b}$$
$$\therefore b = 2$$
$$\therefore y = 2^x - 2$$

5. $m = \frac{5 - (-3)}{-3 - 1} = -2$

6. $f(x) = -3x + 6$ and $g(x) = x + 2$

6.1 $-3x + 6 = x + 2$

$$-3x - x = 2 - 6$$
$$-4x = -4$$
$$\therefore x = 1$$
$$\therefore y = g(1) = 1 + 2 = 3$$
$$\therefore \text{ Point of intersection is (1; 3)}$$

6.2 f : x-intercept, let $y = 0$:

$$0 = -3x + 6$$
$$3x = 6$$
$$\therefore x = 2$$

g : x-intercept, let $y = 0$:

$$0 = x + 2$$
$$\therefore x = -2$$

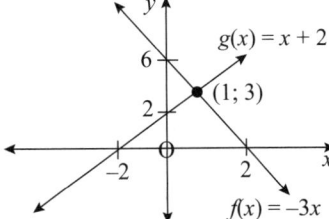

6.3 $y = -3x$

6.4 $y = -x$

6.5 $f(5) = -3(5) + 6 = -9$

Test B (page 135)

1.1   At P and Q:

$$-x + 2 = -\frac{3}{x}$$
$$-x^2 + 2x + 3 = 0$$
$$x^2 - 2x - 3 = 0$$
$$(x - 3)(x + 1) = 0$$
$$\therefore x = 3 \text{ or } -1$$
$$\therefore Q(3; -1) \text{ and } P(-1; 3)$$

1.2   B, x-intercept of f:

$$0 = -x + 2$$
$$\therefore x = 2$$
$$\therefore B(2; 0)$$

1.3   At D: $x = 2$ ∴ $y = -\frac{3}{2}$ ∴ BD $= \frac{3}{2}$ units

1.4   At F, $x = -6$ and $y = \frac{1}{2}$

At G, $x = -6$; $y = 8$

$$\therefore \text{GF} = 7\frac{1}{2} \text{ units}$$

2.1   At A, $y = 0$

$$\therefore \frac{4}{x} - 1 = 0$$
$$\therefore x = 4$$
$$\therefore A(4; 0)$$

2.2 $g(x) = x - 1$ (asymptote of hyperbola with vertical shift of 1 unit downwards)

2.3   At P and Q, $f(x) = g(x)$:

$$\frac{4}{x} - 1 = x - 1$$
$$4 - x = x^2 - x$$
$$0 = x^2 - 4$$
$$0 = (x - 2)(x + 2)$$
$$x = 2 \text{ or } -2$$
$$\therefore P(2; 1) \text{ and } Q(-2; -3)$$

2.4 $h(x) = -\frac{4}{x} + 1$ (the sign of the y-values changes)

3. $y = -x + 4$ and $2y - x - 2 = 0$

3.1   For $y = -x + 4$:

x-intercept, let $y = 0$:
$$0 = -x + 4$$
$$\therefore x = 4$$
$$\therefore B(4; 0)$$

For $2y - x - 2 = 0$:

x-intercept, let $y = 0$:
$$\therefore x = -2$$
$$\therefore A(-2; 0)$$
$$\therefore \text{AB} = 4 + 2 = 6 \text{ units}$$

3.2   For $y = -x + 4$

y-intercept, let $x = 0$:
$$y = -(0) + 4$$
$$\therefore y = 4$$
$$\therefore C(0; 4)$$
$$\therefore \text{CD} = 4 - 1 = 3 \text{ units}$$

3.3   For $y = -x + 4$, at $y = -1$:
$$-1 = -x + 4$$
$$\therefore x = 5$$
$$\therefore Q(5; -1)$$

For $2y - x - 2 = 0$, at $y = -1$:
$$2(-1) - x - 2 = 0$$
$$\therefore x = -4$$
$$\therefore P(-4; -1)$$
$$\therefore \text{PQ} = 4 + 5 = 9 \text{ units}$$

3.4   For G, at $x = 3$:
$$y = -3 + 4 = 1$$
$$\therefore G(3; 1)$$

For F, at $x = 3$:
$$2y - 3 - 2 = 0$$
$$2y = 5$$
$$\therefore y = \frac{5}{2}$$
$$\therefore F\left(3; \frac{5}{2}\right)$$
$$\therefore \text{FG} = \frac{5}{2} - 1 = \frac{3}{2} \text{ units}$$

4.1 g: $m = \tan \theta$
$$= \tan 135°$$
$$= -1$$
$$\therefore g(x) = -x$$

f: $y = \frac{a}{x}$ ... ①

Substitute $C\left(4; \frac{3}{2}\right)$ into ①:
$$\frac{3}{2} = \frac{a}{4}$$
$$\therefore a = 6$$
$$\therefore f(x) = \frac{6}{x}, x > 0$$

h: $A\left(\sqrt{6}; \sqrt{6}\right)$
$$\therefore x^2 + y^2 = r^2$$ ... ①

Substitute $A\left(\sqrt{6}; \sqrt{6}\right)$ into ①:
$$6 + 6 = r^2$$
$$\therefore r^2 = 12$$
$$\therefore h(x) = \sqrt{12 - x^2}$$

m: $y = ax^2 - 4$ ... ①

Substitute $(2\sqrt{3}; 0)$ into ①:

$0 = a(2\sqrt{3})^2 - 4$

$4 = 12a$

$\therefore a = \frac{1}{3}$

$\therefore m(x) = \frac{1}{3}x^2 - 4$

4.2 $A(\sqrt{6}; \sqrt{6})$

4.3 $m = \frac{\frac{3}{2} - (-4)}{4 - 0}$

$= \frac{\frac{11}{2}}{4}$

$= \frac{11}{8}$

$\therefore y = \frac{11}{8}x - 4$

Test C (page 136)

1. $f(x) = 3x + 9$, $g(x) = -x^2 + 9$

1.1 f: x-intercept, let $y = 0$:

$0 = 3x + 9$

$3x = -9$

$\therefore x = -3$

y-intercept: $y = 9$

g: x-intercept, let $y = 0$:

$0 = -x^2 + 9$

$x^2 - 9 = 0$

$(x - 3)(x + 3) = 0$

$\therefore x = 3$ or $x = -3$

y- intercept: $y = 9$

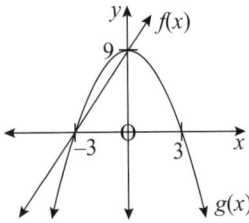

1.2 $m = -\frac{1}{3}$

y-intercept: $y = -6$

$\therefore h(x) = -\frac{1}{3}x - 6$

1.3 $h(x) = f(x)$

$\therefore -\frac{1}{3}x - 6 = 3x + 9$

$-15 = \frac{10}{3}x$

$-45 = 10x$

$x = -\frac{9}{2}$

$y = f\left(-\frac{9}{2}\right) = 3\left(-\frac{9}{2}\right) + 9 = -\frac{9}{2}$

$\therefore$ Point of intersection $\left(-\frac{9}{2}; -\frac{9}{2}\right)$

1.4 Range: $y \in (-\infty; 9]$

1.5 $x < 0; x \in \mathbb{R}$

1.6 $-3 \le x \le 0; x \in \mathbb{R}$

2. $f(x) = 2x + 3$ and $g(x) = -x + 6$

2.1 $f(x) = g(x)$

$2x + 3 = -x + 6$

$2x + x = 6 - 3$

$3x = 3$

$\therefore x = 1$

$y = f(1)$

$= 2(1) + 3$

$= 5$

$\therefore A(1; 5)$

2.2.1 f: x-intercept, let $y = 0$:

$0 = 2x + 3$

$2x = -3$

$\therefore x = -\frac{3}{2}$

$\therefore B\left(-\frac{3}{2}; 0\right)$

$\therefore f(x) > 0$ for $x > -\frac{3}{2}; x \in \mathbb{R}$

2.2.2 g: x-intercept, let $y = 0$:

$0 = -x + 6$

$\therefore x = 6$

$\therefore g(x) \le 0$ for $x \ge 6; x \in \mathbb{R}$

2.2.3 $f(x) . g(x) > 0$

$-\frac{3}{2} < x < 6; x \in \mathbb{R}$

2.3 $y = 2 . 3^x - 6$

x-intercept, let $y = 0$:

$0 = 2 . 3^x - 6$

$2 . 3^x = 6$

$3^x = 3$

$\therefore x = 1$

y-intercept, let $x = 0$:

$y = 2 . 3^0 - 6 = -4$

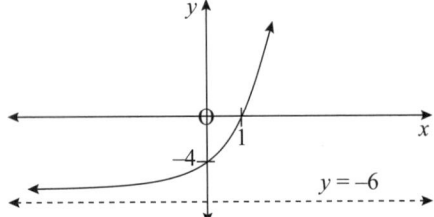

3.1 SW $= y = 5\frac{1}{4} = \frac{17}{4}$

  $\therefore y = \frac{17}{4} - \frac{5}{4}$

  $\quad = 3$

  $\therefore$ W(2; 3)

  $f : y = ax^2 + q$ $\quad\quad\quad\quad\quad$ ... Ⓐ

  Substitute A(4; 6) into Ⓐ:

  $6 = a(4)^2 + q$

  $\therefore q = -16a + 6$ $\quad\quad\quad\quad$ ... ①

  Substitute W(2; 3) into Ⓐ:

  $3 = a(2)^2 + q$

  $3 = 4a + q$ $\quad\quad\quad\quad\quad$ ... ②

  Substitute ① into ②:

  $4a - 16a + 6 = 3$

  $\quad\quad -12a = 3 - 6$

  $\quad\quad -12a = -3$

  $\therefore a = \frac{1}{4}$ $\quad\quad\quad\quad\quad$ ... ③

  Substitute ③ into ①:

  $q = -16\left(\frac{1}{4}\right) + 6 = 2$

  $\therefore f(x) = \frac{1}{4}x^2 + 2$

3.2 $y = ax^2 + q$

  $\therefore a = \frac{1}{4}$

  $\therefore$ T(0; 2)

  $\therefore q = 2 - 4\frac{1}{4} = -\frac{9}{4}$

  $\therefore g(x) = \frac{1}{4}x^2 - \frac{9}{4}$

3.3 M(−2; 3)

3.4 MW $= 2 + 2 = 4$ units

3.5 OT $= 2$ units

3.6 $f(4) - g(4)$

  $= \frac{1}{4}(4)^2 + 2 - \left[\frac{1}{4}(4)^2 - \frac{9}{4}\right]$

  $= 6 - \frac{7}{4}$

  $= \frac{17}{4}$

3.7 Range: $y \in \left[-\frac{9}{4}; \infty\right)$

Topic 8   Euclidean geometry (page 138)

Exercise 1 (page 144)

1.	$a = 140°$	**2.**	$b = 30°$	**3.**	$c = 53°$	**4.**	$d = 40°$
5.	$e = 60°$	**6.**	$f = 35°$	**7.**	$g = 140°$	**8.**	$h = 40°; i = 115°$
9.	$j = 20°$	**10.**	$k = 30°$	**11.**	$m = 18°; l = 72°$	**12.**	$n = 117°$
13.	$p = 74°$	**14.**	$q = 70°$	**15.**	$s = 53°; r = 53°$	**16.**	$t = 116°$
17.	$u = 131°; v = 58°$	**18.**	$w = 120°$	**19.**	$a = 65°$	**20.**	$b = 50°$
21.	$c = 28°$	**22.**	$d = 50°$	**23.**	$e = 53°; f = 60°$	**24.**	$g = 65°; h = 115°$
25.	$i = 60°; j = 120°$	**26.**	$k = 45°; l = 135°$	**27.**	$m = 13$ mm	**28.**	$n = 15$ mm
29.	$p = \sqrt{72}$ mm $= 6\sqrt{2}$ mm			**30.**	$q = 90°$		

Exercise 2 (page 146)

1.1 $\hat{B} = \hat{A} = x$ (Angles opposite equal sides $\triangle$ABC; AC = BC.)

  $\therefore D\hat{C}B = \hat{B} = x$ (Angles opposite equal sides $\triangle$DBC; DC = DB.)

  $\therefore A\hat{D}C = 2x$ (Exterior angle equals sum of two interior opposite angles.)

  $\therefore A\hat{C}D = 2x$ (Angles opposite equal sides $\triangle$ADC; AD = DC.)

  $\hat{A} + A\hat{D}C + A\hat{C}D = 180°$ (Sum of interior angles of a triangle.)

  $\therefore 5x = 180°$

  $\therefore x = 36°$

1.2 $B\hat{D}C = x + 12°$ (Exterior angle of $\triangle$ADC equals sum of interior opposite angles.)

  $\hat{B} = B\hat{D}C = x + 12°$ (Angles opposite equal sides $\triangle$DBC; DC = BC.)

  $A\hat{C}B = \hat{B} = x + 12°$ (Angles opposite equal sides $\triangle$ABC; AB = AC.)

  $\hat{A} + \hat{B} + A\hat{C}B = 180°$ (Sum of interior angles in $\triangle$ABC.)

  $\therefore x + x + 12° + x + 12° = 180°$

  $\therefore 3x = 156°$

  $\therefore x = 52°$

1.3 $A\hat{C}B = 180° - (2x + 20°)$ (Straight line BCE.)

  $= 160° - 2x$

  $D\hat{B}A = \hat{A} + A\hat{C}B$ (Exterior angle of $\triangle$ABC equals sum of interior opposite angles.)

  $\therefore x + 80° = 70° + 160° - 2x$

  $\therefore x + 2x = 230° - 80°$

  $\therefore 3x = 150°$

  $\therefore x = 50°$

1.4 $2a + 2b = 180°$ (Co-interior angles, AD $\parallel$ BC.)

  $\therefore a + b = 90°$

  $x + a + b = 180°$ (Sum of interior angles in $\triangle$ABK.)

  $\therefore x + 90° = 180°$

  $\therefore x = 90°$

1.5 $\text{E}\hat{\text{D}}\text{B} = \hat{\text{A}} = x + 15°$ (Corresponding angles, AC ∥ DE.)

   $\text{D}\hat{\text{E}}\text{C} = \hat{\text{D}} + \hat{\text{B}}$ (Exterior angle of △DEB equals sum of interior opposite angles.)

   $\therefore 110° - x = x + 15° + 55°$

   $\therefore 110° - 70° = x + x$

   $\therefore 40° = 2x$

   $\therefore x = 20°$

1.6 $a + b + 140° = 180°$ (Sum of interior angles of △DBC.)

   $\therefore a + b = 40°$

   $\therefore 2a + 2b = 80°$

   $x + 2a + 2b = 180°$ (Sum of interior angles of △ABC.)

   $\therefore x + 80° = 180°$

   $\therefore x = 100°$

1.7 $a + b + 70° = 180°$ (Sum of interior angles of △BCF.)

   $\therefore a + b = 110°$

   $\therefore 2a + 2b = 220°$

   $\text{A}\hat{\text{B}}\text{C} = 180° - 2a$ (Straight line ABD.)

   $\text{A}\hat{\text{C}}\text{B} = 180° - 2b$ (Straight line ACE.)

   $\hat{\text{A}} + \text{A}\hat{\text{B}}\text{C} + \text{A}\hat{\text{C}}\text{B} = 180°$ (Sum of interior angles in △ABC.)

   $\therefore x + 180° - 2a + 180° - 2b = 180°$

   $\therefore x = 180° - 180° - 180° + 2a + 2b$

   $= -180° + 220°$

   $= 40°$

2.1 $\text{A}\hat{\text{B}}\text{C} + \text{A}\hat{\text{C}}\text{B} = 110°$ (Sum of interior angles in △ABC.)

   $\therefore \text{A}\hat{\text{B}}\text{C} = \text{A}\hat{\text{C}}\text{B} = 55°$ (Given.)

   $\therefore \text{A}\hat{\text{C}}\text{D} = 125°$ (Straight line BCD.)

   $\therefore \text{A}\hat{\text{C}}\text{E} = \text{E}\hat{\text{C}}\text{D} = 62,5°$ (CE bisects AĈD, given.)

2.2 $\text{A}\hat{\text{E}}\text{C} = \text{E}\hat{\text{C}}\text{D}$ (Alternate angles, AE ∥ BD.)

   $= 62,5°$

3. $2\hat{\text{B}} = \hat{\text{A}}$ (Given.)

   $\therefore \hat{\text{B}} = \frac{1}{2}\hat{\text{A}}$

   $2\hat{\text{C}} = \hat{\text{A}}$ (Given.)

   $\therefore \hat{\text{C}} = \frac{1}{2}\hat{\text{A}}$

   $\hat{\text{A}} + \hat{\text{B}} + \hat{\text{C}} = 180°$ (Sum of interior angles in △ABC.)

   $\therefore \hat{\text{A}} + \frac{1}{2}\hat{\text{A}} + \frac{1}{2}\hat{\text{A}} = 180°$

   $\therefore 2\hat{\text{A}} = 180°$

   $\therefore \hat{\text{A}} = 90°$

   $\therefore \hat{\text{B}} = \frac{1}{2}\hat{\text{A}}$

   $= 45°$

   and $\hat{\text{C}} = \frac{1}{2}\hat{\text{A}} = 45°$

4. PR̂S = P̂ + Q̂   (Exterior angle of △PQR equals sum of interior opposite angles.)

∴ 2P̂ = P̂ + Q̂   (PR̂S = 2P̂, given.)

∴ 2P̂ – P̂ = Q̂

∴ P̂ = Q̂

5. B̂ = Ĉ   (Angles opposite equal sides △ABC, AB = BC.)

DÂC = B̂ + Ĉ   (Exterior angle of △ABC equals sum of interior opposite angles.)

= 2Ĉ   (B̂ = Ĉ, proved)

EÂC = $\frac{1}{2}$DÂC = Ĉ   (AE bisects DÂC, given.)

∴ AE ∥ BC   (EÂC = alternate Ĉ)

6.1 $AB^2 = AE^2 + BE^2$ (Pythagoras.)

$= (AD^2 – DE^2) + BE^2$ (Pythagoras.)

$= AD^2 – DE^2 + (BD + DE)^2$ (Pythagoras.)

$= AD^2 – DE^2 + BD^2 + 2BD \cdot DE + DE^2$

$= AD^2 + BD^2 + 2BD \cdot DE$

6.2 $AC^2 = AE^2 + EC^2$ (Pythagoras.)

$= AE^2 + (DC – DE)^2$

$= AE^2 + DC^2 – 2DC \cdot DE + DE^2$

6.3 $AB^2 + AC^2 = AD^2 + BD^2 + 2BD \cdot DE + AE^2 + DC^2 – 2DC \cdot DE + DE^2$

$= AD^2 + BD^2 + 2BD \cdot DE + (AE^2 + DE^2) + BD^2 – 2BD \cdot DE$ (BD = DC, given.)

$= AD^2 + 2BD^2 + AD^2$ (Pythagoras.)

$= 2AD^2 + 2BD^2$

Exercise 3   (page 153)

1. The diagonals of a rectangle are equal.

2. All the sides are equal.
The diagonals intersect at right angles.
The diagonals bisect the angles.

3.1 Rectangle and square

3.2 Rhombus, square and kite

3.3 Parallelogram, rectangle, rhombus and square

3.4 Rhombus and square

3.5 Parallelogram, rectangle, rhombus and square

4.1 $a = 35°$ **4.2** $b = 115°$ **4.3** $c = 100°$ **4.4** $d = 36°$

4.5 $e = 80°$ **4.6** $f = 106°$ **4.7** $g = 60°$ **4.8** $h = 50°$

4.9 $i = 40°$ **4.10** $j = 135°$

Exercise 4   (page 155)

1.1 AB̂C = AD̂C = 70°   (Opposite angles of a parallelogram are equal.)

∴ ABO = 50°

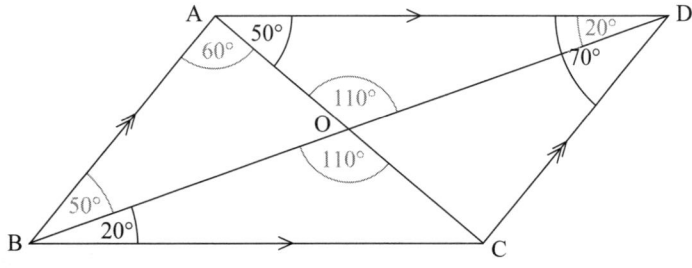

1.2   A$\hat{D}$B = D$\hat{B}$C = 20°   (Alternate angles, AD ∥ BC.)

     A$\hat{O}$D = 110°   (Sum of interior angles of △AOD.)

     B$\hat{O}$C = A$\hat{O}$D = 110°   (Vertically opposite angles.)

1.3   B$\hat{A}$O + 50° = 110°   (Exterior angle of △ABO equals the sum of interior opposite angles.)

     ∴ B$\hat{A}$O = 60°

2.   A$\hat{B}$E = A$\hat{E}$B = 62°   (Angles opposite equal sides △ABE; AB = AE.)

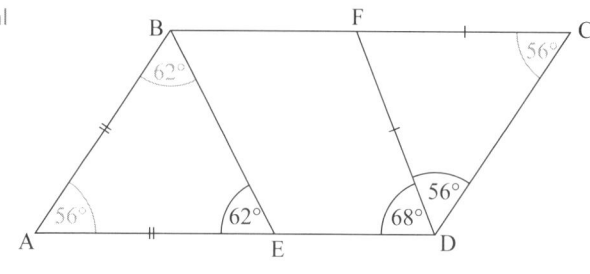

     $\hat{A}$ = 56°   (Sum of interior angles of △BAE.)

     $\hat{C}$ = F$\hat{D}$C = 56°   (Angles opposite equal sides △FDC; FC = FD.)

     $\hat{A}$ + A$\hat{D}$C = 56° + 68° + 56°

             = 180°

     ∴ AB ∥ DC   (Co-interior angles are supplementary.)

     $\hat{C}$ + C$\hat{D}$A = 56° + 56° + 68°

             = 180°

     ∴ BC ∥ AD   (Co-interior angles are supplementary.)

     ∴ BADC is a parallelogram   (Both pairs of opposite sides are parallel.)

3.   A$\hat{B}$D = C$\hat{B}$D = 36°   (Diagonals of a rhombus bisect angles.)

     $\hat{C}$ + A$\hat{B}$C = 180°   (Co-interior angles, AB ∥ DC.)

     ∴ $\hat{C}$ = 108°

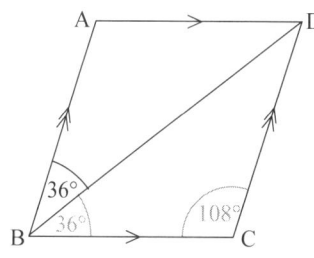

4.1   A$\hat{B}$E = A$\hat{E}$B   (Angles opposite equal sides △ABE; AB = AE.)

     = $\dfrac{180° - 110°}{2}$ (Interior angles of △ABE.)

     = 35°

     ∴ C$\hat{E}$D = 70°   (Straight line AED.)

4.2   DC = AB = 50 mm   (Opposite sides of a parallelogram are equal.)

     ∴ $\hat{D}$ = 70°   (Co-interior angles, AB ∥ DC.)

     ∴ EC = DC = 50 mm   (Sides opposite equal angles △EDC.)

     AD = BC = 80 mm   (Opposite sides of a parallelogram are equal.)

     ED = AD − AE

          = 80 − 50

          = 30 mm

5.1   PR = QS   (Diagonals of a rectangle are equal.)

     PO = OR and QO = OS   (Diagonals of a rectangle bisect each other.)

     ∴ PO = OS

     ∴ S$\hat{P}$O = 20°   (Angles opposite equal sides △POS; PO = OS.)

     P$\hat{O}$Q = S$\hat{P}$O + P$\hat{S}$O   (Exterior angle △POS equals sum of interior opposite angles.)

     ∴ P$\hat{O}$Q = 40°

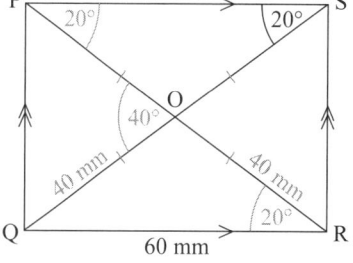

5.2	$P\hat{R}Q = S\hat{P}O$	(Alternate angles, PS ∥ QR.)
	$= 20°$	
5.3	$PS = QR$	(Opposite sides of a parallelogram are equal.)
	$= 60$ mm	
5.4	$PO = OR$	(Diagonals of a parallelogram bisect each other.)
	$= 40$ mm	
5.5	$OQ = 40$ mm	(Diagonals of a rectangle are equal and bisect each other.)

6.	$M\hat{K}P = A\hat{K}P$	(Diagonals of a square bisect the angles.)	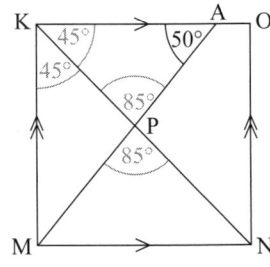
	$\therefore A\hat{K}P = 45°$		
	$\therefore A\hat{P}K = 85°$	(Sum of interior angles of △KPA.)	
	$M\hat{P}N = K\hat{P}A$	(Vertically opposite angles.)	
	$= 85°$		

7.1	$D\hat{C}O = O\hat{A}B$	(Alternate angles, AB ∥ DC.)	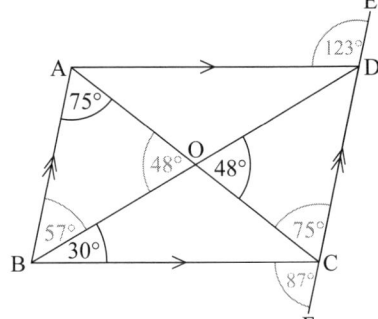
	$= 75°$		
	$B\hat{D}E = D\hat{O}C + D\hat{C}O$	(Exterior angle of △OCD equals sum of interior opposite angles.)	
	$= 48° + 75°$		
	$= 123°$		
7.2	$A\hat{O}B = D\hat{O}C = 48°$	(Vertically opposite angles.)	
	$A\hat{B}O = 57°$	(Sum of interior angles of △ABO.)	
	$B\hat{C}F = A\hat{B}C$	(Alternate angles, AB ∥ EF.)	
	$= 87°$		

8.	$D\hat{E}C = E\hat{C}B$	(Alternate angles, AD ∥ BC.)	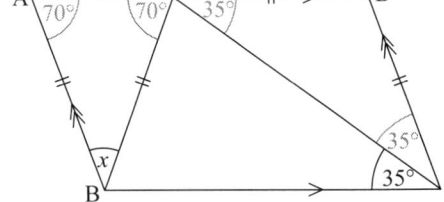
	$= 35°$		
	$E\hat{C}D = D\hat{E}C$	(Angles opposite equal sides △EDC; ED = DC.)	
	$= 35°$		
	$\hat{A} = D\hat{C}B$	(Opposite angles of parallelogram are equal.)	
	$= 70°$		
	$AB = DC$	(Opposite sides of parallelogram are equal.)	
	$= BE$	(Given.)	
	$A\hat{E}B = \hat{A}$	(Angles opposite equal sides △ABE; AB = BE.)	
	$= 70°$		
	$x = 40°$	(Sum of interior angles of △ABE.)	

9.	$R\hat{S}T = 30°$	(Angles opposite equal sides △TRS; TR = SR.)	
	$P\hat{S}T = 30°$	(Alternate angles, PS ∥ QR.)	
	$\hat{Q} = P\hat{S}R$	(Opposite angles of parallelogram are equal.)	
	$= 60°$		
	$P\hat{T}Q = \hat{Q}$	(Angles opposite equal sides △PQT; PQ = PT.)	
	$= 60°$		

$\hat{P} = 60°$ (Sum of interior angles of △PQT.)

∴ QT = PQ   (Equilateral △PQT; equiangular.)

PQ = SR   (Opposite sides of a parallelogram are equal.)

∴ QT = 60 mm

10.   UT = RS   (Opposite sides of a rectangle are equal.)

     = 16 mm

∴ WT = 9 mm

SV = VT   (V the midpoint of ST, given.)

     = 12 mm

RU = ST   (Opposite sides of a rectangle are equal.)

     = 24 mm

$VW^2 = WT^2 + VT^2$ (Pythagoras, $\hat{T} = 90°$.)

     = 144 + 81

     = 225

$RV^2 = 16^2 + 12^2$ (Pythagoras, $\hat{S} = 90°$.)

     = 400

$RW^2 = 24^2 + 7^2$ (Pythagoras, $\hat{V} = 90°$.)

     = 625

$RV^2 + VW^2 = 225 + 400$

     = 625

     $= RW^2$

∴ $R\hat{V}W = 90°$ (Converse of Pythagoras.)

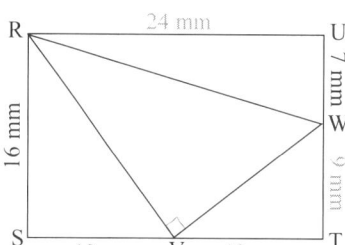

11.   BO = OD   (Diagonals of a rhombus bisect each other.)

     = 24 mm

$A\hat{O}B = 90°$ (Diagonals of a rhombus intersect at right angles.)

$AO^2 = AB^2 - BO^2$ (Pythagoras, $A\hat{O}B = 90°$.)

     = 676 − 576

     = 100

∴ AO = 10 mm

AO = OC   (Diagonals of a rhombus bisect each other.)

∴ AC = 20 mm

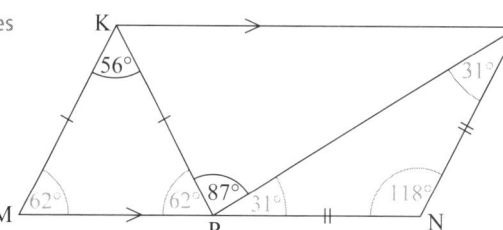

12. $\hat{M} = K\hat{P}M$ (Angles opposite equal sides △KMP; KM = KP.)

     $= \dfrac{180° - 56°}{2}$ (Sum of interior angles of △KMP.)

     = 62°

∴ $R\hat{P}N = 180° - (87° + 62°)$ (Straight line MPN.)

     = 31°

$P\hat{R}N = R\hat{P}N$ (Angles opposite equal sides △RPN; RN = PN.)

     = 31°

$\hat{N} = 118°$ (Sum of interior angles of △RPN.)

∴ KM ∥ RN   ($\hat{M}$ and co-interior $\hat{N}$ are supplementary.)

∴ KMNR is a parallelogram   (Both pairs of opposite sides are parallel.)

13. $\quad$ S$\hat{\text{R}}$P = Q$\hat{\text{P}}$R $\qquad$ (Alternate angles, PQ ∥ SR.)

$\qquad\quad$ = 50°

$\qquad$ S$\hat{\text{O}}$R = 90° $\qquad\qquad$ (Sum of interior angles of △SOR.)

$\qquad$ ∴ PQRS is a rhombus $\qquad$ (Parallelogram with diagonals that intersect at 90°.)

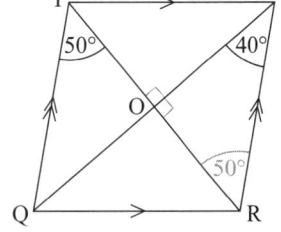

Or:

In △POS and △ROS:

$\qquad$ OS = OS $\qquad\qquad\qquad$ (Common.)

$\qquad$ P$\hat{\text{O}}$S = S$\hat{\text{O}}$R $\qquad\qquad$ (Proved.)

$\qquad$ PO = RO $\qquad\qquad\qquad$ (Diagonals of a parallelogram bisect each other.)

$\qquad$ ∴ △POS ≡ △ROS $\qquad\quad$ (SAS.)

$\qquad$ ∴ PS = SR $\qquad\qquad\qquad$ (△POS ≡ △ROS.)

$\qquad$ ∴ PQRS is a rhombus $\qquad$ (Parallelogram with two adjacent sides that are equal.)

14. $\quad$ A$\hat{\text{B}}$E = A$\hat{\text{E}}$B $\qquad\qquad$ (Angles opposite equal sides △AEB; AB = AE.)

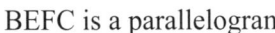

$\qquad\quad = \dfrac{180° - 48°}{2}$ $\qquad\qquad$ (Sum of interior angles of △AEB.)

$\qquad\quad = 66°$

$\qquad$ ∴ E$\hat{\text{B}}$C = 180° − 66° $\qquad$ (Straight line ABC.)

$\qquad\qquad = 114°$

$\qquad$ BEFC is a parallelogram $\qquad$ (BC = EF and BC ∥ EF.)

$\qquad$ ∴ BE ∥ CF

$\qquad$ F$\hat{\text{C}}$D = E$\hat{\text{B}}$C $\qquad\qquad$ (Corresponding angles, BE ∥ CF.)

$\qquad\qquad = 114°$

$\qquad$ ∴ $\hat{\text{D}}$ = 46° $\qquad\qquad\qquad$ (Sum of interior angles of △CFD.)

15.1 $\quad$ A$\hat{\text{B}}$C + D$\hat{\text{C}}$B = 180° $\qquad$ (Co-interior angles, AB ∥ DC.)

$\qquad$ E$\hat{\text{B}}$C = $\frac{1}{2}$A$\hat{\text{B}}$C $\qquad\qquad$ (BE bisects A$\hat{\text{B}}$C, given.)

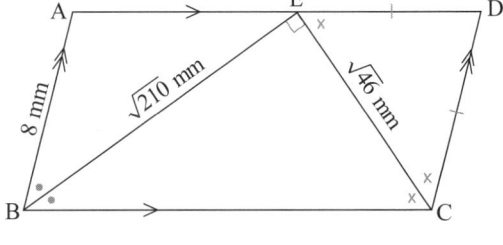

$\qquad$ E$\hat{\text{C}}$B = $\frac{1}{2}$D$\hat{\text{C}}$B $\qquad\qquad$ (CE bisects B$\hat{\text{C}}$D, given.)

$\qquad$ ∴ E$\hat{\text{B}}$C + E$\hat{\text{C}}$B = $\frac{1}{2}$A$\hat{\text{B}}$C + $\frac{1}{2}$D$\hat{\text{C}}$B

$\qquad\qquad\qquad\qquad\quad = \frac{1}{2}$(A$\hat{\text{B}}$C + D$\hat{\text{C}}$B)

$\qquad\qquad\qquad\qquad\quad = 90°$

$\qquad$ ∴ B$\hat{\text{E}}$C = 90° $\qquad\qquad$ (Sum of interior angles of △EBC.)

$\qquad$ ∴ BC^2 = BE^2 + EC^2 $\qquad$ (Pythagoras, B$\hat{\text{E}}$C = 90°.)

$\qquad\qquad = 210 + 46$

$\qquad\qquad = 256$

$\qquad$ ∴ BC = 16 mm

15.2 $\quad$ D$\hat{\text{E}}$C = E$\hat{\text{C}}$B $\qquad\qquad$ (Alternate angles, AD ∥ BC.)

$\qquad$ ∴ ED = DC $\qquad\qquad\qquad$ (Sides opposite equal angles △EDC; D$\hat{\text{E}}$C = D$\hat{\text{C}}$E.)

$\qquad$ DC = AB $\qquad\qquad\qquad$ (Opposite sides of a parallelogram.)

$\qquad$ ∴ ED = DC = 8 mm

16. $P\hat{T}Q = 60°$ (Straight line QTS.)

$\quad$ $T\hat{Q}P = T\hat{P}Q$ (Angles opposite equal sides △PQT; PQ = PT.)

$\quad\quad\quad = 60°$ (Sum of interior angles of △PQT.)

$\quad$ ∴ PT = TQ   (Sides opposite equal angles △PTQ; $T\hat{P}Q = P\hat{Q}T$.)

$\quad$ PT = TR and QT = TS   (Diagonals of a parallelogram bisect each other.)

$\quad$ ∴ PT = TR = QT = TS

$\quad$ $S\hat{P}T = P\hat{S}T$ (Angles opposite equal sides △PTS; PT = ST.)

$\quad\quad = \dfrac{180° - 120°}{2}$ (Sum of interior angles of △PTS.)

$\quad\quad = 30°$

$\quad$ ∴ $Q\hat{P}S = 60° + 30°$

$\quad\quad\quad = 90°$

$\quad$ ∴ PQRS is a rectangle   (The parallelogram has one right angle.)

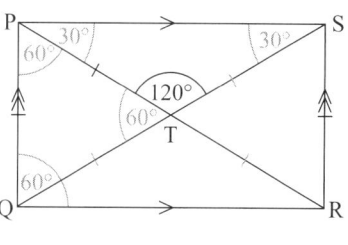

17. Let $\hat{B} = x$

$\quad$ ∴ $A\hat{F}B = x$ ($A\hat{F}B = \hat{B}$, given.)

$\quad$ ∴ AB = AF   (Sides opposite equal angles △ABF; $\hat{B} = A\hat{F}B$.)

$\quad$ AB = DC   (Opposite sides of a parallelogram are equal.)

$\quad$ ∴ FC = DC   (AF = FC, given.)

$\quad$ $D\hat{F}C = 180° - (72° + x)$ (Straight line BFC.)

$\quad\quad\quad = 108° - x$

$\quad$ $F\hat{D}C = D\hat{F}C$ (Angles opposite equal sides △DFC; DC = FC.)

$\quad\quad\quad = 108° - x$

$\quad$ $\hat{C} = 180° - (108° - x + 108° - x)$ (Sum of interior angles of △DFC.)

$\quad\quad = 2x - 36°$

$\quad$ $\hat{B} + \hat{C} = 180°$ (Co-interior angles, AB ∥ CD.)

$\quad$ ∴ $x + 2x - 36° = 180°$

$\quad\quad\quad\quad ∴ 3x = 216°$

$\quad\quad\quad\quad\quad ∴ x = 72°$

$\quad\quad\quad\quad ∴ \hat{C} = 2(72°) - 36°$

$\quad\quad\quad\quad\quad\quad = 108°$

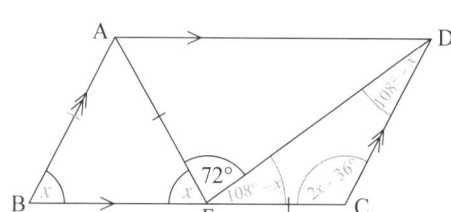

18. ABCD is a parallelogram   (Both pairs of opposite sides are parallel.)

$\quad$ AB = BC   (Given.)

$\quad$ ∴ ABCD is a rhombus   (A parallelogram with one pair of adjacent sides equal.)

$\quad$ $D\hat{A}B + A\hat{D}C = 180°$ (Co-interior angles, AB ∥ CD.)

$\quad$ ∴ $D\hat{A}B = 68°$

$\quad$ $D\hat{A}C = C\hat{A}B$ (Diagonals of rhombus bisect angles.)

$\quad\quad = 34°$

$\quad$ $C\hat{A}P = A\hat{P}Q$ (Alternate angles, AC ∥ QP.)

$\quad\quad = 20°$

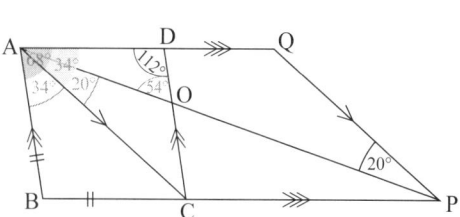

$$\therefore \ D\hat{A}O = 34° - 20°$$
$$= 14°$$
$$A\hat{O}D = 54° \qquad \text{(Sum of interior angles of } \triangle AOD.)$$

Exercise 5 (page 158)

1.1 $K\hat{N}M + P\hat{M}N = 180°$ (Co-interior angles, KN ∥ PM.)

$O\hat{N}M = \frac{1}{2}K\hat{N}M$ (ON bisects $K\hat{N}M$, given.)

$O\hat{M}N = \frac{1}{2}N\hat{M}P$ (OM bisects $N\hat{M}P$, given.)

$\therefore \ O\hat{N}M + O\hat{M}N = \frac{1}{2}K\hat{N}M + \frac{1}{2}P\hat{M}N$
$$= \frac{1}{2}(K\hat{N}M + P\hat{M}N)$$
$$= 90°$$

$\therefore \ N\hat{O}M = 90°$ (Sum of interior angles of $\triangle EBC$.)

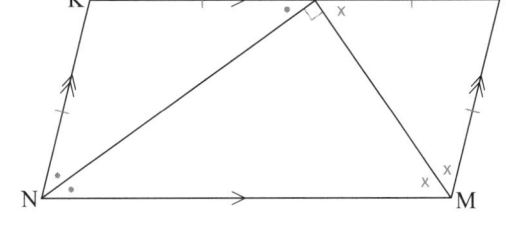

1.2 $K\hat{O}N = O\hat{N}M$ (Alternate angles, KP ∥ NM.)

$\therefore \ KO = KN$ (Sides opposite equal angles $\triangle KNO$; $K\hat{N}O = K\hat{O}N$.)

$\therefore \ \triangle KNO$ is isosceles (Two sides are equal.)

1.3 $KN = PM$ (Opposite sides of a parallelogram are equal.)

$P\hat{O}M = O\hat{M}N$ (Alternate angles, KP ∥ NM.)

$\therefore \ OP = PM$ (Sides opposite equal angles $\triangle OPM$; $P\hat{O}M = P\hat{M}O$.)

$\therefore \ OP = KN = KO$

$\therefore \ KP = 2KN$

2. $x + 5x + 4x + 2x = 360°$ (Sum of interior angles of a quadrilateral.)

$\therefore \ 12x = 360°$

$\therefore \ x = 30°$

$\therefore \ \hat{B} = 5x$
$$= 150°$$

$\therefore \ AD \parallel BC$ ($\hat{A}$ and co-interior $\hat{B}$ are supplementary.)

$\therefore \ ABCD$ is a trapezium

3.1 In $\triangle ABN$ and $\triangle CDM$:

$AB = DC$ (Opposite sides of parallelogram equal.)

$B\hat{A}N = D\hat{C}M$ (Alternate angles, AB ∥ DC.)

$AN = CM$ (Given.)

$\therefore \ \triangle ABN \equiv \triangle CDM$ (SAS.)

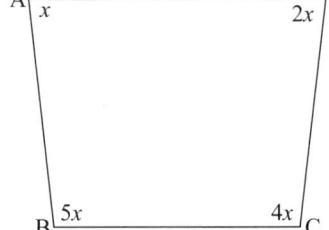

3.2 $A\hat{N}B = D\hat{M}C$ ($\triangle ABN \equiv \triangle CDM$.)

$\therefore \ B\hat{N}M = D\hat{M}N$ (Straight line ANMC.)

$\therefore \ BN \parallel MD$ ($B\hat{N}M =$ alternate $D\hat{M}N$.)

$BN = MD$ ($\triangle ABN \equiv \triangle CDM$.)

$\therefore \ NBMD$ is a parallelogram (One pair of opposite sides is equal and parallel.)

4. $\hat{AOD} = 90°$ (Diagonals of a rhombus intersect at right angles.)

$\hat{ADB} = x$ (Given.)

$\therefore \hat{ODC} = x$ (Diagonals of a rhombus bisect the angles.)

$\hat{AOD} = \hat{ODC} + \hat{OCD}$ (Exterior angle of $\triangle DOC$ equals sum of interior opposite angles.)

$\therefore 90° = x + 2x + 30°$

$\therefore 60° = 3x$

$\therefore x = 20°$

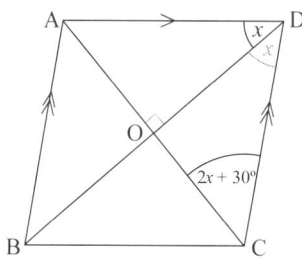

5. In $\triangle POQ$ and $\triangle ROS$:

QO = SO   (Given.)

$\hat{QPO} = \hat{ORS}$ (Alternate angles, PQ || SR.)

$\hat{POQ} = \hat{ROS}$ (Vertically opposite.)

$\therefore \triangle POQ \equiv \triangle ROS$ (AAS.)

$\therefore$ PQ = SR

$\therefore$ PQRS is a parallelogram   (One pair of opposite sides is equal and parallel.)

6. $\hat{BAD} + \hat{ADC} = 180°$ (Co-interior angles, AB || DC.)

$\hat{DAE} = \frac{1}{2}\hat{BAD}$ (AE bisects $\hat{DAB}$, given.)

$\hat{ADE} = \frac{1}{2}\hat{ADC}$ (DE bisects $\hat{ADC}$, given.)

$\hat{DAE} + \hat{ADE} = \frac{1}{2}\hat{BAD} + \frac{1}{2}\hat{ADC}$

$\qquad\qquad = \frac{1}{2}(\hat{BAD} + \hat{ADC})$

$\qquad\qquad = 90°$

$\therefore \hat{AED} = 90°$ (Sum of interior angles of $\triangle AED$.)

AEDF is a parallelogram   (Both pairs of opposite sides are parallel.)

$\therefore$ AEDF is a rectangle   (Parallelogram has one right angle.)

7.1 $\hat{AQP} = 60°$ ($\triangle PAQ$ is equilateral, given.)

$\hat{BSA} = 60°$ ($\triangle BPS$ is equilateral, given.)

PQRS is a parallelogram   (PS || QR and PQ || SR.)

$\therefore \hat{PQR} = \hat{PSR}$ (Opposite angles of parallelogram are equal.)

$\therefore \hat{PQR} + \hat{AQP} = \hat{PSR} + \hat{BSA}$

$\therefore \hat{AQR} = \hat{BSR}$

7.2 In $\triangle AQR$ and $\triangle RSB$:

$\hat{AQR} = \hat{BSR}$ (Proved.)

AQ = PQ   (Given.)

PQ = SR   (Opposite sides of parallelogram are equal.)

$\therefore$ AQ = SR

QR = PS   (Opposite sides of parallelogram.)

PS = BS   (Given.)

$\therefore$ QR = BS

$\therefore \triangle AQR \equiv \triangle RSB$ (SAS.)

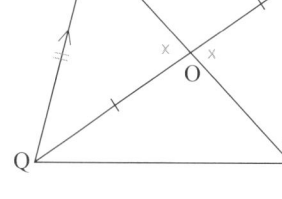

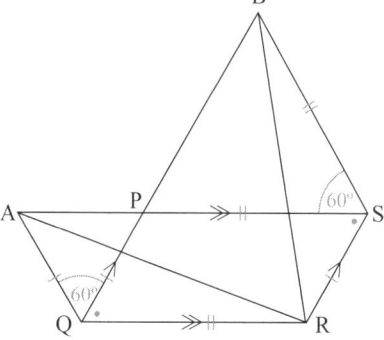

7.3 QR = PS   (Proved.)

  = 8 cm

PQ = SR   (Proved.)

  = 5 cm

∴ Perimeter of △AQP = 15 cm

8. In △APO and △CQO:

AO = CO   (Given.)

PÂO = QĈO   (Alternate angles, AB ∥ DC.)

AÔP = CÔQ   (Vertically opposite angles.)

∴ △APO ≡ △CQO   (AAS.)

∴ AP = CQ

AB = DC   (Opposite sides of a parallelogram
                                     are equal.)

∴ AB − AP = DC − CQ

  ∴ PB = DQ

∴ PBQD is a parallelogram   (One pair of opposite sides is parallel and equal.)

9.1 In △AMK and △NBH:

AM = NB   (Given.)

NÂK = MN̂H   (Corresponding angles, AC ∥ NH.)

NM̂K = MB̂H   (Corresponding angles, MK ∥ BC.)

∴ △AMK ≡ △NBH   (AAS.)

9.2 MK = BH   (△AMK ≡ △NBH.)

∴ MBHK is a parallelogram   (One pair of opposite sides is equal and parallel.)

10. EA = OB   (Opposite sides of
                                     parallelogram EABO
                                     are equal.)

OB = DO   (Diagonals of parallelogram
                                     ABCD bisect each other.)

∴ EA = DO

∴ EAOD is a parallelogram   (One pair of opposite sides is
                                     equal and parallel.)

11. Let OĈE = x

∴ OĈF = x (Diagonal of rhombus ADCB bisects angle.)

FÔC = FĈO   (Angles opposite equal sides △OFC;
                                     OF = FC.)

  = x

EÔC = OĈE   (Angles opposite equal sides
                                     △OEC; OE = EC.)

  = x

OF ∥ EC   (FÔC = alternate OĈE = x.)

OE ∥ FC   (EÔC = alternate FĈO = x.)

∴ OECF is a parallelogram   (Both pairs of opposite sides are parallel.)

∴ OECF is a rhombus   (Parallelogram with two adjacent sides equal.)

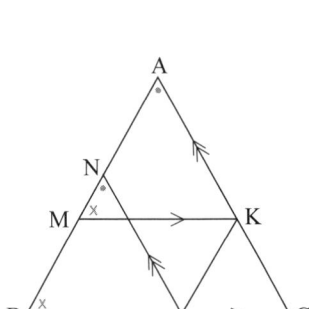

12. PQ = TU and PQ ∥ TU   (PTUQ is a parallelogram.)

RS = TU and RS ∥ TU   (RTUS is a parallelogram.)

∴ PQ = RS and PQ ∥ RS

∴ PQSR is a parallelogram   (One pair of opposite sides is equal
and parallel.)

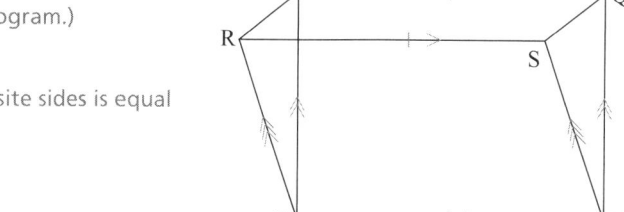

13. AB̂E = EB̂C   (BE bisects AB̂C, given.)

AÊB = EB̂C   (Alternate angles, AD ∥ BC.)

∴ AB̂E = AÊB

∴ AB = AE   (Sides opposite equal angles
△ABE; AB̂E = AÊB.)

EĈD = EĈB   (CE bisects BĈD, given.)

DÊC = EĈB   (Alternate angles, AD ∥ BC.)

∴ EĈD = DÊC

∴ ED = DC   (Sides opposite equal angles △EDC; EĈD = DÊC.)

BC = AD   (Opposite sides of parallelogram ABCD are equal.)

    = AE + ED

    = BA + CD

14. ABRC is a parallelogram   (AB ∥ CR and AC ∥ BR.)

AB = CR   (Opposite sides of parallelogram are equal.)

CR = QB   (Given.)

∴ AB = QB

Let QÂB = x

∴ AQ̂B = x   (Angles opposite equal sides △AQB; AB = QB.)

∴ AB̂D = 2x   (Exterior angle of △ABQ equals sum of interior
opposite angles.)

P̂ = AB̂D   (Alternate angles, AB ∥ PR.)

    = 2x

∴ QP̂R = 2AQ̂P

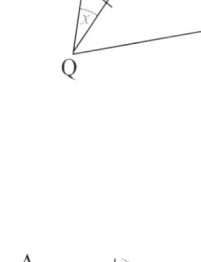

15. Let AB̂E = x

∴ AÊB = x   (Angles opposite equal sides △ABE; AB = AE.)

EB̂C = AÊB   (Alternate angles, AD ∥ BC.)

    = x

D̂ = AB̂C   (Opposite angles of parallelogram are equal.)

    = 2x

CÊD = D̂   (Angles opposite equal sides △EDC; EC = DC.)

    = 2x

BĈE = CÊD   (Alternate angles, AD ∥ BC.)

    = 2x

BÊC = BĈE   (Angles opposite equal sides △BEC; BE = BC.)

    = 2x

AÊB + CÊD + BÊC = 180°   (Straight line AED.)

∴ 5x = 180°

∴ x = 36°

∴ BÊC = 72°

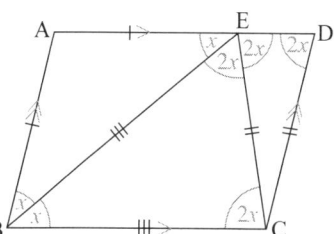

16.1 $CO = OE$ (Diagonals of parallelogram BCDE bisect each other.)

$EF = FO$ (Diagonals of parallelogram AODE bisect each other.)

$\therefore CO = 2EF$

16.2 $BO = OD$ (Diagonals of parallelogram BCDE bisect each other.)

$OD = AE$ (Opposite sides of parallelogram AODE are equal.)

$\therefore BO = AE$

$\therefore$ ABOE is a parallelogram (One pair of opposite sides is equal and parallel.)

17. $\hat{BAF} = \hat{AFD}$ (Alternate angles, AB ∥ DC.)

$= 3x$

$\therefore \hat{DAB} = 4x$

$\hat{DAC} = \hat{CAB}$ (Diagonals of rhombus bisect angles.)

$= 2x$

$\therefore \hat{FAC} = x$

$\therefore$ AF bisects $\hat{DAC}$

18. $\hat{AEB} = \hat{FAE}$ (Alternate angles, AF ∥ BE.)

$= x$

$\therefore AB = BE$ (Sides opposite equal angles △ABE; $\hat{AEB} = \hat{BAE} = x$.)

$\hat{DFO} = \hat{FBE}$ (Alternate angles, AF ∥ BE.)

$= y$

$AB = AF$ (Sides opposite equal angles △ABF; $\hat{F} = \hat{ABF} = y$.)

$\therefore BE = AF$ (Both equal AB.)

AFEB is a parallelogram (One pair of opposite sides is equal and parallel.)

$\therefore AB = FE$ (Opposite sides of parallelogram AFEB are equal.)

19. Let $\hat{P} = x$

$\hat{PSM} = x$ (Angles opposite equal sides △PSM: PM = MS.)

$\hat{SMN} = \hat{PSM} + \hat{P}$ (Exterior angle △SPM equals sum of interior opposite angles.)

$= 2x$

$\hat{TNQ} = \hat{SMN}$ (Corresponding angles, SM ∥ TN.)

$= 2x$

$\hat{Q} = \hat{NTQ}$ (Angles opposite equal sides △TNQ; TN = NQ.)

$= \dfrac{180° - 2x}{2}$ (Sum of interior angles of △TNQ.)

$= 90° - x$

$\hat{R} + \hat{P} + \hat{Q} = 180°$ (Sum of interior angles of △RPQ.)

$\therefore \hat{R} + x + 90° - x = 180°$

$\therefore \hat{R} = 90°$

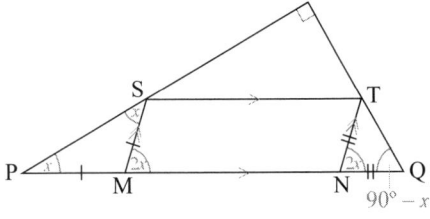

20. In △ABE and △DAF:

$AB = AD$ (All sides of a square are equal.)

$AE = DF$ (E and F midpoints, given.)

$E\hat{A}B = F\hat{D}A$ (Angles of a square 90°.)

∴ $\triangle ABE \equiv \triangle DAF$ (SAS.)

Let $A\hat{O}B = x$

∴ $E\hat{A}O = x$ ($\triangle ABE \equiv \triangle DAF$.)

∴ $B\hat{A}O = 90° - x$

$A\hat{O}B + B\hat{A}O + A\hat{B}O = 180°$ (Sum of interior angles of $\triangle AOB$.)

∴ $A\hat{O}B + 90° - x + x = 180°$

∴ $A\hat{O}B = 90°$

∴ $BE \perp AF$

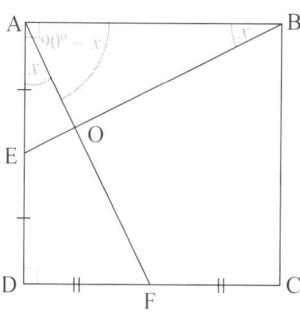

Exercise 6   (page 162)

1.   FG ∥ BC   (In $\triangle ABC$, AF = FB, AG = GC)

∴ FG ∥ BD

CH = HE   (In $\triangle CAE$, AG = FB, GH ∥ AE)

HI ∥ CD   (In $\triangle ECD$, CH = HE, EI = ID)

∴ HI ∥ BD

∴ FG ∥ HI   (Both lines ∥ BD.)

2.1   MN ∥ QR AND MN = $\frac{1}{2}$QR   (In $\triangle PQR$, PM = MQ and PN = NR.)

∴ MN = 5 mm

2.2   MS ∥ QR   (MN ∥ QR.)

∴ MQRS is a parm   (Both pairs of opposite sides are parallel.)

MQ = SR   (Opposite sides of parallelogram are equal.)

     = 4 mm

∴ PQ = 8 mm   (PM = MQ, given.)

3.1   AO = OG   (In $\triangle AGC$, AE = EC and OE ∥ GC.)

3.2   DO ∥ BG   (In $\triangle ABG$, AD = DB and AO = OG.)

∴ OC ∥ BG

∴ OBGC is a parm   (BO ∥ GC and OC ∥ BG.)

3.3   OF = FG = 15 cm   (Diagonals of parallelogram bisect one another.)

∴ OG = 30 cm

AO = OG   (Proved.)

     = 30 cm

AF = 30 + 15 = 45 cm

4.1 $B\hat{O}C = 90°$ (Diagonals of rhombus intersect at right angles.)

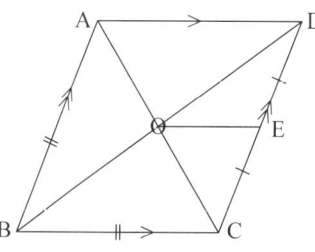

4.2   BO = OD = 40 mm   (Diagonals of rhombus bisect one another.)

and AO = OC = 30 mm

$BC^2 = BO^2 + CO^2$ (Pythagoras, $B\hat{O}C = 90°$.)

     = 1 600 + 900

$BC = \sqrt{2\,500} = 50$ mm

4.3   OE ∥ BC and OE = $\frac{1}{2}$BC   (In $\triangle DBC$, DO = OB and DE = EC.)

∴ OE = 25 mm

5.1   ZC = CB   (In △ZAB, ZY = YA and YC ∥ AB.)

   DABC is a parm   (DC ∥ AB and DA ∥ CB.)

   ∴ DA = CB   (Opposite sides of parallelogram DABC.)

   ∴ DA = ZC   (CB = ZC.)

   ∴ DACZ is a parm.   (One pair of opposite sides is equal and parallel.)

   ∴ DZ ∥ AC   (Opposite sides of parallelogram DACZ.)

5.2   Area △DAC = area △ZAC   (Same base AC, AC ∥ DZ.)

   Area △DAC + area △ABC = Area △ZAC + area △ABC

   ∴ Area ABCD = area △ZAB

6.   AG = GC   (In △ABC, AF = FB and FG ∥ BC.)

   AE = ED   (In △ADC, AG = GC and EG ∥ DC.)

   ∴ EF ∥ DB   (In △ADB, AE = ED and AF = FB.)

7.   MN ∥ BC and MN = $\frac{1}{2}$BC   (In △ABC, AM = MB and AN = NC.)

   $\hat{O}_1 = \hat{P}_1$ (Corresponding angles MN ∥ BC.)

   AO = OP   (In △ABP, AM = MB and MO ∥ BP.)

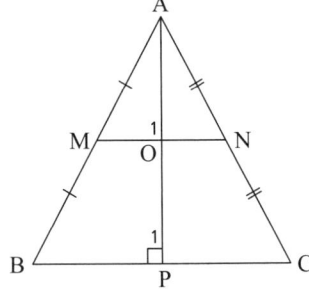

   Area △AMN = $\frac{1}{2}$MN × AO

   = $\frac{1}{2}\left(\frac{1}{2}BC\right)\left(\frac{1}{2}AP\right)$

   = $\frac{1}{4}\left(\frac{1}{2}BC \cdot AP\right)$

   = $\frac{1}{4}$ area ABC

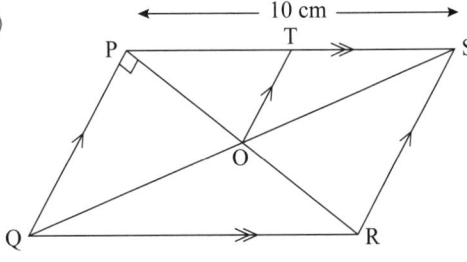

8.1   QR = PS   (Opposite sides of parallelogram.)

   = 10 cm

   $PR^2 = QR^2 - PQ^2$ (Pythagoras, ∠QPR = 90°.)

   = 100 − 36

   ∴ PR = $\sqrt{64}$ = 8 cm

8.2   PO = OR   (Diagonals of parallelogram bisect one another.)

   PT = TS   (In △PRS, PO = OR and OT ∥ RS.)

   = 5 cm

8.3   OT = $\frac{1}{2}$RS   (In △PRS, PO = OR and PT ∥ TS.)

   = $\frac{1}{2}$PQ   (RS = PQ, opposite sides of parallelogram.)

   = 3 cm

8.4   Area of parallelogram PQRS = base × ⊥ height

   = PQ × PR

   = 6 × 8

   = 48 cm^2

9.1   AM = MD   (M is midpoint of AD.)

   ∴ MB ∥ YD   (In △AYD, AB = BY and AM = MD.)

   AB = AM   (AD = 2AB.)

9.2   Let $\hat{B}_1 = \hat{M}_1 = x$ (In △ABM, AB = AM.)

   $\hat{B}_2 = \hat{M}_1 = x$ (Alternate angles AD ∥ BC.)

   ∴ BM bisects A$\hat{B}$C

9.3 In △BYF and △CDF:

(a) BY = AB   (Given.)

AB = CD   (Opposite sides parallelogram ABCD.)

∴ BY = DC

(b) $\hat{F}_1 = \hat{F}_2$ (Vertically opposite angles.)

(c) $\hat{Y} = \hat{D}_1$ (Alternate angles, AY ∥ DC.)

∴ △BYF ≡ △CDF   (SAA)

∴ BF = FC

AD = BC   (Opposite sides parallelogram ABCD.)

∴ MD = FC

∴ MFCD is a parm   (MD = FC and MD ∥ FC)

∴ MO = OC   (Diagonals of parallelogram bisect one another.)

∴ O is the midpoint of MC

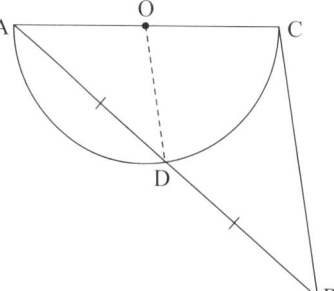

10. BE = EA   (In △BAC, BF = FC and EF ∥ AC.)

CG = GD   (AE = EB and AD ∥ EG ∥ BC.)

11. Construct OD.

OD ∥ BC and OC = $\frac{1}{2}$BC   (In △ABC, AD = DB and AO = OC, radii.)

∴ BC = 2 × OD

    = 2 × radius of the circle

    = AC

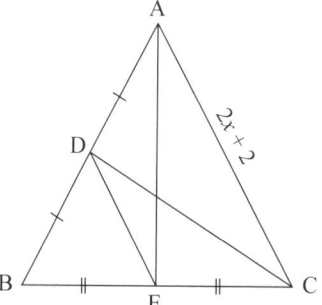

12. DE ∥ AC and DE = $\frac{1}{2}$AC   (In △ABC, AD = DB and BE = EC.)

∴ DE = $\frac{2x+2}{2} = x + 1$

Test A   (page 164)

1.1 AP̂Y = 40°   (Vertically opposite angles.)

$\hat{C} = 50°$ (Sum of interior angles in △POC.)

$\hat{B} = \hat{C} = 50°$ (Angles opposite equal sides △ABC; AB = AC.)

YÂP = $\hat{B} + \hat{C}$ (Exterior angle of △ABC equals sum of interior opposite angles.)

     = 100°

$\hat{Y} = 40°$ (Sum of interior angles of △APY.)

1.2 AY = AP   (Sides opposite equal angles △APY, $\hat{Y}$ = YP̂A = 40°.)

2. In △ABC and △EDC:

BC = DC   (Given.)

AC = EC   (Given.)

BĈA = CÂE   (Alternate angles, AE ∥ BD.)

$$C\hat{A}E = C\hat{E}A$$ (Angles opposite equal sides △ACE; AC = CE.)

$$C\hat{E}A = E\hat{C}D$$ (Alternate angles, AE ∥ BD.)

$$\therefore B\hat{C}A = E\hat{C}D$$

$$\therefore \triangle ABC \equiv \triangle EDC$$ (SAS.)

3.   AD ∥ BC and AB ∥ DC

AD = BC and AB = DC

AE = EC and BE = ED

AD = BC and AD ∥ BC

4.1 $$A\hat{B}D = B\hat{D}C$$ (Alternate angles, AB ∥ DC.)

$$= 3x$$

4.2 $$3x + 4x + 75° = 180°$$ (Sum of interior angles of △ABO.)

$$\therefore 7x = 105°$$

$$\therefore x = 15°$$

4.3 $$B\hat{C}D + A\hat{D}C = 180°$$ (Co-interior angles, AD ∥ BC.)

$$\therefore B\hat{C}D + 5(15°) = 180°$$

$$\therefore B\hat{C}D = 105°$$

5.1   In △AOD and △COB:

AO = OC (Given.)

$$D\hat{A}O = O\hat{C}B$$ (Alternate angles, AD ∥ BC.)

$$A\hat{O}D = C\hat{O}B$$ (Vertically opposite angles.)

$$\therefore \triangle AOD \equiv \triangle COB$$ (AAS.)

5.2   AD = BC (△AOD ≡ △COB.)

∴ ABCD is a parallelogram (One pair of opposite sides is equal and parallel.)

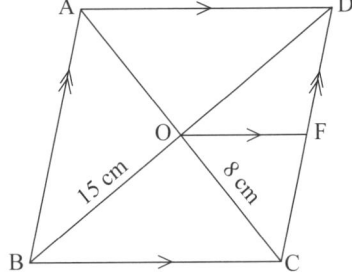

6.1   DC = 20 mm   **6.2**   AE = 30 mm   **6.3**   OD = 19 mm   **6.4**   BF = 62 mm

7.   BO = OD = 15 cm and AO = OC = 8 cm (Diagonals of rhombus bisect one another.)

$$B\hat{O}C = 90°$$ (Diagonals of rhombus intersect at right angles.)

$$BC^2 = BO^2 + OC^2$$ (Pythagoras, B\hat{O}C = 90°.)

$$= 15^2 + 8^2$$

$$\therefore BC = 17 \text{ cm}$$

$$DF = FC \text{ and } OF = \tfrac{1}{2}BC$$ (In △DBC, DO = OB and OF ∥ BC.)

$$\therefore OF = 8,5 \text{ cm}$$

Test B   (page 166)

1. $$C\hat{D}A = \hat{A} = 50°$$ (Alternate angles, AB ∥ CE.)

$$A\hat{D}B = 60°$$ (Straight line CDE.)

$$B\hat{F}D = A\hat{D}B = 60°$$ (Angles opposite equal sides △BFD; BF = BD.)

$$B\hat{F}D = x + \hat{A}$$ (Exterior angle of △ABF equals sum of interior opposite angles.)

$$\therefore 60° = x + 50°$$

$$\therefore x = 10$$

2.

$$\hat{R} = P\hat{Q}R \qquad \text{(Angles opposite equal sides } \triangle PQR; PQ = PR.)$$

$$P\hat{A}Q = \hat{R} + 30° \qquad \text{(Exterior angle of } \triangle AQR.)$$

$$\hat{P} = P\hat{A}Q = \hat{R} + 30° \quad \text{(Angles opposite equal sides } \triangle QPA; QA = QP.)$$

$$\hat{R} + P\hat{Q}R + \hat{P} = 180° \qquad \text{(Sum of interior angles in } \triangle PQR.)$$

$$\therefore \hat{R} + \hat{R} + \hat{R} + 30° = 180°$$

$$\therefore \hat{R} = 50°$$

3. Let $A\hat{B}E = x$

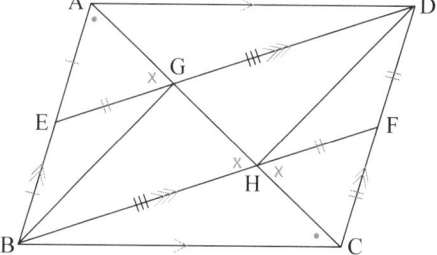

$$\therefore A\hat{E}B = x \qquad \text{(Angles opposite equal}$$
$$\text{sides } \triangle BAE; AB = AE.)$$

$$E\hat{B}C = A\hat{E}B \qquad \text{(Alternate angles,}$$
$$\text{AD } \| \text{ BC.)}$$

$$= x$$

$$\hat{D} = A\hat{B}C \qquad \text{(Opposite angles of a parallelogram are equal.)}$$

$$= 2x$$

$$C\hat{E}D = \hat{D} \qquad \text{(Angles opposite equal sides } \triangle EDC; EC = DC.)$$

$$= 2x$$

$$x + 90° + 2x = 180° \qquad \text{(Straight line AED.)}$$

$$\therefore 3x = 90°$$

$$\therefore x = 30°$$

$$\therefore E\hat{B}C = 30°$$

4.1

$$AB = DC \qquad \text{(Opposite sides of parallelogram}$$
$$\text{ABCD are equal.)}$$

$$\therefore \tfrac{1}{2}AB = \tfrac{1}{2}BC$$

$$\therefore EB = DF$$

$$\therefore \text{EBFD is a parallelogram.} \quad \text{(One pair of opposite sides is}$$
$$\text{equal and parallel.)}$$

4.2 In $\triangle AEG$ and $\triangle CFH$:

$$AE = CF \qquad \text{(Proved.)}$$

$$E\hat{A}G = F\hat{C}H \qquad \text{(Alternate angles, AB } \| \text{ DC.)}$$

$$ED \| BF \qquad \text{(EBFD is a parallelogram, proved.)}$$

$$\therefore A\hat{G}E = G\hat{H}B \qquad \text{(Corresponding angles, ED } \| \text{ BF.)}$$

$$G\hat{H}B = F\hat{H}C \qquad \text{(Vertically opposite angles.)}$$

$$\therefore A\hat{G}E = F\hat{H}C$$

$$\therefore \triangle AEG \equiv \triangle CFH \qquad \text{(AAS.)}$$

$$\therefore EG = HF \qquad (\triangle AEG \equiv \triangle CFH.)$$

4.3

$$ED = BF \qquad \text{(Opposite sides of parallelogram EBFD are equal.)}$$

$$\therefore ED - EG = BF - HF$$

$$\therefore GD = BH$$

$$ED \| BF \qquad \text{(EBFD is a parallelogram.)}$$

$$\therefore GD \| BH$$

$$\therefore \text{GBHD is a parallelogram.} \quad \text{(One pair of opposite sides is equal and parallel.)}$$

5. $AB \parallel QR$ and $AB = \frac{1}{2}QR$ (In $\triangle PQR$, $PA = AQ$ and $PB = BR$.)

$DE \parallel QR$ and $DE = \frac{1}{2}QR$ (In $\triangle OQR$, $OD = DQ$ and $OE = ER$.)

$\therefore AB \parallel QR \parallel DE$ and $AB = \frac{1}{2}QR = DE$

$\therefore ADEB$ is a parm   (One pair of opposite sides is equal and parallel.)

6. In $\triangle PTC$ and $\triangle PAC$:

  (a) $\hat{P}_1 = \hat{P}_2$ (Given.)

  (b) $PC = PC$ (Common.)

  (c) $P\hat{C}T = P\hat{C}A = 90°$ (Straight line TCA.)

    $\therefore \triangle PTC \equiv \triangle PAC$ (AAS.)

    $\therefore TC \equiv CA$

    $AB = BP$ (In $\triangle APT$, $TC = CA$ and $BC \parallel PT$.)

    $AO = OV$ (In $\triangle ATV$, $TC = CA$ and $CO \parallel TV$.)

    $BO = BA + AO$

      $= \frac{1}{2}PA + \frac{1}{2}AV$

      $= \frac{1}{2}(PA + AV)$

      $= \frac{1}{2}PV$

Test C   (page 167)

1.1 $D\hat{C}B = E\hat{D}C = 20°$ (Alternate angles, $DE \parallel BC$.)

1.2 $B\hat{D}C = 92°$ (Sum of interior angles of $\triangle DBC$.)

1.3 $A\hat{D}C = \hat{B} + D\hat{C}B$ (Exterior angle of $\triangle DBC$ equal sum of interior opposite angles.)

    $= 88°$

1.4 $D\hat{C}E = \hat{A}$ (Angles opposite equal sides $\triangle ADC$; $AD = DC$.)

    $= \frac{180° - 88°}{2}$ (Sum of interior angles in $\triangle ADC$.)

    $= 46°$

1.5 $A\hat{E}D = E\hat{C}D + D\hat{C}E$ (Exterior angle of $\triangle DEC$ equals sum of interior opposite angles.)

    $= 66°$

2.1 $B\hat{E}C = \hat{A} + A\hat{C}E$ (Exterior angle of $\triangle ACE$ equals sum of interior opposite angles.)

    $130° = x + 50°$

    $\therefore x = 80°$

2.2 $E\hat{C}G = 2x - 110°$

    $= 2(80°) - 110°$

    $= 50°$

    $\therefore AB \parallel FG$ ($B\hat{E}C$ and co-interior $E\hat{C}G$ are supplementary.)

3. $Q\hat{M}R = Q\hat{R}M = x$ (Angles opposite equal sides $\triangle MQR$; $MQ = RQ$.)

  $P\hat{M}S = 180° - 54° - x$ (Straight line PMQ.)

    $= 126° - x$

  $\hat{P} = P\hat{M}S$ (Angles opposite equal sides $\triangle PMS$; $SP = SM$.)

    $= 126° - x$

  $\hat{Q} = 180° - 2x$ (Sum of interior angles of $\triangle MQR$.)

  $\hat{P} + \hat{Q} = 180°$ (Co-interior angles, $PS \parallel QR$.)

  $126° - x + 180° - 2x = 180°$

  $\therefore x = 42°$

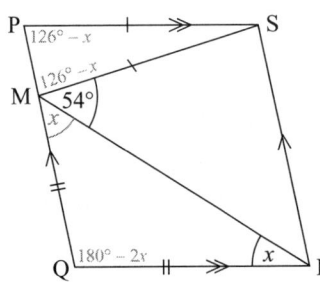

4.1 A$\hat{\text{B}}$O = O$\hat{\text{B}}$C   (Diagonals of a rhombus bisect angles.)

       = 40°

   A$\hat{\text{O}}$B = 90°   (Diagonals of a rhombus intersect at right angles.)

   B$\hat{\text{A}}$O = 50°   (Sum of interior angles of △ABO.)

4.2 A$\hat{\text{D}}$C = A$\hat{\text{B}}$C   (Opposite angles of a rhombus are equal.)

      = 80°

4.3 D$\hat{\text{C}}$E = A$\hat{\text{D}}$C

      = 80°   (Alternate angles, AD ∥ BE.)

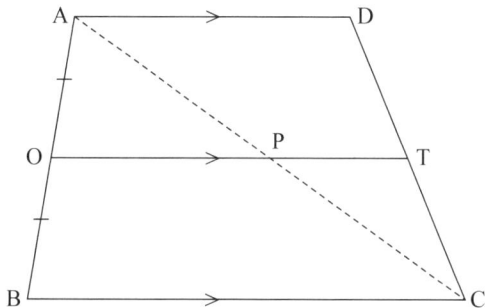

5.1 AQ = QE   (In △ABE, AF = FB and FQ ∥ BE.)

     = $\frac{1}{2}$AE

  AE = EC   (Given.)

     = $\frac{1}{2}$AC

  AQ = $\frac{1}{4}$AC

5.2 AP = PR   (In △ABR, AF = FB and FP ∥ BR.)

  PE ∥ RC   (In △ARC, AP = PR and AE = EC.)

  ∴ PE ∥ FR

  ∴ FPER is a parallelogram   (PE ∥ FR and FP ∥ RE.)

5.3 Area △AFR = area △BFR   (Base AF = FB, common apex R ∴ same height.)

6.

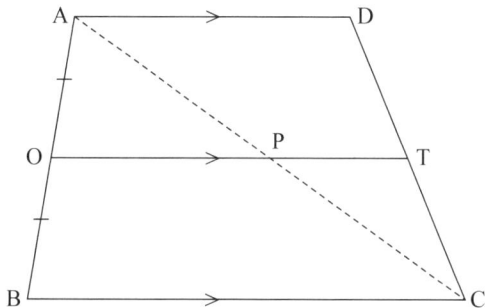

6.1 DT = TC   (AO = OB and AD ∥ OT ∥ BC.)

6.2 Draw AC to intersect OT at P.

  AP = PC and OP = $\frac{1}{2}$BC   (In △ABC, AO = OB and OP ∥ BC.)

        PT = $\frac{1}{2}$AD   (In △ADC, AP = PC and DT = TC.)

      OT = OP + PT

        = $\frac{1}{2}$BC + $\frac{1}{2}$AD

     2OT = BC + AD

Topic 9   Analytical geometry (page 169)

Exercise 1   (page 174)

1.1 $AB = \sqrt{(2-14)^2 + (5-9)^2}$

$= \sqrt{(-12)^2 + (-4)^2} = \sqrt{160} = 4\sqrt{10}$

1.2 $m_{AB} = \frac{9-5}{14-2} = \frac{4}{12} = \frac{1}{3}$

1.3 $M\left(\frac{2+14}{2}; \frac{5+9}{2}\right) = M(8; 7)$

1.4 $m_{DE} = m_{AB} = \frac{2}{5}$

1.5 $m_{RT} \times m_{AB} = -1 \therefore m_{RT} = -\frac{5}{2}$

2.1 $PQ = \sqrt{(-4-1)^2 + (7+5)^2}$

$= \sqrt{(-5)^2 + (12)^2} = \sqrt{169} = 13$ units

2.2 $m_{PQ} = \frac{-5-7}{1+4} = -\frac{12}{5}$

2.3 $M\left(\frac{-4+1}{2}; \frac{7-5}{2}\right) = M\left(-\frac{3}{2}; 1\right)$

2.4 $m_{RT} = m_{PQ} = -\frac{12}{5}$

2.5 $m_{ME} \times m_{PQ} = -1 \therefore m_{ME} = \frac{5}{12}$

3.1 $RS = \sqrt{(3+1)^2 + (\sqrt{2+\sqrt{2}})^2}$

$= \sqrt{4^2 + (2\sqrt{2})^2} = \sqrt{24} = 2\sqrt{6}$ units

3.2 $m_{RS} = \frac{-\sqrt{2}-\sqrt{2}}{-1-3} = \frac{-2\sqrt{2}}{-4} = \frac{\sqrt{2}}{2}$

3.3 $M\left(\frac{3-1}{2}; \frac{\sqrt{2}-\sqrt{2}}{2}\right) = M(1; 0)$

3.4 $m_{TZ} = m_{RS} = \frac{\sqrt{2}}{2}$

3.5 $m_{ZB} \times m_{RS} = -1 \therefore m_{ZB} = -\frac{2}{\sqrt{2}} = -\frac{2\sqrt{2}}{2} = -\sqrt{2}$

4.1 $m_{AB} = \frac{5-3}{5-1} = \frac{2}{4} = \frac{1}{2}$

$m_{CD} = \frac{4-2}{7-3} = \frac{2}{4} = \frac{1}{2}$

AB $\parallel$ CD $m_{AB} = m_{CD}$

4.2 $m_{AB} = \frac{2-0}{8-2} = \frac{2}{6} = \frac{1}{3}$

$m_{CD} = \frac{-4-2}{2-4} = \frac{-6}{-2} = \frac{3}{1}$

$\therefore$ Neither

4.3 $m_{AB} = \frac{2-8}{7-4} = \frac{-6}{3} = -\frac{2}{1}$

$m_{CD} = \frac{2+4}{0+3} = \frac{6}{3} = \frac{2}{1}$

$\therefore$ Neither

4.4 $m_{AB} = \frac{-7-5}{3-5} = \frac{-12}{-2} = \frac{6}{1}$

$m_{CD} = \frac{1-3}{5+7} = \frac{-2}{12} = -\frac{1}{6}$

AB $\perp$ CD $m_{AB} \times m_{CD} = -1$

5.1 Distance $= \sqrt{(-4-0)^2 + (-7-0)^2}$

$= \sqrt{(-4)^2 + (-7)^2} = \sqrt{65} \approx 8{,}06$

5.2 $m = \frac{-7-0}{-4-0} = \frac{-7}{-4} = \frac{7}{4}$

5.3 $M\left(\frac{0-4}{2}; \frac{0-7}{2}\right) = M\left(-2; -\frac{7}{2}\right)$

6. $OA = \sqrt{(3-0)^2 + (5-0)^2}$

$= \sqrt{(3)^2 + (5)^2} = \sqrt{34}$

$OB = \sqrt{(-4-0)^2 + (-4-0)^2}$

$= \sqrt{(-4)^2 + (-4)^2} = \sqrt{32}$

Point B is closer to the origin.

7. **A.** $x = 5$ **B.** $y = 2x - 3$

C. $y = -3$ **D.** $2y + 3x = 10$

$\therefore y = -\frac{3}{2}x + 5$

E. $3y - 2x = 1$ **F.** $2y + 3x + 1 = 0$

$\therefore y = \frac{2}{3}x + 1$ $\therefore y = -\frac{3}{2}x - 1$

G. $2y = 5$ **H.** $x + 4 = 0$

$\therefore y = \frac{5}{2}$ $\therefore x = -4$

7.1 A $\parallel$ H   (Both vertical lines are parallel to the y-axis.)

C $\parallel$ G   (Both horizontal lines are parallel to the x-axis.)

D $\parallel$ F   (Both lines have the same gradient.)

7.2 A and H are both $\perp$ to C and G,

$\therefore$ E $\perp$ D, E $\perp$ F

8.1 **(a)** $M\left(\frac{4p+2p}{2}; \frac{-2n+6n}{2}\right) = M\left(\frac{6p}{2}; \frac{4n}{2}\right)$

$= M(3p; 2n)$

(b) $m_{AB} = \frac{-2n-6n}{4p-2p} = \frac{-8n}{2p} = -\frac{4n}{p}$

8.2 **(a)** $M\left(\frac{a-a}{2}; \frac{3b-3b}{2}\right) = M(0; 0)$

(b) $m_{AB} = \frac{-3b-3b}{-a-a} = \frac{-6b}{-2a} = \frac{3b}{a}$

8.3 **(a)** $M\left(\frac{2n^2+n^2}{2}; \frac{-2p-p}{2}\right) = M\left(\frac{3n^2}{2}; \frac{-3p}{2}\right)$

(b) $m_{AB} = \frac{-p+2p}{n^2-2n^2} = \frac{p}{-n^2}$

9. $PS = \sqrt{(3-2)^2 + (6-3)^2}$

$= \sqrt{(1)^2 + (3)^2} = \sqrt{10}$

$QS = \sqrt{(-1-2)^2 + (4-3)^2}$

$= \sqrt{(-3)^2 + (1)^2} = \sqrt{10}$

PS = QS $\therefore$ Points P and Q are equidistant from the point S.

10. $m_{AB} = \frac{5-2}{4-1} = \frac{3}{3} = 1$

$m_{BC} = \frac{8-5}{7-4} = \frac{3}{3} = 1$

$m_{AB} = m_{BC} = 1$. Points are therefore collinear.

Exercise 2 (page 178)

1. $\left(\frac{x_1 + x_2}{2}; \frac{y_1 + y_2}{2}\right) = (1; -2,5)$

$\therefore \frac{t+4}{2} = 1$

$t + 4 = 2$

$\therefore t = -2$

2.1 $m_{AB} = \frac{0-p}{-3-2} = \frac{1}{10}$

Points A(2; p) and B(−3; 0) are given.

$\therefore \frac{-p}{-5} = \frac{1}{10}$

$\therefore -10p = -5$

$\therefore p = \frac{1}{2}$

2.2 If AB $\perp$ TM and $m_{TM} = -5$, find the value(s) of p.

$m_{TM} \times m_{AB} = -1 \therefore m_{AB} = \frac{1}{5}$

$m_{AB} = \frac{0-p}{-3-2} = \frac{1}{5}$

$\therefore \frac{-p}{-5} = \frac{1}{5}$

$\therefore -5p = -5$

$\therefore p = 1$

3. CD $= \sqrt{53}$; C(1; −2); D(x; 5)

$\sqrt{53} = \sqrt{(1-x)^2 + (-2-5)^2}$

$53 = 1 - 2x + x^2 + 49$

$0 = x^2 - 2x - 3$

$0 = (x-3)(x+1)$

$x = 3$ or $x = -1$

In this case, D lies in the 2nd quadrant, therefore $x = -1$.

4.1 $\left(\frac{x_1 + x_2}{2}; \frac{y_1 + y_2}{2}\right) = M(3; 5)$

$\therefore \frac{4+p}{2} = 5$

$\therefore 4 + p = 10$

$\therefore p = 6$

4.2 $\left(\frac{x_1 + x_2}{2}; \frac{y_1 + y_2}{2}\right) = M(1; t)$

$\therefore \frac{-4+p}{2} = 1$ and $\frac{6+5}{2} = t$

$\therefore -4 + p = 2$ and $\frac{11}{2} = t$

$\therefore p = 6$ and $t = 5,5$

5.1 AC $= \sqrt{(2-5)^2 + (5-1)^2}$

$= \sqrt{(-3)^2 + (4)^2} = \sqrt{25} = 5$

AB $= \sqrt{(2-2)^2 + (5-t)^2} = \sqrt{25 - 10t + t^2}$

AB = AC (given)

$\therefore$ AB^2 = AC^2

$\therefore 25 - 10t + t^2 = 25$

$\therefore t^2 - 10t = 0$

$\therefore t(t - 10) = 0$

$\therefore t = 0$ or $t = 10$

5.2 B(2; 0), D(5; 6)   **5.3**   B(2; 10), D(5; −4)

6.1 $m_{AB} = \frac{-6-p}{3-2} = -\frac{2}{3}$

$\frac{-6-p}{1} = -\frac{2}{3}$

$-18 - 3p = -2$

$-3p = 16$

$\therefore p = -\frac{16}{3}$

6.2 AB $= \sqrt{(2-3)^2 + (p+6)^2}$

$= \sqrt{1 + p^2 + 12p + 36}$

$\sqrt{17} = \sqrt{p^2 + 12p + 37}$

$\therefore 17 = p^2 + 12p + 37$

$\therefore 0 = p^2 + 12p + 20$

$0 = (p + 10)(p + 2)$

$\therefore p = -10$ or $p = -2$

6.3 $\left(\frac{x_1 + x_2}{2}; \frac{y_1 + y_2}{2}\right) = M(t; 10)$

$\left(\frac{2+3}{2}; \frac{p-6}{2}\right) = M(t; 10)$

$\frac{2+3}{2} = t$ and $\frac{p-6}{2} = 10$

$\therefore \frac{5}{2} = t$ and $p - 6 = 20$

$\therefore p = 26$

7. $\left(\frac{x_1 + x_2}{2}; \frac{y_1 + y_2}{2}\right) = M(3; 2)$

$\left(\frac{-3+x}{2}; \frac{4+y}{2}\right) = M(3; 2)$

$\frac{-3+x_2}{2} = 3$ and $\frac{4+y_2}{2} = 2$

$-3 + x = 6$ and $4 + y = 4$

$x = 9$ and $y = 0$

T(9; 0)

8.1 PQ $= \sqrt{(-2-a)^2 + (3+7)^2}$

$= \sqrt{4 + 4a + a^2 + 100}$

$$2\sqrt{41} = \sqrt{a^2 + 4a + 104}$$
$$4 \times 41 = a^2 + 4a + 104$$
$$0 = a^2 + 4a - 60$$
$$0 = (a + 10)(a - 6)$$
$$\therefore a = -10 \text{ or } a = 6$$

8.2 $m_{PQ} = 2$

$m_{PQ} = \frac{3 + 7}{-2 - a} = 2$

$10 = -4 - 2a$

$2a = -14$

$\therefore a = -7$

$m_{QR} = \frac{-7 - b}{-7 + 8} = 2$ (Points are collinear.)

$-7 - b = 2$

$\therefore b = -9$

8.3 $\left(\frac{x_1 + x_2}{2}; \frac{y_1 + y_2}{2}\right) = Q(a; -7)$

$\left(\frac{-2 - 8}{2}; \frac{3 - b}{2}\right) = Q(a; -7)$

$\frac{-2 - 8}{2} = a$ and $\frac{3 - b}{2} = -7$

$a = -5$ and $3 - b = -14$

$\therefore b = 17$

9.1 If MA = MB it means that M is the midpoint of AB.

$\left(\frac{-3 + p}{2}; \frac{1 + t}{2}\right) = M(2; 3)$

$\frac{-3 + p}{2} = 2$ and $\frac{1 + t}{2} = 3$

$-3 + p = 4$ and $1 + t = 6$

$p = 7$ and $t = 5$

9.2 $\left(\frac{p + 10}{2}; \frac{t + 2}{2}\right) = M(8; 5)$

$\frac{p + 10}{2} = 8$ and $\frac{t + 2}{2} = 5$

$p + 10 = 16$ and $t + 2 = 10$

$p = 6$ and $t = 8$

10.1 $C\left(\frac{1 + 5}{2}; \frac{1 + 7}{2}\right) = C(3; 4)$

10.2 radius $= CB = \sqrt{(5 - 3)^2 + (7 - 4)^2}$

$\qquad = \sqrt{(2)^2 + (3)^2} = \sqrt{13}$

10.3 $CD = \sqrt{(3 - 5)^2 + (4 - 1)^2}$

$\qquad = \sqrt{(-2)^2 + (3)^2} = \sqrt{13}$

CD = radius, therefore D lies on the circle.

11. PQ = QA   (given)

$\qquad \therefore$ Q is the midpoint of PA.

$\left(\frac{2 + x}{2}; \frac{7 + y}{2}\right) = Q(3; 4)$

$\frac{2 + x}{2} = -3$ and $\frac{7 + y}{2} = 4$

$2 + x = -6$ and $7 + y = 8$

$x = -8$ and $y = 1$

$\therefore$ A(–8; 1)

12. $m_{CA} = \frac{7 - 1}{-1 - 5} = \frac{6}{-6} = -1$

$m_{AN} = \frac{1 - y}{5 + 4} = -1$

$\frac{1 - y}{9} = -1$

$1 - y = -9$

$y = 10$

13. AB $\perp$ DE $\therefore m_{AB} \times m_{DE} = -1 \therefore m_{AB} = -\frac{3}{2}$

$m_{AB} = \frac{y - 5}{-4 + 2} = -\frac{3}{2}$

$\frac{y - 5}{-2} = -\frac{3}{2}$

$y - 5 = 3$

$y = 8$

Exercise 3   (page 182)

1.1 D(4; –2), B(1; 0)

Midpoint of AC: $\left(\frac{4 + 1}{2}; \frac{-2 + 0}{2}\right) = \left(\frac{5}{2}; -1\right)$

A(–1; –3), C(6; 1)

Midpoint of AD: $\left(\frac{-1 + 6}{2}; \frac{-3 + 1}{2}\right) = \left(\frac{5}{2}; -1\right)$

Therefore the diagonals bisect each other.

1.2 $AB = \sqrt{(-1 - 1)^2 + (-3 - 0)^2}$

$\qquad = \sqrt{13}$ units

$DC = \sqrt{(6 - 4)^2 + (1 - (-2))^2}$

$\qquad = \sqrt{13}$ units

$\therefore$ AB = DC

$m_{AB} = \frac{-3 - 0}{-1 - 1} = \frac{-3}{-2} = \frac{3}{2}$

$m_{CD} = \frac{-2 - 1}{4 - 6} = \frac{-3}{-2} = \frac{3}{2}$

$\therefore$ AB $\parallel$ DC

1.3 ABCD is a parallelogram (One pair of opposite sides is both equal and parallel and the diagonals bisect each other.)

2. Points A(2; 4), B(6; 4) and C(6; 7) are given.

2.1 $m_{AB} = \frac{4 - 4}{6 - 2} = \frac{0}{4} = 0$

$m_{AC} = \frac{7 - 4}{6 - 2} = \frac{3}{4}$

$m_{BC} = \frac{7 - 4}{6 - 6} = \frac{3}{0} \therefore$ undefined

2.2 $AB \perp BC$

2.3 $\triangle ABC$ is a right-angled triangle with $\hat{B} = 90°$.

2.4 $AB = \sqrt{(2-6)^2 + (4-4)^2}$

$\quad = \sqrt{(-4)^2 + (0)^2} = \sqrt{16} = 4$ units

$\quad AC = \sqrt{(2-6)^2 + (4-7)^2}$

$\quad\quad = \sqrt{(-4)^2 + (-3)^2} = \sqrt{25} = 5$ units

$\quad BC = \sqrt{(6-6)^2 + (4-7)^2}$

$\quad\quad = \sqrt{(0)^2 + (-3)^2} = \sqrt{9} = 3$ units

2.5 $AC^2 = AB^2 + BC^2$ (Pythagoras)

$\quad \therefore \triangle ABC$ is a right-angled triangle with $\hat{B} = 90°$.

3. The points P(2; 7), E(–6; –9), and T(6; –5) in a Cartesian plane form an triangle.

3.1 $m_{PE} = \frac{-9-7}{-6-2} = \frac{-16}{-8} = \frac{2}{1}$

3.2 $M\left(\frac{2-6}{2}; \frac{7-9}{2}\right) = M(-2; -1)$

3.3 $m_{TM} = \frac{-5+1}{6+2} = \frac{-4}{8} = -\frac{1}{2}$

3.4 $PE \perp TM$ $\quad m_{PE} \times m_{TM} = -1$

3.5 $PE = \sqrt{(2+6)^2 + (7+9)^2}$

$\quad\quad = \sqrt{(8)^2 + (16)^2} = 8\sqrt{5}$

$\quad TM = \sqrt{(6+2)^2 + (-5+1)^2}$

$\quad\quad = \sqrt{(8)^2 + (-4)^2} = \sqrt{80} = 4\sqrt{5}$

$\quad$ Area of $\triangle PET = \frac{bh}{2} = \frac{PE \times TM}{2} = \frac{8\sqrt{5} \times 4\sqrt{5}}{2}$

$\quad\quad\quad = \frac{160}{2}$

$\quad\quad\quad = 80$ units

4.1 $m_{AB} = \frac{-3-3}{1-5} = \frac{-6}{-4} = \frac{3}{2}$

$\quad m_{BC} = \frac{t+3}{-1-1} = \frac{3}{2}$ (gradients = for collinear points)

$\quad \therefore \frac{t+3}{-2} = \frac{3}{2}$

$\quad \therefore 2t + 6 = -6$

$\quad \therefore t = -6$

4.2 For $\hat{A} = 90°$, $AD \perp AB$

$\quad m_{AD} = -\frac{2}{3}$ $\quad m_{AB} \times m_{AD} = -1$

$\quad \therefore m_{AD} = \frac{1-3}{p-5} = \frac{-2}{3}$

$\quad \frac{-2}{p-5} = \frac{-2}{3}$

$\quad\quad -6 = -2p + 10$

$\quad\quad 2p = 16$

$\quad\quad p = 8$

5.1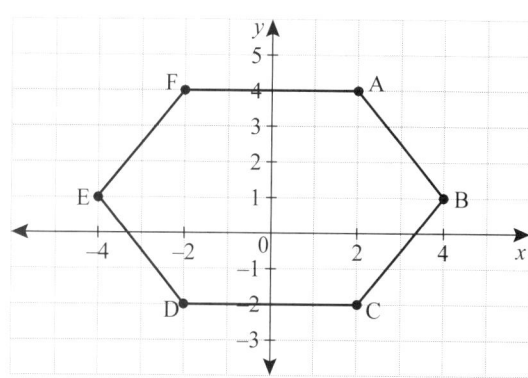

5.2 $AB = \sqrt{(2-4)^2 + (4-1)^2}$

$\quad\quad = \sqrt{(-2)^2 + (3)^2} = \sqrt{13}$

$\quad BC = \sqrt{(2-4)^2 + (-2-1)^2}$

$\quad\quad = \sqrt{(2)^2 + (-3)^2} = \sqrt{13}$

$\quad CD = 4$ units

5.3 Perimeter $= 4 \times \sqrt{13} + (2 \times 4) = 22{,}42$ units

5.4 FA: $y = 4$

$\quad DC: y = -2$

5.5 $m_{BC} = \frac{1+2}{4-2} = \frac{3}{2}$

$\quad m_{EF} = \frac{4-1}{-2+4} = \frac{3}{2}$

$\quad BC \parallel EF$ $m_{BC} = m_{EF}$

5.6 $AF \parallel DC$ $m_{AF} = m_{DC} = 0$

$\quad m_{ED} = \frac{1+2}{-4+2} = \frac{3}{-2} = -\frac{3}{2}$

$\quad m_{AB} = \frac{4-1}{2-4} = \frac{3}{-2} = -\frac{3}{2}$

$\quad \therefore ED \parallel AB$ $\quad m_{ED} = m_{AB}$

5.7 Hexagon

6.1 $m_{PA} = \frac{3+1}{2+4} = \frac{4}{6} = \frac{2}{3}$

$\quad m_{AT} = \frac{-3-3}{6-2} = \frac{-6}{4} = -\frac{3}{2}$

$\quad PA \perp AT$ $\quad m_{PA} \times m_{AT} = -1$

$\quad \therefore \triangle PAT$ is right angled with $\hat{A} = 90°$.

6.2 $PA = \sqrt{(-4-2)^2 + (-1-3)^2}$ units

$\quad\quad = \sqrt{(-6)^2 + (-4)^2} = \sqrt{52}$

$\quad AT = \sqrt{(6-2)^2 + (-3-3)^2}$

$\quad\quad = \sqrt{(4)^2 + (-6)^2} = \sqrt{52}$ units

$\quad \therefore AT = PA$

$\quad \therefore \triangle PAT$ is an isosceles right-angled triangle.

$\quad \therefore A\hat{P}T = A\hat{T}P = 45°$

  (∠s opp. = sides.)

6.3   Area of $\triangle$PAT $= \frac{bh}{2} = \frac{PA \times AT}{2} = \frac{\sqrt{52} \times \sqrt{52}}{2}$

$= \frac{52}{2} = 26$ units^2

Exercise 4   (page 185)

1.1 $m_{PQ} = \frac{y_2 - y_1}{y_2 - y_1} = \frac{-4 - 2}{4 - 2} = \frac{-6}{2} = -\frac{3}{1}$

$m_{RS} = \frac{y_2 - y_1}{y_2 - y_1} = \frac{4 - (-2)}{8 - 10} = \frac{6}{-2} = -\frac{3}{1}$

$\therefore$ PQ $\parallel$ RS $m_{PQ} = m_{RS}$

$m_{QR} = \frac{y_2 - y_1}{y_2 - y_1} = \frac{-2 - (-4)}{10 - 4} = \frac{2}{6} = \frac{1}{3}$

$m_{PS} = \frac{y_2 - y_1}{y_2 - y_1} = \frac{4 - 2}{8 - 2} = \frac{2}{6} = \frac{1}{3}$

$\therefore$ QR $\parallel$ PS $m_{QR} = m_{PS}$

$\therefore$ PQRS is a parallelogram.

Both pairs of opposite sides are parallel.

1.2 $\therefore$ PQ $\perp$ QR $m_{PQ} \times m_{QR} = -1$

PQRS is a parallelogram (proved in 1.1 above) with $\hat{Q} = 90°$.

$\therefore$ PQRS is a rectangle.

1.3   PQ $= \sqrt{(2 + 4)^2 + (2 - (-4))^2}$

$= \sqrt{(-2)^2 + (6)^2} = \sqrt{40} = 2\sqrt{10}$

QR $= \sqrt{(4 - 10)^2 + (-4 - (-2))^2}$

$= \sqrt{(-6)^2 + (-2)^2} = \sqrt{40} = 2\sqrt{10}$

$\therefore$ PQ $=$ QR

PQRS is a rectangle with adjacent sides equal
$\therefore$ PQRS is a square.

2.   PT $= \sqrt{(-2 - 4)^2 + (3 + 3)^2}$

$= \sqrt{(-6)^2 + (6)^2} = \sqrt{72} = 6\sqrt{2}$

PE $= \sqrt{(4 - 8)^2 + (-3 - 5)^2}$

$= \sqrt{(-4)^2 + (-8)^2} = \sqrt{80} = 4\sqrt{5}$

TE $= \sqrt{(-2 - 8)^2 + (3 - 5)^2}$

$= \sqrt{(-10)^2 + (-2)^2} = 2\sqrt{26}$

$\triangle$PET is a scalene triangle

(No sides are equal in length.)

3.1 $m_{JL} = \frac{7 - 0}{1 - 2} = -\frac{7}{1}$

$m_{KM} = \frac{8 + 1}{8 + 5} = \frac{9}{13}$

$m_{JL} \times m_{LM} \neq -1$

$\therefore$ Diagonals are not perpendicular.

3.2 $m_{JK} = \frac{8 - 7}{8 - 1} = \frac{1}{7}$

$m_{ML} = \frac{0 + 1}{2 + 5} = \frac{1}{7}$

$\therefore$ JK $\parallel$ ML $m_{JK} = m_{ML}$

$m_{JM} = \frac{7 + 1}{1 + 5} = \frac{8}{6} = \frac{4}{3}$

$m_{KL} = \frac{8 - 0}{8 - 2} = \frac{8}{6} = \frac{4}{3}$

$\therefore$ JM $\parallel$ KL $m_{JM} = m_{KL}$

JKLM is a parallelogram. Both pairs of opposite sides are parallel.

4.1   Midpoint of diagonal BN:

$\left(\frac{-10 - 2}{2} ; \frac{0 - 8}{2} \right) = (-6; -4)$

Midpoint of diagonal DA:

$\left(\frac{-3 - 9}{2} ; \frac{-1 - 7}{2} \right) = (-6; -4)$

$\therefore$ Diagonals bisect each other.

$m_{BN} = \frac{0 + 8}{-10 + 2} = \frac{8}{-8} = -1$

$m_{DA} = \frac{-1 + 7}{-3 + 9} = \frac{6}{6} = 1$

$\therefore$ BN $\perp$ DA $m_{BN} \times m_{DA} = -1$

$\therefore$ The diagonals bisect each other at right angles.

4.2   square, rhombus

4.3   BN $= \sqrt{(-10 + 2)^2 + (0 + 8)^2}$

$= \sqrt{(8)^2 + (8)^2} = \sqrt{128} = 8\sqrt{2}$

DA $= \sqrt{(-3 + 9)^2 + (-1 + 7)^2}$

$= \sqrt{(6)^2 + (6)^2} = \sqrt{72} = 6\sqrt{2}$

Diagonals are of unequal lengths.
$\therefore$ BAND is not a square.

4.4   ME $=$ 2MA and points D, M, A and E are collinear, $\therefore$ A must be the midpoint of ME.

$\therefore \left(\frac{-6 + x}{2} ; \frac{-4 + y}{2} \right) = (-3; -1)$

$\frac{-6 + x}{2} = -3$ and $\frac{-4 + y}{2} = -1$

$-6 + x = -6$ $-4 + y = -2$

$x = 0$ and $y = 2$

$\therefore$ E(0; 2)

4.5   BAND is a rhombus   (Unequal diagonals bisect each other at right angles.)

4.6   Area of a rhombus $= \frac{BN \times DA}{2}$

$= \frac{8\sqrt{2} \times 6\sqrt{2}}{2} = 48$ units^2

5.1   T(6; 1)

5.2 $m_{UR} = \frac{5-1}{2+2} = \frac{4}{4} = 1$

$m_{US} = \frac{1+3}{-2-2} = \frac{4}{-4} = -1$

$\therefore UR \perp US$ $\quad\quad\quad\quad m_{UR} \times m_{US} = -1$

$\therefore$ The adjacent sides meet at right angles, therefore RUST is at least a rectangle.

$UR = \sqrt{(5-1)^2 + (2+2)^2}$

$\quad\quad = \sqrt{(4)^2 + (4)^2} = \sqrt{32} = 4\sqrt{2}$ units

$US = \sqrt{(1+3)^2 + (-2-2)^2}$

$\quad\quad = \sqrt{(4)^2 + (-4)^2} = \sqrt{32} = 4\sqrt{2}$ units

$UR = US$

$\therefore$ RUST is a square (Adjacent sides are equal, and they meet at right angles.)

6.1 A(2; 4,5), B(5,5; –1), C(–1,5; –4,5), D(–5; 1)

6.2 $m_{AC} = \frac{4,5+4,5}{2+1,5} = \frac{16}{7}$

$m_{DB} = \frac{-1-1}{5,5+5} = -\frac{4}{21}$

Diagonals do not intersect at right angles.

$m_{AB} = \frac{4,5+1}{2-5,5} = -\frac{11}{7}$

$m_{DC} = \frac{1+4,5}{-5+1,5} = -\frac{11}{7}$

$\therefore AB \parallel DC$ $\quad\quad\quad\quad m_{AB} = m_{DC}$

$m_{AD} = \frac{4,5-1}{2+5} = \frac{1}{2}$

$m_{BC} = \frac{-1+4,5}{5,5+1,5} = \frac{1}{2}$

$\therefore AD \parallel BC$ $\quad\quad\quad\quad m_{AD} = m_{BC}$

$\therefore$ ABCD is a parallelogram (Both pairs of opposite sides are parallel.)

7.1 $AB = \sqrt{(-4)^2 + (-3)^2} = \sqrt{25} = 5$ units

$AC = \sqrt{(-4)^2 + (-3)^2} = \sqrt{25} = 5$ units

$\therefore \triangle ABC$ is an isosceles triangle

(Two sides equal in length.)

7.2 D(–4; –6)

7.3 D(–4; 6)

8. $KT = \sqrt{(4-1)^2 + (9-3)^2}$

$\quad\quad = \sqrt{(3)^2 + (6)^2} = \sqrt{45}$

$EI = \sqrt{(1-5)^2 + (8-6)^2}$

$\quad\quad = \sqrt{(-4)^2 + (2)^2} = \sqrt{20}$

Area KITE $= \frac{\sqrt{45} \times \sqrt{20}}{2} = 15$ units^2

Test A (page 186)

1. Let R(2; 5) = $(x_1; y_1)$ and T(–1; 3) = $(x_2; y_2)$.

1.1 Distance $= \sqrt{(x_2 - x_1)^2 + (y_2 - y_1)^2}$

$RT = \sqrt{(-1-2)^2 + (3-5)^2}$

$\quad\quad = \sqrt{(-3)^2 + (-2)^2} = \sqrt{13}$

1.2 $m = \frac{y_2 - y_1}{x_2 - x_1}$

$m_{RT} = \frac{3-5}{-1-2} = \frac{-2}{-3} = \frac{2}{3}$

1.3 Midpoint $= \left(\frac{x_1 + x_2}{2}; \frac{y_1 + y_2}{2}\right)$

$\left(\frac{2-1}{2}; \frac{5+3}{2}\right) = \left(\frac{1}{2}; 4\right)$

2.1 $m_{DE} = \frac{y_2 - y_1}{x_2 - x_1} = \frac{-4-5}{-4-2} = \frac{-9}{-6} = \frac{3}{2}$

$m_{EF} = \frac{y_2 - y_1}{x_2 - x_1} = \frac{2-(-4)}{0-(-4)} = \frac{6}{4} = \frac{3}{2}$

Points D, E and F are collinear.

2.2 $m_{DG} = \frac{y_2 - y_1}{x_2 - x_1} = \frac{9-5}{-4-2} = \frac{4}{-6} = -\frac{2}{3}$

$DG \perp DE$ $\quad\quad\quad\quad m_{DG} \times m_{DE} = -1$

3.1

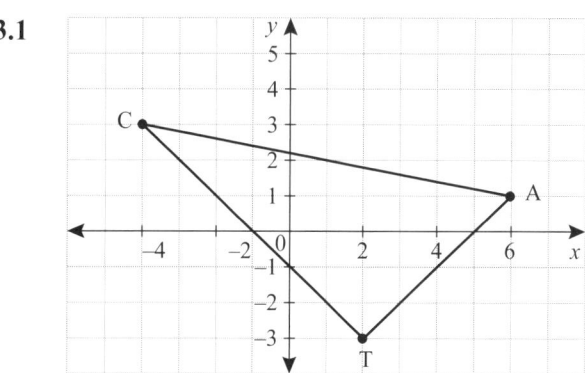

3.2 $m_{CT} = \frac{y_2 - y_1}{x_2 - x_1} = \frac{-3-3}{2-(-4)} = \frac{-6}{6} = -1$

$m_{TA} = \frac{y_2 - y_1}{x_2 - x_1} = \frac{1-(-3)}{6-2} = \frac{4}{4} = 1$

$CT \perp TA$ $\quad\quad\quad\quad m_{CT} \times m_{TA} = -1$

$\therefore \hat{T} = 90°$

$\therefore \triangle CAT$ is a right-angled triangle.

3.3 S(0; 1) (by inspection)

4. Midpoint of QA: $\left(\frac{-3+3}{2}; \frac{-5+5}{2}\right) = (0; 0)$

Midpoint of UD: $\left(\frac{2-2}{2}; \frac{-3+3}{2}\right) = (0; 0)$

Line segments QA and UD have a common midpoint.

$\therefore$ QA and UD bisect each other.

5. $m_{ST} = \frac{p-(-2)}{6-4} = 1$

$\frac{p+2}{2} = 1$

$p + 2 = 2$

$\quad p = 0$

6. M(3; 2) is the midpoint of AB.

$\therefore \left(\frac{4+a}{2}; \frac{7+b}{2}\right) = (3; 2)$

$\therefore \frac{4+a}{2} = 3$ and $\frac{7+b}{2} = 2$

$4 + a = 6$ and $7 + b = 4$

$a = 2$ and $b = -3$

7. Distance $= \sqrt{(x_2 - x_1)^2 + (y_2 - y_1)^2}$

$\quad\quad 5 = \sqrt{(2-6)^2 + (3-y)^2}$

$\therefore 25 = (2-6)^2 + (3-y)^2$

$\therefore 25 = 16 + 9 - 6y + y^2$

$\therefore 0 = y^2 - 6y$

$\therefore 0 = y(y-6)$

$\therefore y = 0$ or $y = 6$

8.1 Paula is correct.

Distance $= \sqrt{(x_2 - x_1)^2 + (y_2 - y_1)^2}$

PT $= \sqrt{(-2-1)^2 + (-2-2)^2}$

$\quad = \sqrt{(-3)^2 + (-4)^2} = \sqrt{25} = 5$

TE $= \sqrt{(1-1)^2 + (2-(-2)^2}$

$\quad = \sqrt{(0)^2 + (4)^2} = \sqrt{16} = 4$

EP $= \sqrt{(-2-1)^2 + (-2-(-2)^2}$

$\quad = \sqrt{(-3)^2 + (-0)^2} = \sqrt{9} = 3$

Triangle is scalene (All sides have different lengths.)

$PT^2 = TE^2 + EP^2$

$\therefore \hat{E} = 90°$

$\therefore \triangle PET$ is a scalene right-angled triangle.

8.2 Area $= \frac{lb}{2} = \frac{TE \times EP}{2} = \frac{4 \times 3}{2} = 6$ units^2

Test B (page 187)

1.1 S(8; 0)

1.2 In trapezium PQRT, RS = ST.

S(8; 0) is the midpoint of RT.

$\therefore \left(\frac{3+x}{2}; \frac{-3+y}{2}\right) = (8; 0)$

$\therefore \frac{3+x}{2} = 8$ and $\frac{-3+y}{2} = 0$

$3 + x = 16$ and $-3 + y = 0$

$x = 13$ and $y = 3$

$\therefore T(13; 3)$

2.1 Midpoint for AB: $P\left(\frac{1+4}{2}; \frac{1+5}{2}\right) = P(2,5; 3)$

2.2 Distance $= \sqrt{(x_2 - x_1)^2 + (y_2 - y_1)^2}$

AB $= \sqrt{(1-4)^2 + (1-5)^2}$

$\quad = \sqrt{(-3)^2 + (-4)^2} = \sqrt{25} = 5$

Diameter of small circle $= 5 - 2 \times 2 = 1$ unit

2.3

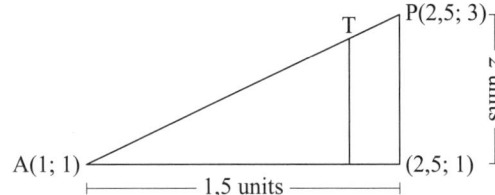

AT : TP $= 2 : 0,5 = 4 : 1$

$\therefore$ AP : TP $= 5 : 1$

$\frac{1,5}{5} = 0,3$ and $\frac{2,0}{5} = 0,4$

$\therefore T(2,5 - 0,3; 3 - 0,4)$

$\therefore T(2,2; 2,6)$

3.1 Distance $= \sqrt{(-2-10)^2 + (12-2)^2} = \sqrt{244}$

$\quad\quad\quad\quad = 2\sqrt{61} \times 2$ km $= 31,24 = 31$ km

3.2 (10; –8)

3.3 $m = \frac{-12+10}{2-c} = -\frac{1}{6}$

$\frac{-2}{2-c} = -\frac{1}{6}$

$-12 = -2 + c$

$\quad c = -10$

3.4 $10\sqrt{2} = \sqrt{(-2-d)^2 + (12-2)^2}$

$\quad 200 = 4 + 4d + d^2 + 100$

$\quad\quad 0 = d^2 + 4d - 96$

$\quad\quad 0 = (d-8)(d+12)$

$\quad\quad d = -12$ or $d = 8$

In this case, $d = 8$ is inadmissible since checkpoint 6 lies in the 2nd quadrant,

$\therefore d = -12$

3.5 A(1; 9,5), B(4; 7), C(7; 4,5)

4.

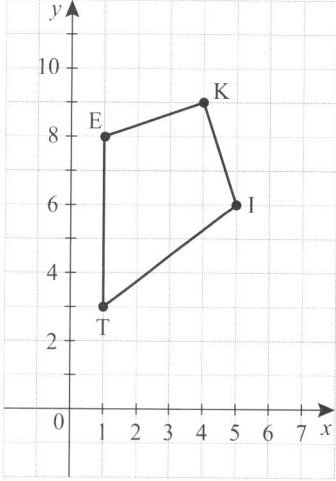

K(4; 9), I(5; 6), T(1; 3) and E(1; 8)

$m_{KT} = \frac{9-3}{4-1} = \frac{6}{3} = \frac{2}{1}$

$m_{EI} = \frac{8-6}{1-5} = \frac{2}{-4} = -\frac{1}{2}$

$\therefore$ KT $\perp$ EI $m_{KT} \times m_{EI} = -1$

Diagonals intersect at right angles.

Let the midpoint of EI be M, then M(3; 7).

$m_{KM} = \frac{7-9}{3-4} = \frac{-2}{-1} = \frac{2}{1}$

$m_{MT} = \frac{3-7}{1-3} = \frac{-4}{-2} = \frac{2}{1}$

Points K, M and T are collinear.

Therefore the midpoint of diagonal EI lies on KT.

The diagonals intersect at right angles and the diagonal KT bisects the diagonal EI.

KITE is therefore a kite.

Test C   (page 188)

1.1 $M\left(\frac{-3+7}{2}; \frac{3-3}{2}\right) = M(2; 0)$

1.2 $m_{AD} = \frac{0-3}{2+3} = -\frac{3}{5}$

$m_{DC} = \frac{0-5}{2-5} = \frac{-5}{-3} = \frac{5}{3}$

AD $\perp$ DC $m_{AD} \times m_{DC} = -1$

1.3 $DC = \sqrt{(2-5)^2 + (0-5)^2}$

$= \sqrt{(-3)^2 + (-5)^2} = \sqrt{34}$

$AD = \sqrt{(-3-2)^2 + (3-0)^2}$

$= \sqrt{34}$

$\therefore AB = 2\sqrt{34}$

Area $= \frac{\sqrt{34} \times 2\sqrt{34}}{2} = 34$ units^2

2.1 $m_{BC} = 2$ (given)

$\frac{-2-6}{3-p} = 2$

$\frac{-8}{3-p} = 2$

$-8 = 6 - 2p$

$p = 7$

2.2 $AB = \sqrt{(-5-3)^2 + (6+2)^2} = \sqrt{128}$

$\sqrt{128} = \sqrt{(3-p)^2 + (-2-6)^2}$

$128 = (3-p)^2 + (-2-6)^2$

$128 = 9 - 6p + p^2 + 64$

$0 = p^2 - 6p - 55$

$0 = (p+5)(p-11)$

$p = -5$ or $p = 11$

3.1 $DE = \sqrt{(-3-3)^2 + (3+5)^2}$

$= \sqrt{(-6)^2 + (8)^2} = \sqrt{100} = 10$

3.2 $m_{DE} = \frac{3+5}{-3-3} = \frac{8}{-6} = -\frac{4}{3}$

3.3   For $D\hat{E}F = 90°$, $m_{DE} \times m_{EF} = -1 = 3$ and

$m_{EF} = \frac{3}{4}$

$m_{EF} = \frac{-5-k}{3+1} = \frac{3}{4}$

$4(-5-k) = 3 \times 4$

$-20 - 4k = 12$

$k = -8$

3.4 $M\left(\frac{-3-1}{2}; \frac{3-8}{2}\right) = M(-2; -2,5)$

3.5 $G(-7; 0)$

4. $C(12; 8)$

$m_{AC} = \frac{8-4}{12-a} = \frac{1}{7}$

$\frac{4}{12-a} = \frac{1}{7}$

$28 = 12 - a$

$a = -16$

5.   Given: E(0; 0) and D(12; 15)

5.1   Midpoint of ED = (6; 7,5)

5.2   (3; 3,75), (6; 7,5) and (9; 11,25)

5.3   (4; 5), (8; 10)

Topic 10   Finance (page 190)

Exercise 1 (page 191)

1.1 $4,5\% = \frac{4,5}{100} = 0,045$

1.2 $5,1\% = \frac{5,1}{100} = 0,051$

1.3 $4,9\% = \frac{4,9}{100} = 0,049$

1.4 $5,6\% = \frac{5,6}{100} = 0,056$

2. $P = 2\,000(1 + 0,055(3)) = 2\,330$

3. $P = \frac{5\,000}{(1 + 0,045(3))} = 4\,405,29$

4. $n = \frac{\left(\frac{4\,500}{2\,650} - 1\right)}{0,04} = 17,45$

5. $i = \frac{\left(\frac{4\,500}{2\,650} - 1\right)}{5} = 0,1396$

6. $P = 20\,000\left(1 + \frac{0,0485}{12}(4 \times 12)\right) = 23\,880$

7. $P = 20\,000(1 + 0,051)^5 = 25\,647,41$

8. $P = \frac{7\,500}{(1 + 0,048)^3} = 6\,515,95$

9. $i = \sqrt[3]{\frac{10\,000}{7\,800}} - 1 = 0,0863$

10. $P = 2\,000\left(1 + \frac{0,05}{4}(3 \times 4)\right) = 2\,300$

11. $P = 2\,000\left(1 + \frac{0,05}{12}(3 \times 12)\right) = 2\,300$

12. $P = 20\,000\left(1 + \frac{0,051}{12}\right)^{5 \times 12} = 25\,795,29$

13. $P = \frac{7\,500}{\left(1 + \frac{0,048}{12}\right)^{3 \times 12}} = 6\,496,02$

Exercise 2 (page 193)

1. $F = P(1 + in)$, where $P = R10\,000$,

$i = \frac{9}{100} = 0,09$ *i represents the interest rate per period as a decimal, given that r = 9% p.a.*

and $n = 5$ *n represents the number of interest periods.*

$F = 10\,000(1 + 0,09(5)) = R14\,500$

Interest paid = R4 500

2.1 Interest rate (r) per day $= \frac{12}{365}$

$= 0,0329\%$ per day

2.2 $F = P(1 + in)$, where $P = R7\,000$,

$i = \frac{0,12}{365}$ *i represents the interest rate per day as a decimal, where r = 12% p.a.*

and $n = 120$ *n represents the number of interest periods.*

$F = 7\,000\left(1 + \frac{0,12}{365}(120)\right) = R7\,276,16$

2.3 Interest paid = R276,16

3.1 $F = 25\,500(1 + 0,08(3)) = R31\,620$

3.2 Interest = R6 120

3.3.1 $F = 25\,500(1 + 0,08(6)) = R37\,740$

3.3.2 Interest = R12 240

> **Note:**
> In calculations using simple interest, the interest doubles if the time period is doubled.

4. $F = 15\,000\left(1 + \frac{0,054}{12}(4)\right) = R15\,270$

Take the annual interest rate as a decimal and divide by 12 as the money is in the account for 4 months.

Exercise 3 (page 194)

1.1 $P = \frac{F}{(1 + in)}$, F = R20 000,

$i = 0,05$ and $n = 2$

$P = \frac{20\,000}{(1 + 0,05(2))} = R18\,181,82$

1.2 $P = \frac{F}{(1 + in)}$, F = R100 000, $i = \frac{0,07}{365}$ and $n = 120$

$P = \frac{100\,000}{\left(1 + \frac{0,07}{365}(120)\right)} = R97\,750,40$

1.3 $P = \frac{F}{(1 + in)}$, F = R55 000, $i = 0,06$ and $n = 5$

$P = \frac{55\,000}{(1 + 0,06(5))} = R42\,307,69$

1.4 $P = \frac{F}{(1 + in)}$, F = R5 000, $i = 0,02$ and $n = 12$

note the interest rate is per month and n = 1 × 12 = 12

$P = \frac{5\,000}{(1 + 0,02(12))} = R4\,032,26$

2. $i = \frac{\frac{F}{P} - 1}{n} = \frac{\frac{8\,000}{6\,000} - 1}{2} = 0,167$

The rate of interest, $r = 16,7\%$ p.a.

3.1 $n = \frac{\frac{F}{P} - 1}{i} = \frac{\frac{8\,000}{5\,000} - 1}{0,08} = 7,5 = 7$ years 6 months

3.2 $n = \frac{\frac{F}{P} - 1}{i} = \frac{\frac{75\,000}{65\,000} - 1}{0,12} = 1,28 = 1$ year 3 months

> **Note:**
> To calculate the number of months with an answer of 1,28… subtract 1 and then multiply by 12 i.e. 12 months in a year.

3.3 $n = \frac{\frac{F}{P} - 1}{i} = \frac{\frac{15\,000}{10\,000} - 1}{0,14} = 3,57 = 3$ years 7 months

4.1 $F = P(1 + in) = 36\,255(1 + 0,06 \times 1)$
$= R38\,430,30$

4.2 $F = P(1 + in) = 36\,255(1 + 0,06 \times 2)$
$= R40\,605,60$

4.3 $F = P(1 + in) = 36\,255(1 + 0,06 \times 3)$
$= R42\,780,90$

4.4 $F = P(1 + in) = 36\,255(1 + 0,06 \times 4)$
$= R44\,956,20$

4.5 Interest earned $= R40\,605,60 - R36\,255$
$= R4\,350,60$

4.6 Interest earned $= R44\,956,20 - R36\,255$
$= R8701,20$

4.7 The future values calculated have the same (constant) interest added from year to year. This is a property related to simple interest, the rate of change is constant.

4.8 When the time period is doubled, the interest earned is also doubled. This is a property related to simple interest where the rate of change is constant.

Exercise 4   (page 195)

1.1 Balance $= R7\,500 - R1\,200 = R6\,300$

1.2 Interest rate per period $= 9\% \div 12$
$= 0,75\%$ per month.

1.3 $F = P(1 + in)$
$= R6\,300(1 + 0,0075(20))$

i is the interest as a decimal.

$F = R7\,245$
Monthly payments $= R7\,245 \div 20 = R362,25$

1.4 Interest paid $=$ total amount paid $-$ marked price
$= R1\,200 + R362,25 \times 20 - R7\,500 = R945$

2.1 Deposit $= \frac{20}{100} \times 4\,450 = R890$

2.2 Balance outstanding $= R4\,450 - R890$
$= R3\,560$

2.3 $F = P(1 + in) = 3\,560\left(1 + \frac{0,12}{12}(15)\right)$

i is the interest as a decimal per month.

$F = R4\,094$

2.4 Monthly instalment $= R4\,094 \div 15 = R272,93$

2.5 Interest paid $= R890 + R4\,094 - R4\,450$
$= R534$

3.1 $F = R174,31 \times 36 = R6\,275,16 + R1\,200$
$= R7\,475,16$

3.2 $i = \frac{\frac{F}{P} - 1}{n}$

$i = \frac{\frac{7\,475,16}{5\,000} - 1}{36} = 0,013754222$ per month

i is the interest as a decimal per month
Per annum ($\times 12$) $= 0,1650$, rate $r = 16,5\%$

4. $F = P(1 + in)$
$F = 16\,500(1 + 0,10(2)) = R19\,800$
Monthly instalments $= R19\,800 \div 24 = R825$

Exercise 5   (page 197)

1.1 $F = P(1 + i)^n$, $P = R120\,000$, $i = 0,06$, $n = 3$

the interest rate per period as a decimal

$F = R120\,000(1 + 0,06)^3 = R142\,921,92$

1.2 Interest earned $= R142\,921,92 - R120\,000$
$= R22\,921,92$

2.1 $F = P(1 + i)^n$, $P = R4\,500$, $i = \frac{0,056}{12}$

the interest rate per period as a decimal

$n = 18$ *number of interest periods*

$F = 4\,500\left(1 + \frac{0,056}{12}\right)^{18} = R4\,893,37$

2.2 Interest earned $= R4\,893,37 - R4\,500$
$= R393,37$

3.1 **Option A:** $F = P(1 + i)^n$
$= R200\,000\left(1 + \frac{0,0575}{12}\right)^{2 \times 12}$
$= R224\,313,07$

Option B: $F = P(1 + in)$
$= 200\,000\left(1 + \frac{0,062}{12}\right)(2 \times 12)$
$= R224\,800$

Over a short period of time, Option B is the better choice.

3.2 **Option A:** $F = P(1 + i)^n$
$= R200\,000\left(1 + \frac{0,0575}{12}\right)^{10 \times 12}$
$= R354\,938,36$

Option B: $F = P(1 + in)$
$= R200\,000\left(1 + \frac{0,062}{12}\right)(10 \times 12)$
$= R324\,000$

Over a longer period of time Option A is the better choice.

4.1 $F = P(1 + i)^n$
$= R2\,800\left(1 + \frac{0,048}{4}\right)^{5 \times 4} = R3\,554,42$

4.2 $F = P(1 + i)^n$
$= R2\,800\left(1 + \frac{0,048}{12}\right)^{5 \times 12} = R3\,557,79$

4.3 $F = P(1 + i)^n$
$= R2\,800\left(1 + \frac{0,048}{12}\right)^{10 \times 12} = R4\,520,68$

5.1 The future value increased from questions 4.1 to 4.2. There are more periods of interest in a year.

5.2 The future value increased from questions 4.2 to 4.3. As the number of years increase so do the number of interest periods.

5.3 Interest earned in question 4.2 = R757,79
Interest earned in question 4.3 = R1 720,68
And R757,79 × 2 = R1 515,58; this amount is less than the interest earned in question 4.3. When the number of years is doubled and all else remains the same, the interest does not double when using compound interest.

Exercise 6 (page 198)

1.1 (a) $P = \frac{F}{(1+i)^n} = \frac{16\,585}{(1+0,048)^3} = R14\,408,93$

(b) Interest paid = R2 176,07

1.2 (a) $P = \frac{F}{(1+i)^n} = \frac{10\,000}{\left(1+\frac{0,165}{365}\right)^{120}} = R9\,472,10$

(b) Interest paid = R527,90

1.3 (a) $P = \frac{F}{(1+i)^n} = \frac{5\,800}{\left(1+\frac{0,06}{4}\right)^{5\times 4}} = R4\,306,33$

(b) Interest paid = R1 493,67

1.4 (a) $P = \frac{F}{(1+i)^n} = \frac{1\,000}{(11+0,02)^{12}} = R788,49$

> **Note:**
> This interest rate is PER MONTH and not per annum compounded monthly. This indicates a very high rate of interest.

(b) Interest paid = R211,51

1.5 (a) $P = \frac{F}{(1+i)^n} = \frac{1\,000}{\left(1+\frac{0,02}{12}\right)^{12}} = R980,21$

(b) Interest paid = R19,79

2. $i = \sqrt[n]{\frac{F}{P}} - 1 = \sqrt[2]{\frac{10\,000}{7\,500}} - 1 = 0,1547.$
Annual rate of interest = 15,47%

3.1 $i = \sqrt[n]{\frac{F}{P}} - 1 = \sqrt[3]{\frac{8\,000}{5\,000}} - 1 = 0,1696.$
Annual rate of interest = 16,96%

3.2 $i = \sqrt[n]{\frac{F}{P}} - 1 = \sqrt[4]{\frac{75\,000}{65\,000}} - 1 = 0,03642.$
Annual rate of interest = 3,64%

3.3 $i = \sqrt[n]{\frac{F}{P}} - 1 = \sqrt[2]{\frac{15\,000}{10\,000}} - 1 = 0,22474.$
Annual rate of interest = 22,47%

4.1 $F = P(1 + in)$
$= 15\,000(1 + 0,09(10)) = R28\,500$

4.2 $F = P(1 + i)^n$
$= 15\,000(1 + 0,09)^{10} = R35\,510,46$

5. $F = P(1 + i)^n$
$= 35\,000\left(1 + \frac{0,09}{2}\right)^{10} = R54\,353,93$

6. $P = \frac{F}{(1+i)^n} = \frac{68\,575}{\left(1+\frac{0,055}{52}\right)^{3\times 52}} = R58\,149,38$

7. $n = \frac{\frac{F}{P}-1}{i} = \frac{\frac{25\,768,45}{15\,000}-1}{0,058} = 12,377$
$= 12$ years and $4\frac{1}{2}$ months

8.1 $F = P(1 + in)$
$= 350\,000(1 + 0,056(5)) = R448\,000$

8.2 $F = P(1 + i)^n$
$= 350\,000\left(1 + \frac{0,053}{12}\right)^{5\times 12} = R455\,934,73$

8.3 The investment using compound interest is better. Even though the interest rate appears to be lower, interest is calculated more frequently.

Exercise 7 (page 200)

1.1 The South African inflation rate appears to be less stable than the Great Britain as there are many more peaks and troughs on the graph. The inflation rate was rather erratic pre-2001, with the highest rate occurring during 1985 at approximately 20% per annum. The inflation rate for Great Britain was rather erratic pre-1983, with the highest rate during 1975 at 25%. The South African graph during 2001 to 2011 follows a similar pattern to the Great Britain graph during 1973 to 1983.

1.2 The graph of the Great Britain rate indicates that the inflation rate has been relatively stable since 1993, after a small peak in 1991. Should we interpolate from these two graphs, perhaps the South African inflation rate may follow the graph of Great Britain with another small peak and then reach a period of time where it would become more stable.

2.1 $F = P(1 + i)^n$
$= 168\,000(1 + 0,053)^3 = R196\,152,75$

2.2 $F = P(1 + i)^n$
$= 3\,599(1 + 0,053)^5 = R4\,659,33$

2.3 $F = P(1 + i)^n$
$= 12\,000(1 + 0,053)^2 = R13\,305,71$

2.4 $F = P(1 + i)^n$
$= 36\,000(1 + 0,053)^3 = R42\,032,73$ per annum

1.1 US\$1 = R10,66

US\$27 589 = R294 098,74

1.2 ฿1 = R0,33

฿4 375 = R1 443,75

1.3 £1 = R17,88

£19 000 = R339 720

1.4 J\$1 = R0,09

J\$19 000 = R1 710

1.5 €1 = R14,30

€55 725 = R796 867,50

2.1 R17,88 = £1

$R20\,000 = \frac{20\,000}{17,88} = £1\,118,57$

2.2 R0,33 = ฿1

$R10\,000 = \frac{10\,000}{0,33} = ฿30\,303,03$

2.3 R10,66 = US\$1

$R15\,000 = \frac{15\,000}{10,66} = US\$1\,407,13$

2.4 R0,09 = J\$1

$R5\,000 = \frac{5\,000}{0,09} = J\$55\,555,56$

2.5 R14,30 = €1

$R15\,000 = \frac{15\,000}{14,30} = €1\,048,95$

Test A (page 202)

1.1 Straight line method:

$F = P(1 + in)$

$F = 32\,000(1 + 0,12(5))$

$F = R51\,200$

1.2 Exponential method:

$F = P(1 + i)^n$

$F = 32\,000(1 + 0,12)^5$

$F = R56\,394,93$

2.1 $\text{Deposit} = \frac{12}{100} \times 26\,800 = R3\,216$

Present value = R26 800 − R3 216 = R23 584

2.2 $F = P(1 + in)$

$= 23\,584(1 + 0,095(3)) = R30\,305,44$

2.3 Monthly instalments = R30 305,44 ÷ (3 × 12)

= R841,82

2.4 Interest paid = money paid − price of furniture

= R30 305,44 + R3 216 − R26 800

= R6 721,44

3.1 $F = P(1 + i)^n$

$= 168\,000(1 + 0,07)^5 = R235\,628,69$

3.2 $\text{Deposit} = \frac{20}{100} \times 168\,000 = R33\,600$

Present value = R168 000 − R33 600

= R134 400

4.1 **Option A:** $F = P(1 + in)$

$= 100\,000(1 + 0,085(1))$

$= R108\,500$

Option B: $F = P(1 + i)^n$

$= 100\,000\left(1 + \frac{0,082}{12}\right)^{12}$

$= R108\,515,31$

Over 1 year, option B is better.

4.2 **Option A:** $F = P(1 + in)$

$= 100\,000\left(1 + \frac{0,085}{2}(1)\right)$

$= R104\,250$

Option B: $F = P(1 + i)^n$

$= 100\,000\left(1 + \frac{0,082}{12}\right)^6$

$= R104\,170,68$

Over 6 months, option A is better.

5.1 US\$1 = R10,66

US\$12 000 = R12 000 × 10,66 = R127 920

5.2 ฿1 = R0,33

฿6 000 = R6 000 × 0,33 = R1 980

5.3 £1 = R17,88

£7 500 = R7 500 × 17,88 = R134 100

5.4 J\$1 = R0,09

J\$7 500 = R7 500 × 0,09 = R675

6.1 R17,88 = £1

$R20\,000 = \frac{20\,000}{17,88} = £1\,118,57$

6.2 R0,33 = ฿1

$R10\,000 = \frac{10\,000}{0,33} = ฿30\,303,03$

6.3 R14,30 = €1

$R15\,000 = \frac{15\,000}{14,30} = €1\,048,95$

6.4 R0,09 = J\$1

$R5\,000 = \frac{5\,000}{0,09} = J\$55\,555,56$

Test B (page 203)

1.1 $i = \frac{\frac{F}{P} - 1}{n} = \frac{\frac{11\,238,50}{9\,500} - 1}{2} = 0,0915$

This is a rate of 9,15% p.a.

1.2 $i = \sqrt[n]{\frac{F}{P}} - 1 = \sqrt[2]{\frac{11\,238,50}{9\,500}} - 1 = 0,0877$

This is a rate of 8,77% p.a.

2.1 Thandi's net salary $= R13\,650 - \frac{20}{100} \times 13\,650$

$= R10\,920$

Xolani's net salary $= R12\,890 - 0,22 \times 12\,890$

$= R10\,054,20$

2.2 Thandi's 30% = $0{,}30 \times 10\,920$ = R3 276
Xolani's 30% = $0{,}30 \times 10\,054{,}20$ = R3 016,26
Maximum value of the bond they can afford
= R3 276 + R3 016,26 = R6 292,26

2.3.1 $F = P(1 + i)^n$
$F = 650\,000\left(1 + \frac{0{,}08}{12}\right)^1 = R654\,333{,}33$

Amount of the bond after 1 month.

Interest after 1 month
= R654 333,33 – R650 000 = R4 333,33

2.3.2 They can pay a monthly payment of
R6 292,26, which is more than the interest.
Yes, they can afford this bond.

3.1 Deposit = $0{,}20 \times (3\,299 + 3\,699)$ = R1 399,60

3.2 Money owed = R3 299 + R3 699 – R1 399,60
= R5 598,40

3.3 $F = P(1 + in)$
$= 5\,598{,}40\left(1 + \frac{0{,}18}{12} \times 36\right)$
= R8 621,54

Monthly payments = $\frac{8\,621{,}54}{36}$ = R239,49

3.4 Interest paid
= payments made – original price
= R1 399,60 + R8 621,54 – R6 998
= R3 023,14

4.1 $F = P(1 + i)^n$
$= 64\,000(1 + 0{,}075)^8$
= R114 142,58

4.2.1 $F = P(1 + i)^n$
$= 165\,000\left(1 + \frac{0{,}098}{12}\right)^5$
= R171 848,45

4.2.2 Interest paid = R171 848,45 – R165 000
= R6 848,45

4.2.3 Balance outstanding at start
= R165 000 – R20 000
= R145 000
$F = P(1 + i)^n$
$= 145\,000\left(1 + \frac{0{,}098}{12}\right)^5 = R151\,018{,}33$

Test C (page 204)

1.1 $F = P(1 + i)^n$
$2P = P(1 + i)^6$
$i = \sqrt[6]{\frac{2P}{P}} - 1 = 0{,}1225 \Rightarrow$ rate 12,25% p.a.

1.2.1 $F = P(1 + in)$
$= 20\,400(1 + 0{,}052 \times 4)$
$= R24\,643{,}20$

1.2.2 $F = P(1 + i)^n$
$= 20\,400(1 + 0{,}052)^4$
$= R24\,985{,}79$

1.2.3 Interest = R4 243,20

1.2.4 Interest = R4 585,79

2. Ben: $F = P(1 + in)$
$= 42\,500(1 + 0{,}065(7)) = R61\,837{,}50$
He also receives a bonus = $0{,}04 \times 42\,500$
$= R1\,700$

Total value of investment
= R61 837,50 + R1 700
= R63 537,50

Josh: $F = P(1 + i)^n$
$= 42\,500\left(1 + \frac{0{,}06}{12}\right)^{7 \times 12} = R64\,615{,}71$
Josh receives the bigger amount.

3.1 Deposit = $0{,}15 \times 250\,000$
= R37 500

3.2 Amount = R250 000 – R37 500
= R212 500

3.3 $F = P(1 + i)^n$
$= 212\,500\left(1 + \frac{0{,}08}{12}\right)^1 = R213\,916{,}67$
Interest charged = R213 916,67 – R212 500
= R1 416,67

4.1.1 $F = P(1 + in)$
$= 85\,000(1 + 0{,}09(6))$
= R130 900

4.1.2 Interest = R130 900 – R85 000
= R45 900

4.2 $i = \sqrt[n]{\frac{F}{P}} - 1 = \sqrt[3]{\frac{6\,850}{5\,000}} - 1 = 0{,}1106.$
Interest rate = 11,1%

5.1 R17,88 = £1
R320 000 = $\frac{320\,000}{17{,}88}$ = £17 897,09

5.2 €1 = R14,30
€58 450 × R14,30 = R835 835

5.3 US$1 = R10,66
US$1 325 = R1 325 × 10,66
= R14 124,50

5.4 R0,09 = J$1
R15 000 = $\frac{15\,000}{0{,}09}$ = J$166 666,67

5.5 R$1 = R4,66
R$225 × R4,66 = R1 048,50

Topic 11   Statistics (page 205)

Exercise 1   (page 214)

1.1
$$
\begin{array}{c|llllllllllll}
4 & 2 & 5 & 8 \\
5 & 1 & 1 & 1 & 1 & 2 & 4 & 4 & 5 & 6 & 6 & 8 \\
6 & 0 & 1 & 2 & 4 & 6 & 9
\end{array}
$$
Key: $\frac{4}{2} = 42$

1.2   Range = 69 – 42 = 27

1.3   Median mass = $\frac{54 + 55}{2} = 54,5$

1.4 $Q_3 = 61$

1.5 $Q_1 = 51$
IQR = 61 – 51 = 10
Semi-IQR = $\frac{10}{2} = 5$

2.1   25

2.2   Lowest score: 60
Highest score: 98
Range: 98 – 60 = 38

2.3   Median score: 74

2.4 $Q_1 = 65$
$Q_3 = \frac{83 + 84}{2} = 83,5$
IQR = 83,5 – 65 = 18,5
Semi-IQR = $\frac{18,5}{2} = 9,25$

2.5
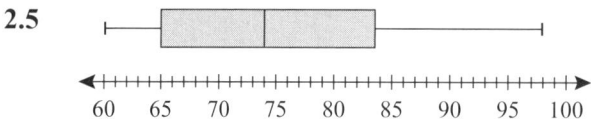

```
  60   65   70   75   80   85   90   95   100
```

3.1   Highest score: 100
3.2   Range: 100 – 55 = 45
3.3   Median: 80
3.4   Not possible to calculate the mean score
3.5   Not possible to calculate how many learners wrote the test
3.6   Upper quartile: 85
3.7   50% of the learners scored above 80.
3.8   25% of the learners scored below 70.
3.9   0% of the learners scored below 55.
3.10 50% of the learners scored between 70 and 85.

4.1

	5.00–7.00 p.m.	7.00–9.00 p.m.	

$$
\begin{array}{rr|c|l}
 & 3\ 3 & 2 & \\
 & 8\ 8\ 7\ 4 & 3 & 8\ 9 \\
 9\ 8\ 8\ 7\ 7\ 5\ 4\ 3 & & 4 & 8\ 9 \\
 8\ 8\ 7\ 6\ 6\ 5\ 2\ 1\ 0\ 0 & & 5 & 0\ 0\ 4\ 5\ 8\ 8\ 9\ 9 \\
 & 6\ 4\ 2\ 0 & 6 & 0\ 1\ 3\ 4\ 4\ 6\ 7\ 7\ 7\ 8 \\
 & & 7 & 3\ 4\ 4\ 5\ 6\ 8\ 8 \\
 & & 8 & 0
\end{array}
$$
Key: $\frac{2}{3} = 23$

4.2.1 Range: 66 – 23 = 43

4.2.2 Range: 80 – 48 = 32

4.3.1 Median: $\frac{49 + 50}{2} = 49,5$

4.3.2 Median: 64

4.4.1 $Q_1 = \frac{1}{4}(n + 1)$
$= \frac{1}{4}(28 + 1) = 7,25$th score
$= 43$

$Q_3 = \frac{3}{4}(n + 1)$
$= \frac{3}{4}(28 + 1) = 21,75$th score
$= 57$

4.4.2 $Q_1 = 58$
$Q_3 = \frac{3}{4}(n + 1) = \frac{3}{4}(28 + 1)$
$= 21,75$th score $= 74$

4.5.1–2
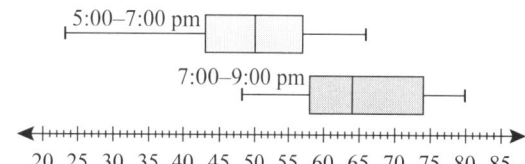

```
5:00–7:00 pm
                7:00–9:00 pm

20  25  30  35  40  45  50  55  60  65  70  75  80  85
```

Exercise 2   (page 219)

1.1

Intervals (%)	Tally	Frequency	Cumulative frequency
20–29	||||	4	4
30–39	|	1	5
40–49	||||	4	9
50–59	‖‖‖	5	14
60–69	‖‖‖ |||	8	22
70–79	||||	4	26
80–89	‖‖‖ ‖‖‖	10	36
90–99	||||	4	40
Total			**40**

1.2   4 learners
1.3   5 learners
1.4   Median interval: 60–69
1.5   Modal interval: 80–89

1.6

Intervals (%)	Tally	Frequency (*f*)	Midpoint for interval (*x*)	*f* × *x*
20–29	||||	4	24,5	4 × 24,5 = 98
30–39	|	1	34,5	1 × 34,5 = 34,5
40–49	||||	4	44,5	4 × 44,5 = 178
50–59	ⴕ	5	54,5	5 × 54,5 = 272,5
60–69	ⴕ |||	8	64,5	8 × 64,5 = 516
70–79	||||	4	74,5	4 × 74,5 = 298
80–89	ⴕ ⴕ	10	84,5	10 × 84,5 = 845
90–100	||||	4	95	4 × 95 = 380
Total		**40**		**∑(*fx*) = 2 622**

Mean (approx.) = $\frac{\Sigma(fx)}{n} = \frac{2\,622}{40} = 65{,}55\%$

1.7

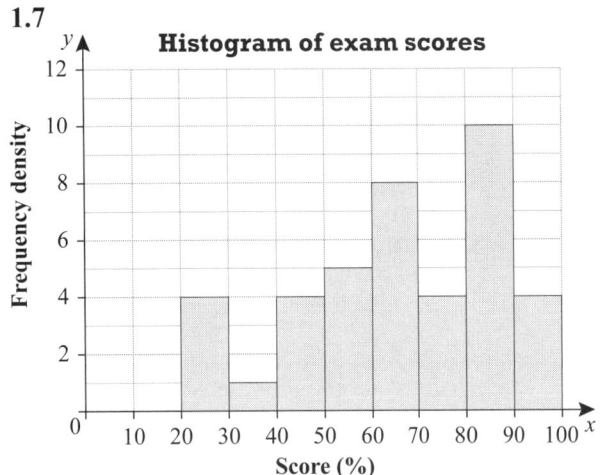

Histogram of exam scores

2.

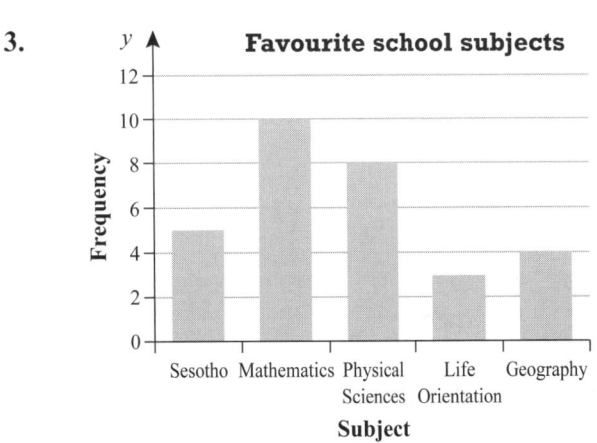

Population 2003–2007

3.

Favourite school subjects

4.1 55 learners

4.2 30 boys

4.3 80 – 50 = 30 girls.
There are 20 more boys than girls.

4.4 Grade 11 has 35 girls and 35 boys. Grade 9 has 30 girls and 30 boys.

4.5 330 learners

4.6 Grade 8 and Grade 9: 30 girls each. Grade 11 and Grade 12: 35 girls each

5.1

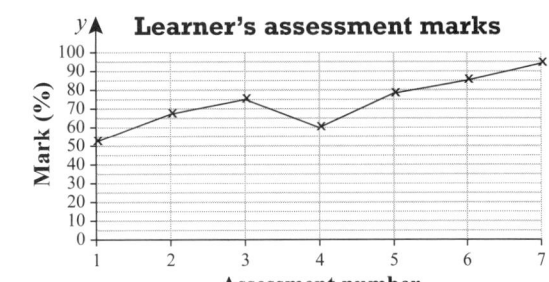

Learner's assessment marks

5.2 Mean = $\frac{510}{7}$
= 72,9%

5.3 Assessments 3, 5, 6 and 7

6.1

Flavour	Tally	Frequency
Radical relish (R)	ⴕ ⴕ ||	12
Hot stuff (H)	ⴕ	5
Nibble mania (N)	ⴕ |	6
Cheesy corn (C)	||||	4
Salted mix (S)	ⴕ	5
Original flip (O)	||||	4
Total		**36**

6.2 **Flavour preferences**

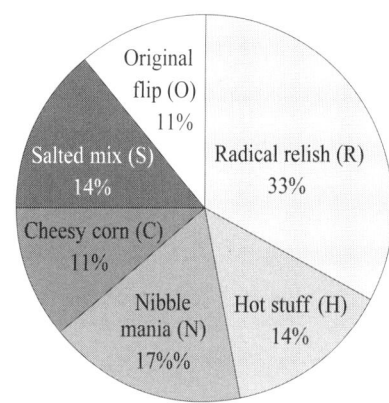

6.3 Radical relish

7.1 5 5 5 5 6 10 10 12 12 14

7.1.1 Mean = $\frac{84}{10}$ = 8,4

7.1.2 Mode = 5

7.1.3 Median = 5,5th score = $\frac{6+10}{2}$ = 8

7.1.4 Range = 14 − 5 = 9

7.2 1 1 3 4 6 8 9

7.2.1 Mean = $\frac{32}{7}$ = 4,6

7.2.2 Mode = 1

7.2.3 Median = 4th value = 4

7.2.4 Range = 9 − 1 = 8

7.3 5 5 5 6 6 6 6 8 8 9 9 9

7.3.1 Mean = $\frac{82}{12}$ = 6,83

7.3.2 Mode = 6

7.3.3 Median = 6,5th value = 6

7.3.4 Range = 9 − 5 = 4

8.1

Score	Frequency	Frequency × score
5	3	15
7	8	56
9	2	18
10	3	30
12	4	48
Total	**20**	**167**

Mean = $\frac{167}{20}$ = 8,35

Median = 10,5th value = 7

Mode = 7

8.2

Score	Frequency	Frequency × score
10	3	30
15	5	75
20	2	40
25	4	100
30	1	30
Total	**15**	**275**

Mean = $\frac{275}{15}$ = 18,$\dot{3}$

Median = 8th value = 15

Mode = 15

9.

Interval	Frequency (f)	Midpoint (x)	Frequency × interval $f \times x$
10–19	2	14,5	2 × 14,5 = 29
20–29	3	24,5	3 × 24,5 = 73,5
30–39	7	34,5	7 × 34,5 = 241,5
40–49	1	44,5	1 × 44,5 = 44,5
50–59	2	54,5	2 × 54,5 = 109
Total	**15**		**$\sum fx$ = 497,5**

Mean (approx.) = $\frac{\Sigma(fx)}{n}$ = $\frac{497,5}{15}$ = 33,17

10.1

Interval	Tally	Frequency	Mid-interval	Frequency × mid-interval
130–137	||||	4	133,5	534,0
138–145	|||| ||||	9	141,5	1 273,5
146–153	|||| |||| |	11	149,5	1 644,5
154–161	|||| ||	7	157,5	1 102,5
162–169	||||	5	165,5	827,5
170–177	||||	4	173,5	694,0
Total		**40**		**6 076,0**

10.2.1

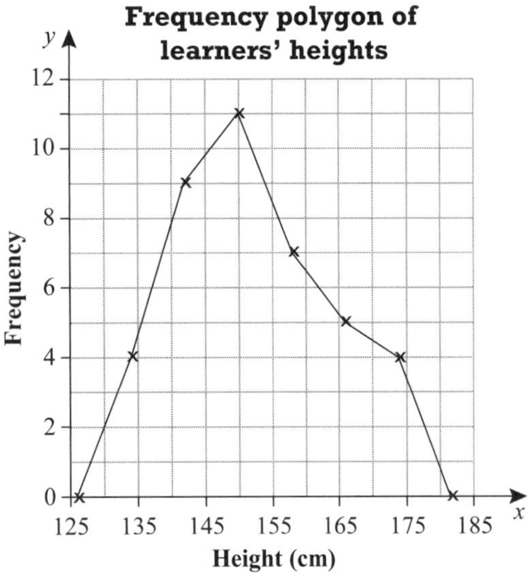
Frequency polygon of learners' heights
y
12
10
8
6
4
2
0
Frequency
125
135
145
155
165
175
185
x
Height (cm)

10.2.2 Approximate mean height = $\frac{6\,076}{40}$ = 151,9 cm

10.3 Actual mean height = $\frac{6\,062}{40}$ = 151,55 cm

The answer differs by 0,35 cm.

10.4

13	1	2	2	2	8	8	9				
14	0	1	2	2	5	5	6	6	7	9	
15	0	0	0	1	2	2	2	4	6	7	9
16	0	1	1	2	3	6	8	9			
17	0	0	1	3							

Key: $\frac{15}{4}$ = 154 cm

10.5 Median = $\frac{150 + 151}{2}$ = 150,5

Mode: There are three modes for this set of data: 132, 150 and 152.

10.6 20th percentile is found at score number 8. Therefore 32 learners lie above the 20th percentile.

10.7 60th percentile is found between score number 24 and score number 25. Therefore 16 learners lie between the 20th and the 60th percentiles.

Test A (page 222)

1.1 Mean = $\frac{49}{7}$ = 7

Median = 8

Mode = 8

1.2 Mean = $\frac{4\,758}{12}$ = 396,5

Median = $\frac{395 + 396}{2}$ = 395,5

Mode = 392

1.3 Mean = $\frac{165}{8}$ = 20,625

12 15 17 18 20 25 26 32

Median = $\frac{18 + 20}{2}$ = 19

No mode exists for this data.

2.1

1	1	2	2	2	3	3	3	3	3
4	4	5	6	6	7	7	8	9	9
9	9	9	9	9	9	10			

Range = 10 – 1 = 9

Q_3: $k = \frac{p(n+1)}{100}$

$\therefore k = \frac{75(27+1)}{100}$

= 21st value is 9 $\therefore Q_3 = 9$

Q_1: $k = \frac{p(n+1)}{100}$

$\therefore k = \frac{25(27+1)}{100}$

= 7th value is 3 $\therefore Q_1 = 3$

IQR = $Q_3 - Q_1$ Semi-IQR = $\frac{\text{IQR}}{2}$

= 9 – 3 = $\frac{6}{2}$

= 6 = 3

2.2 Minimum: 1

$Q_1 = 3$

Median = 6

$Q_3 = 9$

Maximum: 10

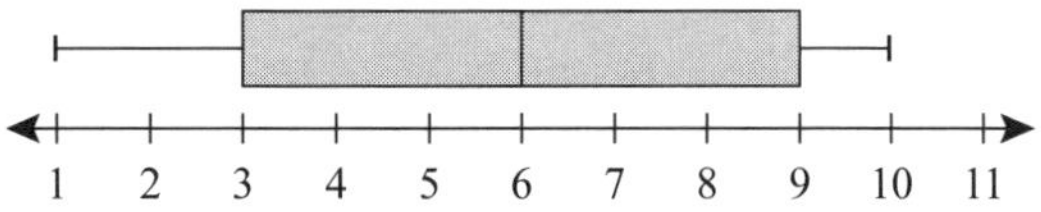
1
2
3
4
5
6
7
8
9
10
11

3.1

Sport	Tally	Frequency
Soccer	卌 ||||	9
Netball	||||	4
Cricket	卌 ||	7
Rugby	卌 ||	7
Hockey	卌 ||	7
Basketball	卌 |	6

3.2

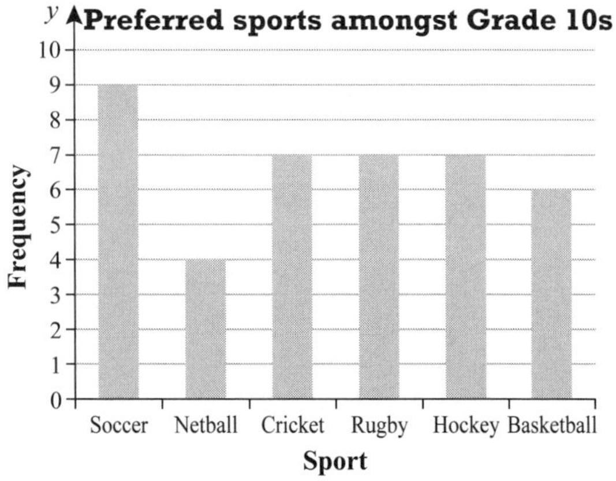
Preferred sports amongst Grade 10s
y
10
9
8
7
6
5
4
3
2
1
0
Frequency
Soccer
Netball
Cricket
Rugby
Hockey
Basketball
Sport

3.3 Soccer

3.4 The information is categorical and not continuous (there are categories and not numbers) so you cannot perform calculations on them.

4.1 $x = 79\ 490$ km^2

4.2

17 010	79 490	92 100	116 320	123 910
129 370	129 480	169 580	361 830	

Median $= \left(\frac{9+1}{2}\right)$th value

$= $ 5th value

$= 123\ 910$ km^2

4.3 Mean $= \frac{1\ 219\ 090}{9}$

$= 135\ 454{,}44\ \ldots$

$= 135\ 454$ km^2 to the nearest km^2

4.4 The median is lower than the mean, and as seven provinces are smaller than the mean, we can conclude that there is one (or more) unusually large province (an outlier). (In this case it is Northern Cape.) The median would probably be a better indicator of central tendency than the mean.

4.5

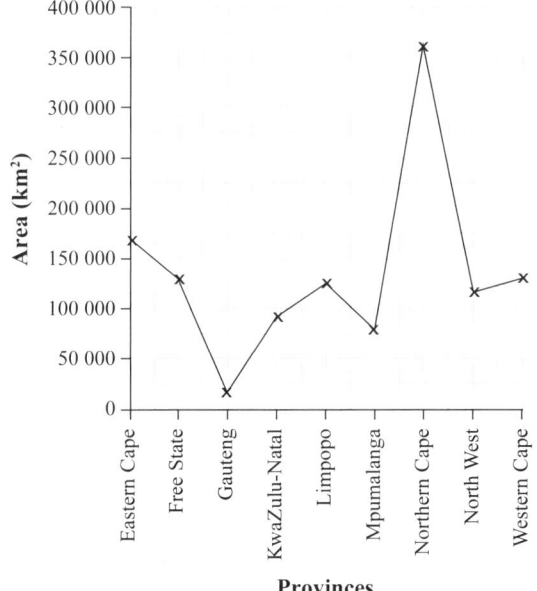

Relative sizes of South Africa's provinces

4.6.1 $\frac{129\ 370}{1\ 219\ 090} \times 360 = 38{,}2°$

4.6.2 $\frac{27{,}2}{360} \times 1\ 219\ 090 = 92\ 109{,}02\ \ldots$

∴ it would represent KwaZulu-Natal.

Test B (page 224)

1.1 (You may find it useful to use a stem-and-leaf plot to organise this many values.)

```
2 | 7 8
3 | 3 5 7 7 7 8
4 | 0 0 1 3 3 3 3 4 4 6 6 6 7 7 7 8
5 | 0 0 2 2 3 4 4 4 4 4 5 5 6 6 8 8 9
6 | 0 0 1 2 2 4 4 5
7 | 4
```

Range $= 74 - 27 = 47$

$Q_1: k = \frac{p(n+1)}{100}$

$\therefore k = \frac{25(50+1)}{100}$

$= 12{,}75$

So, we calculate the number that is three-quarters (0,75) of the way between the 12th and 13th values.

The 12th and 13th values are both 43, so $Q_1 = 43$.

$Q_3: k = \frac{p(n+1)}{100}$

$\therefore k = \frac{75(50+1)}{100}$

$= 38{,}25$

So, we calculate the number that is one-quarter (0,25) of the way between the 38th and 39th values.

The 38th value is 56 and the 39th value is 58.

∴ the 38,25th value Q_3 is $56 + 0{,}25(58 - 56)$

$= 56{,}5$

Interquartile range $= Q_3 - Q_1$

$= 56{,}5 - 43$

$= 13{,}5$

Semi-interquartile range $= \frac{\text{IQR}}{2}$

$= \frac{13{,}5}{2}$

$= 6{,}75$

1.2 The interquartile range is a more reliable measure because it is not affected by extreme scores (outliers).

2.1

Gross vehicle mass (GVM) (kg)	Frequency	Frequency × mid-interval
$3\ 000 \leq x < 5\ 000$	49	196 000
$5\ 000 \leq x < 7\ 000$	72	432 000
$7\ 000 \leq x < 9\ 000$	123	984 000
$9\ 000 \leq x < 11\ 000$	135	1 350 000
$11\ 000 \leq x < 13\ 000$	109	1 308 000
$13\ 000 \leq x < 15\ 000$	86	1 204 000
Total	**574**	**5 474 000**

Estimated mean $= \frac{5\,474\,000}{574}$

$= 9\,537$ kg

2.2

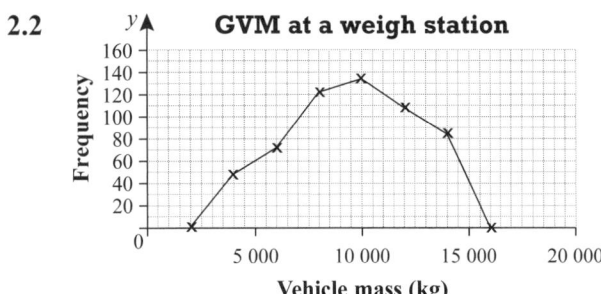

GVM at a weigh station

2.3 The modal score cannot be found because actual scores (masses of vehicles) are not given. The mode is the most common score. Also, it is unlikely that any vehicles will have the same mass, so there will probably not be a mode.

2.4 $k = \frac{p(n+1)}{100}$

$= \frac{25(574+1)}{100}$

$= 143{,}75$th score

This score falls in the 7 000–9 000 kg category.

3.1 Mean $= \frac{18\,742}{30} = $ R624,73

3.2

Turnover (R)	Frequency
$100 \le R < 300$	1
$300 \le R < 500$	8
$500 \le R < 700$	10
$700 \le R < 900$	8
$900 \le R < 1\,100$	3

3.3

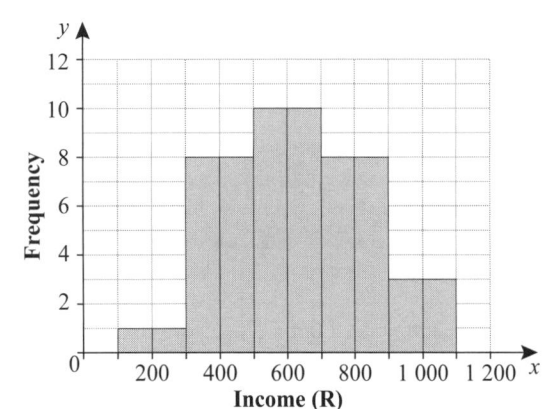

Income (R)

3.4 Need to calculate the 60th percentile.

$k = \frac{p(n+1)}{100}$

$= \frac{60(30+1)}{100}$

$= 18{,}6$

The 18,6th value falls in the R500 –R699$^+$ category.

There are 10 values in this category, so we need the number that is $\frac{8,6}{10}$ of the way through the category.

$500 + \frac{8,6}{10}(200)$ The category width is 200.

$= $ R672

The owner should therefore keep the business.

4.1 Five-number summaries for leagues A and B:

	League A	League B
Minimum	73	94
1st quartile	95,5	114
Median	121,5	126,5
3rd quartile	128	131
Maximum	147	145

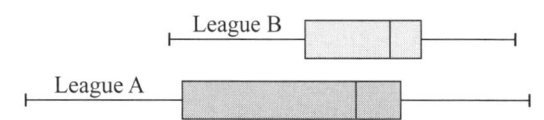

70 75 80 85 90 95 100 105 110 115 120 125 130 135 140 145 150

4.2.1 50% , i.e. 7 of the players in league B scored above the median of 126,5.

25%, i.e. 4 of the players in league A scored 128 and above (the third quartile).

Therefore 11 teams scored 127 or more home runs.

4.2.2 League B performed more consistently. Both their range and their IQR are smaller than that of league A, which means that as a league they scored more consistently. League A has a larger range and a larger IQR which implies that the overall performance is less consistent.

Test C (page 225)

1.1

Boys		Girls
	3	9
9 8 4 4 1	4	2 3 5 7 8 9
7 6 6 1	5	4 5 6 7 7
9 5 2	6	1 2 5 7 7 7 8 8
9 7 2	7	2
9 7 5 1 0 0	8	0
2 2 1 0	9	
1	10	

Key: $\frac{7}{2} = 72$ kg

1.2 Lowest: 41

Q_1: 56 (score no. 7)

Median = $\frac{72+77}{2}$ = 74,5   midpoint of scores 13 and 14

Q_3: 87 (score no. 20)

Highest: 101

1.3 Mean = $\frac{1\,838}{26}$ = 70,69 kg

1.4 The mean is the average of all the scores. The median is the middle most score of a ranked set of scores.

1.5 In this case, the median is possibly the better measure since 50% of all the boys weigh above 74,5 kg. The mean seems too low despite its being the arithmetic average.

1.6 New mean = $\frac{1\,838 + 10 \times 76,3}{36}$ = 72,25 kg

2.1.1 25%

2.1.2 48 kg (the exact value for Q_1 = 47,75 kg)

2.2 50% of the girls' weights lie between 48 kg and 67 kg.

2.3 50%

2.4 50% of 300 = 150 girls

2.5 IQR = 67 – 48 = 19 kg

Semi-IQR = 9,5 kg

2.6 The IQR informs us about the spread of the data about the median.

In this case, the weights are tightly packed around the median, meaning that 50 % of the girls weigh within a 19 kg band about the median weight of 57 kg.

3.1 and 3.2

Interval (Weight in kg)	Tally	Frequency	Midpoint of class interval	
$30 \leq w \leq 39$	|	1	34,5	$1 \times 34,5 = 34,5$
$40 \leq w \leq 49$	⦚⦚ ⦚⦚ |	11	44,5	$11 \times 44,5 = 489,5$
$50 \leq w \leq 59$	⦚⦚ ||||	9	54,5	$9 \times 54,5 = 490,5$
$60 \leq w \leq 69$	⦚⦚ ⦚⦚ |	11	64,5	$11 \times 64,5 = 709,5$
$70 \leq w \leq 79$	||||	4	74,5	$4 \times 74,5 = 298$
$80 \leq w \leq 89$	⦚⦚ ||	7	84,5	$7 \times 84,5 = 591,5$
$90 \leq w \leq 99$	||||	4	94,5	$4 \times 94,5 = 378$
$100 \leq w \leq 109$	|	1	104,5	$1 \times 104,5 = 104,5$
				$\sum fx = 3\,096$

Mean = $\frac{\Sigma(fx)}{n} = \frac{3\,096}{48}$ = 64,5 kg

3.3

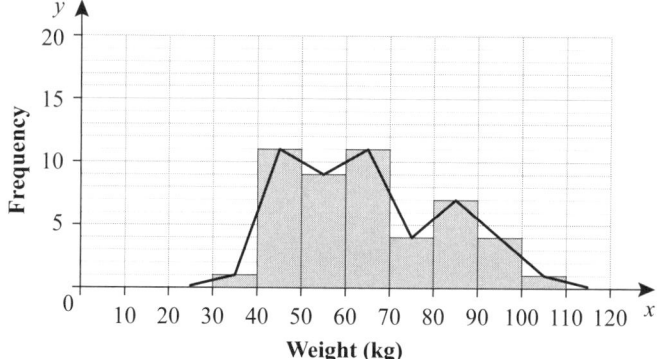

3.4 The data has two clear modes because the weights of both the girls and the boys are mixed. The lower mode will be for the girls and the higher mode for the boys.

Topic 12 Trigonometry in all four quadrants and word problems (page 227)

Exercise 1 (page 230)

1.

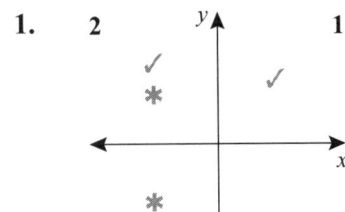

$\sin \theta > 0$ ✓
$\theta \in [90°; 270°]$ ✱
Answer: 2nd quadrant

2.

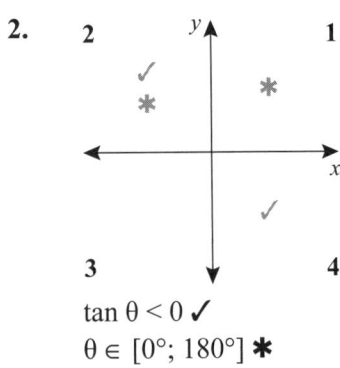

$\tan \theta < 0$ ✓
$\theta \in [0°; 180°]$ ✱
Answer: 2nd quadrant

3.

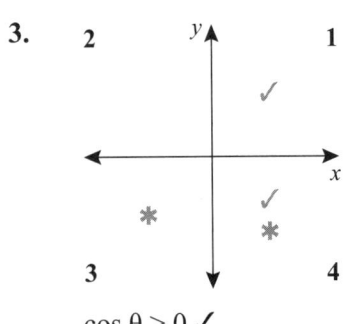

$\cos \theta > 0$ ✓
$\theta \in [180°; 360°]$ ✱
Answer: 4th Quadrant

4.

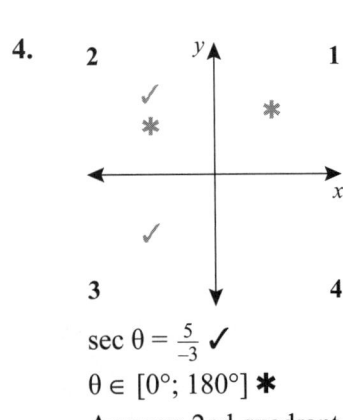

$\sec \theta = \frac{5}{-3}$ ✓
$\theta \in [0°; 180°]$ ✱
Answer: 2nd quadrant

5.

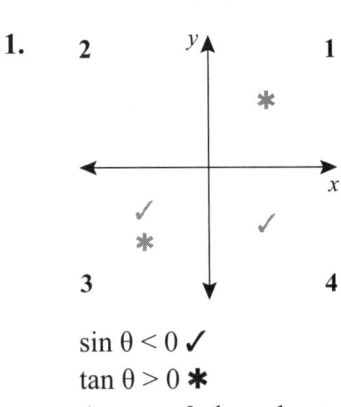

$\cot \theta < 0$
$\theta \in [\theta°; 180°]$ ✱
Answer: 2nd quadrant

Exercise 2 (page 230)

1.

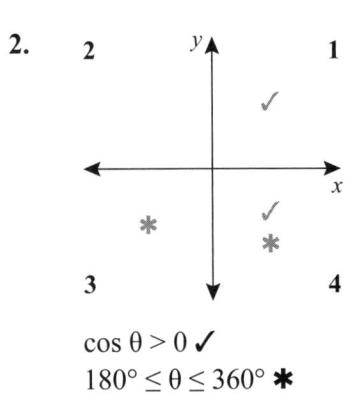

$\sin \theta < 0$ ✓
$\tan \theta > 0$ ✱
Answer: 3rd quadrant

2.

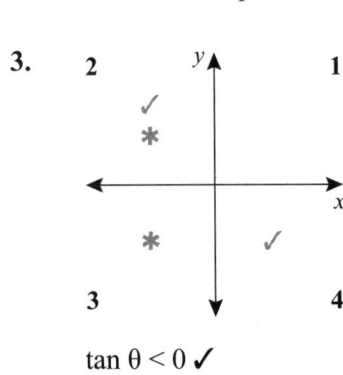

$\cos \theta > 0$ ✓
$180° \leq \theta \leq 360°$ ✱
Answer: 4th quadrant

3.

$\tan \theta < 0$ ✓
$90° \leq \theta \leq 270°$ ✱

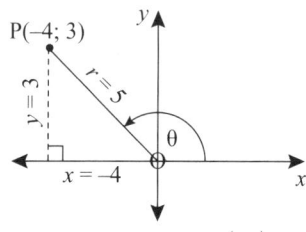

P(–4; 3)
$r = 5$
$y = 3$
θ
$x = –4$

$\sin θ - \cos θ = \frac{3}{5} - \left(\frac{-4}{5}\right) = \frac{7}{5}$

4.1 $\tan θ = \frac{y}{x} = \frac{-12}{-5} = \frac{12}{5}$

4.2 $r = 13$ (Pyth) triple 5: 12: 13

$\sin θ = \frac{y}{r} = \frac{-12}{13}$

4.3 $1 - \cos^2 θ$

$= 1 - \left(\frac{-5}{13}\right)^2$

$= 1 - \frac{25}{169}$

$= \frac{144}{169}$

4.4 $\frac{\sin θ}{\cos θ} = \frac{\frac{-12}{13}}{\frac{-5}{13}} = \frac{-12}{-5} = \frac{12}{5}$

4.5 The answers are the same $∴ \tan θ = \frac{\sin θ}{\cos θ}$.

5.1 $r = \sqrt{x^2 + y^2}$ (Pythagoras)

$= \sqrt{(2)^2 + (-2\sqrt{3})^2}$

$= \sqrt{4 + 4 \cdot 3}$

$= 4$

$\sin θ = \frac{y}{r} = \frac{-2\sqrt{3}}{4} = \frac{\sqrt{3}}{2}$

5.2 $\cos^2 θ + \sin^2 θ = \left(\frac{2}{4}\right)^2 + \left(\frac{-2\sqrt{3}}{4}\right)^2$

$= \frac{1}{4} + \frac{3}{4} = \frac{4}{4} = 1$

5.3 $\sqrt{3} \tan θ = \sqrt{3}\left(\frac{-2\sqrt{3}}{2}\right) = -3$

5.4 $\cos θ = \frac{2}{4} = \frac{1}{2}$

$θ = 60°$ (special ∠s)

6.1

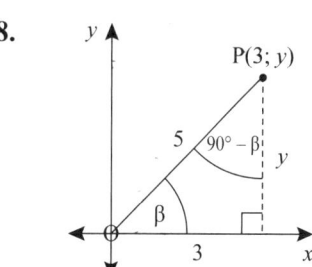

P(–5; $\sqrt{11}$)
6
$\sqrt{11}$
α
M   –5
x

$r^2 = (-5)^2 + \left(\sqrt{11}\right)^2$ (Pyth, M̂ = 90°)

$= 25 + 11$

$r = \sqrt{36} = 6$

6.2 $1 - \cos^2 α$

$= 1 - \left(\frac{-5}{6}\right)^2$

$= 1 - \frac{25}{36}$

$= \frac{11}{36}$

6.3 $\sqrt{11}(\tan α + 2 \sin α)$

$= \sqrt{11}\left(\frac{\sqrt{11}}{-5}\right) + 2 \cdot \frac{\sqrt{11}}{6}$

$= \frac{11}{-5} + \frac{11}{3}$

$= \frac{22}{15}$

7.

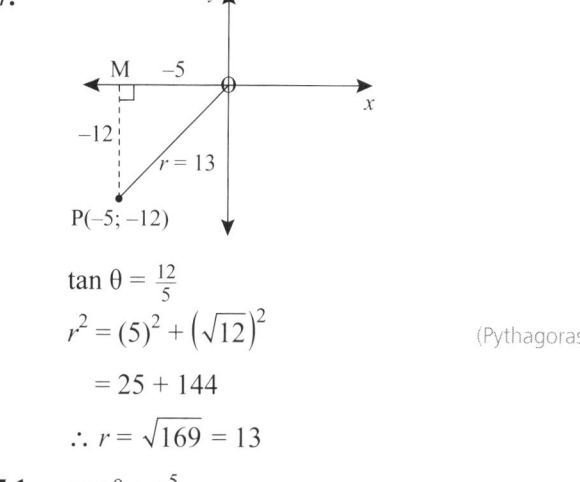

M   –5
x
–12
$r = 13$
P(–5; –12)

$\tan θ = \frac{12}{5}$

$r^2 = (5)^2 + \left(\sqrt{12}\right)^2$ (Pythagoras)

$= 25 + 144$

$∴ r = \sqrt{169} = 13$

7.1 $\cos θ = \frac{-5}{13}$

7.2 $\tan θ \sin θ$

$= \frac{12}{5} \cdot \frac{-12}{13}$

$= \frac{-144}{65}$

7.3 $\cos^2 θ + \sin^2 θ$

$= \left(\frac{-5}{13}\right)^2 + \left(\frac{-12}{13}\right)^2$

$= \frac{25}{169} + \frac{144}{169}$

$= \frac{169}{169}$

$= 1$

7.4 $13 \sin θ - 5 \tan θ$

$= 13\left(\frac{-12}{13}\right) - 5\left(\frac{12}{5}\right)$

$= -12 - 12$

$= -24$

8.

P(3; y)
5
90° – β
y
β
3
x

$5 \cos β - 3 = 0$

$∴ \cos β = \frac{3}{5}$

$y^2 = r^2 - x^2$ (Pythagoras)

$= 25 - 9$

$y = \sqrt{16} = 4$

8.1 $\tan \beta = \frac{4}{3}$

8.2 $\cos (90° - \beta) = \frac{4}{5}$

8.3 $\tan \beta \cos \beta = \frac{4}{3} \times \frac{3}{5} = \frac{4}{5}$

8.4 $\frac{\sin \beta}{\cos \beta} = \frac{4}{5} \div \frac{3}{5} = \frac{4}{5} \times \frac{5}{3} = \frac{4}{3}$

Exercise 3   (page 235)

1.

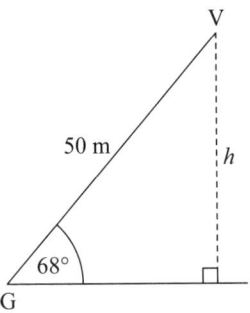

$\frac{h}{50} = \sin 68°$

$h = 50 \sin 68°$

$\quad = 46,4$ m

2.

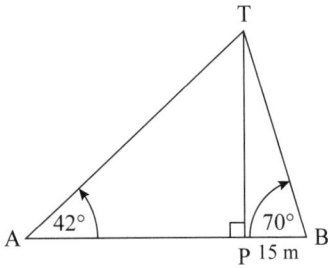

2.1 $\frac{TP}{PB} = \tan 70°$

$TP = 15 \tan 70°$

$\quad = 41,2$ m

2.2 $\frac{AP}{TP} = \frac{1}{\tan 42°}$

$AP = \frac{41,2}{\tan 42°} = 45,8$ m

3.

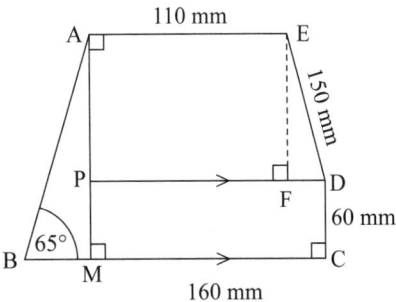

3.1 $\frac{AM}{BM} = \tan 65°$

$AM = 80 \tan 65° = 171,6$ mm

3.2 $EF = AM - 60 = 111,6$

Area trapezium APDE $= \frac{1}{2}$ (sum ‖ sides) × h

$= \frac{1}{2}(110 + 160) \times 111,6$

$= 15\ 066$ mm^2

3.3 $\sin E\hat{D}P = \left(\frac{111,6}{150}\right)$

$E\hat{D}P = \sin^{-1}\left(\frac{111,6}{150}\right) = 48°$

4.

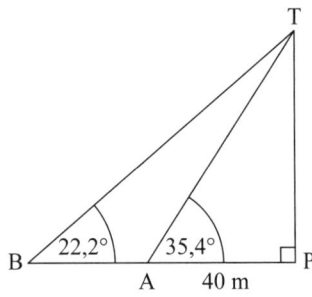

4.1 $\frac{TP}{40} = \tan 35,4°$

$TP = 40 \tan 35,4°$

$\quad = 28,4$ m

4.2 $\frac{BP}{TP} = \frac{1}{\tan 22,2°}$

$BP = \frac{28,4}{\tan 22,2°} = 69,7$ m

$\therefore$ BA $= 69,7 - 40 = 29,7$ m

5.

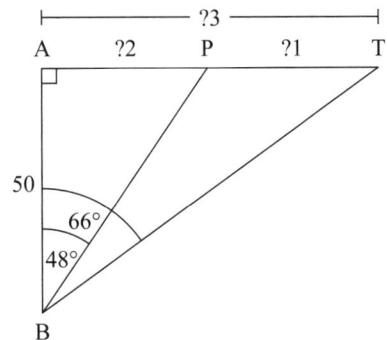

In $\triangle$BAT:

$\frac{AT}{50} = \tan 66°$

AT $= 50 \tan 66° = 112,3$ m

In $\triangle$PAB

$\frac{AP}{50} = \tan 48°$

AP $= 50 \tan 48° = 55,5$ m

PT $= 112,3 - 55,5 = 56,8$ m

6.1   In $\triangle$BAM:

$\frac{AB}{80} = \cos 33,4°$

AB $= 80 \cos 33,4° = 66,8$ mm

6.2 AB = BC = DC   (ABCD square)

$\frac{FC}{BC} = \tan 27{,}2°$

$FC = 66{,}79 \tan 27{,}2°$

$\quad = 34{,}3$ mm

$\therefore DF = 66{,}79 - 34{,}3$

$\quad\quad = 32{,}5$ mm

6.3 Area ABFD = area square ABCD – area $\triangle$BFC

$= 66{,}79^2 - \frac{1}{2}(66{,}79)(34{,}3)$

$= 3\ 315{,}5$ mm^2

7.

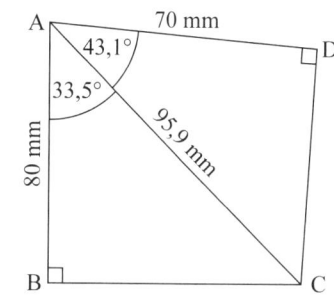

7.1 In $\triangle$BAC: $\frac{AC}{AB} = \frac{1}{\cos 33{,}5°}$

$AC = \frac{80}{\cos 33{,}5°} = 95{,}9$ mm

7.2 In $\triangle$CAD: $\cos C\hat{A}D = \frac{70}{95{,}9°}$

$C\hat{A}D = \cos^{-1}\frac{70}{95{,}9°} = 43{,}2°$

7.3 $CD = \sqrt{AC^2 - AD^2}$ (Pythagoras)

$\quad = 65{,}6$ mm

8.1 In $\triangle$PAM:

$\frac{AP_1}{MA} = \tan 64{,}9°$

$AP_1 = 750 \tan 64{,}9° = 1\ 601{,}1$ m

8.2 $\frac{P_2A}{750} = \tan 58°$

$P_2A = 750 \tan 58° = 1\ 200{,}3$ m

$P_1P_2 = 1\ 601{,}1 - 1\ 200{,}3 = 400{,}8$ m

She drops 400,8 m in 60 seconds.

$\therefore$ She drops 1200,3 m in $60 \times \frac{1\ 200{,}3}{400{,}8}$

$= 179{,}69$ seconds $\approx$ 3 minutes.

9.

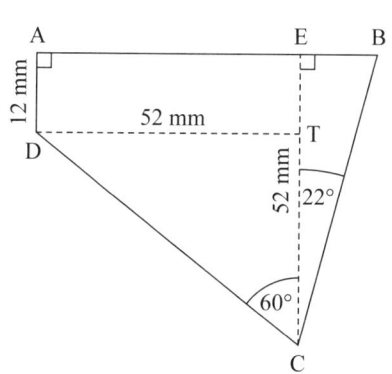

In $\triangle$BEC:

$\frac{EB}{52} = \tan 22°$

$EB = 52 \tan 22° = 21$ mm

$\therefore$ AE = DT and AD = ET

$\therefore$ TC = 52 – 12 = 40 mm

In $\triangle$DTC:

$\frac{DT}{TC} = \tan 60°$

$DT = 40 \tan 60° = 69{,}3$ mm

$\therefore$ AB = DT + EB = 69,3 + 21 = 90,3 mm

Test (page 237)

1.1

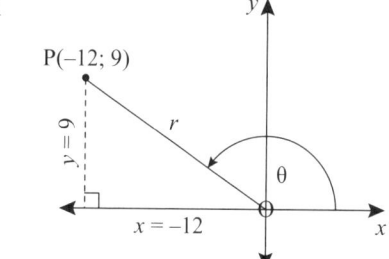

1.1.1 $\tan \theta = \frac{y}{x} = \frac{9}{-12} = -\frac{3}{4}$

1.1.2 $r = \sqrt{(12)^2 + (9)^2} = 15$

$1 - \cos^2 \theta = 1 - \left(\frac{-12}{15}\right)^2 = \frac{9}{25}$

1.2 $\tan \theta = \frac{p}{1}$

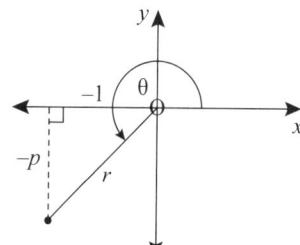

1.2.1 $r = \sqrt{p^2 + 1}$ (Pythagoras)

$\sin \theta = \frac{y}{r} = \frac{-p}{\sqrt{p^2 + 1}}$

1.2.2 $\cos \theta = \frac{x}{r}$

$= \frac{-1}{\sqrt{p^2 + 1}}$

$= \frac{-1}{\sqrt{\tan^2 \theta + 1}}$

2.

$\cos A = \frac{x}{r} = \frac{2m}{m^2+1}$

$y = \sqrt{(m^2+1)^2 - (2m)^2}$ (Pythagoras)

$\quad = \sqrt{m^4 + 2m^2 + 1 - 4m^2}$

$\quad = \sqrt{m^4 - 2m^2 + 1}$

$\quad = \sqrt{(m^2-1)^2}$

$\quad = m^2 - 1$

$\sin A = \frac{m^2-1}{m^2+1}$

3. Draw two separate triangles:

Triangle 1

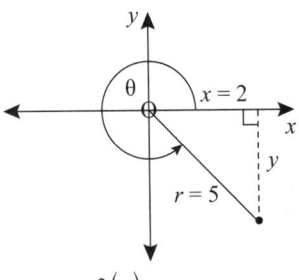

$\cos\theta = \frac{2}{5}\left(\frac{x}{r}\right)$

$y = -\sqrt{25-4} = -\sqrt{21}$

Triangle 2

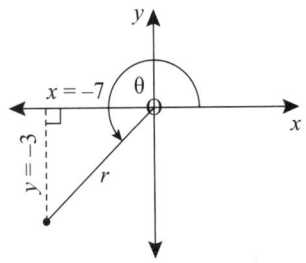

$\tan\alpha = \frac{y}{x} = \frac{3}{7}$

$r = \sqrt{49+9}$ (Pythagoras)

$\quad = \sqrt{58}$

$\therefore \frac{2}{21}\tan^2\theta + 58\sin^2\alpha = \frac{2}{21}\left(\frac{\sqrt{21}}{2}\right)^2 + 58\left(\frac{-3}{\sqrt{58}}\right)^2$

$\quad = \frac{2}{21} \times \frac{21}{4} + 58 \times \frac{9}{58}$

$\quad = \frac{1}{2} + 9 = 9\frac{1}{2}$

4.1

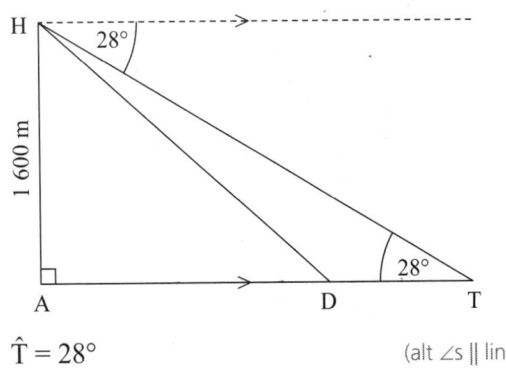

$\hat{T} = 28°$ (alt ∠s ∥ lines)

4.2 $\frac{AT}{1\,600} = \frac{1}{\tan 28°}$

$AT = \frac{1\,600}{\tan 28°} = 3\,009{,}16 \text{ m} \approx 3\,009 \text{ m}$

4.3 DT : AD = 1 : 5

$DT = \frac{1}{6}(AT) = \frac{1}{6}(3\,009) = 501{,}5 \approx 502 \text{ m}$

4.4 AD = $3\,009 - 502 = 2\,507$

$\tan H\hat{D}A = \frac{1\,600}{2\,507}$

$H\hat{D}A = \tan^{-1}\left(\frac{1\,600}{2\,507}\right)$

$\quad\quad = 32{,}5°$

5. *Note:* To determine EC, we need EG and GC. To find GC we need DG ∴ work in △DEG.

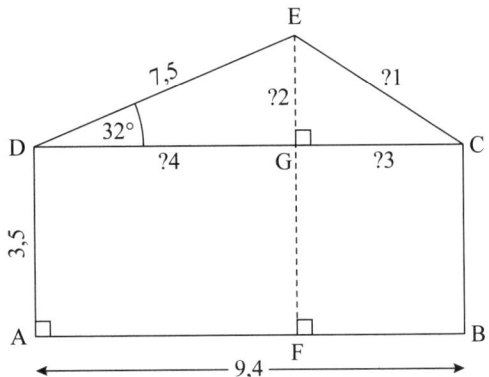

5.1 ?4: $\frac{DG}{DE} = \cos 32°$

DG = $7{,}5 \cos 32° = 6{,}36$ m

?3: GC = $9{,}4 - 6{,}36 = 3{,}04$ m

?2: $\frac{EG}{7{,}5} = \sin 32°$

EG = $7{,}5 \sin 32° = 3{,}97$ m

?1: EC = $\sqrt{EG^2 + GC^2}$ (Pythagoras)

$\quad\quad = \sqrt{3{,}97^2 + 3{,}04^2} = 5{,}00$ m

5.2 $\tan D\hat{C}E = \frac{3{,}97}{3{,}04}$

$D\hat{C}E = \tan^{-1}\left(\frac{3{,}97}{3{,}04}\right) = 52{,}6°$

5.3 Area △DEC $= \frac{1}{2}$ base × height

$\quad\quad = \frac{1}{2}(9{,}4)(3{,}97)$

$\quad\quad = 18{,}66 \text{ m}^2$

5.4 EF = GF + EG (GF = AD)

$\quad\quad = 3{,}5 + 3{,}97 = 7{,}47$ m

Topic 13 Measurement (page 239)

Exercise 1 (page 243)

1.1 **(a)** Volume = base area × height

$$\therefore V = xbl$$

(b) Surface area = 2 × base area + perimeter
× height

$$= 2xb + 2(x + b)l$$
$$= 2(xb + xl + bl)$$

1.2 **(a)** Volume = base area × height

$$\therefore V = \tfrac{1}{2}abl$$

(b) Surface area = 2 × base area + perimeter
× height

$$= 2\left(\tfrac{1}{2}ab\right) + (a + b + c)l$$
$$= ab + l(a + b + c)$$

1.3 **(a)** Volume = base area × height

$$= \pi a^2 l$$

(b) Surface area = 2 × base area + perimeter
× height

$$= 2\pi a^2 + 2\pi al$$
$$= 2\pi a(a + l)$$

1.4 **(a)** Volume = base area × height

$$= a \times a \times l$$
$$= a^2 l$$

(b) Surface area = 2 × base area + perimeter
× height

$$= 2a^2 + 4al$$
$$= 2a(a + 2l)$$

2.1 **(a)** Volume = base area × height

$$= \tfrac{1}{2}bHh$$
$$= \tfrac{1}{2} \times 4 \times 3 \times 7$$
$$= 42 \text{ cm}^3$$

(b) $AB^2 = 4^2 + 3^2$ (Pythagoras)

$$= 16 + 9$$
$$= 25$$
$$\therefore AB = 5 \text{ cm}$$

Surface area
= 2 × base area + perimeter × height

$$= 2 \times \tfrac{1}{2} \times 4 \times 3 + (3 + 4 + 5) \times 7$$
$$= 12 + 84$$
$$= 96 \text{ cm}^2$$

2.2 Volume = base area × height

$$\therefore V = \pi r^2 h$$
$$= \pi \times 3{,}5^2 \times 7{,}5$$
$$= 288{,}6 \text{ cm}^3$$

Surface area
= 2 × base area + perimeter × height
$$= 2 \times \pi \times 3{,}5^2 + 2 \times \pi \times 3{,}5 \times 7{,}5$$
$$= 241{,}9 \text{ cm}^2$$

2.3 Volume = base area × height

$$= lbh$$
$$= 2{,}5 \times 6{,}5 \times 3$$
$$= 48{,}75 \text{ cm}^3$$

Surface area = 2 × base area + perimeter ×
height
$$= 2 \times 2{,}5 \times 6{,}5 + 2(2{,}5 + 6{,}5) \times 3$$
$$= 32{,}5 + 54$$
$$= 86{,}5 \text{ cm}^2$$

2.4 $AB^2 = 8^2 + 8^2$ (Pythagoras)

$$= 128$$
$$\therefore AB = 11{,}3 \text{ cm}$$

Volume = base area × height
= (area of rectangle + area of triangle)
× height
$$= (5 \times 11{,}3 + \tfrac{1}{2} \times 8 \times 8) \times 24$$
$$= 88{,}5 \times 24$$
$$= 2\,124 \text{ cm}^3$$

Surface area = 2 × base area + perimeter ×
height
$$= 2(5 \times 11{,}3 + \tfrac{1}{2} \times 8 \times 8) + (5 +$$
$$11{,}3 + 5 + 8 + 8) \times 24$$
$$= 1\,072{,}2 \text{ cm}^2$$

2.5 $AB^2 = 17^2 - 15^2$ (Pythagoras)

$$= 64$$
$$\therefore AB = 8 \text{ cm}$$

Volume = base area × height
$$= lbh$$
$$= 15 \times 8 \times 20$$
$$= 2\,400 \text{ cm}^3$$

Surface area = 2 × base area + perimeter
× height
$$= 2 \times 15 \times 8 + 2(15 + 8) \times 20$$
$$= 1\,160 \text{ cm}^2$$

2.6 $AB^2 = 34^2 - 16^2$ (Pythagoras)

$$= 900$$
$$\therefore AB = 30 \text{ cm}$$

Volume = base area × height

$$= \text{area } \triangle ABC \times \text{height}$$
$$= \tfrac{1}{2} \times 30 \times 16 \times 12$$
$$= 2\,880 \text{ cm}^3$$

Surface area
$$= 2 \times \text{base area} + \text{perimeter} \times \text{height}$$
$$= 2 \times \tfrac{1}{2} \times 30 \times 16 + (34 + 30 + 16) \times 12$$
$$= 960 + 480$$
$$= 1\,440 \text{ cm}^2$$

2.7 Volume = base area × height
$$= \tfrac{1}{2} \times 4a \times 3a \times 8a$$
$$= 48a^3 \text{ cm}^3$$

AB = 5a   (Pythagoras)

Surface area = 2 × base area + perimeter
$$\times \text{ height}$$
$$= 2 \times \tfrac{1}{2} \times 4a \times 3a + (3a + 4a + 5a) \times 8a$$
$$= 12a^2 + 12a \times 8a$$
$$= 12a^2 + 96a^2$$
$$= 108a^2 \text{ cm}^2$$

2.8 Volume = base area × height
$$= (\text{area of square} + \text{area of semi-circle}) \times \text{height}$$
$$= \left(28 \times 28 + \tfrac{1}{2} \times \pi \times 14^2\right) \times 85$$
$$= 92\,809{,}5 \text{ cm}^3$$

Surface area
$$= 2 \times \text{base area} + \text{perimeter} \times \text{height}$$
$$= 2\left(28 \times 28 + \tfrac{1}{2} \times \pi \times 14^2\right) + \left(28 + 28 + 28 + \tfrac{1}{2} \times 2 \times \pi \times 14\right) \times 85$$
$$= 13\,062{,}2 \text{ cm}^2$$

2.9 Radius of semi-circle: $\dfrac{15 - (14 + 4)}{2} = \dfrac{7}{2}$
$$= 3{,}5$$

Volume = base area × height
$$= (\text{area of rectangle} - \text{area of semi-circle}) \times \text{height}$$
$$= \left(15 \times 8 - \tfrac{1}{2} \times \pi \times 3{,}5^2\right) \times 60$$
$$= 6\,045{,}5 \text{ cm}^3$$

Surface area
$$= 2 \times \text{base area} + \text{perimeter} \times \text{height}$$
$$= 2\left(15 \times 8 - \tfrac{1}{2} \times \pi \times 3{,}5^2\right) + \left(4 + 8 + 15 + 8 + 4 + \tfrac{1}{2} \times 2 \times \pi \times 3{,}5\right) \times 60$$
$$= 3\,201{,}2 \text{ cm}^2$$

2.10 Volume = base area × height
$$= \tfrac{1}{4} \times \pi \times 17^2 \times 72$$
$$= 16\,342{,}6 \text{ cm}^3$$

Surface area
$$= 2 \times \text{base area} + \text{perimeter} \times \text{height}$$
$$= 2\left(\tfrac{1}{4} \times \pi \times 17^2\right) + \left(17 + 17 + \tfrac{1}{4} \times 2 \times \pi \times 17\right) \times 72$$
$$= 4\,824{,}6 \text{ cm}^2$$

3. Volume of cylinder = $\pi r^2 h$
$$= \pi \times 3^2 \times 10$$
$$= 282{,}74 \text{ cm}^3$$

Volume of square hole = lbh
$$= 4 \times 4 \times 10$$
$$= 160 \text{ cm}^3$$

Volume of remaining wood = $282{,}74 - 160$
$$= 122{,}74 \text{ cm}^3$$

4.1 $x^2 = 2^2 - 1{,}6^2$
$$= 4 - 2{,}56$$
$$x = 1{,}44$$
$$\therefore x = 1{,}2$$
$$y^2 = 13^2 - 12^2$$
$$= 169 - 144$$
$$= 25$$
$$\therefore y = 5 \text{ mm}$$

4.2 Volume of A = base area × height
$$= \tfrac{1}{2} bHh$$
$$= \tfrac{1}{2} \times 1{,}6 \times 1{,}2 \times 5$$
$$= 4{,}8 \text{ cm}^3$$

Volume of B = base area × height
$$= \tfrac{1}{2} bHh$$
$$= \tfrac{1}{2} \times 5 \times 12 \times 150$$
$$= 4\,500 \text{ mm}^3$$
$$= 4{,}5 \text{ cm}^3$$

∴ the volume of A is greater.

4.3 Doubling the height will double the volume.
$$\therefore 4{,}8 \times 2 = 9{,}6 \text{ cm}^3$$

4.4 Doubling the base dimensions of B will increase the base area by four times. The volume will therefore increase by four times.
$$\therefore 4{,}5 \times 4 = 18 \text{ cm}^3$$

5. $AD^2 = 5^2 - 4^2$ (Pythagoras)

    $= 9$

    $\therefore AD = 3$ m

    Surface area $= 2 \times$ base area $+$ perimeter
                       $\times$ height

$$= 2 \times \tfrac{1}{2} \times 6 \times 4 + (5 + 5 + 6) \times 9$$
$$= 168 \text{ m}^2$$

6. Volume $=$ base area $\times$ height

$$= \tfrac{1}{2}\pi r^2 h$$
$$= \tfrac{1}{2} \times \pi \times 0{,}06^2 \times 0{,}18$$
$$= 0{,}00102 \text{ m}^3$$

7.1 Volume $=$ base area $\times$ height

$$= \pi r^2 h$$
$$= \pi \times 10^2 \times 15$$
$$= 4\,712{,}4 \text{ cm}^3$$

7.2 Open-ended container (without lid)

    Surface area $=$ base area $+$ perimeter $\times$ height

$$= \pi r^2 + 2\pi rh$$
$$= \pi \times 10^2 + 2 \times \pi \times 10 \times 15$$
$$= 1\,256{,}6 \text{ cm}^2$$

7.3 Open-ended container

    Surface area $=$ base area $+$ perimeter $\times$ height

$$= \pi r^2 + 2\pi rh$$
$$= \pi \times 10^2 + 2 \times \pi \times 10 \times 1$$
$$= 377 \text{ cm}^2$$

8. Surface area

    $= 2 \times$ base area $+$ perimeter $\times$ height

    $= 2($area of semi-circle $+$ area of rectangle$)$
      $+$ perimeter $\times$ height

$$= 2\left(\tfrac{1}{2} \times \pi \times 0{,}4^2 + 1{,}7 \times 0{,}8\right) + \left(1{,}7 + 0{,}8 + 1{,}7\right.$$
$$\left. + \tfrac{1}{2} \times 2 \times \pi \times 0{,}4\right) \times 5$$
$$= 30{,}5 \text{ m}^2$$

9.1 Area of vertical carpet $= 3(l \times b)$

$$= 3 \times 60 \times 15$$
$$= 2\,700 \text{ cm}^2$$

    Area of horizontal carpet $= 3(l \times b)$

$$= 3 \times 60 \times 20$$
$$= 3\,600 \text{ cm}^2$$

    $\therefore$ Total area $= 6\,300 \text{ cm}^2$

9.2 Volume $=$ volume of three rectangular prisms

                 (Each step can be treated as a prism.)

$$= 15 \times 20 \times 80 + 30 \times 20 \times 80 + 45 \times$$
$$20 \times 80$$
$$= 24\,000 + 48\,000 + 72\,000$$
$$= 144\,000 \text{ cm}^3$$

10. Volume of box $=$ base area $\times$ height

$$= l \times b \times h$$
$$= 16 \times 12 \times 4$$
$$= 768 \text{ cm}^3$$

    Volume of one can $=$ base area $\times$ height

$$= \pi r^2 h$$
$$= \pi \times 2^2 \times 4$$
$$= 50{,}3 \text{ cm}^3$$

    Volume of 12 cans $= 12 \times 50{,}3$

$$= 603{,}2 \text{ cm}^3$$

    Remaining volume $= 768 - 603{,}2$

$$= 164{,}8 \text{ cm}^3$$

Exercise 2   (page 254)

1.1 Volume $= \tfrac{1}{3}bh\text{H} = \tfrac{1}{3}(10)(10)(12) = 400 \text{ cm}^3$

    $\text{TSA} = b^2 + 4\left(\tfrac{bh}{2}\right) = 10^2 + 4\left(\tfrac{10 \times 13}{2}\right) = 360 \text{ cm}^2$

1.2 Volume $= \tfrac{1}{3}bh\text{H} = \tfrac{1}{3}(8)(8)(9) = 192 \text{ cm}^3$

    Slant height for triangular face $= \sqrt{9^2 + 4^2}$

$$= \sqrt{97}$$
$$\therefore \text{TSA} = b^2 + 4\left(\tfrac{bh}{2}\right) = 8^2 + 4\left(\tfrac{8\sqrt{97}}{2}\right)$$
$$= 221{,}58 \text{ cm}^2$$

1.3 Volume $= \tfrac{1}{3}\pi r^2 \text{H} = \tfrac{1}{3}\pi(5)^2(12) = 100\pi$

$$= 314{,}16 \text{ units}^3$$

    $\text{TSA} = \pi r^2 + \pi rs$

    $s = \sqrt{12^2 + 5^2} = 13 \text{ units}$

    $\therefore \text{TSA} = \pi(5)^2 + \pi(5)(13) = 90\pi$

$$= 282{,}74 \text{ units}^2$$

2. $\text{TSA} = 4\pi r^2 = 4\pi(4)^2 = 64\pi = 201 \text{ cm}^2$

    Volume $= \tfrac{4}{3}\pi r^3 = \tfrac{4}{3}\pi(4)^3 = 268{,}08 = 268 \text{ cm}^3$

3. Circumference $= 2\pi r$

    $\therefore 50\pi = 2\pi r$

    $\therefore r = 25$ m

Volume for a hemisphere

$= \frac{2}{3}\pi r^3 = \frac{2}{3}\pi(25)^3$

$= 32\,724,92 \text{ m}^3$

$\text{TSA} = 2\pi r^2 = 2\pi(25)^2 = 1\,250\pi = 3\,926,99 \text{ m}^2$

4. $V = \frac{4}{3}\pi r^3 = \frac{4}{3}\pi(3,75)^3 = 220,89 \text{ cm}^3$

$\text{TSA} = 4\pi r^2 = 4\pi(3,75)^2 = 176,71 \text{ cm}^2$

5. $V = \frac{1}{3}\pi r^2 H = \frac{1}{3}\pi(12)^2(28) = 4\,222,30 \text{ mm}^3$

$s = \sqrt{28^2 + 12^2} = 4\sqrt{58} \text{ mm}$

$\text{TSA} = \pi r(r + s) = \pi(12)(12 + 4\,158) = 480\pi$

$= 1\,600,82 \text{ mm}^2$

6. Height of the base triangle

$= \sqrt{\left(10\sqrt{3}\right)^2 - \left(5\sqrt{3}\right)^2} = 15 \text{ m}$

$\therefore V = \frac{1}{6}bhH = \frac{1}{6}(10\sqrt{3})(15)(12) = 519,62 \text{ m}^3$

7. Height for triangular face:

$h = \sqrt{(45)^2 - (7,5)^2} = 44,37 \text{ cm}$

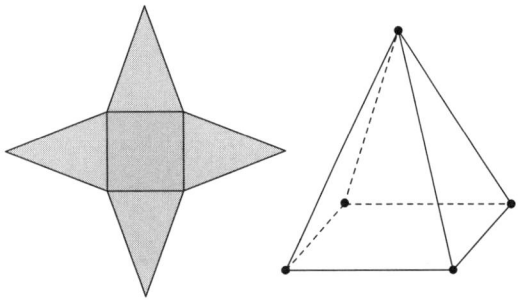

Surface area of pyramid:

$A = b^2 + 4 \times \frac{1}{2}hb$

$= 15^2 + 4 \times \frac{1}{2}(44,37)(15) = 1\,556,1 \text{ cm}^2$

8.1 $V = \frac{1}{6}bhH$

$398 = \frac{1}{6}b(13)(5\sqrt{6})$

$b = \frac{398 \times 6}{(13)(5\sqrt{6})} = 14,998 \text{ cm}$

$\therefore b = 15 \text{ cm}$ (Correct to the nearest cm.)

8.2 Each face has an area of $\frac{1}{2}15 \times 13 = 97,5 \text{ cm}^2$

Area of kite paper needed $= 4 \times 97,5 \text{ cm}^2$

$= 390 \text{ cm}^2$

9. Slant height $= \sqrt{25^2 + 9^2} = \sqrt{706} \text{ cm}$

Surface area $= \pi rs = \pi(9)\left(\sqrt{706}\right) = 751,27 \text{ cm}^2$

10. $V = \frac{1}{3}b^2 H$

$\frac{1}{3}b^2(15) = 500$

$\therefore b^2 = \frac{500 \times 3}{15}$

$\therefore b = \sqrt{500 \times \frac{3}{15}} = 10 \text{ m}$

$\therefore$ Square base has a length of 10 m.

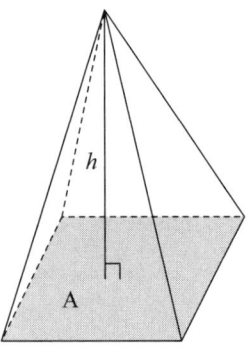

Height of triangular face $= \sqrt{15^2 + 5^2} = 5\sqrt{10}$

Area of one triangular face

$= \frac{1}{2}bh = \frac{1}{2}(10)\left(5\sqrt{10}\right) = 25\sqrt{10} = 79,1 \text{ m}^2$

11. Height of triangular base $= \sqrt{6^2 - 3^2} = 3\sqrt{3}$

$\text{TSA} = \frac{bh}{2} + 3\left(\frac{bh_2}{2}\right) = \frac{6(3\sqrt{3})}{2} + 3\left(\frac{6(10)}{2}\right)$

$= 105,59 \text{ units}^2$

Exercise 3 (page 257)

1. Total volume = volume of cylinder +volume of a sphere (two hemispheres added together form one complete sphere)

$= \pi r^2 H + \frac{4}{3}\pi r^3$

$= \pi(3)^2(10) + \frac{4}{3}\pi(3)^3$

$= 126\pi$

$= 395,84 \text{ mm}^3$

Total surface area $= 2\pi rH + 4\pi r^2$

$= 2\pi(3)(10) + 4\pi(3)^2$

$= 96\pi$

$= 301,59 \text{ mm}^2$

2. Volume of cone $= \frac{1}{3}\pi r^2 h = \frac{1}{3}\pi(40)^2 \cdot 30$

$= 16\,000\pi$

Volume of cylinder $= \pi r^2 h = \pi(40)^2 \cdot 50$

$= 80\,000\pi$

Total volume = volume of cone

$= 16\,000\pi + 80\,000\pi = 96\,000\pi$

$= 301\,592,89 \text{ m}^3$

3. Surface area of band $= 2\pi rh = 2\pi(5)(2)$

$= 20\pi \text{ cm}^2$

Surface area of cone $= \pi r^3 = \pi(5)(13)$

$= 65\pi \text{ cm}^2$

Total exterior surface area $= 20\pi + 65\pi = 85\pi$
$$= 267{,}04 \text{ cm}^2$$

4. Volume of cylinder $= \pi r^2 h = \pi(11)^2(53)$
$$= 6\,413\pi \text{ cm}^3$$

Volume of cone $= \frac{1}{3}\pi r^2 h = \frac{1}{3}\pi(11)^2(53)$
$$= \frac{6\,413}{3}\pi \text{ cm}^3$$

Volume of liquid $= 6\,413\pi + \frac{6\,413}{3}\pi$
$$= 13\,431{,}36 \text{ cm}^3$$

5.1 Each triangular face will have a height of
$\sqrt{2^2 - 1^2} = \sqrt{3}$ cm

Volume for one tetrahedron
$= \frac{1}{6}bh\text{H} = \frac{1}{6}(2)(\sqrt{3}) \times \frac{2\sqrt{6}}{3} = \frac{2\sqrt{2}}{3}$ cm^3

Volume for each sinker
$= 2 \times \frac{2\sqrt{2}}{3}$ cm$^3 = \frac{4\sqrt{2}}{3}$ cm^3
$\approx 1{,}89$ cm^3

5.2 Smallest box: $2 \text{ cm} \times \sqrt{3} \text{ cm} \times 2\left(\frac{2\sqrt{6}}{3}\right)$ cm
$= 2 \text{ cm} \times 1{,}73 \text{ cm} \times 3{,}265 \text{ cm}$
$= 20 \text{ mm} \times 17{,}3 \text{ mm} \times 32{,}7 \text{ mm}$

5.3 To ensure that the sinkers would fit into the box we would have to round our answers UP, and not to the nearest mm.
$\therefore$ Dimensions $= 20 \text{ mm} \times 18 \text{ mm} \times 33 \text{ mm}$
Rounding off to the nearest mm

6. Volume of cylinder $= \pi r^2 h = \pi(2{,}1)^2(4{,}2)$
$$= 58{,}18857 \ldots$$

Volume of glass ball $= \frac{4}{3}\pi r^3 = \frac{4}{3}\pi(2{,}1)^3$
$$= 38{,}7923 \ldots$$

Volume of packaging $= 58\,18857 - 38{,}7923$
$$= 20 \text{ m}^3$$

7.1 $\text{AH}^2 = 0{,}8^2 + 1{,}5^2$
$\text{AH}^2 = 2{,}89$
$\text{AH} = 1{,}7$ m

7.2 Surface area of roof $= 4 \times \frac{1}{2}(3 \times 1{,}7)$
$$= 10{,}2 \text{ m}^2$$

7.3 Surface area of walls $= 4 \times 3 \times 2{,}1$
$$= 25{,}2 \text{ m}^2$$
Total surface area $= 10{,}2 \text{ m}^2 + 25{,}2 \text{ m}^2$
$$= 35{,}4 \text{ m}^2$$

8.1 Volume $= \frac{4}{3}\pi(8)^3$
$$= 2\,144{,}66 \text{ mm}^3$$

8.2 New volume : original volume $= 2^3 : 1$
$$= 8 : 1$$

8.3 Volume including silver $= \frac{4}{3}\pi(9)^3$
$$= 3\,053{,}63 \text{ mm}^3$$

Volume of silver $= 3\,053{,}63 - 2\,144{,}66$
$$= 908{,}97 \text{ mm}^3$$

9. Surface area of open cylinder
$= 2\pi rh = 2\pi(9)(38) = 684\pi \text{ cm}^2$

Surface area of cone $= \pi rs$
$$= \pi(11)(17) = 187\pi \text{ cm}^2$$

Total surface area for one cracker
$= 684\pi + 187\pi = 871\pi \text{ cm}^2$

Surface area for 10 crackers
$= 10 \times 871\pi \text{ cm}^2 = 8\,710 \text{ cm}^2$

Cardboard needed for 10 crackers
$= 0{,}871 \text{ m}^2$

10. *Note:* Bottom and back faces are included.
10.1 44 blocks in total $\therefore$ total volume $= 44 \text{ cm}^3$
10.2 Row 1 = 10 faces, row 2 = 8 faces,
row 3 = 14, row 4 = 15, row 5 = 30

11.1 Total volume $=$ vol. of top cone $+$ vol. of
sphere $+$ vol. of bottom cone
$$= \frac{1}{3}\pi r^2 h + \frac{4}{3}\pi r^3 + \frac{1}{3}\pi R^2 H$$
$$= \frac{1}{3}\pi(2)^2(4{,}5) + \frac{4}{3}\pi(1{,}8)^3$$
$$+ \frac{1}{3}\pi(3)^2(9)$$
$$= 6\pi + 27\pi + 24{,}43$$

Volume for each shape $= 128{,}10 \text{ cm}^3$

11.2 Total surface area of one ball
$= 4\pi r^2 = 4\pi(1{,}8)^2 = \frac{324}{25}\pi$

Surface area for 2 balls each with two coats
$= 2 \times 2 \times \frac{324}{25}\pi = 162{,}86 \text{ cm}^2$

Cost for this coating: $162{,}86 \times \text{R}55$
$$= \text{R}8\,957{,}31$$

11.3 Inside and outside area for top cones
$= 2 \times 2 \times \pi rs = 4\pi(2)\left(\sqrt{1 + 4{,}5^2}\right)$
$= 115{,}86 \text{ cm}^2$
$= 116 \text{ cm}^2$

Outside bottom cone
$= 2 \times \pi rs = 2\pi(3)\left(\sqrt{3^2 + 9^2}\right)$
$= 178{,}82 = 179 \text{ cm}^2$
Total area $= 116 + 179 = 295 \text{ cm}^2$

Test A (page 259)

1.

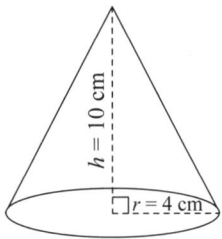

1.1 $V = \frac{1}{3}\pi r^2 h$

1.2 $V = \frac{1}{3}\pi r^2 h = \frac{1}{3}\pi(4)^2(10) = 167{,}6$ cm^3

1.3 TSA $= \pi r^2 + \pi rs$

1.4 TSA $= \pi r^2 + \pi rs = \pi(4)^2 + \pi(4)\left(\sqrt{10^2 + 4^2}\right)$
        $= 185{,}6$ cm^2

1.5 TSA $= 2 \times \pi rs = 2 \times \pi(4)\left(\sqrt{10^2 + 4^2}\right)$
        $= 270{,}69$ cm$^2 = 271$ cm^2

2.1 $V = \frac{4}{3}\pi r^3$

2.2 $V = \frac{4}{3}\pi(25)^3 = 65\ 449{,}85$ m^3

2.3 TSA $= 4\pi r^2$

2.4 TSA $= 4\pi(25)^2 = 2\ 500\pi = 7\ 853{,}98$ m^3

2.5 TSA $= 2\pi r^2 + \pi r^2$

2.6 TSA $= 3\pi r^2 = 3\pi(25)^2 = 1\ 875\pi = 5\ 890{,}49$ m^2

3.1 $V = \frac{1}{3}b^2 \times H$

3.2 $V = \frac{1}{3} \times 4^2 \times 12 = 64$ m^3

3.3 Height of triangular face
        $= \sqrt{12^2 + 2^2} = 2\sqrt{37} = 12{,}166$ m

3.4 TSA $= 4^2 + 4\left(\frac{4 \times 2\sqrt{37}}{2}\right) = 113{,}32 = 113$ cm^2

3.5

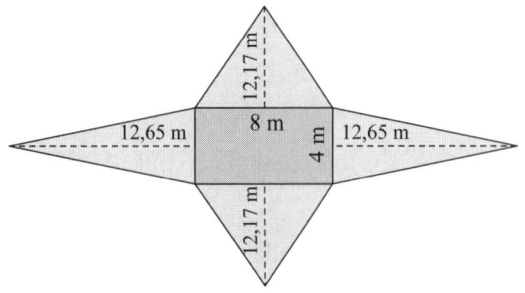

3.6 Height $= \sqrt{12^2 + 4^2} = 4\sqrt{10} = 12{,}65$ m

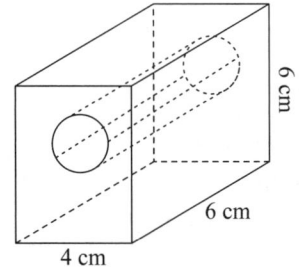

4.1 $V = lbh - \pi r^2 h$
        $= 4 \times 6 \times 6 - \pi(1)^2(6)$
        $= 125{,}15$ cm^3

4.2 Inside surface $= 2\pi rh = 2\pi(1)(6) = 12\pi$
        Cost of paint $= 2 \times 12\pi \times$ R115 $=$ R8 670,80

Test B (page 261)

1.1 Height of the cone $= 140 - 40 = 100$ cm
        Volume of cone $= \frac{1}{3}\pi r^2 h$
        $= \frac{1}{3}(40)^2 - 100$
        $= 167\ 551{,}6082$ cm^3
        Volume of hemisphere $= \frac{1}{2}\left(\frac{4}{3}\pi r^3\right)$
        $= \frac{1}{2} \cdot \frac{4}{3}\pi(40)^3$
        $= 134\ 041{,}2866$ cm^3
        Total volume of model $= 301\ 592{,}89$ cm^3

1.2 $H^2 = 1^2 + (0{,}4)^2 = 1{,}16$
        $H = 1{,}077032961$ m

        Total exterior surface area
        $=$ surface area of hemisphere $+$ surface area
          of cone
        $= \frac{1}{2} \cdot 4\pi r^2 + \pi rH$
        $= 2\pi(0{,}4)^2 + \pi(0{,}4)(1{,}07703961)$
        $= 2{,}358...$ m^2
        $= 2{,}36$ m^2

1.3 Mass $= 2{,}36 \times 2{,}5$
        Mass $= 5{,}90$ kg

2. **Glass 1:** Volume for a cylinder
        $= \pi r^2 h = \pi(2{,}5)^2(6) = 117{,}81$ cm^3

        Glass 2: Volume for a cylinder $+$ volume of a
        hemisphere
        Volume for a cylinder $= \pi r^2 h$
        $= \pi(3)^2(3) = 27\pi$ cm^3

        Volume for a hemisphere $= \frac{2}{3}\pi r^3$
        $= \frac{2}{3}\pi(3)^3 = 18\pi$ cm^3

        Total volume for glass 2: $27\pi + 18\pi$
        $= 45\pi = 141{,}37$ cm^3

        Glass 3: Volume for a cone
        $= \frac{1}{3}\pi r^2 h = \frac{1}{3}\pi(3)^2\left(\sqrt{6^2 - 3^2}\right) = 48{,}97$ cm^3

        Glass 2 holds the most liquid, and glass 3 the
        least.

3.1 The lampshade is the lower half of a cone. Using the midpoint theorem we can deduce that the slant height of the "full" cone is twice the length of the truncated cone.

Therefore the slant height of the original cone is $2 \times 13 = 26$ cm.

Given that the perpendicular height for the full cone is 24 cm, $R_1 = \sqrt{26^2 - 24^2} = 10$ cm

In the smaller cone: $R_2 = \sqrt{13^2 - 12^2} = 5$ cm

3.2 Surface area for the open, full cone:
$\pi R_1 s = \pi(10)(26) = 260\pi$

Surface area for the open, small cone:
$\pi R_2 s = \pi(5)(13) = 65\pi$

$\therefore$ Area of lampshade $= 260\pi - 65\pi = 195\pi$
$= 612{,}61$ cm^2

4. Surface area of one tennis ball $= 100\pi$.
$100\pi = 4\pi r^2$
$\therefore r = 5$ cm

Height of surface for labelling $= 3 \times$ the diameter of one tennis ball $= 30$ cm

Radius for curved labelling surface $= 5$ cm

$\therefore$ Area $= 2\pi r H$
$= 2\pi(5)(30) = 300\pi = 942{,}48$ cm^2

5. Volume of sand in cone:
$V = \frac{1}{3}\pi r^2 H = \frac{1}{3}\pi(5)^2(15) = 125\pi$

Height of the sand in the cylinder:
$125\pi = \pi(6)^2 H$
$\therefore H = \frac{125\pi}{36\pi} = 3{,}472$ cm

Volume that cylinder can hold:
$V = \pi r^2 H = \pi(6)^2(17) = 612\pi$

Percentage of cylinder filled with sand:
$\frac{125\pi}{612\pi} \times 100 = 20{,}42\%$

6. 3 cubes have 5 open faces $= 15$ faces
9 cubes have 4 open faces $= 36$ faces
1 cube has 3 open faces $= 3$ faces
Total number of open faces $= 54$ open faces
Total surface area $= 54$ cubic units^2

Test C   (page 261)

1.1 TSA of geyser $= 2\pi r^2 + 2\pi rh = 2\pi(0{,}31)^2$
$+ 2\pi(0{,}31)(1{,}22) = 2{,}98$ m$^2 = 3$ m^2

1.2 Volume of geyser $= \pi r^2 h = \pi(30)^2(120)$
$= 108\,000\pi = 339\,292{,}0066$ cm$^3 = 339{,}29$ ℓ

2. Volume of concrete beam $= lbh = (0{,}5)(0{,}4)(3)$
$= 0{,}6$ m^3

Volume of one pillar $= \pi r^2 h = \pi(0{,}2)^2 . 2$
$= 0{,}25$ m^3

Volume of two pillars $= 2 \times 0{,}25$ m$^3 = 0{,}5$ m^3
Volume of concrete needed: $0{,}25$ m$^3 + 0{,}5$ m^3
$= 1{,}1$ m^3

3.1 Surface area of cylinder $= 2\pi rh$
$= 2\pi(10)(75)$
$= 4\,712{,}39$ m^2

Surface area of dome $= \frac{1}{2}(4\pi r^2)$
$= 2 . (10)^2 . \pi$
$= 628{,}32$ m^2

Total surface area $= 4\,712{,}39$ m$^2 + 628{,}32$ m^2
$= 5\,340{,}71$ m^2

3.2 Volume of rectangular prism $= lbh$
$= 0{,}6 \times 0{,}5 \times 2$
$= 0{,}6$ m^3

Volume of pyramid $= \frac{1}{3}lbh$
$= \frac{1}{3}(0{,}6)(0{,}5)(0{,}8)$
$= 0{,}08$ m^3

Total volume of 2 pillars $= 2(0{,}6 + 0{,}08)$
$= 1{,}36$ m^3

4.1 Radius$_1 = 2$ cm; Radius$_2 = 1$ cm
Perpendicular height of cone $= \sqrt{12^2 - 2^2}$
$= \sqrt{35}$ cm

Perpendicular height of bottom half of cone
$= \sqrt{6^2 - 1^2} = \sqrt{35}$ cm

4.2 $V = \frac{1}{3}\pi r^2 h = \frac{1}{3}\pi(1)^2(\sqrt{35}) = 6{,}20$ cm^3

4.3 $V = \frac{4}{3}\pi r^3 = \frac{4}{3}\pi(2)^3 = \frac{32}{3}\pi = 33{,}51$ cm^3

4.4 Volume of space left $=$ vol. of cone $-$ vol. of chocolate $-$ vol. of hemisphere of ice cream
$= \frac{1}{3}\pi(2)^2(2\sqrt{35}) - \frac{\pi}{3}(\sqrt{35})^3 - (\frac{32}{3}\pi \div 2)$
$= \frac{8\pi}{3}\sqrt{35} - \frac{\pi}{3}(\sqrt{35}) - (\frac{16}{3}\pi) = 26{,}61$ cm^3

5. Volume of box $= (85)(\frac{85}{2}) = \frac{7\,225}{2}$ mm^2
Volume of one golf ball
$= \frac{4}{3}\pi r^3 = \frac{4}{3}\pi(\frac{85}{4})^3 = \frac{614\,125\pi}{48}$ mm^3

Volume of two golf balls
$= 2 \times \frac{614\,125\pi}{48}$ mm$^3 = \frac{614\,125\pi}{24}$ mm^3

Percentage of air surrounding golf balls
$= (\frac{7\,225}{2} \div \frac{614\,125\pi}{24}) \times 100 = 4{,}49 = 4{,}5\%$

Topic 14   Probability   (page 265)

Exercise 1   (page 267)

1.1   (a) no chance or impossible event
(b) 0
(c) 0
(d) 0%

1.2   (a) a certain event
(b) 1
(c) 1
(d) 100%

2.1   (a) $P(\text{six}) = \frac{1}{6}$

(b) 0,17

2.2   (a) $P(\text{no six}) = \frac{5}{6}$

(b) 0,83

2.3   (a) $P(\text{even number}) = \frac{3}{6} = \frac{1}{2}$

(b) 0,5

2.4   (a) $P(\text{Composite number}) = \frac{2}{6} = \frac{1}{3}$

(b) 0,33

3.1   (a) $P(\text{black card}) = \frac{26}{52} = \frac{1}{2}$

(b) 50%

3.2   (a) $P(\text{face card}) = \frac{12}{52} = \frac{3}{13}$

(b) 23%

3.3   (a) $P(\text{five}) = \frac{4}{52} = \frac{1}{13}$

(b) 7,7%

Exercise 2   (page 270)

1.1

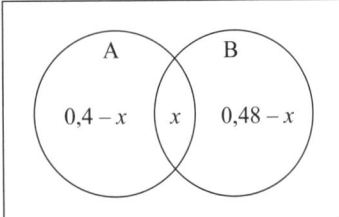

Let $P(A \cap B) = x$

1.2 $P(A \cap B)$ Find the value of x.
$P(A \cup B) = P(A) + P(B) - P(A \cap B)$
$0,73 = 0,4 + 0,48 - x$
$x = 0,88 - 0,73$
$= 0,15$

1.3   A and B are not mutually exclusive because
$P(A \cap B) \neq 0$.

2.1

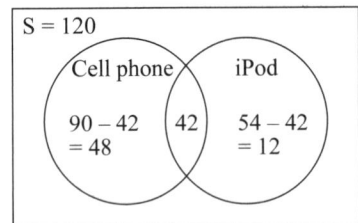

2.2 $48 + 12 = 48$
2.3 $120 - 90 = 30$

3.1

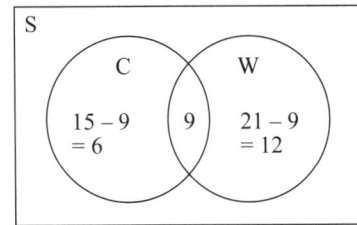

3.2   27 learners
3.3 $P(\text{only cricket}) = \frac{6}{27}$ or 0,22
3.4 $P(C \cap W) = \frac{9}{27} = \frac{1}{3}$ or 0,3
3.5 $P(C \cup W) = P(C) + P(W) - P(C \cap W)$
$= \frac{21}{27} + \frac{15}{27} - \frac{9}{27}$
$= \frac{27}{27} = 1$

3.6   No, $P(C \cap W) \neq \varnothing$
3.7   Yes, $P(C) = 1 - P(W)$

Test A   (page 271)

1.1   Event A: certain event $P(A) = 1$
1.2   Event B: impossible $P(B) = 0$
1.3   Event C: 50-50 chance $P(C) = 0,5$
1.4   Event D: cannot tell from information given
1.5   Event E: 50-50 chance $P(E) = 0,5$
1.6   Event F: above average $P(F) = \frac{36}{52} = 0,69$
1.7   Event G: below average.
$P(G) = \frac{2}{6} = 0,33$ marks

2.1   (a) $P(A) = \frac{3}{6} = \frac{1}{2}$
(b) $P(A) = 0,5$
(c) Event A = 50%

2.2   (a) $P(B) = \frac{5}{6}$
(b) $P(B) = 0,83$
(c) Event B = 83%

2.3 (a) $P(C) = \frac{1}{6}$

(b) $P(C) = 0,17$

(c) Event C = 16,67%

2.4 (a) $P(D) = \frac{5}{6}$ or $P(D) = 1 - P(C) = 1 - \frac{1}{6} = \frac{5}{6}$

Events C and D are complementary events. This means that $P(D) + P(C) = 1$.

(b) $P(D) = 0,83$

(c) Event D = 83%

2.5 Event E: $P(E) = 0$

3.1 two members

3.2

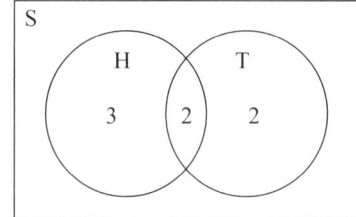

3.3 1

3.4 Yes, $P(T) + P(H) = 1$

4.1

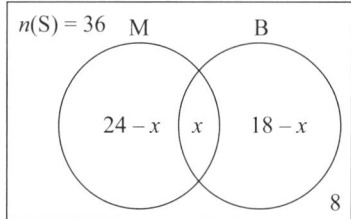

4.2 Solve for x:

$24 - x + x + 18 - x + 8 = 36$

$x = 14$

4.3 $P(\text{only } M) = \frac{10}{36} = 0,28$

4.4 $P(M \cup B) = P(M) + P(B) - P(M \cap B)$

$= \frac{24}{36} + \frac{18}{36} - \frac{14}{36} = \frac{28}{36} = 0,78$

4.5.1 No. $P(M \cap B) \neq \varnothing$

4.5.2 No. $P(M) \neq 1 - P(B)$

Test B (page 272)

1.1 $P(\text{Black card}) = \frac{26}{52} = \frac{1}{2}$

1.2 $P(\text{Face card}) = \frac{12}{52} = \frac{3}{13}$

1.3 $P(\text{King card}) = \frac{4}{52} = \frac{1}{13}$

1.4 $P(\text{King of hearts}) = \frac{1}{12}$

2.1

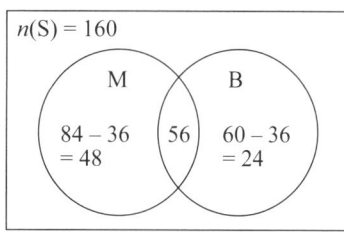

2.2.1 $P(B) = \frac{84}{160} = 0,525$

2.2.2 $P(\text{only } M) = \frac{48}{160} = 0,3$

2.2.3 $P(M \cup B) = P(M) + P(B) - P(M \cap B)$

$= \frac{84}{160} + \frac{60}{160} - \frac{36}{160} = \frac{108}{160} = 0,68$

2.3 No: $P(M \cap B) \neq \varnothing$

2.4 Opinion question. If yes, the answer must mention that the probability of the learners taking Mathematics or Business Studies is 68%, which is highly likely to happen. (*This answer is preferable.*)

If no, the answer must mention that the probability of taking Mathematics is only 65% and the probability that they will take Business is only 53%.

3.1 8 blue pens and 24 green pens

3.2 Yes. $P(B) = 1 - P(G)$

3.3 Yes. $P(B \cap G) = \varnothing$

4.1 Diagram B

4.2 Diagram C

4.3 Diagram A

5.1

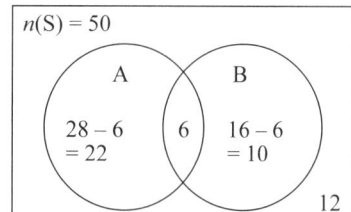

5.2

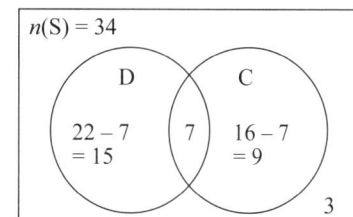

5.3

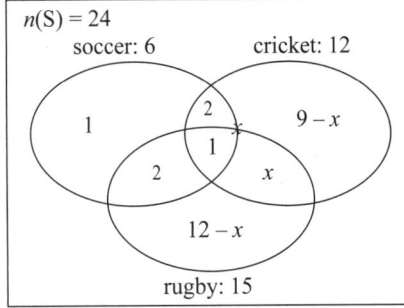

$n(S) = 24$
soccer: 6   cricket: 12
1   2   9 − x
    1
2   x
12 − x
rugby: 15

There were 24 learners who participated in the survey, so adding up all the sections of the diagram's circles gives:
$1 + 2 + 2 + 1 + (12 − x) + (9 − x) + x = 24$
$27 − x = 24$
$−x = −3$
$\therefore x = 3$

There were 3 learners who liked cricket and rugby but not soccer.

Test C (page 274)

1.1 P(not get a ball in the net) $= \frac{8}{40} = 0{,}2$

1.2 Number of times $= 0{,}8 \times 100 = 80$ times

1.3 The approximate answer to question 1.1 can be reduced by testing the probability using many more shots at the hoop.

2.1 P(green) $= \frac{12}{24} = 0{,}5$

2.2 P(red or yellow) $= \frac{10}{24} = 0{,}42$

2.3 P(not red) $= 1 − $ P(red) $= 1 − \frac{6}{24} = \frac{18}{24} = 0{,}75$

2.4 P(not yellow or blue) $= \frac{18}{24} = 0{,}75$

3.1 15

3.2.1 P(E$_1$) $= \frac{7}{15}$

3.2.2 P(E$_4$) $= \frac{4}{15}$

3.2.3 P($\bar{E}_2$) $= \frac{10}{15}$

3.2.4 P($\bar{E}_4$) $= \frac{11}{15}$

3.2.5 P(E$_1$ or E$_4$) = P(E$_1$) + P(E$_4$) − P(E$_1$ and E$_4$)
$= \frac{7}{15} + \frac{4}{15} − \frac{0}{15}$
$= \frac{11}{15}$

3.2.6 P(E$_2$ and E$_3$) = P(E$_2$) . P(E$_3$)
$= \frac{5}{15} \times \frac{3}{15}$
$= \frac{15}{225} = \frac{1}{15}$

4. For mutually exclusive events:
$P(A \cup B) = P(A) + P(B) − P(A \cap B)$,
where $P(A \cap B) = \varnothing$
$0{,}75 = 4x + x − 0$
$5x = 0{,}75$
$x = 0{,}15$

5.1 $100 − 94 = 6$

5.2

$n(S) = 100$
C   A
$50 − 31 − x$ 27 $66 − 44 = 22$
    4
x   13
$40 − 17 − x$
S

5.3 $94 = 50 − 31 − x + 27 + 4 + x + 22 + 13$
$+ 40 − 17 − x$
$= 108 − x$
$x = 14$

5.4.1 P(A) $= \frac{66}{100} = 0{,}66$

5.4.2 P(only A) $= \frac{22}{100} = 0{,}22$

5.4.3 P(only A and S) $= \frac{13}{100} = 0{,}13$

5.4.4 P(liked all 3) $= \frac{4}{100} = 0{,}04$